What is Cultural Factor

Biology Beyond Darwin

Vipin Gupta

Copyright © 2022 Vipin Gupta

All rights reserved.

All rights reserved by the author. No part of this publication may be reproduced, stored in a retrieval system, or transmitted in any form, or by any means, electronic, mechanical, photo-copying, recording, or otherwise, without the author's prior permission.

ISBN: 9798421648970

Black & White Paperback

Independently Published Imprint

ISBN: 9798421649458

Color Hardcover

Independently Published Imprint

First Published: February 2022

Cover Design & Illustrations::
Kiran SJ

Page Setup:
Surajit Paul

Index:
Liya Jayabalan and Salini Kurup

EBook Editor:
Suresh Velu

Project Manager:
Bhakti Gupta

Contents

Project Motivation		1
List of Abbreviations		10
About the Author		11
Preface: Understanding the Cultural Factor—Biology Beyond Darwin		13
Chapter 1	**The Entity Tensor** : *Exchanging the God Within*	31
1.1	How Culture Transforms an Entity	34
1.2	How an Entity Transforms Culture	36
1.3	How Energy Transforms Mass Consciousness	38
1.4	How Mass Consciousness Transforms Energy	41
Chapter 2	**The Time Tensor** : *Exchanging the God Without*	46
2.1	Forming Time with Negative Nine Finite Points	50
2.2	Norming Time with Positive Nine Finite Points	50
2.3	Transforming Time with Zero Finite Points	51
2.4	Transcending Time with Six Hundred Finite Points	54
2.5	Enjoying Time with Fifteen Finite Points	61
	2.5.1 The Sixteen-wheel system	62
	2.5.2 The Three-Wheel System	65
	2.5.3 The Two-Wheel System	66
2.6	Exchanging Time with Seven Thousand Finite Points	71
2.7	Developing Time as a Child with One-Hundred-And-Twenty-Eight Finite Points	75
2.8	Reversing Time with Sixteen Finite Points	77
2.9	Acculturing Time with Twenty-Seven Finite Points	82
2.10	Working to be Free from Time with Negative Sixteen Finite Points	87
	2.10.1 First, Shape the Twelfth House with Goal	87
	2.10.2 Second, Shape the Tenth House with Willfulness	90
	2.10.3 Third, Shape the Eighth House with Infinite Breeding	92
	2.10.4 Fourth, Shape the Sixth House with Entanglement	94
	2.10.5 Fifth, Shape the Fourth House with Culture	95
	2.10.6 Sixth, Shape the Second House with Materialism	97

2.11	Knowing the Value of Time with Negative Twelve Finite Points	100
	2.11.1 First, Shape the Eleventh House into a Leader	100
	2.11.2 Second, Shape the Ninth House into a Consciousness System	103
	2.11.3 Third, Shape the Seventh House into a Conscious System	103
	2.11.4 Fourth, Shape the Fifth House into a Circular System	105
	2.11.5 Fifth, Shape the Third House into a Circle	106
	2.11.6 Sixth, Shape the First House into an Energy System	107
2.12	Feeding the Value of Time with Negative One Finite Points	111
Chapter 3	**The Space Tensor: Exchanging the Entity Within**	**118**
3.1	Forming Space for the Universe of Entities with Five Finite Points	123
3.2	Transforming Space in Five Ways	125
	3.2.1 Transforming Space with One Finite Point	125
	3.2.2 Transforming Space with Two Finite Points	127
	3.2.3 Transforming Space with Three Finite Points	128
	3.2.4 Transforming Space with Four Finite Points	129
	3.2.5 Transforming Space with Five Finite Points	130
3.3	Norming Space with Five Forming and Five Transforming Finite Points	131
3.4	Transcending Space with Twenty Finite Points	134
3.5	Enjoying Space with Thirty-Three Finite Points	138
3.6	Exchanging Space with Sixty Finite Points	143
3.7	Developing Space with One Hundred Eighty Finite Points	148
3.8	Reversing Space with Eighteen Thousand Finite Points	152
3.9	Acculturing Space to the Triple Copy Culture with Ninety Thousand Finite Points	159
3.10	Working to be Free from the Triple Copy Culture with Negative Sixteen Finite Points	163
3.11	Knowing the Value of Freedom with Negative Eight Points	165

3.12	Feeding the Value of Freedom with Negative Nineteen Finite Points	167
Chapter 4	**The Causation Tensor : Exchanging the Entity Without**	**173**
4.1	Forming Causation at the First Infinity Level: The Infinite Council and The Intellectual Body	178
	4.1.1 Role of the Earth Wheel	179
	4.1.2 Role of the Moon Wheel	179
	4.1.3 Role of the Sun Wheel	180
	4.1.4 Role of the Wheel of Wheels	182
	4.1.5 Symmetry of the Intrinsic Wheels with the Extrinsic Cycles	184
4.2	Norming Causation at the Second Infinity Level: The Divine Council and The Mental Body	187
	4.2.1 Role of the Time Wheel	187
	4.2.2 Role of the White Matter	188
	4.2.3 Role of the Serotonin-Melatonin Production Cycle	190
	4.2.4 Role of the Divine Council	193
	4.2.5 Role of the Serotonin	195
	4.2.6 Role of the Homozygote	197
	4.2.7 Role of the Heterozygote	200
	4.2.8 Role of the Melatonin	202
4.3	Transforming Causation at the Third Infinity Level: The Council of Elements and The Astral Body	207
	4.3.1 Transforming Causation at the Cellular Level	207
	4.3.2 Transforming Causation Within Body	210
	4.3.3 Transforming Causation Without Body	214
	4.3.4 Transforming Causation Within Oneself	216
	4.3.5 Transforming Causation Without Oneself	225
4.4	Transcending Causation at the Fourth Infinity Level: The Council of Seven Light Forces and The Etheric Body	237
4.5	Enjoying Causation at the Fifth Infinity Level: The Supreme Council and the Causal Body	249
4.6	Exchanging Causation at the Sixth Infinity Level: The Council of Twelve and the Self-Luminous Entity	253
4.7	Developing Causation at the Seventh Infinity Level: The Ring Council and the First Eden	259

	4.7.1 Dividing Causation. Divided Causal Body and Soul Mates Within the Soul Family	266
	4.7.2 Reversing Causation. Divided Etheric Body and Souls Within the Oversoul Family	277
	4.7.3 Acculturing Causation. Divided Astral Body and Flamemates Within the Flame Family	282
	4.7.4 Working Causation. Divided Physical Body and the Self Within the Soul Universe	300
	4.7.5 Knowing Causation. Divided Mental Body and the Twin flame Without Flame Family	309
	4.7.6 Feeding Causation. Divided Intellectual Body and the Flame Without the Human Being	324
Chapter 5	**Transcending the Causation Tensor: Exchanging I AM**	**333**
5.1	Imperative for Transcending the Causation Tensor	341
	5.1.1 Breeding Causation. Divided Body and Gender Differentiation Within Group	342
	5.1.2 Freeriding Causation. Divided Consciousness and Generation Differentiation Within Gender	344
	5.1.3 Objectifying Causation. Divided Entity and Group Differentiation Within Geography	346
	5.1.4 Spreading Causation. Divided Geography and Class Differentiation Within Time	351
	5.1.5 Compounding Causation. Divided Time and Absolute Differentiation Within Space	354
	5.1.6 Espousing Causation. Divided Space and Speciation within Goal	357
	5.1.7 Gene Stacking. Divided Causation and Genetics within Nature	360
	5.1.8 Infinite Adding. Divided Nature and Gene Flow within Ego	362
	5.1.9 Eventual State. Divided Divinity Making One Zero	365
5.2	Imperative for Transcending the Cultural Factor	368
Acknowledgments		372
English Index		373
Hindi Index		414

Project Motivation

Project VIPIN, literally *Vastly Integrated Processes Inside Nature*, is a project about myself. Myself of each person, inanimate or animate, is the same. Myself is a beautiful garden without walls. Therefore, myself becomes fragmented and diffused, lost within the universe. Life's journey is to discover myself, unite the fragmented parts, and fuse oneself as a conscious system conscious of one's natural value. Our natural value perpetuates the value of our essence as our essential nature. Each fragmented part operates as a heart system, forming a circle without the circle of our life as the creator of infinite circles. "I" becomes the heart of the universe that beats with the sound of "my" reproduction.

Project VIPIN stands for an investigation into the nature of reality by disentangling the vastly integrated processes that make it appear like an illusion. It is important to develop a clarified consciousness of reality, how it forms, what it norms, and why it transforms. We believe we have the tools to know the reality. Believing is not reality. For example, if one person believes God exists and another believes God does not exist, it does not mean the reality of God is contingent on the varying beliefs. Modern science states that reality is relative and specific to a person who is observing and forming a belief. By observing and believing that we know reality, we only create an illusion of the omnipresent reality within our mental realm. Since the illusion is our creation, we assume that it is real. The illusion is not real; the creation is.

Our energy transforms the illusion into the imaginary and the real into the complex. The complex we create reproduces our reality polluted by a belief in the imaginary. Therefore, the creation comprises the mass of our polluted consciousness. As the creators, we develop a charm attraction towards the creation and strangely repel the illusion that we are the creatures blessed with life by an Almighty Creator for enjoying the creation. After destroying our creation, we become that Almighty Creator blessing everyone with our present consciousness. Thus, everyone becomes the producer of our present consciousness as the absolute, generating an ascending entropy. Suppose we wish to stop the entropy at the individual, local, national, and international levels. In that case, we need to develop our gifted potentialities, not depend on the divinity of the transcendental forces. That requires us to transcend diverse religious ideologies, including using science as secular religion and spirituality as insular religion.

Project VIPIN is educative, illuminating, and revolutionary. It educates the readers about the metaphysics of reality. It illuminates the cause, the sequence, and the consequence of reality. It revolutionizes our understanding of the cause and empowers us to manage the sequence cost-effectively for realizing the desired consequence. It is a consequence of a sequence of research projects seeking to discern the cultural origins of the variations in development at the individual, local, national, and international levels.

The culture originates from transforming our essential nature with I am a deity consciousness. Our "I am a deity consciousness" contaminates our essential nature and accultures everyone with our consciousness of reality. Have you ever reflected upon why people seek to become cultural icons who limit the child's imagination about the realm of possibility by making themselves larger-than-life? How does larger-than-life fill the hole to become our mental space and close the mental space from enjoying the possibilities of life? How can one free one's mind from the present reality and exchange that with the conscious imagination of the impossible to make that possible?

The cultural factor transforms the human factor into the ego by making our reality discordant from the evolving time. Our ego diffuses our energy to form networking linkages that make everyone a consumer of our ego. After consuming our ego, everyone becomes a creator of our copies by idealizing us as a divine entity producing what they are consuming. The copies begin consuming the creation as they believe that they are worthy of the blessings of us, the Almighty Creator of the ego. After consuming the creation, the copies become the creature reproducing the ego energy. Thus, the Almighty Creator, the creator, the creation, and the creature become a four-body system that forms the essence of reality.

The life cycle originates from our weak psychic linkages as an Almighty Creator with our future as the creator, our present as the creation, and our past as the creature. Everyone following us consumes our creation, thus making us dependent on the consciousness they are producing. Our creation is a part of our extended self. Everyone trades a part of our self, thus becoming dependent on the conscious decisions we are taking. Everyone begins personifying our conscious decisions as their conscious decisions programmed within their body as the genetic sequences. Thus, the original gets lost in transition. The copy is bred as the illusion of the original's reality. We all become copies when we identify ourselves as social animals.

Ego is energy in descending motion, leading to the outflow of the thermodynamically self-steaming air in ascending motion. Emotion is energy in descending motion, following the inflow of the consciousness-cooling air in descending motion. We become emotional when we experience the cooling-effect of someone's conscious energy. We experience the cooling effect when someone has empathy with us and is willing to make us a part of their extended self. We make others a part of our extended self by becoming invested in their conscious well-being.

Conscious energy cools down the anxiety in our mind to produce the flow of our consciousness of the anxiety in a horizontal motion. The consciousness flow heats our body as we become mentally agitated and begin perspiring with the fear of the unknown. The unknown is our extended self. We do not know how the others we are devoted to will behave when we show concern for their conscious well-being. Others may imagine that to be an intrusion into their personal space. Since you have already decided to invest, their false imagination of your intentions steams the air you are breathing. The self-steaming air you breathe is diffused by the one seeking space by repelling you out of his space.

When one is not concordant with the social others, one becomes a discordant entity and produces discordant energy. The producer of the discordant energy is also the consumer of the discordant energy. We make others a part of our extended self because of our illusionary I am a deity consciousness. We believe in our divinity, either inherent or borrowed from another divine entity. Therefore, we seek to make others feel they cannot make conscious decisions and superposition our conscious decisions on them. Thus, ego, emotion, and anxiety form a circle guiding our life from birth to death.

Mother Nature gifts each of her children the potential to be limitless. The limiting factors are not natural—they are supernormal mediation of the human beings both while living within the physical body and after that. Supernormal mediation leads to thermodynamic entropy. That's why we are experiencing ascending entropy over time in the mental, intellectual, and physical realms.

I invite you to open your mind to grasp the reasoning given in the book. Use reason as the light guiding your conscious decisions in life. In our lives, we use theories, ideas, notions, opinions, emotions, egos, conceptions, perceptions, experiences, and references to guide our para-conscious decisions. The decisions are conscious when we have certainty of fulfilling the purpose of life through them. The decisions are para-conscious when we lack clarity about the purpose of life and therefore rely on guides to guide us to live our lives. We live life by experiencing its diverse colors and pairs of opposites the purpose of life. We believe we have lived a wholesome life if we have enjoyed everything and matured over time. However, maturity does not lead to freedom. On the contrary, it entangles us in teaching others our life lessons, wishing to accelerate everyone's maturity.

Everyone wishes to diffuse their consciousness system for guiding everyone's conscious system. Thus, life becomes a puzzle we need to disentangle for enjoying our conscious system without the weight of the infinite consciousness system. The infinite consciousness system is the para-conscious system whose weight pollutes our consciousness and limits our power to make conscious decisions with our conscious consciousness. Instead, we gravitate towards spirituality guided by the spirit. The spirit motivates us to fill the darkness within our minds with the light of the universe. When we run after the light of the universe, we lose sight of the light of our inner self. We become our foe, always negating our intuition with the false reasoning, fears, and hopes. We begin seeking God without us instead of enjoying oneness with God within us.

Our essence is immanent within the reality of everything we manifest. Thus, we become entangled with the goals we pursue to manifest diverse realities. A need emerges for introspection into what we are introverting through the extroversion of our potential. We introvert the para-consciousness of the things we create in diverse forms. Those things remain within our consciousness and attract us towards the reality of everything others manifest, motivating us to trade that reality for becoming the foundation of everything. As others follow, the thermodynamic force transforms the foundation into the quantum photon. As we follow others, the quantum photon transforms into quantum light.

As everyone reproduces the quantum light, they produce the light at the speed of light. The speed of light is the conscious system making us conscious of the light we were before becoming a photon and transforming into the diverse forms in the physical realm. The light illuminates the path for realizing our essence by cultivating and developing our essential nature into its undivided form of the greeter essence that greets the reality as a primordial greeter. As a primordial greeter, we enjoy absolute freedom from the cultural factor and develop the power to create energy, transcending the law of limitation of the theory of energy conservation and the scientific paradigm.

Our actions are reactions to voids in our consciousness about reality. We act when we wish to know our reality. Our actions are physical events with biological effects and chemical sequences. They generate organic resonances of the reality we illuminate with our conscious actions and inorganic dissonances of the reality we destroy with our psychological reactions. Organic resonances form the molecules perpetuating the reality we create at each moment. Inorganic dissonances transform the molecules into atoms manifesting the reality of our knowing at the moment. Every moment we make our knowing a part of our subconscious mind. Our subconscious mind reproduces our actions in reaction to the entropy in our conscious knowing of reality.

We substitute the reality of entropy with an illusion of God knowing what we don't. For scientists, that God is Einstein. Everyone else complements the illusion of the social, general God with the imagination of their personal, special God. The reality of both general and special Gods is relative to the institutional, parabolic God that a dominant religion institutionalizes. We are living in an age where science is the dominant religion. Unless you have faith in Einstein, you are not fit to be called a scientist and indeed a rational human.

How scientific is the reality that makes you a hero if you reference Einstein as the authority substantiating your knowing and zero otherwise? Since the present paradigm of reality is not scientific, there is a need to discover the metaphysics of physical sciences and illuminate the dynamics transcending the limits of known reality.

The first nine books under Project VIPIN show that Einstein's theories are correct only under the ideal condition of zero mass of consciousness. If there is zero mass of consciousness, energy is zero because consciousness breeds the energy. When we are conscious of a thing in the mental realm, our consciousness illuminates that thing as a part of our conscious system.

Our conscious system forms the speed the light of our consciousness takes to make us conscious of that thing in the physical realm. Now, if in an equation, $m = 0$, $e = 0$, and c is the initial spatial state that gets illuminated at the speed of light, then any form of e will be equal to any transformation of m. That is a trivial finding. What is important is the consequential entity state illuminated by the initial spatial state and how its reproduction leads one to the entity whose initial spatial state channeled within one's consciousness of the reproductive reality for transcending the limitations of the speed of light.

Natural science is the science of illusion. It is about making sense of the real by imagining an ideal mathematical representation. The ideal that represents the real is not real, but an illusion of the real's copy within our mind. We believe we know the real after observing the reality, and we manifest our belief in the form of a theory.

Twelve Books in Project VIPIN
1. What is divine energy
2. What is present reality
3. Is present reality
4. Is divine energy
5. What is consciousness
6. What is para-consciousness
7. What is self-awareness
8. What is human factor
9. What is trading factor
10. What is cultural factor
11. What is exchange factor
12. What is technological growth

A theory is always falsifiable because our beliefs change every moment. Even in a greenhouse, you believe it is hot at one moment. At the next moment, you believe it is cold. When you perceive the space as closed, you feel hot and wish to have your personal space. When you perceive the space as open, you feel cold and wish to have the walls to protect you from the uncertainty of a danger intruding in the space. The society concerned about our sentient well-being bears the escalating technological cost of managing our perceptual reality. If we stop relying on our beliefs about reality and instead focus on reality, we can enjoy the stillness in our conscious well-being. Then, society can focus on the growth of our conscious well-being instead of managing the entropy costs in our conscious well-being.

This book intends to help one gain freedom from the notions based on ideals, opinions based on theories, and numbers based on dust-ball empiricism purified through mathematical manipulation to give an elite space to the physicists. The prominence of physicists makes us materialistic, racing to be a millionaire, billionaire, trillionaire. One accumulates wealth through unsustainable control over Mother Nature's resources intended for every child. On the one hand, physicists claim that energy is not creatable, only transformable. But, on the other hand, they endorse the claim of the economists that the trillionaires are creating the energy that makes them wealthy and not taking that energy from anyone else. According to them, the energy belongs rightfully to the wealthy; it is private property, not a public good. The hypocrisy of the scientists is evident in the present paradigm. It is called heads I win, tails you lose. One claims that energy is constant to acculture everyone into the appropriation of energy. The claimant is rewarded for illuminating the limitation and the actor with the sanctification of the appropriation as the private property right. And, if you decide not to believe the claimant, you lose out as it is your share of the energy that is now institutionally sanctioned as the believer's property right.

Many philosophers have advocated spirituality as the new paradigm for discerning the unknown reality. Spirituality and science have specific roles in our life. They are tools for us to enjoy life. In practice, people let spirituality and science enjoy them. Everybody has become a producer, a consumer, or a trader of either spirituality or science, if not both.

One produces spirituality by sharing one's spiritual experiences. Everyone's spiritual experiences are different. When we trade someone's experiences, we develop a wish to enjoy the same experiences by becoming a copy of that someone. For example, suppose we are devoted to a divine entity and experience the divinity of that divine entity. In that case, we wish to reproduce that experience personally and socially and eventually make that an institutional experience the whole community shares for their salvation, i.e., shared becoming of a copy of the one they are experiencing. As the next step in the sequence, one produces science by investigating everyone's shared becoming and servicing our discovery for making everyone conscious of the alternatives to shared becoming.

When we trade someone's ideas, we become bound by their ideal-effect and take them as our ideal. We begin conceiving the ideal as the holy spirit breeding like an original within our light. We begin breeding their ideas as our ideas about the purpose of life. The breeding system becomes a fact of life, contaminating the breathing system. With the pollution of our mental space, we begin behaving like an immortal entity. Even after our death, we decide to stay with our devoted followers as their guiding spirit, generating an illusion of making conscious decisions for their conscious well-being. Consequently, we circle back to produce spirituality as a parallel universe.

Human development becomes an artificial element formed due to the reversal of the development already within us. When we lack confidence in ourselves and our capability to manifest the reality we wish, we augment the reality with the complexity of our unrealized imagination of natural reality. That creates a need for confidence-complementing development. Similarly, when we lack awareness of ourselves and our investment in the reality of others, we destroy the reality with the illusion of our realized imagination of a supernatural reality beyond us. That creates a need for awareness-supplementing development. The kith complements the confidence of the kin who share the genetics of confidence-deficit in their self, without us. The aliens supplement the awareness of the kith who share the genetics of awareness-munificence of their extended self within us.

The kin is like a child. The kith is like a sister courageous to gain confidence and convince the child to move fast without doubting his capability. Trading the sister's courage becomes the key to a child servicing the value of his capability for his conscious well-being. The alien is like a brother with conviction in the sister's investment in him. He blocks the sister's awareness of her extended self within him, thus making the sister dependent on his social conscious well-being. Self-developed personal conscious well-being is more than the extended self-developed social conscious well-being. When we seek to develop ourselves as a social animal whose well-being is conditional on the well-being of others, we expect our well-being to exceed the well-being of social others. Thus, it leads to everyone's entropy and eventually makes us zero.

List of Abbreviations

DIVINE   d = determination, I = imagination, v = virtue, I = intuition, n = natural, e = excellence

GUIDER   g = global, u = unique, i = inclusive, d = diverse, e = engagement, r = responsibility

SHEENY   s = social, h = human, e = ecological, e = economic, n = national, y = psychological

About the Author

Vipin Gupta, Ph.D., is a professor of management, and co-director of the Center for Global Management at the Jack H. Brown College of Business and Public Administration, California State University, San Bernardino. He has a Ph.D. from the Wharton School of the University of Pennsylvania. He has been a gold medalist for outstanding academic performance in the Indian Institute of Management post-graduate program, Ahmedabad; a top rank holder in the B.Com. (Hons) Program of Delhi University from Sri Ram College of Commerce; and an all-India rank holder at the graduate program of the Institute of Cost and Works Accountants of India.

Vipin Gupta was previously at Simmons University, Boston, Grand Valley State University, and Fordham University. He has offered several training programs and workshops on strategic planning and cross-cultural management to senior executives, administrators, defense personnel, and research methods to doctoral students and faculty in India and the US. He has been a visiting or guest faculty at more than thirty business schools in India. His workshops and lectures have been covered by several leading national and regional newspapers and television channels.

Professor Gupta has authored more than 250 journal articles and book chapters, including in leading journals such as Journal of Business Venturing, Family Business Review, Research in Organizational Behavior, Asia-Pacific Journal of Management, Multinational Business Review, Journal of World Business, Advances in Global Leadership, and Management Review. Besides delivering lectures and keynotes in several nations, he has presented at international academic conferences in more than sixty nations, including the Academy of Management, IFSAM, EGOS, Society of Industrial Organization Psychologists, Global Entrepreneurship Conference, and Family Enterprise Research Conference. He has been on the governing board and organizing committee of several international conferences. In 2017, he served as the academic program chair for the 52nd CLADEA Assembly.

Dr. Gupta is the co-editor of the seminal GLOBE (Global Leadership and Organizational Behavior Effectiveness Program) book Culture, Leadership, and Organizations – The GLOBE Study of 62 Societies (Sage Publications, 2004). He is the principal investigator of the path-breaking CASE (Culturally-sensitive Assessment Systems and Education) Project on family businesses. He edited two critically acclaimed books on the theme of strategic management, performance, and leadership in the emerging markets: Creating Performing Organizations (Sage Publications, 2003) and Transformative Organizations (Sage Publications, 2004). He is the author of the Strategic Management and Business Policy: Concepts and Applications (PHI Learning, 2003 and 2005). He has been the principal editor of ten books on family business models in ten different regional clusters and the eleventh book on family businesses' gender dimension (ICFAI University Press, 2004). He has also co-authored a research manuscript MNC Subsidiaries in China: An empirical study of growth and development strategy (Information Age Publishing; 2015), and a textbook Leadership Across the Globe (Routledge USA, 2015).

Vipin Gupta has been a recipient of the coveted 2005 Scott Myers Award for Applied Research in the Workplace from Society for Industrial Organization Psychologists, USA. As a 2015-16 American Council of Education fellow, he visited sixty-two universities, colleges, and higher education institutions in nine European nations, the USA, and India

Preface

Understanding the Cultural Factor
Biology Beyond Darwin

God's Dilemma: I need space for my transformation as "God" (*Ishvara*, 5) into the "creator" (*Brahma*, 59) to "greet" (*Palala*, 64 = 59 + 5) my "energy fragment" (*Yogihridaya*, 64) as an "Immortal Almighty Creation" (*Palala*, 64).

How the Cultural Factor Transforms the Potential into the Space for Transformation?

The cultural factor transforms the potential into the space for transformation after an entity activates its potential to conceive a population that needs space for its growth beyond the primordial space that gives birth to the entity.

"Biology" (*Jivavijnana*, 90,000) is a science of the transformation of the "population" (*Abadi*, 18) into divergent groups, norming a "genus" (*Varga*, 9) of "species" (*Prajati*, 17) that enjoys a life of "eternity" (*Ananta*, 90,000) by "metamorphosing" (*Pudgala*, ½) into the diverse "genera" (*Varga*, 9) evolving with time. Biologists hold Charles Darwin as the Father of biology.

According to Charles Darwin, a "population" (*Abadi*, 18) "perpetuates" (*Ramakatha*, 9) its para-conscious "spontaneous fragmentation" (*Khandana*, 42) into the "divergent groups" (*Varga*, 9) due to the "density" (*Sandra*, 60) of growth over time. Mother Nature selects the divergent groups with "perfect adaptation" (*Anurupana*, 13) to the "local conditions" (*Yavadartha*, 0) for norming the species that live and transforms them into the genera that form the "convergent groups" (*Parivar*, 10), norming the "families" (*Parivar*, 10) of species livable in future. Father Nature selects the convergent groups with "adaptation" (*Anukulana*, 16) to the "global conditions" (*Sthanata*, 119) for "ordering" (*Dishta*, 6) the "growth" (*Dishta*, 6) of the species into the asymmetric "classes" (*Varna*, 869) bound for extinction because of the struggle for existence and the survival of only the fittest species within each family. The "taxonomical divergence" (*Parindartha*, -3) rises over time as the groups increasingly diverge from the symmetric "kinds" (*Prakara*, 17) bound for life within the domain of a selected kingdom, breaking away from their primordial genera.

In reality, a future convergent group is a reproduction of the past divergent group. The reproduction force of the reproductive endogenous behavior, not exogenous selection, is responsible for the present extinction group. A divergent group reproduces its copies because it perceives the local conditions as unhealthy for its survival. Over time the copies converge with the divergent group because the reproduction makes the global conditions unhealthy for the survival of the divergent copies. Therefore, reproduction does not lead to a divergent group of copies.

The natural selection of the lower-order "divergent groups" (*Varga*, 9) as fitter than the higher-order "convergent groups"

(*Parivar*, 10) is not the origin of species that live under the "global condition" (*Sthanata*, 119) necessary for substantiating a "theory" (*Niyama*, 127) of nurturing "Mother Nature" (*Kudrat*, 8). Similarly, the "supernatural fragmentation" (*Khandana*, 42) of the higher-order "convergent groups" (*Parivar*, 10) into the lower-order "divergent groups" (*Varga*, 9) for "perfect adaptation" (*Anurupana*, 13) to the "local condition" (*Yavadartha*, 0) substantiating the "theory-effect" (*Maha dasha*, 0) of a theory. A theory takes "Father Nature" (*Yahoodi*, 1) as the "ideal" (*Adarsha*, 1) agent for managing the "adverse selection" (*Sakriya-Niskriya*, 10^{1000}) of a "child" (*Arcisa*, 128) Mother Nature loves.

The "evolutionary force" (*Tushti*, 85) of time is the "perfect guider" (*Raja-rajesvari*, $85 = 50 + 17 + 9 + 9$) of the "descending-order development" (*Rajo Guna*, 50) once the "species" (*Prajati*, 17) "emerge" (*Ummujjati*, 9) as the "convergent group" (*Parivar*, 10) and "diverge" (*Varga*, 9) into a "divergent group" (*Varga*, 9).

The "revolutionary force" (*Uttama*, 85) of space "curves" (*Vakra*, 85) the "backward value" (*Pramod*, 85) of time for an "ascending-order development" (*Sato Guna*, 200). It develops new "kind" (*Prakara*, 17) of "species" (*Prajati*, 17) with the "convergent energy" (*Samvat shakti*, 28) of the "present growth" (*Sva*, 11) of species.

Time as a "twin entity" (*Ghatika*, 24) "copies" (*Nakala*, 0) the "future value" (*Anagata*, 0) of space as an "illusionary force" (*Mahendrani*, 0) for the "horizontal-order development" (*Tamo Guna*, 90,000) of a diverse group of species within a class of entities normed by the space over time. The time-invariant "imagination force" (*Siddhi*, $57 = 19 * 3$) of space as a "para entity" (*Ashtavakra*, 19) orders each group of species into heterogeneous families as a function of their "backward-order development" (*Divyata*, 57) with the forward-moving time. An "ideal self-luminous entity" (*Purusha*, 12), acculturated to a "finite coordinate" (*Pradesha*, 9) of space and time, orders the heterogeneous family into a "homogeneous genus" (*Prakara*, 17) through a "replication" (*Samalekha*, 90) of its "forward-order development" (*Samalekha*, 90) beyond the finite coordinate that perpetuates the "cultural factor" (*Sugriva*, 5) as the "genetic code" (*Somapa*, 180).

Thus, the cultural factor transforms the "potential" (*AUM*, 18) within an "entity" (*Hasti*, 24) into the "population" (*Abadi*, 18) bred by the entity for forming the "space" (*Vaas*, 18,000) for transformation. It norms the "conscious energy" (*Amrit*, 1000 = 900 + 10) of the ascending, descending, horizontal, backward, and forward order developments as the "five genetic codes" (*Pasha*, 900 = 180 * 5). Five genetic codes "loop" (*Pasha*, 900) time as "quantum time" (*Pasha*, 900) reproducing the "primordial space" (*Anataramsa*, 100) for the growth of each group into "infinity" (*Tamo guna*, 90,000 = 900 * 100), sustaining the "horizontal-order development" (*Tamo guna*, 90,000).

The Cultural Factor is About Our Acculturation

The "cultural factor" (*Sugriva*, 5) acculturates us to the perpetuating value of reality. Our body seeks to excrete the "perpetuating value" (*Saranyu*, 5) for illuminating our reality. The "illuminating value" (*Prakasha*, 7) of our reality gives us the freedom to perpetuate the reality we wish as the culture. The goal of shaping the "culture" (*Sadakhya*, 9) leads to our "entanglement" (*Kula*, 9) with the "universe" (*Brahman*, 2), whose perpetuating value we shape. We fulfill a goal by following the "path of action" (*Karma Marga*, 86) and producing a "fruit of action" (*Karma Phal*, -10^{19}). The fruit of action is the "path of knowing" (*Jnana Marga*, -10^{19}) that generates a "fruit of knowing" (*Jnana Phal*, 1). The fruit of knowing makes us the "worker" (*Karmi*, 1) for everyone who does not know how to fulfill their wishes. It makes us a "deity" (*Deva*, 1) whose "knowing" (*Jnana*, 19) makes him the "denominator" (*Shudra*, 1) of the "path of devotion" (*Bhakti Marga*, 1). After making everyone's path of devotion our fruit of knowing, we enjoy everyone's "fruit of devotion" (*Bhakti Phal*, 96).

The fruit of devotion makes us a "living entity" (*Sushupti*, 96) living in everyone's "consciousness" (*Chaithanya*, 4) of our "impact" (*Siddhi*, 57) and everyone a "luminous entity" (*Kali*, 96) servicing us their "being energy" (*Kali Shakti*, 96) for enjoying the impact of our "divinity" (*Divyata*, 57). It motivates everyone to follow our "path of divinity" (*Siddhi Marga*, 96) for making a "difference" (*Faraka*, 8) in the "world" (*Duniya*, -2). Thus, everyone enjoys the "fruit of impact" (*Siddhi Phal*, 25) that we are "making" (*Karin*, 5000) with our "multiplying" (*Eke*, 5000), and they are "taking" (*Lena*, 19) with their "energy" (*Shakti*, 19) and "shaping" (*Tarashana*, 0) into their "own" (*Atmiya*, 1). It leads everyone to a "path of extroversion" (*Prabhasa Marga*, 25) of their "organizational entropy" (*Prabhasa*, 4). It leads us to a "path of introversion" (*Siddha marga*, 25) as a "sentient entity" (*Siddha*, 7) conscious of our "reality" (*Vastavikta*, 7).

Everyone enjoys our "consciousness system" (*Sara Kalpa*, 10^{10}) as the "fruit of joyful life" (*Prabhasa Phal*, 10^{10}), extroverted by the organizational entropy of their reality. Everyone becomes the "enjoyer" (*Rasi*, 9) of our "culture" (*Sadakhya*, 9). We enjoy everyone's "conscious system" (*Shunya Kalpa*, 8×10^{15}) as the "fruit of reality" (*Siddha Phal*, 8×10^{15}), introverted by our life as a sentient entity. Consequently, we become the "gravitational center" (*Guru Sthana*, 629) of everyone's "conscious system" (*Shunya Kalpa*, 8×10^{15}). We become the "organization" (*Sangathan*, $29 = 629$), trading everyone's "creative force" (*Uma*, 6). We become the "star" (*Tara*, 2), servicing everyone the "goal" (*Lakshya*, 9) of becoming a "thing" (*Vastu*, 9) that "matters" (*Sadhibhuta*, 158) to us. When one "entangles" (*Uljhana*, 158) their "consciousness" (*Chaithanya*, 4) within the "perpetuating value" (*Saranyu*, 5) of our "reality" (*Vastavikta*, 7), they enjoy the "convergence" (*Avadhana*, 158) with our "cycle of divinity" (*Siddhi chakra*, 158). The cycle of divinity is the "cycle of conscious life" (*KLIM shakti chakra*, 158). It makes everyone's "happiness" (*Hriti*, 158) conditional on the "illusionary joy" (*Vastu*, 9) of "breeding" (*Janana*, 15) the "growth" (*Dishta*, 6) of the "wheel of the sentient entity" (*Kaal Sarpa chakra*, 39).

Therefore, a need emerges for everyone to seek rebirth seeking the "human body" (*Karana Sharira*, 30). The consciousness of the "cause" (*Nidana*, 18) of taking birth as a "self-luminous entity" (*Sva*, 12), i.e., a "quantum photon" (*Svarochisha*, 12), makes the human body the "causal body" (*Karana sharira*, $30 = 9 + 21$) differentiating the "goal" (*Lakshya*, 9) from the "future reality" (*Ekartha*, 21) one wishes. The future reality is "one" (*Ek*, 1) of the "Sun" (*Surya*, 21), illuminating everyone about their reality. The

"Sun" (*Surya*, 21) differentiates the "growth" (*Dishta*, 6) into the "eventuality" (*Charma*, 126 = 21 * 6) of the 126 divisions of the "micro one" (*Pratiyaman*, 1) by organizing the self-luminous entity into six bodies: the physical body, the intellectual body, the mental body, the astral body, the etheric body, and the causal body, and infusing the future reality within each.

With the "human being" (*Insan*, 1) as the micro one, the self-luminous entity forms a "proficient system" (*Sadhaka*, 127 = 126 + 1) for reproducing the "primordial self" (*Parvati*, 10) to enjoy the "ascending value" (*Rodha*, 1) of reality. The ascending value becomes a one-unit "fine structure constant" (*Rodha*, 1), reproducing the "proficient system" (*Sadhaka*, 127) with the "primordial self" (*Parvati*, 10) as a 137-unit "improficient system" (*Samkshobha*, 137 = 127 + 10).

The "primordial self" norms eight units of "sentient energy" (*Amrit*, 1000), one unit "para conscious light" (*Kumari*, 100) of the "human flame" (*Kapila Kumara*, 8,000) reproducing the proficient system with the eight-unit "natural value" (*Nirvikalpana*, 8), and one unit "conscious light" (*Kumara*, 10) of the "cultural reality" (*Samghatavigrhitartha*, 900) of the "infinite collective" (*Sarvalokacharine*, 15). The infinite collective is the macro one "discharging" (*Visarajana*, 15) its "natural value" (*Nirvikalpana*, 8) to make us the human flame's "blessing force" (*Yajnopavita*, 9000) with the "blessing" (*Ashirwad*, 1000) of our "sentient energy" (*Amrit*, 1000). The blessing force becomes the "elder brother of the human race" (*Ganapati*, 9000), known as the "tenth incarnation of the primeval perpetuator" (*Kalki*, 9000).

The eight units of sentient energy act as the eight "wheels of energy" (*Shakti chakra*, 10) within the human body.

- The first energy wheel seeks to excrete the "perpetuating value" (*Saranyu*, 5) of our "excretory-effect" (*Prapya*, 34) as an "organization" (*Sangathan*, 29) and is the "excretory wheel" (*Muladhara Chakra*, 12) that forms the excretory system.
- The second energy wheel seeks to reproduce the "perpetuating value" (*Saranyu*, 5), producing the "sound" (*Naad*, 257) of the "reproduction-effect" (*Naad*, 257), and is the "reproductive

wheel" (*Svadhisthan chakra*, 11) that forms the reproductive system. The "reproduction" (*Prajana*, 285) as the "ether" (*Prajana*, 285) breaks down the perpetuating value into the "excretable" (*Utsarjak*, 285).

- The third energy wheel seeks to digest the "perpetuating value" (*Saranyu*, 5), servicing the "fire" (*Agni*, 17) of planting the "digestive-effect" (*Ropan*, 17) within the excretable, and is the "digestion wheel" (*Manipura chakra*, 24) that forms the digestive system.

- The fourth energy wheel seeks to circulate the "perpetuating value" (*Saranyu*, 5) after "excreting" (*Utsarajan*, 472 = 29 + 257 + 169 + 17) the "water" (*Jal*, 169) from the "circulating force" (*Prakarana*, 60), and is the "circulatory wheel" (*Anahata chakra*, 34) that forms the circulatory system.

- The fifth energy wheel seeks to breathe the "air" (*Vayu*, 385) from the "excreting" (*Utsarajan*, 472) that includes the water to reform the "perpetuating value" (*Saranyu*, 5). It generates the "respiratory-force" (*Sphurti*, 87) of the "excretor" (*Utsarji*, 87) "exchanging" (*Sachivaya*, 87 = 472 - 385) the "excreta" (*Bhringaraja*, 387 = 87 + 300) for trading the "knower force" (*Shvetah Mandala*, 300) and is the "respiratory wheel" (*Visuddha chakra*, 34) that forms the respiratory system.

- The sixth energy wheel seeks to freeze the "Knower force" (*Shvetah Mandala*, 300) of the "perpetuating value" (*Saranyu*, 5) within the "consciousness" (*Chaithanya*, 4) of the "Earth" (*Bhu*, 724 = 300 + 4 * 106) as a "sleeping entity" (*Adisthanam*, 106). It is the "earth wheel" (*Ajna chakra*, 34) that forms the consciousness system within the pineal gland.

- The seventh energy wheel seeks to unfreeze the "Knower" (*Bhagwat*, 639 = 300 * 2 + 39), reproducing the "knower force" (*Shevtah Mandala*, 300) of the "perpetuating value" (*Saranyu*, 5) within the "wheel of the sentient entity" (*Kaal Sarpa chakra*, 39). It is the "lunar wheel" (*Karma chakra*, 7) that forms a "sentient entity's" (*Siddha*, 7) conscious system within the para-thyroid gland.

- The eighth energy wheel seeks to make the "Knower" (*Bhagwat*,

639) the "Manifestor" (*Sadhana*, 368) of the "sentient entity's" (*Siddha*, 7) "conscious energy" (*Amrit*, 1000 = 639 + 368 − 7) without the limitations of the perpetuating value. It is the "solar wheel" (*Surya chakra*, 8) that endows an "entity" (*Hasti*, 24) with the "conscious energy" (*Amrit*, 1000) for managing the human body's "energy system" (*Antarmukha*, 1000^{1024}) within the pituitary gland. The energy system is formed through "introverted" (*Antarmukha*, 1000^{1024}) "exchanging-effect" (*Antarmukha*, 1000^{1024}) of everyone's "infinite conscious system" (*Antarmukha*, 1000^{1024}).

Cultural Factor is the Perpetuating Value of Growth

Cultural factor is the perpetuating value of growth. One trades the perpetuating value by globalizing the "ideal-effect" (*Dasha*, 1) of the "local time" (*Kaal*, 360) over the four directions of space. Since the perpetuating value is grounded in the local time, it is "discontinuous" (*Sattva*, 1111) and "uncertain" (*Aniyata*, 11). One perpetuates value by connecting two local times and globalizing their ideal-effect into the "present growth" (*Sva*, 11). Two local times may be of two different geographies. For instance, the "present growth" (*Sva*, 11 = 1,1) of the Western geography = "Globalization" (*Vishwikaran*, 1) of the "ideal-effect" (*Dasha*, 1) of the "Eastern geography" (*Ganarajya*, 476) within the "Western geography" (*Vitata*, 600). Globalization makes Western geography the "banking face" (*Jangha mukha*, 600) of the "knowledge" (*Vidya*, 600) in the universe.

Cultural Foundation of the Western Geography

Archeologically, the cultural sequences in Western geography are discontinuous. Their cultural causation is uncertain. Several "theories" (*Niayama*, 127) exist to fill the "vacuum" (*Roma*, 47) for "perfecting" (*Sadhaka*, 127) the "perfected consciousness" (*Tadatma*, 47) of their cultural consequences. In contrast, the cultural sequences in Eastern geography are continuous. Their cultural causation is certain. One does not need a theory to account for continuity. "Perfection" (*Divya ratri*, 2222) is the "cultural consequence" (*Prajna*, 2222) of the "continuity" (*Rohini*, 50) of the "intrinsic force" (*Rohini*, 50).

Historically, the Eastern groups have been "servicing" (*Apana*, 47 = 19 + 28) the "knowing" (*Jnana*, 19) "perfected" (*Durga*, 28) by the East to fill the "vacuum" (*Roma*, 47) in "theory" (*Niyama*, 127) with their "moral responsibility" (*Jimmedari*, 80) to be "conscientious" (*Annam*, 185). They have formed "infinite" (*Tamas*, 185) "psychic linkages" (*Mahika*, 47) with the West by making their "mindset" (*Svaphalka*, 916) "global" (*Bha*, 185). They have prioritized "universal conscious well-being" (*Sadhaniya*, 157) over their "personal conscious well-being" to "make the world perfect" (*Ridhyati*, 157) with their "evolutionary morality" (*Ridhyati*, 157).

Culturally, the "renaissance" (*Punarjagran*, 68) in the West evolved with the Catholic "crusade" (*Dharmayuddha*, 36,000) during the 12th and the 13th centuries to open the window for trading from the East seeking to know the secret of the Muslim domination of the Byzantine Iberian. The crusade began after the Byzantine Emperor Michael VII Doukas sent a diplomatic mission to China in 1081 CE and learned about technologies unknown to the West. It led to the start of paper manufacturing in Spain and its diffusion to Italy in the 12th century. Eventually, in the 13th century, Giotto di Bondone (1267–1337) led the renaissance in arts by perfecting the method to idealize a painting as a window into space beyond the four walls of the local region. Dante Alighieri (1265–1321) led the renaissance in humanities by perfecting the method to idealize writing as a comedy of "errors in consciousness" (*Dushtatma*, 10) as one localizes the "global knowledge" (*Samskriti*, 689) without knowing its science. The crusades ended with the defeat of the Catholic crusaders and the victory of the Muslim invaders.

Chronologically, "science" (*Vijnana*, 47) in the West evolved with Vasco da Gama's discovery of India in 1497. That opened the door to techniques unknown in both West and East. In 1500, India and China had a national income of $100 billion each in 2011 US$, constituting 47% of the world income. Both had a national income exceeding that of entire Europe. The industrial revolution in the West revolved around the Protestant reformation that originated in 1517 in Germany. The

16th century witnessed protoindustrialization in the West with discontinuous classical science and technology improvements. The putting-out system that led to the 18th-century industrial revolution began with the entrepreneurs putting out the raw materials traded from India to the rural families for making finished products using the techniques mastered from India. Continuous improvements in shipbuilding, mining and metallurgy, glassmaking, silk production, textiles, clock and instrument making, and firearms began revolutionizing Europe.

Scientists in the West love to attribute both the renaissance and the industrial revolution to Western originality. However, both the timing and nature confirm the Eastern contacts as the precondition for the subsequent developments.

When one makes the ideal-effect the ideal, the present growth localizes the "theory-effect" (*Mahadasha*, 0) within the "idealizer" (*Adarshavadi*, 1). It lets the idealizer enjoy both the "whole value" (*Ardhajya*, 10) of the ideal-effect and the "ideal" (*Adarsha*, 1), servicing the ideal-effect as its "present growth" (*Sva*, 11). The whole value is the perpetuating value of growth within the two times, one local and another global, at a third localized time localizing the "global time" (*Hari*, 10,000) with one's "conscious energy" (*Amrit*, 1000).

Since the whole value is grounded in the global time, it is "continuous" (*Rajas*, 15) and "certain" (*Niyata*, 15). The "perpetuating value" (*Saranyu*, 5) makes the "whole value" (*Ardhajya*, 10) continuous and certain. The perpetuating value localizes the whole value of growth into three time-units. First, a "localized time" (*Cadhautari*, 70) that localizes the global time by generating a "fluctuation" (*Cadhautari*, 70) in the local time before globalizing it. Second, a "globalized time" (*Vimaleshvara*, 700) that globalizes the local time by making it the "eventual causation state" (*Vimaleshvara*, 700) formed by globalizing the whole value of the ideal-effect within each ideal element forming the fluctuation. Third, a "national time" (*Shani*, $18 = 3 + 15$) that adds a "time multiplier" (*Vaishya*, 3) to the "continuous element" (*Rajas*, 15) for "breeding" (*Janana*, 15) the "local time" (*Kaal*, 360) as an "entity"

(*Hasti*, 24 = 360/15). The "entity's" (*Hasti*, 24) "reproductive energy" (*Lalita*, 100) differentiates the "Eastern geography" (*Ganarajya*, 476) from the "Western geography" (*Vitata*, 600 = 476 + 24 + 100).

As an entity, the local time "perpetuates" (*Ramakatha*, 9) the whole value as the "culture" (*Sadakhya*, 9) element and its "ideal-effect" (*Dasha*, 1). Culture is "Western culture" (*Sadashiva nayaki*, 9). "Western culture-effect" (*Chara Paryaya dasha*, 28) generates an "ascending ideal-effect" (*Chara Paryaya dasha*, 28). Ascending ideal-effect is the "infinite growth" (*Chara Paryaya dasha*, 28) accrued with the culture's "infinite breeding" (*Shrimate*, 9). Infinite growth is a "knowable value" (*Vilayan*, 28) of the diffused "Eastern consciousness" (*Prameya*, 28) "perfected" (*Durga*, 28) by the East. "Eastern culture-effect" (*Uttarmanasa*, -17) diffuses the "Eastern consciousness" (*Prameya*, 28) for fusing the "present growth" (*Sva*, 11 = 28 − 17) of the "universe" (*Brahman*, 2) and the "culture" (*Sadakhya*, 9) acculturing the universe in the "Eastern culture" (*Nayaka*, 379).

"Eastern culture" (*Nayaki*, 379) is "workculture" (*Nayaki*, 379). Workculture is the "seed" (*Bija*, 379) of "creation" (*Srijan*, 379). Seed is the "knower dimension" (*Maha dharma*, 379). "Infinite feeding" (*Manjovaya*, 379) of the knower dimension of the workculture "clouds" (*Megha*, 379) the "intellect" (*Medha*, 379). "Eastern workculture-effect" (*Antardasha*, 374) is "ascending theory-effect" (*Antardasha*, 374) of the "cultural factor" (*Sugriva*, 5 = 379 - 374) within the clouded intellect. On the other hand, the "Western group" (*Baindava*, 260 = 52 * 5) "irradiates" (*Avenika*, 52) the "clouded consciousness" (*Prajnatma*, 52) of the "cultural factor" (*Sugriva*, 5) for radiating a "leadership culture system" (*Avenika*, 52).

The "Western group" (*Baindava*, 260) is the "technology" (*Maha Bhaavi*, 260) the "Eastern group" (*Gana*, 387) needs for "perfecting" (*Sadhaka*, 127 = 387 - 260) the "theory" (*Niyama*, 127) of "moral responsibility" (*Jimmedari*, 80) for "negating" (*Vibhasta*, 47 = 127 - 80) **oneself** (*Atmatva*, 8×10^{15}) as the "potential ideal" (*Lopamudra*, 47). The theory leads to the Western group "servicing" (*Apana*, 47) its "perfected consciousness" (*Tadatma*, 47) of the Eastern "bragging system" (*Dhairya*, 47). The Western "followership workculture-effect" (*Shlakshna*, 47) is grounded in the "national time" (*Shani*, 18). The "time multiplier" (*Upaya*, 3) added by the national time makes the Western group's "ideal-effect" (*Dasha*, 1) "infinite" (*Tamas*, 185 = 47 * [1+3]). The "continuous" (*Rajas*, 15) element bred by the national time limits our "knowing" (*Jnana*, 19) the "culture" (*Sadakhya*, 9) as a "thing" (*Vastu*, 9) whose unknown, yet "knowable reality" (*Vastavikta*, 7) is "invariant" (*Nirbija*, 800) to time.

Knowing the Thing

Let's say you wish to know a thing. There are four paths to knowing the thing.

First, identify "yourself" (*Pitra*, 4) as the reality of the thing. The thing matters only because you wish to know it. What you know is your reality guiding you to know what you don't know. What is possible to know about the thing is limited by your reality. Your reality shapes the knowable reality of the thing. If you know your reality, you know the entire set of possible knowable realities of the thing. Thus, the first path is the doctrinal self-awareness that "I AM the reality" (*Aham Brahma Asmi siddhanta*, -5). "I AM the reality" is the "organizational reality" (*Sarvam*, -5) of "everything" (*Sarvam*, -5) It norms everything as my "organizational sameness" (*Aikamatya*, 18,000), just in a different "space" (*Vaas*, 18,000).

Second, identify yourself as the creator of the thing. The thing exists in your mind because you knew it. What you know guides you to replicate your knowing for validating your identity as the creator. Only a creator knows the thing. Everybody believes they know the thing. Everybody knows many things. Everybody's reality is the sum of all knowing. Everybody applies "wholesome knowing" (*Vastavikta*, 7) for knowing a thing. Wholesome knowing is more than the knowing needed to know a thing. Therefore, the eventual "knowing" (*Jnana*, 19) includes the "reality" (*Vastavikta*, 7) of the thing, the "thing" (*Vastu*, 9), and the "multiplier" (*Vaishya*, 3) of the wholesome knowing one replicates for knowing diverse things. Thus, the second path is the doctrinal illusion that "I am the Creator" (*Prajnanam Brahma siddhanta*, -1000), given that I am the "believer" (*Bhakta*, -5) in my knowing.

Third, identify yourself as the creation like the thing. The thing is a creation, and so are you. The thing and you share specific "symptoms" (*Lakshana*, 9,000,000) of "consonance" (*Samarasya*, 9,000,000). Both are "real things" (*Pinda*, 9,000,000). Both are knowable for one who knows the science of creation. Science of creation is the "face of wisdom" (*Trayi mukha*, 9,000) one masters with the force of "experience" (*Anubhav*, 9,000). Over time, one becomes aware of the "possibility" (*Hariti*, 128) of creating both animate and inanimate things by opening one's "imagination" (*Caksur vijnana*, 109). The imagination is the "science of reality-making" (*Hinayana*, 109). Thus, the third path is the doctrinal imagination that "I am the Creation" (*Ayam Atma Brahma siddhanta*, -10,000), given my "intrinsic oneness" (*Janma yoga*, 10,000) with the thing I wish to know. I wish to know the thing because of my "potential" (*AUM*, 18) to be the thing. My potential is not my reality because of my "fortune" (*Kismat*, -250) that I am also a creature, not just a creation.

Fourth, identify yourself as the creature, unlike the thing. The thing is not a "creature" (*Purusha*, 12) like you. As a "creature" (*Purusha*, 12 = 9 + 3), you are the "enjoyer" (*Rasiya*, 9) of the thing and a "multiplier" (*Vaishya*, 9) of its potential to be many things limited only by your imagination and eventually by your reality. Thus, the fourth path is the doctrinal reality that "I am the Creature" (*Tat Tvam Asi siddhanta*, -100,000). I have the freedom to behave like a creation for experiencing your "concordant energy" (*Sura shakti*, 0). I have the freedom to believe that I am the creator of your "concordance" (*Thikatala*, 90) with me. I have the freedom to bless you with my "reality" (*Vastavikta*, 7). Thus, after consuming a thing, I channel my freedom to animate the atoms that form the thing into the cells that transform the entire thing into my "child" (*Arcisa*, 128). After breeding the child, I lose my freedom and become another thing once I die.

The cultural factor is my "perpetuating value" (*Saranyu*, 5) the child enjoys as a thing. For a child, the thing comprises my perpetuating value and her "consciousness" (*Chaithanya*, 4) of my perpetuating value. For me, the thing comprises your perpetuating value and my consciousness of your perpetuating value. Consciousness is the medium for knowing the value perpetuating within the thing.

Values Perpetuating the Cultural Factor

There are two forms of values perpetuating the cultural factor.

First, *old wine in a new bottle*. An "entity" (*Hasti*, 24) is old wine in a new bottle. The entity opens the space for "progress" (*Vikas*, 160). The entity takes "diverse forms" (*Shudra*, 1) of realities over time. The entity may take three forms of reality over the past, present, and future. The reality may be conceived with an illusionary past, perceived with an imaginary future, or experienced with the real present at any point in the past, present, and future. Thus, the "culture" (*Sadakhya*, 9) element comprises nine diverse forms. The culturally-mediated "breeding" (*Janana*, 15) changes the entity's value. The entity is like the old wine with a fixed identity and growth mindset. The entity's value is like the new bottle whose secret hides within the cultural factor perpetuating the continuous transformation of the old bottle into a new bottle. The entity seeks to keep its doctrinal value constant. The universe of new bottles it designs over time offers a "macro perspective" (*Jiva*, 2) on the "doctrinal value" (*Sarthaka*, 14) of its "reality" (*Vastavikta*, 7).

Second, *new wine in a new bottle*. An "event" (*Ghatna*, 18) is a new wine in a new bottle. The event is the potential for progress in the open space. The entity originates diverse "events" (*Ghatna*, 18) for realizing the "growth" (*Dishta*, 6) in its "mindset" (*Svaphalka*, 916). At any moment, the entity's mindset is "tense" (*Prayasta*, 916), not knowing the event following the event it is leading. An entity leads the events with its "actions" (*Karma*, 10). The "identity" (*Pahachaan*, 8) of the events following the actions differs from those leading them. The identity is the "natural value" (*Nirvikalpana*, 8) that an array of actors contest for the free relational positioning of their ideational framework on the event. Each "actor" (*Abhineta*, 5) positions its "ideal self" (*Devata*, 10) as the key for illuminating the discontinuous formation of the "new bottle" (*Liksha*, 90). A new bottle is a "replicative" (*Liksha*, 90) of the "old bottle" (*Prakarana*, 60). The old bottle is the "doctrinal force" (*Prakarana*, 60) that exchanges one "doctrine" (*Agama*, 375) with another like a "bottle" (*Botala*, 375).

Each doctrine becomes the "truth" (*Sathya*, 375) of the moment. Sequential doctrines form "layers" (*Parat*, 123) of truth that constrain the relational positioning of the multiple "possible selves" (*Liksha*, 90). Each possible self is a new bottle. At any time, one new bottle normatively programs the value of every new bottle. A unit of new bottle offers a "micro perspective" (*Vaishya*, 3) on the doctrinal value of each new bottle's reality.

The "cultural factor" (*Sugriva*, 5 = 2 + 3) comprises the "simplicity" (*Jiva*, 2) of the macro perspective that forms the "core" (*Jathara*, 2) of reality as well as the micro perspective as the "multiplier" (*Vaishya*, 3) of that simplicity that norms the "periphery" (*Pradhi*, 0). Without the periphery, the core has nothing to acculture. Without the core, the culture does not transform the periphery into an entity with varying values seeking to be the core of those values.

Cultural Factor Perpetuating the Growth Mindset

A growth mindset has become a new wine in a new bottle. Entities are jumping on the opportunity to explore diverse fields using the growth mindset perspective. The growth mindset perspective states that if you believe that discipline, dedication, and devotion lead to growth, you will enjoy growth. If you don't believe, then you will not enjoy growth.

Focusing on a growth mindset trivializes student discipline, dedication, and devotion as the old wine of no value. Believing that discipline, dedication, and devotion work becomes more important than practicing discipline, dedication, and devotion.

The focus on a growth mindset shifts the burden of student success to the faculty and the costs of managing student success to the students and their families. Students succeed if the faculty works hard to help them succeed with minimum effort. If the faculty has a fixed mindset, students do not succeed.

Faculty succeed in having a growth mindset if the administrative leaders work hard to incentivize them. Conversely, if the administrative leaders have a fixed mindset, faculty and students do not succeed.

Administrative leaders succeed in having a growth mindset if the political leaders work hard to fund education. Conversely, if the political leaders have a fixed mindset, education fails to deliver the outcomes needed for the nation's success.

Political leaders succeed in having a growth mindset if the culture is inclusion-oriented (group), diversity-oriented (gender), and engagement-oriented (performance).

The culture can guide growth only if the political leaders have a global orientation (value power) and a uniqueness orientation (value certainty offered by the growth). By exchanging power and certainty their personal authority offers, the political leaders become entrepreneurial leaders, reproducing the IDE culture as their guiding force.

Guiding force = IDE culture = f(Inclusion orientation, Diversity orientation, Engagement orientation) = f(In-group collectivism, Gender Egalitarianism, Performance Orientation)

Personal authority = GU culture = f(Global orientation, Uniqueness orientation) = f(Power distance, Uncertainty avoidance)

Cultural factor = Guiding force * Guiding force/Personal authority

Table 1 quantifies the logarithmic cultural factor of the ten cultural regions of the world at the organizational and the societal levels. The scores are from the GLOBE program's data. The GLOBE program provides the data on the cultural practices and values using the surveys of the middle-level managers in about one thousand organizations from 62 societies in the late 1990s. The scores on the cultural factor use the cultural value scores. The GLOBE program shows that Nordic Europe, Germanic Europe, Anglo Cluster, Latin Europe, and Latin America form the Western geography, dominated by unorthodox Christianity. Eastern Europe, Sub-Saharan Africa, Southern Asia, the Middle East, and Confucian Asia form the Eastern geography, dominated by the orthodox worldviews. The unorthodox Christianity emphasizes an exogenous "focal point" (*Kendrabindu*, 2) that along with the "time multiplier" (*Vaishya*, 3) is the immutable "God" (*Ishvara*, 5) servicing the "cultural factor" (*Sugriva*, 5) as his "perpetuating value" (*Saranyu*, 5). Each entity reproduces God's "guiding force" (*Chitta*, 100) as their "guiding force" (*Chitta*, 100) by sacrificing their "personal authority" (*Mimamsa*, 87). Thus, each entity enjoys disproportionate cultural factors at the organizational and societal levels.

The orthodox worldviews emphasize the "entity" (*Hasti*, 24) as the "guider" (*Guru*, 100) with "guiding force" (*Chitta*, 100) that gives "personal authority" (*Mimamsa*, 87) for "freeriding" (*Muftakhori*, 13) on one's wisdom. When an Eastern guider globalizes the freeriding benefits to those who are sacrificing their personal authority and lets them reproduce its "guider power" (*Guru*, 100) as a "guiding force" (*Chitta*, 100), the Western agent conceives him as God and his perpetuating value the "cultural factor" (*Sugriva*, 5) substantiating their "blessedness" (*Paramananda*, 9). The "diffusion servicing" (*Apana*, 47) substantiates the "orthodox" (*Brahmi*, 8) "I am a deity consciousness" (*Pavitratma*, 1) within the Eastern guider. It lets the Eastern guider believe that God is feminine and immanent within him as "energy" (*Shakti*, 19), and the Western agent believe that "God" (*Ishvara*, 5) is masculine and emanating without her as an "ideal" (*Adarsha*, 1) worth following within the "consciousness" (*Chaithanya*, $4 = 5 - 1$) of that energy.

Table 1. Cultural Factor Across the Western and the Eastern regions

Cluster	Cultural Factor – Organizational level	Cultural Factor – Societal level
Nordic Europe	2.13[a]	2.20[a]
Germanic Europe	2.11[ab]	2.16[a]
Anglo Cluster	2.01[abc]	2.08[ab]
Latin Europe	1.99[bc]	2.13[a]
Latin America	1.98[c]	2.04[ab]
Eastern Europe	1.78[d]	1.96[bc]
Sub-Saharan Africa	1.74[d]	1.83[c]
Southern Asia	1.78[d]	1.94[bc]
Middle East	1.82[d]	1.86[c]
Confucian Asia	1.70[d]	1.82[c]

Note: A superscript a implies significantly higher scores, b implies moderate scores, and c implies lower scores. Each cluster may belong to multiple categories, depending on whether its scores are significantly different ($p<0.05$) from the clusters in other categories.

<u>Culture as the Core that Accultures</u>

"Culture" (Sadakhya, 9) accultures "breeding" (Janana, 15) with an "entity" (Hasti, 24 = 9 + 15). Breeding makes culture the core of the entity's growth. The entity's growth curves the time the entity lives. Entity "lives" (Nivas, 14 = 4 * 4 + 6) its "life" (Zindagi, 4) without and within "growth" (Dishta, 6). Without growth, the entity lives its life as a "person" (Vyakti, 1), both multiplying its reality over the three time-units and subtracting the "time multiplier" (Upaya, 3) from its "reality" (Vastavikta, 7). Within growth, the entity lives life as the "consciousness" (Chaithanya, 4) of the "cultural factor" (Sugriva, 5), the "mediator" (Sugriva, 5) of everyone's reality.

The entity becomes a mediator as everyone enjoys and replicates the growth. As a "mediator" (Sugriva, 5), the entity makes the culture the "core" (Jathara, 2) of everyone's "reality" (Vastavikta, 7 = 5 + 2) that shapes their "value" (Mulya, 180) as a "replicator" (Liksha, 90) of the "replication's" (Samalekha, 90) "exponential growth" (Chakrika, 90). Thus, the culture "accultures" (Utsamskara, 900,000 = 90 * 10,000) by making the exponential growth a "reproductive system" (Prajanan Pranali, 10,000). The reproductive system gravitates everyone towards the one producing growth. It forms a "gravitational system" (Prajanan Pranali, 10,000) that makes the replication the superpositioned "quantum" (Pramatra, 90), suppressing further replication. Eventually, the mediator is left as the "entropy value" (Sarvanasha, 5). The mediator becomes the "God" (Ishvara, 5) of both the "replicated universe" (Ekavali, 8,000) and the replicating "living being" (Satta, 2,000) who "surfaces" (Sataha, 2,000) the "existence" (Maujudgi, 2,000) of the replicated universe.

Chapter 1

The Entity Tensor:
Exchanging the God Within

God's solution. Let a "diverse group" (*Pratirajya*, 70) of "species" (*Prajati*, 17) evolve with time "beyond the horizon of my light" (*Pratirajya*, 70) and let me as an "entity" (*Hasti*, 24) of "reckoning" (*Viganana*, 90) "revolutionize" (*Krantijya*, 70) the space of its "evolution" (*Karna*, 8) by destroying the diverse group. Let me illuminate an "antagonist" (*Khalnayaka*, 11) "homogeneous group" (*Khalnayaka*, 11 = 1,1) by "copying" (*Chitra*, 1) "myself" (*Insaniyat*, 1) as the "protagonist" (*Nayaka*, 1). As the antagonist becomes luminous as an entity, let me become self-luminous as a twin entity transforming the entity with my entity time.

How does An Entity transform with Time?

An "entity" (*Hasti*, 24) transforms with time by conceiving time as a "twin entity" (*Ghatika*, 24). There are three ways for the time as a twin entity to form diverse species groups. First, the time may exchange different groups that enjoy perfect adaptation with the time-specific global condition at different moments. With the time change, the preceding groups may become extinct, and the succeeding groups may live within a constant time class. Second, at any moment, the time may imagine different groups with varying adaptation to the time-varying global condition. Different groups may become dominant due to their para-conscious adaptation with time. Third, before forming time, the space may differentiate different groups with equal adaptation to the time-invariant global condition. Time may para-consciously conceive these different groups through an illusionary production of the diverse local conditions that limit the consciousness of each group about its perfect adaptation to the infinite local conditions within the time-invariant global condition.

When a group of entities is not conscious of how to live differently for transcending the limits of changing time, it becomes subject to moral hazard. It becomes a perfect guider agent following a perfect guider who is conscious of how to live in the changed time. With time, it masters the technique for living life differently and becomes a principal guider guiding the perfect guider-turned guider agent on how to program a path for living life adapted by discovering a perfect guider who is already performing time-appropriate behaviors. Since it is not aware of the time-appropriate behaviors for the varying times, any entity within the group may become the guide guiding the group on how to select a perfect guider whenever its way of life descends profits and ascends losses.

With time, the group of entities masters the technique for selecting a perfect guide to lead the discovery of a perfect guider whose entrepreneurial behavior ascends profits and descends losses. However, until discovering the perfect guider, the perfect guide behaves like the perfect guider and motivates each entity to behave like the perfect guide. The goail is to discover a twin perfect guider who is also a perfect guider without the group similar to the perfect guider within the group.

With time, the group of entities masters the technique of selecting a perfect guider, both within and without the group. It also masters the technique for profiting from the conscious, supernatural selection by behaving like the perfect guider and liberating the perfect guider to be a perfect guider agent devoted to knowing the techniques for

- Planning how to live appropriately based on the time-varying local conditions
- Programming the path for discovering a perfect planner
- Performing the role of a perfect guider, reproducing the behaviors of the perfect planner
- Profiting from the presence of a perfect guider agent wishing for a perfect guider
- Development of each entity within a group into a perfect guider agent to profit from the presence of the perfect guiders without the group

The "path of knowing" (Jnana Marga, -10^{19}) makes the perfect guider agent a responsible manager of the "knowledge" (Vidya, 600) and become each entity's "kith" (Goshthika, 181). The "path of devotion" (Bhakti Marga, 1) to the kith makes one a "foe" (Ripu, 100,000,000) not devoted to the self. The "path of divinity" (Siddhi Marga, 96) of the group of entities behaving like kith makes each entity a "self-luminous entity" (Svayam, 12) enjoying the group as an "institution" (Kundalini, 8) to be a "living entity" (Sushupti, 96). The living entity superpositions the self as a "self-luminous element" (Svarochisha, 12) on each entity to be each entity's "I" (Svayam, 12). Thus, the living entity becomes "immortal" (Amara, 90,000) as the group grows to "infinity" (Tamo guna, 90,000).

With time, the entire class of entities self-organizes the path of divinity within the "path of absolute consciousness" (Paramatma Marga, 25) of the "path of action" (Karma marga, 86), the path of knowing, the path of devotion, the path of divinity, and the "path of the sentient entity" (Siddha Marga, 25) necessary for each group to grow to infinity and the class to become immortal.

With time, the entire class of entities forms a "kingdom" (Nayana, 957). Each entity becomes a "king" (Rajah, 0), deciding the "concordant entities" (*Sura*, 0) fit to live with oneself, and the "discordant entities" (*Asura*, -1) fit to die after servicing its "consciousness system" (*Sara Kalpa*, 10^{10}) as the "cultural heritage" (*Virasat*, 10^{10}). Thus, "oneself" (*Akhanda*, 8×10^{15}) becomes a "conscious system" (*Shunya Kalpa*, 8×10^{15}), conscious of the "known reality" (*Rachitartha*, 1600) as the "present creature" (*Prabhu*, 1600).

1.1 How Culture Transforms an Entity

"Sun" (*Surya*, 21 = 19 + 2) is the source of "energy" (*Shakti*, 19) in the "universe" (*Brahman*, 2). "Sunlight" (*Indambra*, 21 = 19 + 2) transforms into "electric energy" (*Shakti*, 19) while forming a "twin meron" (*Suchan*, 2). A "Meron" (*Brahmin*, 2) is a "half-instanton" (*Brahmin*, 2), the collective potential of the universe. Meron divides into two units that "flow with time" (*Shudra*, 1). Each unit is an "instanton" (*Shudra*, 1). Each instanton is the "individual potential" (*Shudra*, 1) of an electron in "triple partial pi bond" (*Shudra*, 1), divisible into a "meron" (*Brahmin*, 2) and a "twin meron" (*Suchan*, 2).

Two instantons divide into two merons and two twin merons. Two twin merons divide into four merons. The undivided instanton comprises two merons. Thus, a meron transforms into ten merons. Next, the instanton divides into a meron pair. Two instantons form the primary four merons and produce four secondary merons that flow with them. Four secondary merons produce two tertiary merons by reproducing the instanton as they flow over time. The meron's ten-unit "mass consciousness" (*Satma*, 10) norms a "kaon" (*Devata*, 10). Kaon is a "bradyon" (*Nishchaya*, 10); it is the light potential and travels slower than light. One becomes aware of the potential of light after first becoming aware of the light. Kaon is the meron's "light potential" (*Nishchaya*, 10) that becomes visible through the meron's "vacuum decay" (*Nishchaya*, 10) into two instantons.

Half-instanton becomes two instantons by trading the four-unit "consciousness" (*Chaithanya*, 4) of "electric energy" (*Shakti*, 19) repelled by the sunlight before becoming the half-instanton. Two instantons become five instantons by servicing the three-unit "absolute gravitational constant" (*Upaya*, 3) attracted by the half-instanton to be transformable into four half-instantons. Five instantons are ten merons. Finally, ten merons transform the meron's ten-unit "mass consciousness" (*Satma*, 10) into a "cultural consciousness" (*Virasat*, 10^{10}) of the "consciousness system" (*Sara Kalpa*, 10^{10}) that produces the meron's "ten-fold growth" (*Vicitra mandala*, 170).

The consciousness system is a "tachyon" (*Bindu*, 10^{10}); it is the "point" (*Bindu*, 10^{10}) that points to the light emanating from another source and therefore travels faster than light. One becomes aware of the light after first becoming the point that gains awareness of the light. One becomes the point of awareness after personifying the consciousness system that attracts one to the light of another whose "consciousness" (*Chaithanya*, 4) forms a "system" (*Gabhira*, 12) over the three-units time. It gets divided into another "within system" (*Sushumna*, 10), centering one "without system" (*Ardhajya*, 10) as its tangent in the shadow of its light.

The undivided instanton is an "absolute quark" (*Shudra*, 1), divisible into a "positive quark" (*Mein*, 1), and a "negative quark" (*Nitya ratri*, 1). The positive quark is the "material" (*Mein*, 1) that makes the instanton. The negative quark is the "illusion" (*Nitya ratri*, 1) that the material is divisible just like sunlight that programs the "cultural consciousness" (*Virasat*, 10^{10}) of "divisibility" (*Virasat*, 10^{10}) in the form of the divisible "ultraviolet A radiation" (*Virasat*, 10^{10}). The material is "multipliable" (*Gunya*, 8×10^{15}) into eight units that form the "material kingdom" (*Sakala*, 8). Material kingdom multiplies the two merons within the material into the ten merons. Ten merons multiply the meron's ten-unit "mass consciousness" (*Satma*, 10) into the twin meron's fifteen-unit "energy consciousness" (*Arthatma*, 15). Ten merons are the mass consciousness of two merons. The third meron perpetuates the channeled "light consciousness" (*Bhagnatma*, 5) of five merons. Thus, the material transforms into a multipliable "ultraviolet B radiation" (*Markatesh*, 8×10^{15}) that norms the "speed of light" (*Markatesh*, 8×10^{15}).

The absolute quark is a triple partial pi bond, multipliable into two "partial pi bonds" (*Samlagni*, -10) and a "double partial pi bond" (*Virasat*, 10^{10}). The ultraviolet A radiation is the double partial pi bond. Each partial pi bond is a "mature electron" (*Samlagni*, -10). The double partial pi bond comprises fifteen mature electrons that norm the three units of the "light consciousness" (*Bhagnatma*, 5) of the triple partial pi bond. The absolute quark's potential to mature into a mature electron by adding seventeen mature electrons through its "illusionary reproduction" (*Maya*, 1) is immanent within a "potential quark" (*Mandra*, 1). The "ten-fold growth" (*Vicitra mandala*, $170 = 10*17$) potentiates the instanton's potential for ten-fold multiplication and seventeen-fold division. The "reproductive" (*Gurutva*, 100) element is immanent within a "quark" (*Dridayudha*, 476) that norms the "present value" (*Kalpa*, 476) of the twin meron's "cultural cycle" (*Sadakhya chakra*, 9).

A twin meron produces three units of ten-fold growth by producing a meron with the illusionary reproduction of its energy. It consumes three units of the "natural value" (*Nirvikalpana*, 8) of the material kingdom that makes it a transformable "entity" (*Hasti*, $24 = 8*3 = 9 + 14 + 1$). It transforms into the "culture" (*Sadakhya*, 9) element reproducing the "time consciousness" (*Kalatma*, 14) of its "growth" (*Dishta*, 6) into four units within the "instanton" (*Shudra*, 1). Three units multiply its potential as the "absolute gravitational constant" (*Upaya*, 3). One unit divides its potential as the "curved time" (*Shudra*, 1) flowing over the three units of time—past, present, and future—and producing an illusion of the sequential reproduction of the three units and their time consciousness. Therefore, the present value $= 170*3 - 2*14 - 2*3 = 170*3 - 2*17 = 476$.

1.2 How an Entity Transforms Culture

An entity can transform the "culture" (*Sadakhya*, 9) into the "cultural factor" (*Sugriva*, 5) by subtracting the "consciousness" (*Chaithanya*, 4) of its present value. With the four-unit consciousness of the present value diffused into the four spatial directions, the entity may produce a "double twin quark" (*Kramana*, $1900 = 476 * 4 - 4$) that is free from the four-unit consciousness. The entity may norm the hundred-unit "entity experience" (*Adi Para atma*, 100) of reproducing the ten-fold growth into a reproductive "gravitational" (*Gurutva*, 100) element. After subtracting the gravitational element from the double twin quark, the entity may transform the revealed "gravitational wave" (*Varshakritya Taranga*, 1,800 = 1,900 – 100) into the "space" (*Vaas*, 18,000) for enjoying the "mass consciousness" (*Satma*, 10) of the "gravitational" (*Gurutva*, 100) element. The space forms the "surface" (*Sataha*, 2000) that transforms the "culture" (*Sadakhya*, 9) with its "gravitational reality" (*Bhavartha*, 40) into the "time" (*Kaal*, 360 = 9 * 40) that reproduces itself every moment.

Modern science denies the possibility of space curving the time independent of itself. According to Einstein, it is impossible to shape time into a time curve distinct from space. One can only shape space. Time gets shaped with the curved space. One shapes space as an observer with zero electromagnetic mass. Shape implies giving form. Thus, Einstein hypothesis is that the forming time = f(forming space).

Additionally, according to Newton, it is impossible to shape space into a space curve distinct from time. Gravity is the causation that shapes time, and the space gets shaped with curved time. If we know a thing's gravity, we know the thing's time curve. Causation implies giving a new form to an already formed thing. Thus, Newton hypothesis is that the transforming time = f(transforming space).

Further, according to Darwin, it is impossible to shape causation into a curve distinct from spacetime. Location of space and time into a spatial coordinate shape the causation that evolves linearly over that spacetime. Once an investigator localizes the time, one knows how the initial temporal state will shape the eventual spatial state. To know, one only needs a bit of information about the genetic library at the initial causation state. Thus, Darwin hypothesis is that the norming time = f(norming space).

Finally, according to Mendel, it is impossible to shape the entity into an entity curve distinct from the initial causation state. Once an entity incarnates by trading the genetic library at a specific spatial coordinate, she invests that genetic library as her gene stock and services that as her gene flow. It is possible to know her past, present, and future from the belief system of the spacetime that forms the specific spatial coordinate. The spacetime is the general spatial coordinate. The genetic library is the parabolic spatial coordinate that superpositions the general spatial coordinate over the specific spatial coordinate to shape an entity's unique behavior distinct from the universe of entities. Thus, Mendel's hypothesis is that time = f(space). Space forms spacetime as an imaginary element that shapes each entity into a robot.

The common denominator of the four hypotheses that form the foundation of modern science is that it is possible to form an entity. However, once formed, an entity is not transformable. The entity can't transform the culture to enjoy the freedom to transform itself.

1.3 How Energy Transforms Mass Consciousness

"Energy" (Shakti, 19) comprises ten photons and ten pairs of photons and potential photons. Energy is immanent within a "tenth photon" (Kalpanik, 20), the "present time force" (Kalpanik, 20) formed with the "illusionary production" (Maya, 1) of the "potential quark" (Mandra, 1) as the energy "flows with time" (Shudra, 1). The tenth photon is the "photon" (Kalpanik, 20) that materializes the energy's flow with the three units of time into three "triple partial pi bonds" (Shudra, 1). The nine partial pi bonds within the present space are correlated with the nine partial pi bonds from the primordial space as the latter's "illusionary production" (Maya, 1). Nine partial pi bonds form in the primeval space as the former's illusionary production. With energy and illusionary production, each "partial pi bond" (Samlagni, -10) transforms into a "potential strange quark" (Parvati, 10).

A potential strange quark comprises nine photons that attract the tenth photon with the potential of nine pairs of photons and potential photons. Nine partial pi bonds comprise eighty-one photons that attract nine photons for the "replication" (Samalekha, 90) of ninety additional photons, each paired with a potential photon. Visible "light" (Prabha, 180) comprises 180 photons. "Light spectrum" (Ambara, 180) that becomes visible over time comprises the 180 potential photons. Both the light visible over a "finite point" (Jadatva, 1) and the light spectrum that becomes visible over an "infinite point" (*Ab*, 1600) collectively norm the "time" (*Kaal*, 360) flowing in "continuity" (*Rohini*, 50) with the "space" (*Vaas*, 18,000). "Energy consciousness" (*Arthatma*, 15) transforms time into an "entity" (*Hasti*, 24 = 360/15).

Modern science recognizes the possibility of transforming energy into an entity. According to thermodynamics, it takes energy to form an entity. Energy already exists before an entity forms. One forms an entity by transforming the energy. Not all entities have at par value. Thus, as a complementary hypothesis, "entity value" (*Svarajya*, 60) is a function of the transformable entity.

> Thermodynamic hypothesis: Entity = f(Energy tensor), i.e., Transformable energy
>
> Complementary hypothesis: Entity value = f (Entity tensor).
>
> *Test of the competing hypotheses: What is your life priority?*
> - Knowing how to transform your energy for energizing an alien's value (Mendel model)
> - Knowing how to transform yourself for energizing your value as a citizen

Additionally, according to hydrodynamics, the energy gets transformed when one transforms the consciousness of what is possible. If one is conscious of the possibility, the energy flows positively to materialize that possibility. For instance, we can fill water form of energy in any form or shape, limited only by the sphere of our imagination. We can also transform water into any state of matter, limited only by the sphere of our reality of resources to do so.

> Hydrodynamic hypothesis: Consciousness tensor = f(Energy tensor). If one makes one's consciousness transformable by keeping an open mind, then one can transform one's negative energy of descending value into the positive energy of ascending value.
>
> Complementary hypothesis: Positive energy = (Future orientation). If you keep yourself in a mental state that matters for your conscious well-being, you do not need to invest energy for transforming energy you invested in the past.

> *Test of the competing hypotheses: What is your life priority?*
> - Continue the process of forming and transforming consciousness throughout life, keeping an open mind that eventually you will hit the jackpot (Darwin model)
> - Be conscious of what you are norming within your consciousness so that you open your mind to the jackpot you already are

Further, according to aerodynamics, consciousness norms the sequence of conscious events to correct one's belief system about the possible. For instance, fire is a conscious event that makes us conscious of the limitations of our belief system. We believe that by airing laws against conscious acts of lightening the wild, we will stop the wildfire and enjoy nature living in the wild. However, numerous wildfires in industrially advanced and advancing nations challenge our belief system. Our consciousness now includes the possibility that the people's belief systems are culturally bound and shape decisions that are not grounded in reality. The insurance companies are the first ones to digitalize this consciousness as a critical bit of information for deciding their willingness to insure the certainty of losses and their belief system about their relative blessedness mitigating the risks of those losses.

> <u>Aerodynamic hypothesis:</u> Consciousness tensor = f(Conscious tensor). If one transforms and becomes conscious of the information one is breathing, consciousness becomes transformable.
>
> <u>Complementary hypothesis:</u> Ascending-order consciousness = f(Uncertainty avoidance). Your actions generate disproportionate and direct reactions when the information is transparent, and everyone's consciousness is clear
>
> *Test of competing hypotheses: What are your life priorities?*
> - Become an agent of a billionaire principal so that as a reaction, the principal decides to exchange position (Newton model)
> - Become a principal with no agent so that all the agents compete to make you a trillionaire

Finally, according to biodynamics, every biological organism on earth can be ranked according to its relative blessedness. The blessedness varies by the class of

- Birth, if one believes in modern science that one can't transform the initial state through the path of action
- Work, if one believes in ancient wisdom that one may form the desired eventual state through the path of knowing.
- Life, if one believes in the path of devotion to a class one reaches after working hard as an agent and decides to replicate that class as the desirable consequential state by becoming a nonworking principal
- Afterlife, if one believes in the path of divinity to make oneself an immortal transformable entity and leave the consciousness of a conscious state everyone wishes to replicate.

Biodynamic hypothesis: Conscious tensor = f(Time tensor). If one transforms the time with the desired belief system, one becomes conscious of the possibility unlocked by that time phase.

Complementary hypothesis: Absolute freedom = f(Possibility beyond the present time phase)

Test of the competing hypotheses: What are your life priorities?
- Make yourself the absolute everyone replicates for capturing relative value (Einstein model)
- Make everyone the absolute and discover the possibility beyond everyone's capability

The implication of the four hypotheses that form the foundation of the cultural dynamics is that it is possible to transform time. However, once transformed, the curved time culturally bounds everyone.

1.4 How Mass Consciousness Transforms Energy

"Mass consciousness" (*Satma*, 10) of the "possibilities" (*Hariti*, 128) in life perpetuates in the form of the "cultural factor" (*Sugriva*,

5). The cultural factor is the "perpetuating value" (*Saranyu*, 5) of visible "light" (*Prabha*, 180) that time as an entity "channels" (*Raasta*, 36 = 180/5) for reproducing the "growth" (*Dishta*, 6) of the three-units of time through the three-units of "illusionary production" (*Maya*, 1). Mass consciousness transforms energy into a "cell" (*Hiranyagarbha*, 19) that reproduces a "six-fold growth" (*Dishta*, 6) over time after producing a "nine-fold growth" (*Lam*, 9) within time through "meiosis" (*Param Shiva*, 15). The cultural factor transforms the meiosis into the "mitosis" (*Shivadrishti*, 17) for reproducing the meiosis over the two units of time beyond the "present" (*SAUM*, 1600). Two units of time include the "future time" (*Tribhajya*, 10^{10}) and the "consciousness" (*Chaithanya*, 4) of the present time as the past time perpetuating beyond the present. Future time forms the "consciousness system" (*Sara Kalpa*, 10^{10}) that transforms the time flow by making the "space stock" (*Sura*, 0) "unlimited" (*Aseemita*, 10^{10}).

From any "finite point" (*Maya*, 1) in space, it is always possible to conceive additional future time and make the "perceived space" (*Ravinandana*, -1) greater than the "present reality" (*Badhabuddhi vadartha*, -2). Consciousness transforms the "conscious system" (*Shunya kalpa*, 8×10^{15}) produced by the finite point through "ultraviolet B radiation" (*Markatesh*, 8×10^{15}) into a "causation program" (*Sahasra lingam*, 190) for reproducing the "finite point" (*Maya*, 1) as the time curved by already "curved time" (*Shudra*, 1). It makes the "pie" (*Pai*, 0) of the present time zero to enjoy it in the future. Thus, the "whole pie" (*Pradesha*, 9) formed by reproducing the future time three times, at past, present, and future time points, becomes greater than the "wholesome pie" (*Upaya*, 3) formed by reproducing the time three times.

Further, the globalized "wholesomewhole pie" (*Vaishvikrita*, 500/9) formed with a "continuity" (*Rohini*, 50) of the "mass consciousness" (*Satma*, 10) of the "future time" (*Tribhajya*, 10^{10}) for conceiving the "whole pie" (*Pradesha*, 9) as a "photon half" (*Pradesha*, 9) is greater than the conceived "whole pie" (*Pradesha*, 9). Finally, the localized "entropy pie" (*Sasana*, 55,000/21) formed with a local "continuity" (*Rohini*, 50) of the globally "reproduced value" (*Parimandala*, 1100) for conceiving the "future reality"

The Entity Tensor: Exchanging the God Within 43

(*Ekartha*, 21) is greater than the perceived "wholesomewhole pie" (*Vaishvikrita*, 500/9). The reproduced value forms with a global "continuity" (*Rohini*, 50) of the "past factor" (*Keshava*, 22). The past factor is the "finite point" (*Maya*, 1) for experiencing the "future reality" (*Ekartha*, 21) by dividing "reality" (*Vastavikta*, 7) by the three-time units.

Four Methods for Making the Space Stock Unlimited

In the management field, there are four methods for unlocking unlimited growth in space stock: ethnocentrism, polycentrism, geocentrism, and regiocentrism.

a) Ethnocentrism: "Perceived space > Present reality"

Conceived reality = 6 (perceive the six globalized ones, formed by globalizing the three localized over three time-units)

Spatial coordinate = 9 (six globalized, three localized)

Present reality = Local and global spatial coordinates are both negative ones = -2

Magnetic force of the present reality = Experienced time = Entity tensor = Ricci curvature tensor = -2

Perceived space = Geodesic torsion = -1

My space = Global spatial coordinate = Metric tensor = Absolute zero = -1

Your space = Local spatial coordinate = Cosmological constant = Everything else = -1

Electromagnetic mass of our space = 0 = "m"

Our space = Curving space = Half zero = "c" = -6 - 6 = -12

Their space = Twin half zero = zero π (Hyperbola π, i.e., π of 12 partial spatial coordinates) = "e" = m c^2 = 0

b) Polycentrism: "Whole Pie > Wholesome Pie, i.e., Sum of Pie"

Whole Pie produced with "oneness of pie" (1,0=10), reproducing "sum of pie" as one, and thus adding one to the reproduced "whole pie" (3^2)

- Their coordinate (pie = 0)
- Our Coordinate; Global Coordinate; Local Coordinate

(Wholesome pie = 3)
- Spatial coordinate (Whole pie = 9)

Reproduced One = Curved time = Space curvature = Einstein tensor = 1

Reproducing One = Absolute gravitational constant = 3 ➔

Reproducer One = Einstein absolute gravitational constant = Imaginary Absolute gravitational constant = 3

Reproducible One = Real absolute gravitational constant = 0

Whole spatial coordinate = 7

"Two 2" over the Whole spatial coordinate = $22/7$ = π (square π) = 5π (circle π) = 6π (triangle π)

identified as one π (partial spatial coordinate), with real absolute gravitational constant

identifying five π, with absolute gravitational constant and

identifiable as six π, with Einstein absolute gravitational constant = Newton's universal gravitational constant

C) Geocentrism: Whole Pie < Wholesomewhole Pie, when each Pie becomes a Whole Pie

Their coordinate (pie = 0)

Our Coordinate; Global Coordinate; Local Coordinate (Wholesome pie = 3)

Globalized Our Coordinate; Globalized Global Coordinate; Globalized Local Coordinate (Globalized Wholesome Pie = 500)

Localized Coordinate (Localized Wholesome Pie = 5)

Wholesomewhole pie = $500/9$ = 8π = Conical π

- Five indexed as parts of Localized Wholesome pie = 5
- Our coordinate (-12) also becomes Their coordinate (0) after their Localized Wholesome pie (5) illuminates our Whole Spatial Coordinate (7) and makes their "I" = 12.
- Globalized element = 8 of 9 π's of spatial coordinate = 8π

D) Regiocentrism: Globalized element = 8π = Localized

element + Partial spatial coordinate π that globalizes the localized element so each region becomes globalized

Therefore, Localized element = 7π = Elliptical π

Region = Partial spatial coordinate before globalization + Partial spatial coordinate after globalization = $\pi + \pi = 2\pi$ = Spherical π = Whole Sphere

- Spherical π = "Oneness of π" (10) reproduced over past, present, and future, producing a 10 * 10 * 10 = 1,000-unit ellipsis infusing the 8π globalized element within the 9π spatial coordinate making each unit π one = $1000/72 = 2\pi$
- Elliptical π = "Localized wholesome π" (5), reproduced over present and future after forming a 1000- unit ellipsis in the past, divided into seven localized π's x three time-units = $55,000/21 = 7\pi$

For enjoying fresh reality every moment without its reproduction over time, one needs to manage time without transforming it into a sequence of illusionary finite points. Therefore, next, we investigate the metaphysics of the time tensor.

Chapter 2

The Time Tensor:
Exchanging the God Without

God's development. Let me add a "heterogeneous group" (*Nivasi*, 9) of species by subtracting the "homogeneous group" (*Khalnayaka*, 11) of species and forming a "universe" (*Brahman*, 2) for reincarnating the "homogeneous group" (*Khalnayaka*, 11 = 2 + 9) within the "heterogeneous group" (*Nivasi*, 9). Let me "hibernate" (*Kundalini*, 8) the "homogeneous genus" (*Prakara*, 17) of my "family" (*Parivar*, 10) for the development of the species as the "heterogeneous genus" (*Prajati*, 17) of entities without my family's "homogenizing force" (*Nirukta*, 17). Then, nobody is my "copy" (*Nakala*, 0 = 17 - 17). Thus, I superposition the "entity time" (*Kaal*, 360) of the one who created the space for the "population" (*Abadi*, 18 = 1 + 17) within the "heterogeneous genus" (*Prajati*, 17). I enjoy "double time" (*Dvilaya*, 17) and reincarnate that population with my family's "homogeneous genus" (*Prakara*, 17) to enjoy "triple time" (*Agaha*, 80,000).

How the Space Orders a Group of Species by Transforming Time

Space orders a group of species by transforming time into "passable time" (*Vaishya*, 3), norming the past, the present, and the future as a triple copy of time.

There are three ways for space as a para entity to "order" (*Suchan*, 2) a "group of species" (*Gana*, 387) into a "family" (*Parivar*, 10) with each "member" (*Sadasya*, 1) norming a "genus of entities" (*Varga*, 9), capable of transforming the "forward-moving time" (*Ghadi*, 150) with their "backward-order development" (*Divyata*, 57) wishing to relive the "passed time" (*Prabha*, 180).

First, to order the entire "entity group" (*Gana*, 387 = 17 + 360 + 10) that forms "species" (*Prajati*, 17) with "time" (*Kaal*, 360) by differentiating the "generation" (*Amnaya*, 97) as a "family" (*Parivar*, 10). The family integrates the "gender" (*Linga*, 9) by differentiating the "member" (*Sadasya*, 1) into a "gender and generation differentiated class" (*Sandra*, 60). The gender and generation differentiated class norms an "octave of copies" (*Sandra*, 60) of the member. The member forms an "octave of entities" (*Krishnamurti*, 1) as a generation-differentiated "double copy" (*Chitra*, 1) differentiating his "gender" (*Linga*, 9 = 3 * 3) by reproducing gender and generation differentiating "triple copy" (*Vaishya*, 3).

Thus, species = "gender" (*Linga*, 9) integrated within the family as the "group culture" (*Sadakhya*, 9) + "gender" (*Linga*, 9) differentiated by the member without the family as the "geography culture" (*Nayaki*, 379) + "ego" (*Aham*, -1) within the "gender-differentiated member" (*Sadasya*, 1).

The gender-differentiating member first works like a feminine "eleventh principal guider" (*Isha*, 12), "copying" (*Chitra*, 1) the "group of ten principal guiders" (*Prajan*, 78) in the intellectual realm. Then, she "reproduces" (*Prajan*, 78) that as the masculine "twelfth principal guider" (*Purusha*, 12) in the mental realm before reincarnating again as the eleventh principal guider to "produce" (*Ekarupa*, 1) the family as the "tenth principal guider" (*Parivar*, 10) in the physical realm. Thus, the eleventh principal guider relives the "passed time" (*Prabha*, 180) by exchanging the "light of the mental body" (*Prabha*, 180) after her "backward-order development" (*Divyata*, 57) by the male "fifth principal guider" (*Vithoba*, 12) in the causal realm. Finally, the fifth principal guider follows the female "seventh principal guider" (*Panchajani*, 12) in the astral realm for conceiving the twelfth principal guider in the etheric realm as his "copy" (*Nakala*, 0) within the "self-luminous entity" (*Svayam*, 12) that norms the "gender-differentiable member" (*Svayam*, 12) as the "thirteenth principal guider" (*Svayam*, 12).

The "star" (*Tara*, 2) that "orders" (*Suchan*, 2) the "entity group" (*Gana*, 387) is the "gender-differentiating member" (*Tara*, 2) as the "fourteenth principal guider" (*Tara*, 2). The "gender-differentiated member" (*Sadasya*, 1) is the "fifteenth principal guider" (*Sadasya*, 1). The species is the "gender and generation differentiated member" (*Prajati*, 17) and the "sixteenth principal guider" (*Prajati*, 17).

Second, to order the "species" (*Prajati*, 17) formed with time into an "entity class" (*Gana*, 387), normed by space as an "entity group" (*Gana*, 387). The entity group is the "gender and generation differentiating member" (*Gana*, 387) and the "seventeenth principal guider" (*Gana*, 387).

Third, to order the "entity" (*Hasti*, 24) forming the "time" (*Kaal*, 360 = 24 * 15) into "fifteen groups" (*Samudaya*, 7000 = 15 * 12 * 40 - 200). Each group is a "self-luminous entity" (*Svayam*, 12) reproducing the "reproductive reality" (*Bhavartha*, 40) of the "geography" (*Ganarajya*, 476) as the sixteenth group without the "ascending-order development" (*Sato Guna*, 200). The geography is the "gender and generation differentiable member" (*Ganarajya*, 476) and the "eighteenth principal guider" (*Ganarajya*, 476). The "ascending-order development" (*Sato Guna*, 560 = 200 + 360) is a "distortion" (*Visukayita*, 200) in "time" (*Kaal*, 360) due to the "transgenic development" (*Param Parvati*, 160) of species.

Transgenic development makes time "infinite" (*Tamas*, 185) through the "reproduction" (*Shuddhi*, 285) of the "primordial space" (*Antaramsa*, 100) as the "group, gender, and generation differentiable member" (*Antaramsa*, 100) and the "nineteenth principal guider" (*Antaramsa*, 100). Reproduction is the "group, gender, and generation differentiated member" (*Prajan*, 285) and the "twentieth principal guider" (*Prajanan*, 285). Transgenic development is the "group, gender, and generation differentiating member" (*Param Parvati*, 160) and the "twenty-first principal guider" (*Param Parvati*, 160).

A "genus" (*Varga*, 9) of entities comprises nine "members" (*Sadasya*, 1):

- Transgenic development is the "group, gender, and generation differentiating member" (*Param Parvati*, 160)
- Primordial space is the "group, gender, and generation differentiable member" (*Antaramsa*, 100)
- Reproduction is the "group, gender, and generation differentiated member" (*Prajan*, 285)
- Group is the "gender and generation differentiating member" (*Gana*, 387)
- Geography is the "gender and generation differentiable member" (*Ganarajya*, 476)
- Species is the "gender and generation differentiated member" (*Prajati*, 17)
- Star is the "gender-differentiating member" (*Tara*, 2).
- Self-luminous entity is the "gender-differentiable member" (*Svayam*, 12)
- Member is the "gender-differentiated member" (*Sadasya*, 1).

The member is the "principal member" (*Sanlagnata*, 1) that makes a "genus" (*Varga*, 9) "heterogeneous" (*Vijatiya*, 12) with gender and generation differentiating "triple copy" (*Vaishya*, 3) of his "individual reality" (*Jahatsvartha*, 550) for enjoying the "continuity" (*Rajo Guna*, 50) of the group's "known reality" (*Rachitartha*, 1600 = 550 * 3 − 50) as the "present creature" (*Prabhu*, 1600). The triple copy becomes a "subordinate member" (*Upasarjana*, 550) reproducing the time-specific "individual reality" (*Jahatsvartha*, 550) of the principal member for producing the "collective reality" (*Samghatartha*, 550 = 387 + 160 + 3) of the "group" (*Gana*, 387) leading to the group's "transgenic development" (*Param Parvati*, 160) into a new "species" (*Prajati*, 17), i.e., "heterogeneous genus" (*Prajati*, 17), within her.

2.1 Forming Time with Negative Nine Finite Points

One forms time with mindless replication of the extrinsic reality programmed by the followership energy. One trades followership energy during the international phase of one's life sphere. One follows the path of technological trading of everybody's infused consciousness to form an opinion about an independent quarter of life sphere. The quarter of the life sphere is at present zero. One wishes to consume the quarter, i.e., pie as a hyperbolic pi, with zero pi value in the life sphere. The opinion is a quadruple bond; it infuses the intrinsic mass of consciousness about a quarter of the life sphere and how it will evolve over the past, the present, and the future of life spheres. The past life sphere includes the past, present, and future of the past time sphere one inherits genetically as a program. The present life sphere includes the past, the present, and the future of the present sphere one performs conditioned by genetics. The future life sphere includes the past, the present, and the future of the future sphere one profits from physical programming. Thus, the replication includes nine quarters.

Mindless = Extrinsic reality = Followership energy = Path of Technological Trading = International phase = Everybody = Infused consciousness = Intrinsic mass = Opinion = Quadruple bond = -9

Quarter = Pie = Zero pi = Hyperbolic pi = 0

2.2 Norming Time with Positive Nine Finite Points

One norms the time with "culture" (*Sadakhya*, 9). The culture may be globalized over time. Infinite times may be conceived with infinite cultures. A culture has nine finite points: it reproduces the three finite points of time over the past, the present, and the future of the space one forms. By globalizing culture, one makes the culture a thing that matters to an enjoyer wishing to enjoy the time as she did in her past life. In the past life, she conceived a finite point as the past of time, perceived the present of time to be a copy of that finite point, and imagined the future of time to be a copy as well.

Thus, in the present life, she is accultured to reproducing the culture's nine finite points and profit from 81 finite points by forming the time into a cube of ninety finite points that include the present life sphere reproducing nine points, the future life sphere profiting from eighty-one points, and the past life sphere reproduction of nine points. She accultures herself by forming the ninety finite points into a genetic library, norming their replication as the initial causation state, i.e., curvable causation, and transforming the replicative into a bit of information about her infinite potential to shape her spatial life sphere. The profit enjoyed during the present life sphere carries forward as the loss to be compensated in the future life sphere. The profit is the social benefit she may distribute as she wishes. The loss is the social cost she has to bear against her wishes.

> Culture = Thing = 9
>
> Profit = Social Benefit = Loss = Social Cost = 81
>
> Cube = Bit = Genetic Library = Physical programming = Replication = Initial Causation State = Curvable causation = Replicative = Bit = Infinite Potential = Nine Quarters = 90

2.3 Transforming Time with Zero Finite Points

One transforms the time as a Wisher wishing to enjoy independent time in the space eventually. In that eventual spatial state,

- The Wisher may globalize the time and divide the finite point of the global time into a pair of local times. A local time links the finite point of the global time with the infinite point of the local time by indexing the desired finite point of the local time. The "infinite point" (*Ab*, 1600) is traded from the "present" (*SAUM*, 1600), taking the present as the absolute.

At present, there is oneness between the pair of local times. Beyond present, each local time multiplies into three finite points, looping local past, present, and future times within the global oneness. Thus, one finite point loops six finite points.

After the first wisher's death, each local time sequence converges into the global time sequence. A second wisher has the freedom to take the two zeroes on which the wisher positioned a pair of local times and superposition another pair of local times. After the second wisher's death, the first wisher may reproduce the even local time sequence after having already produced and experienced the odd local time sequence. Thus, each zero is a finite quantum point. It is a quarter of the whole sphere of time formed with four pairs of local times. Thus, quarter = quantum finite point = 0. Quarter is each Wisher's "eventual spatial state" (*Sura*, 0). Infinite point = Present = 1,6,00 = 1600. Further, the pair comprises four pairs of local times. Thus, pair = 8. Further, each finite point = 1. Each finite point is a Wisher's illusionary production.

Infinite point = Present = Absolute = Param = 1600

Quarter = Quantum finite point = Wisher = Eventual spatial state = 0

Finite point = Illusionary production = 1

Pair = 8

- The Wisher may nationalize the time by conceiving the three finite points as an "absolute gravitational constant" (*Upaya*, 3), letting the Wisher be a primordial paternal producing an infinity of three finite points of localized time for the infinite child citizens, forming an Infinite Wisher. Thus, Infinite Wisher is a "scale factor" scaling the triple bond of finite points into infinity. A child citizen has conditional freedom to conceive the finite quantum point as the "real absolute gravitational constant" (*Ajitatma*, 0) and three unbonded finite points as the "imaginary absolute gravitational constant" (*Vahana mandapa*, 3). Thus, as the present son, the child citizen may modify the national present and future time points into his local present and future time points.

Since the primordial paternal shaped a quarter sphere of time with his national past time point, the param son shapes a half-sphere of time with his local past time point conditioned by the inherited national past time point. Half sphere comprises a pair of national time points (present and future), followed by a pair of local time points (present and future). It is led by a triple sequence that forms corporate time. First, a finite point of the curvable global time. Second, three finite points of the curving national time. Third, three finite points of the curved local time. If there is a reason, these seven finite points of the uncurved corporate time may be curved. Half sphere circulates past time as a square pi. Square pi is a unit pi because further time transformations are a proportion of square pi. Thus, square pi = pi = Half sphere = 22/7

Infinite Wisher = Primeval Wisher = Triple Bond = -6

Primordial paternal = Real absolute gravitational constant = 0

Absolute gravitational constant = Imaginary absolute gravitational constant = Param son = 3

Half sphere = circulate = Two quarters = Square pi = pi = 22/7

- A grandson alien may enjoy the "corporate time" (*Chitrini*, 10^{11}-1) by taking the past "national time" (*Shani*, 18) and "local time" (*Kaal*, 360) as the spheres and making another sphere with the "global time" (*Hari*, 10,000). As the primeval grandson, the grandson alien realizes that the grandfather has shaped the global time and that he has no freedom to modify the national and local time spheres. The global, national, and local times together as a "rotating absolute gravitational constant" (*Yava*, -10) shape the present and the future he experiences.

The rotating absolute gravitational constant is a "partial pi bond" (*Samlagni*, -10). It is a "ligand" (*Samlagni*, -10) kept away by the "Devotee Wishers" (*Jagatkritsna*, -10) devoted to defining the present and the future of their infinite beloved grandsons. A Devotee Wisher is "somebody" (*Salakegolake*, -10) "keeping away" (*Yava*, -10) the freedom for a "Dreamer" (*Taijaisa*, -10) to go back in time and transform the "coordination complex" (*Taijaisa*, -10) that defines the present. Thus, a partial pi bond makes three triplets of inherited time points negative and one past finite point of the corporate time. The rotating absolute gravitational constant makes six pi the universal gravitational constant for each grandson.

> Devotee Wisher = Ligand = Kept Away = Somebody = Keeping Away = Dreamer = Coordination Complex = Rotating Absolute gravitational constant = Partial Pi bond = -10
>
> Primeval grandson = Universal gravitational constant = Three spheres = Triangle pi = 6 x half Sphere = 6 pi = 22/7

- A grandfather deity may divide the global time into ten quarters. First, three to be curved by him into the national time with the rotating absolute gravitational constant. Second, three to be curved by the primordial paternal into the local time with the real absolute gravitational constant. Third, three to be curved by the param son with the imaginary absolute gravitational constant. Fourth, one to be curved by the primeval grandson with the absolute gravitational constant, formed with the primeval grandson's psychic linkage with the three ancestors. The primeval grandson can let his freedom be conditioned by the circle pi formed by the ten quarters. In that case, the injured primeval grandson circulates only the pi in the half-sphere for his freedom decisions and lets the "circle pi" (*Ajari*, 22/7) reproduce the five pi. Such behavior makes five pi a "para gravitational constant" (*Ajari*, 22/7).

> Grandfather deity = Injured = Para gravitational constant = Five pi = 2.5 spheres = Circle pi = 5 pi = 22/7

2.4 Transcending Time with Six Hundred Finite Points

One transcends time with "knowledge" (*Vidya*, 600). One may localize knowledge within time and "spread up" (*Vitata*, 600) the "knowledge-effect" (*Kaivalyashrama*, 240) beyond the "local time" (*Kaal*, 360). The knowledge-effect comprises the "past system"

The Time Tensor: Exchanging the God Without 55

(*Kaivalyashrama*, 140), and its "gravitational energy" (*Lalita*, 100) reproducing within the "present group" (*Kendrabhi mukha*, 140) in the form of the "infinite international force" (*Ketu*, 140). It transforms time into a "globalized" (*Vaishvikrita*, 500/9) element by making the "time value" (*Pulastya*, 500) "coevolve" (*Sahvaas*, 9) with the "absolute development" (*Maha Vibhu*, 14) of the "future system" (*Maha Vibhu*, 14).

The "co-evolving value" (*Vaishvikrita*, 500/9) is the "zeroth gravitational constant" (*Vaishvikrita*, 500/9) that forms the "conical π" (*Vaishvikrita*, 500/9). A cone holds eight shapes, including itself: ellipse, circle, parabola, hyperbola, square, triangle, cone, and full circle. Therefore, conical pi equals eight pi. Eight pi is the co-evolving value of a "devoted mother" (*Vaishvikrita*, 500/9) devoted to "circulate" (*Artta*, 22/7) her knowledge across "gender" (*Linga*, 9) and "generations" (*Amnaya*, 97).

> Devoted Mother = Co-evolving value = Zeroth gravitational constant = Conical pi = Eight pi = 500/9

The "circulating" (*Chakra Yukti*, 10^{1024}) knowledge limits the "entity reality" (*Upakarana artha*, 10^{1024}) to the "energy system" (*Antarmukha*, 10^{1024}) immanent within the entity at the time of birth as an "animal" (*Pashu*, 10^{1024}). Eventually, the animal evolves into a "human" (*Naran*, 275) with "conscious energy" (*Amrit*, 1000) for making "conscious decisions" (*Sahasra lingam*, 190) to transform the maternal "evolutionary force" (*Tushti*, 85) into the "revolutionary force" (*Uttama*, 85) of a "devotee daughter" (*Piyati*, 1000/72). A devotee daughter "revolves" (*Avrit*, 55) the "time value" (*Pulastya*, 500) for experiencing the "paternal" (*Tejas*, 17) element without the "present system" (*Shuru*, 18) so that she may incarnate as a "greeter grandmother" (*Sasana*, 55,000/21) to materialize the "reality" (*Vastavikta*, 7) of the "param paternal" (*Narada*, 7).

The "revolving value" (*Piyati*, 1,000/72) is the "unit gravitational constant" (*Piyati*, 1000/72). It is the "whole sphere" (*Chaarpai*, 1000/72) of the "curved space" (*Chaarpai*, 1000/72) that transforms the "conscious energy" (*Amrit*, 1000) with the materializable "reality" (*Vastavikta*, 7) and the "universe" (*Brahman*, 2) materialized with that reality's "perpetuating value" (*Saranyu*,

5). The perpetuating value forms the "materializing reality" (*Audavita*, 5) as an "illusionary pi" (*Audavita*, 5). The illusionary pi is the "quantum state" (*Audavita*, 5), superpositioned over another quantum state. The immanent quantum state perpetuates its "value" (*Mulya*, 180) within the emanating quantum state. The emanating quantum state materializes the reality of the immanent quantum state.

The "materializable reality" (*Bahutva*, 7) of the immanent quantum state includes both the "materializing reality" (*Audavita*, 5) and the "materialized reality" (*Nyartha*, 2). The materializable reality is the "real pi" (*Bahutva*, 7); it is the "transient value" (*Bahutva*, 7) that illuminates the varying "potential realities" (*Paramanu*, 160) of the devotee daughter. The real pi trades the "infinite entropy-effect" (*Bhuyobhava*, 160) of the finite potential reality as Letter "P" (*Bhuyobhava*, 12). The materialized reality is the "imaginary pi" (*Nyartha*, 2); it is the "panoramic reality" (*Nyartha*, 2) formed after the "destruction" (*Nyartha*, 2) of the potential reality.

The imaginary pi makes the devotee daughter a "channel" (*Raasta* = 36 = 6 * 6) *for reproducing the* "six-fold growth" (*Dishta*, 6) of the panoramic reality for manifesting the varying "I's" (*Svayam*, 12 = 6 * 2) as "quantum photons" (*Svarochisha*, 12). The space reproduces "two twos" (*Keshava*, 22) to produce triple two twos over the three time-units in oneness with a partial two-two. The partial two-two is the constant "primordial space" (*Dik*, 100) that transforms into varying "primeval spaces" (*Para Ganesha*, 19) with each reproduction of "two twos" (*Keshava*, 22) that forms seven two-twos, each as a "pi" (*Artta*, 22/7). The primeval space is the "energy" (*Shakti*, 19) that materializes the "materializable reality" (*Bahutva*, 7). The energy is the factual "reproducing reality" (*Yathartha*, 19).

The whole sphere is the "spherical pi" (*Chaarpai*, 1000/72) with the "materializable reality" (*Bahutva*, 7) and the "materialized reality" (*Nyartha*, 2) as its two pi. The two pi are immanent within the "ellipse" (*Dirghavritta*, 1000) that "channels" (*Raasta*, 36) seventy-two "whole spheres" (*Chaaripai*, 1000/72). A devotee daughter initially trades thirty-six whole spheres using the past system's whole sphere as a "channel" (*Raasta*, 36 = 17 + 19) for experiencing the "paternal" (*Tejas*, 17) element within her "cells" (*Hiranyagarbha*, 19). Eventually, she services thirty-six whole spheres using the present system's whole sphere as a channel for conceiving the "maternal" (*Matri*, 112 = 36 * 3 + 4) element within her "consciousness" (*Chaithanya*, 4) of the three time-channels. Consequently, she channels seventy-two whole spheres for dividing the "ellipse" (*Dirghavritta*, 1000) by seven twos, where each two is the ellipse within the ellipse.

The macro ellipse precedes the micro ellipse. It is "macro" (*Guru*, 100) because it is within the devotee daughter's "conscious energy" (*Amrit*, 1000). With time, the macro ellipse revolves into a micro ellipse. The micro ellipse is "micro" (*Laghu*, 377) because it is a "mirage" (*Marichi*, 377) until the macro ellipse "completes" (*Purna*, 1600) its "revolution" (*Kranti*, 13) to be "present" (*SAUM*, 1600) as a "macro ellipse" (*Dirghavritta*, 1000) within the "knowledge" (*Vidya*, 600) of the "micro ellipse" (*Visaranem*, 55,000). The micro ellipse "revolves" (*Avrit*, 55) the macro ellipse's "self-perpetuating" (*Udvaha*, ½) "time value" (*Pulastya*, 500). Thus, the whole sphere is a "circular-effect" (*Piyati*, 1000/72) of the past macro ellipse channeled over time by a future micro ellipse.

> Devotee Daughter = Revolving value = Circular-effect = Unit gravitational constant = Whole Sphere = Spherical pi = Four pi = 1000/72

A "greeter grandmother" (*Sasana*, 55,000/21) "revolves" (*Avrit*, 55) the "conscious energy" (*Amrit*, 1000) for transforming the "ellipse" (*Dirghavritta*, 1000) into a "devil granddaughter" (*Sautramani*, 13/2) "self-perpetuating" (*Udvaha*, ½) the "revolution" (*Kranti*, 13) into future. It lets the greeter grandmother "forget" (*Visaranem*, 55,000) the "future reality" (*Ekartha*, 21) her "revolution" (*Kranti*, 13) of the "natural value" (*Nirvikalpana*, 8) creates. Consequently, the "evolving value" (*Sasana*, 55,000/21) of the future reality becomes "localized" (*Susana*, 55,000/21) with her. Future reality evolves because the greeter grandmother becomes the "entropy pi" (*Susana*, 55,000/21), experiencing a "curved causation" (*Susana*, 55,000/21) within her "subconscious mind" (*Avachetan*,

101). Entropy pi is the "consequential causation state" (*Susana*, 55,000/21) that transforms the devotee daughter's "conscious causation" (*Susana*, 55,000/21) into an "elliptical pi" (*Susana*, 55,000/21). An ellipse comprises seven shapes, including itself, within a cone. The cone is the "relation" (*Rishta*, 123) the two ellipses conceive between themselves for "creating" (*Sarjana*, 123) the seven shapes as the seven pi's forming an elliptical pi.

Elliptical pi comprises "fourteen quarters" (*Susana*, 55,000/21) comprising seven shapes and seven potential shapes an ellipse forms with its potential to reproduce itself. Each quarter is the "eventual spatial state" (*Sura*, 0) that gives a form to the consciousness of the preceding quarter and transforms the present consciousness of the succeeding quarter. Thus, the fourteen quarters comprise the present consciousness of the fourteenth quarter, the consciousness of the twelfth quarter, the twelve forms of consciousness that shape the first eleven quarters with the evolving value of the thirteenth quarter. Fourteen quarters constitute a "Mass-stress-energy-momentum density tensor" (*Susana*, 55,000/21). They transform the "mass consciousness" (*Satma*, 10) of the first ten quarters by positioning transformability stress on the eleventh quarter. The eleventh quarter repels its energy for managing the stress and circles the twelfth quarter with the momentum density of the thirteenth quarter formed by the repelled energy. The encircled twelfth quarter conceives the encircling momentum density as the fourteenth quarter, evolving with the formation of the thirteenth quarter.

> Greeter Grandmother = Evolving Value = Curved Causation = Consequential Causation State = Conscious Causation = Entropy pi = Elliptical pi = Seven pi = Fourteen Quarters = Mass-stress-energy-momentum density tensor = 55,000/21

The "thirteenth quarter" (*Vishvabhava*, 10^{13}) is a "temporal ellipse" (*Vishvabhava*, 10^{13}). A temporal ellipse is the "luminous force" (*Vishvabhava*, 10^{13}) of the "devil granddaughter" (*Sautramani*, 13/2) wishing to originate the "future reality" (*Ekartha*, 21) the "greeter grandmother" (*Sasana*, 55,000/21) wishes to evolve. A devil granddaughter is a "parabolic pi" (*Sautramani*, 13/2). A "parabola" (*Anuvritta*, 16) comprises six finite points evolving into six infinite points with the involvement of the thirteenth quarter's "four points" (*Vishvabhava*, 10^{13}). The parabolic pi is a "partial pi" (*Sautramani*, 13/2), i.e., ½ pi, since only half of it is "luminous" (*Rochisha*, 13). It sequences four points into a "pie" (*Pai*, 0), thus shaping the sequence involving their value into a "quarter pie" (*Sautramani*, 13/2) of the four-dimensional "spatial pie" (*Pai*, 0).

> Devil granddaughter = Involving value = Parabolic pi = Partial pi = ½ pi = ¼ pie = 13/2

A "point" (*Bindu*, 10^{10}) is the "ninth point" (*Tilaka*, 10^{10}) with the "cultural consciousness" (*Virasat*, 10^{10}) of the eight points that divide its oneness. Eight points include four points of the imagined future reality and four points of the illusionary past reality. They make the ninth point the "conscious reality" (*Virasat*, 10^{10}) of one's "consciousness system" (*Sara Kalpa*, 10^{10}). At each moment, the imagined future reality and the illusionary past reality form the "two points" (*Chakra*, 1/8). The two points rotate like a "wheel" (*Chakra*, 1/8) to "self-reproduce" (*Sah*, 1/8) the "oneness" (*Yoga*, 48) of the present's conscious reality. The conscious reality is divisible into the natural value and multipliable as the "consciousness" (*Chaithanya*, 4) of the "natural value" (*Nirvikalpana*, 8) to produce the "self" (*Atmatva*, 8×10^{15}) as a "conscious one" (*Akhanda*, 8×10^{15}). Thus, the wheel is the "infinite management factor" (*Moksha Astikaya*, 1/8) that forms a "full circle pi" (*Chakra*, 1/8). It comprises "ten pi" (*Chakra*, 1/8), forming the imagined future reality, the illusionary past reality, and the conscious reality of the present, transforming the first two within two points each, and norming the ten points as the "real present reality" (*Chakraj*, 10).

The real present reality is a "cycloid" (*Chakraj*, 10) servicing the "present reality" (*Badhabuddhi vadartha*, -2) of the "natural value" (*Nirvikalpana*, 8). Ten pi is a "full circle pi" (*Chakra*, 1/8) comprising "twenty quarters" (*Chakra*, 1/8). A "full circle" (*Ghatikamandala*, 53) is a "combination" (*Samavaya*, 53) of "two finite points" (*Ghatikamandala*, 53). The first finite point perpetuates the real present reality as a circle over time. The second manifests the circle's "perpetuating value" (*Saranyu*, 5) as the "real absolute gravitational constant" (*Upaya*, 3) without its division into two. Thus, after "breeding" (*Janana*, 15) fifteen quarters over time, it lets "five quarters" (*Audavita*, 5) perpetuate over space as the "illusionary pi" (*Audavita*, 5). Five quarters make "hundred quarters" (*Vivasvan*, 15) the breeding's "natural conscious value" (*Vivasvan*, 15) that curves time with its "time force" (*Mohini*, 15). Natural conscious value is "one without circle" (*Vivasvan*, 15) that circulates time.

Wheel = Two points = Infinite management factor = Full circle pi = Ten pi = Twenty Quarters = 1/8

A Pi is the Potential Shape of Consciousness

A pi is consciousness's "potential shape" (*Artta*, 22/7). It is a "square pi" (*Artta*, 22/7) that circulates its consciousness of the four potential shapes—triangle, circle, full circle, and square—as "four pi" (*Sara*, 16). A pair of "two pi" (*Chaarpai*, 1000/72) enjoy the four pi's "conscious consciousness" (*Prashantatma*, 16) of the "circular-effect" (*Piyati*, 1000/72). "Three pi" (*Upashanta*, 387) enjoy the "complementary value" (*Upashanta*, 387) of a pi with the first two pi within the "competitive linkage" (*Anuprastha Taranga*, 10) with the second two pi. They triangulate the second two pi into "six pi" (*Nakkhatta*, 22/7). Six pi is the triangle pi formed by triangulating two pi.

A triangle forms six shapes, including itself, square, parabola, full circle, hyperbola, and cone. A square comprises "two triangles" (*Duguna*, 18,000). A parabola comprises "three triangles" (*Anvayika*, 16). A full circle comprises "half triangle" (*Ardha tricone*, 53). A hyperbola is a system of "six triangles" (*Gabhira*, 12) that form twelve full circles. A cone comprises "nine triangles" (*Shanku*, 123) that form three squares and three triangles. The ellipse transforms the "relation" (*Rishta*, 123) between each co-adjacent square and triangle into a "circle" (*Valaya*, 100,000).

A circle forms five shapes, including itself, parabola, full circle, hyperbola, and triangle. A circle forms a triangle by conceiving two ellipses without itself and perceiving them as triangulating itself into an ellipse with their "triangulating force" (*Amrit*, 1000). The triangulating force is an ellipse; it brings oneness between a circle and two ellipses by transforming the "quarter" (*Pai*, 0). The quarter trades and services the triangulating force as the two ellipses and forms a circle of oneness between the two ellipses with the third quarter. Thus, the fourth quarter is the "region" (*Bhubhaga*, 1000/72) of oneness within three zeroes. A quarter is a zero without the "oneness" (*Yoga*, 48) of the "four quarters" (*Chaarpai*, 1000/72) within a full circle.

A full circle forms by "trading" (*Samana*, 53) the "linear value" (*Prashant*, 53) of the four quarters. It transforms into a hyperbola by triangulating the four quarters into six triangles using the third quarter to "triangulate" (*Trikonana*, 3) the other three. After triangulating, the third quarter becomes a parabola, trading the growth of the six triangles to be the cone without the circle. The cone is within the square.

A square forms only one shape: the square. After a unit of time, the square begins "perceiving" (*Vibhavana*, 123) itself as the "two triangles" (*Chakor*, 18,000) triangulated by the third unit of time "yet-to-manifest" (*Shesha*, 18,000). The two triangles begin "conceiving" (*Garbhadharana*, 123) the three units of time as the three circles that circulate the square as the "pi" (*Artta*, 22/7). The square, the two triangles, and the three circles form a "cone" (*Shanku*, 123).

2.5 Enjoying Time with Fifteen Finite Points

"Fifteen finite points" (*Ekarshi*, 1000) is the "perpetuated value" (*Ekarshi*, 1000) of the "natural conscious value's" (*Vivasvan*, 15) "conscious energy" (*Amrit*, 1000). "Five finite points" (*Prakashisu*, 10) trade the "time multiplier" (*Upaya*, 3) to service the perpetuated value. The correlation of the five finite points with the fifteen finite points forms a "sixteen-wheel system" (*Guna chakra*, 18). The correlation of the sixteen-wheel system with the time multiplier forms a "three-wheel system" (*Trichakra*, 10). The correlation of the time transformed by the sixteen-wheel system with the space formed by the three-wheel system forms a "two-wheel system" (*Suryaphani chakra*, 10^{10}). The two-wheel system is the "gravitoelectric cycle" (*Suryaphani chakra*, 10^{10}), popularly known as the Palm wheel.

2.5.1 The Sixteen-wheel system

The sixteen-wheel system is the "key to the asymmetry system" (*Yogini chakra*, 18). It transforms the "potential" (*AUM*, 18) within a "shape" (*Rupa*, 100,000) "yet-to-manifest" (*Shesha*, 18,000) into an "energy sequence" (*Agrata*, 916). The energy sequence unlocks the "symmetry system" (*Vyagrata*, 800), the "urgency" (*Vyagrata*, 800) of creation in the mind of a creature for realizing the "essence" (*Sara*, 16) of the "continuity" (*Rohini*, 50 = 800/16) among the culturally-mediated "diverse forms" (*Shudra*, 1) of the "creation" (*Srijan*, 379). The "culture" (*Sadakhya*, 9) is the creation's self-perpetuating potential that creates its diverse forms, including the "creature" (*Purusha*, 12) and the "creator" (*Brahma*, 59). The "relationship" (*Sambandha*, 10) among the diverse forms transforms the creation into a "system" (*Gabhira*, 12) of twelve finite points.

Each "finite point" (*Jadatva*, 1) is a "zodiac house" (*Jadatva*, 1)—a "convergence point" (*Jadatva*, 1) of the creation's correlation with the system-mediated finite point. The system is the "creature" (*Purusha*, 12) rounded with the "infinite networking system" (*Pravritta*, 12) of the twelve finite points. The full circle indexes the diverse "starting points" (*Prasthan*, 53) for "infinite networking" (*Sitadevimudra Pradayakaya*, 90) by the creature as a system. Infinite networking is the "infinite potential" (*Sitadevimudra Pradayakaya*, 90) a creature "unlocks" (*Khola*, 90) by looping the diverse "start points" (*Ab*, 1600) into a "circle" (*Valaya*, 100,000). The creature becomes a creator by starting to triangulate each start point with a "planning point" (*Tilaka*, 10^{10}) for squaring thirty-six masculine "leader species" (*Taraka*, 36) with thirty-six feminine "follower species" (*Lahari*, 36). The twelve zodiac houses constitute the first twelve wheels. They comprise:

- First, the Aries house of self-identity forms the capability to be a creator by transforming the creature networking the relationship among the diverse forms into a "giant" (*Kuha*, 10^{10}). As a giant, the creature identifies the infinite potential for creation with oneself. It makes oneself the "zodiac system"

(*Shunya kalpa*, 8 x 10^{15}) guided by the giant. The giant is the "astrological system" (*Sara kalpa*, 10^{10}) a creature trades as an "astrological element" (*Svarochisha*, 12) and services as a "system" (*Gabhira*, 12) for managing the "zodiac" (*Bhogi*, 1) element. The zodiac element is the "sensuous" (*Bhogi*, 1) idealization of oneself as the "ideal" (*Adarsha*, 1) for everyone.

- Second, the Taurus house of material growth transforms the giant's "finite temporary form" (*Rochisha*, 13) as a "zodiac entity" (*Rashi*, 13) into the "infinite temporary forms" (*Asthayitva*, 81) to "profit" (*Labha*, 81) from the "identity's" (*Pahachaan*, 8) "enduring" (*Pariksha*, 68 = 81 − 13) "self-idealization" (*Pahachaan*, 8). A creator shapes each "temporary form" (*Asthayin*, 9) through one's variable investment into a diverse relationship.

- Third, the Gemini house of conscious endeavors norms resourceful trading with "kins" (*Parijan*, 90,000) to compensate for the perceived "loss" (*Shreya*, 81) from the enduring self-idealization. It lets one trade the kin's "leadership energy" (*Nayakatva shakti*, 90,000) and service oneself as the reproducible "infinity point" (*Amara*, 90,000) within each kin.

- Fourth, the Cancer house of maternal emotions empowers one to transcend the limitations of pure "reproduction" (*Prajanan*, 285) of oneself by making one conceive empathy for the kith and seek an exchange of their "qualities" (*Guna*, 0). Consequently, the "friend circle" (*Vartula*, 10,000) shapes one's multiform "growth potential" (*Nanarupa*, 10,000) as a "reproductive system" (*Prajanan Pranali*, 10,000).

- Fifth, the Leo house of discriminating faculty makes one sentimentally attached to the "aliens" (*Vakradrishti*, -6). It powers one to enjoy servicing one's qualities beyond the home base for realizing the need to transcend those qualities. It frees one from oneself and makes one conscious of the reality beyond self-idealization. An alien engages in a "quality-free" (*Nirguna*, -6) "reciprocal exchange" (*Artharthi*, -6) of qualities wishing to join the friend circle and move one into the "foe circle" (*Vartula*, 100,000). It lets one enjoy the alien's "animate inequality" (*Nirdharma*, -6) after profiting from the "kith's" (*Goshthika*, 181) "animate equality" (*Ekarajya*, 181). The animate equality is the "profit" (*Labha*, 81) and its "reproductive energy" (*Lalita*, 100) that makes one the alien's kith. The alien is attracted to one due to the perceived "loss" (*Shreya*, 81) from not reproducing the kin's "leadership energy" (*Nayakatva shakti*, 90,000).

- Sixth, the Virgo house of universal work service makes one psychically conscious of oneself as the "infinitesimal" (*Atisukshma*, 90,000), seeking growth of "career" (*Pesha*, 80) for alleviating the foe circle and sacrificing "health" (*Tandarusti*, 13,000) endowed by the friend circle. The career is the "path of materialism" (*Pravrtti marga*, 80) guided by the "outer sense" (*Bahirani*, 80) after "poisoning" (*Jahar*, 80) the "inner sense" (*Ajjhattikani*, 80). The health is "immanent" (*Antara*, 13,000) within the inner sense as the "time" (*Kaal*, 360). Psychically, one knows that it is only a matter of time for the alien to conceive the "worker social cost" (*Indu*, 82) of reproducing the kin's "leadership energy" (*Nayakatva shakti*, 90,000).

- Seventh, the Libra house of compatible relationships opens one to the "normative planning" (*Abadha*, 10) by the "better half" (*Parakarana*, -1/2) for enjoying the "pleasures" (*Rati*, 179) of life in the "available future time" (*Ayati*, 863) for fulfilling the destiny. It makes one realize the "worker social benefit" (*Shubh*, 82) of the "followership energy" (*Pasaka*, -9) for knowing what makes the half better. The better half is the earthly "feminine spirit" (*Parakarana*, -1/2) that trades one's "self-perpetuating" (*Udvaha*, ½) "destiny" (*Niyati*, -1). The half is the fiery "masculine spirit" (*Avyayatman*, 1/2) that "self-perpetuates" (*Amsha*, ½) one as a "divine entity" (*Deva*, 1) without the better half after death.

- Eight, the Scorpio house of afflicted life and conflicted relationships closes one with the "normative programming" (*Dharana*, -8) of the "impregnated life purpose" (*Abhipraya*, 38). It makes one "possessive" (*Swamigata*, 160) about the better half who gifted one the "divinity" (*Divyata*, 57). It motivates one to reincarnate, seeking a life of "eternity" (*Ananta*, 90,000) in "unison" (*Samanjasya*, 16).

- Ninth, the Sagittarius house of spiritual blessedness and

fortune reopens one to a life journey where one consciously reproduces the better half's qualities within oneself by idealizing those qualities in a "guider" (*Guru*, 100) one seeks. It eventually makes one a "guider" (*Guru*, 100), guiding everyone's "normative performing" (*Rachana*, 286) for the sake of "universal sentient benefit" (*Sadhaniya*, 157).

- Tenth, the Capricorn house of purposeful spiritual calling "distances" (*Yojana*, 190) one from everyone. It gives one time to reflect on one's "normative profiting" (*Nanarupa*, 10,000) by reliving the life from "zero" (*Shunya*, 0) to "hero" (*Nayaka*, 1) within one "lifetime" (*Savahita*, 125).

- Eleventh, the Aquarius house of knowing and servicing the method for everyone's "perfection" (*Divya ratri*, 2222) leads to one's "normative development" (*Viparyaya*, 888) as a "perfectionist" (*Purnatavadi*, 90). It gives one the "entrepreneurial opportunity" (*Samanantara*, 80,000) to be a "perfect greeter" (*Udbhava*, 10^{96}), gifting one's "mood" (*Bhava*, 360) for perfection to everyone.

- Twelfth, the Pisces house of "attachment" (*Moha*, 250) to oneself and "liberation" (*Siddha ratri*, 17) from others follows as a consequence of one's "normative organization" (*Garjya*, 10^{1024}) like "everyone" (*Pratyeka*, 180). Everyone seeks to "illuminate" (*Dyut*, 150) everyone with the "gift of consciousness" (*Jinpa*, 10^9) of their "value" (*Mulya*, 180) as the "light" (*Prabha*, 180) that "matters" (*Sadhibhuta*, 158) for materializing the next three wheels as a three-wheel system.

2.5.2 The Three-Wheel System

The three-wheel system is the "creative cycle" (*Trichakra*, 10) that unfolds the following three wheels over time.

- Thirteenth, the creature as the house of alienation from the creation diffuses its "consciousness stock" (*Atma*, 4) about the creation's "truth" (*Sathya*, 375) to form the eight planetary bodies. The "consciousness" (*Chaithanya*, 4) of one's "creation" (*Srijan*, 379 = 4 + 375) ends the "urgency" (*Vyagrata*, 800) to manifest the "symmetry system" (*Vyagrata*, 800). It frees the creature from breeding everyone as its "homolog" (*Svarupanugata*, 19) to perpetuate its value after "death" (*Vira ratri*, 18).

- Fourteenth, the creation as the house of togetherness with the creature fuses the creature's body to form the "New Moon" (*Chandra*, 82) as the ninth body. It lets the creature enjoy the "panoramic reality" (*Nyartha*, 2) of the eight planetary bodies through the togetherness of the moon and the eight planetary bodies within the solar universe. The togetherness of the moon generates thermodynamic entropy of the eight planetary bodies and their eventual "destruction" (*Nyartha*, 2). Consequently, the "new moon" (*Chandra*, 82) enjoys the "diversity" (*Vividthta*, 38) of the creature's consciousness and the "supreme sacrifice" (*Svaha*, 44) of the eight planetary bodies within itself.

- Fifteenth, the creator as the house of otherness infuses the moon within the "Sun" (*Surya*, 21), the tenth body, after "reincarnating" (*Manojava*, 61) as a "white star" (*Yama*, $180 = 61 * 3 - 3$), the eleventh body, by first conceiving a mind-born "time multiplier" (*Upaya*, 3) and then subtracting the mind-born time multiplier.

The white star manifests the next two wheels as the sixteenth wheel within the sixteen-wheel system as the seventeenth wheel.

2.5.3 The Two-Wheel System

The two-wheel system is the "innovative cycle" (*Suryaphani chakra*, 10^{10}) that folds the following two wheels as the "revenue" (*Rajasva*, 10^{10}) accrued from the "creator sequence" (*Ajanma*, 9,000).

- Sixteenth, the "entropy" (*Mahodbhava*, 5) as the house of growth continues the "creator sequence" (*Ajanma*, 9,000). The "Almighty Creator" (*Tvam*, 26) conceives each of the ten bodies as the white star's "light spectrum" (*Ambara*, 180), forming a "gravitational wave" (*Varshakritya Taranga*, $1800 = 180 * 10$). The "Almighty Creation" (*Durga*, 27) perceives the "ten-body system" (*Veerbhadra*, 100) to be the "dark matter's" (*Dakshinashapati*, 1600) "gravitational light" (*Digambara*, 100).

The "Almighty Creature" (*Shri Vishvakarma*, 28) experiences the "dark matter" (*Dakshinashapati*, 1600 = 16 * 100 = 1800 - 100 - 100) as the twelfth body formed with the "gravitational energy" (*Lalita*, 100) of its "essence" (*Sara*, 16). The "entity" (*Hasti*, 24) manifests the "essence" (*Sara*, 16) with the essence's self-perpetuating "natural value" (*Nirvikalpana*, 8). Without "entropy" (*Mahodbhava*, 5), the entity transfuses her "energy" (*Shakti*, 19 = 24 - 5) into an innovative "particle" (*Hemarenu*, 19), the thirteenth body.

- Seventeenth, the "growth" (*Dishta*, 6) as the house of entropy discontinues the creator sequence. The "particle" (*Hemarenu*, 19) refolds its "relationship" (*Sambandha*, 10) with the "ten-body system" (*Veerbhadra*, 100) into a nine-unit "culture" (*Sadakhya*, 9) element by manifesting itself as the "Sun" (*Surya*, 21), the tenth body. The Sun embodies the white star and the dark matter within it as the two-unit "neutron star" (*Tara*, 2). The dark matter "becomes" (*Poshan*, 179 = 6 * [9 + 21] -1) the "dark force" (*Mahayana*, 179) with the "growth" (*Dishta*, 6) of both the "culture" (*Sadakhya*, 9) and the "sun" (*Surya*, 21) without the "white star" (*Yama*, 180) as an "ideal" (*Adarsha*, 1). The "growth" (*Dishta*, 6 = 1/3 * 8) is the "self-reproducing" (*Upanayana*, 1/3) "sixteen-wheel system" (*Guna chakra*, 18) that produces the "time multiplier" (*Upaya*, 3).

The ten bodies trade the white star's "ideal-force" (*Maya*, 1) to service a "continuous stream" (*Rajasva*, 10^{10}) of light by exponentiating the ten-unit "base" (*Kumbhaka*, 10). The dark matter exchanges the continuous stream of light for materializing the ten-unit "base" (*Kumbhaka*, 10) with a "discontinuous stream of photons" (*Satvasva*, 10^9).

Generating Revenues

There are two ways of generating revenues: first, activated by the relationships using the path of devotion, and second, activated by an organization using the path of knowing.

When we are consciously devoted to relationships, we activate the sixteen-wheels of innovation. First, the networking-effect gifts us a consciousness of giant capability, activating

Aries house of the self-identity as a newborn baby. Second, the self-activated path of action transforms our finite temporary giant form into infinite temporary dwarf forms invested into diverse relationships, activating the Taurus house of material growth through childish baby steps. Third, the work raises the percipiency about the kin-mediated resourceful trading, activating the Gemini house of conscious endeavors and paternal ego. Fourth, the trading-effect generates empathy for exchange with kith, activating the Cancer house of the maternal emotions. Fifth, the human-effect evokes a sentimental attachment to aliens for servicing our sisterhood culture, moving us out of our home-base and activating the Leo house of the discriminating faculty for our protection. Sixth, the cultural-effect strangely repels the growth, catalyzing our career as we sacrifice health and activating the Virgo house of the universal service for justifying our brotherhood workculture.

Seventh, the workculture-effect attracts planning from a charming, beautiful, significant other, activating the Libra house of the compatible relationships to balance our life. Eighth, the worker-social-benefit makes us possessive about the significant other, revealing our true ugly self now programmed with impregnated pleasures, activating the Scorpio house of the afflicted life and conflicted relationships. Ninth, the worker-social-cost makes space to reproduce the significant other's quality of togetherness within us, making him a guru within our mental realm, thus energizing our performing and activating the Sagittarius house of spiritual blessedness and fortune. Tenth, the social-benefit shapes time to reflect on our profiting using our quality of autonomous otherness, activating the Capricorn house of the purposeful spiritual calling for freedom from the masculine-effect. Eleventh, the development of social-cost without our feminine-effect generates consciousness of our perfection as a feminine, activating the Aquarius house of knowing one's source of monetizable energy and servicing the method for everyone's perfection. Twelfth, the ascending worker-social-benefit-cost-ratio is the gift of consciousness of our organization as a masculine entity enjoying feminine energy. It activates the Pisces house of attachment to oneself and liberation from others.

The consciousness-guided knowing activates the three-wheels of alienation in a continuing sequence. Thirteenth, the ascending social-benefit-cost-ratio without our polluted consciousness diffuses our consciousness stock to form the eight planetary bodies. It ends the creation urgency within us for breeding everyone as our homologs to enliven our essence after our death. Fourteenth, the descending social-benefit-cost ratio makes us fuse our cleansed consciousness as the ninth body into Full Moon, embodying the departed human entity as a Black Hole, thus activating our creature journey to enjoy everyone's diversity within us. Fifteenth, horizontal social-benefit-cost-ratio infuses our cleansing consciousness as a tenth body into Sun, essentially solid earth. It diffuses our polluting consciousness as the eleventh body into a White star, essentially light. Light continues the creator sequence by blinding us and making us behave like the twelfth body, the Dark Matter. Consequently, we transfuse into an innovative thirteenth body, the Particle, embodying us within the energy of the social-benefit-cost-ratio.

As the sixteenth wheel, the particle unfolds all the sixteen known and three unknown relationships folded as the culture element. Culturally-unfolding relationships appear to generate revenue, essentially the continuous stream of the light of our discontinuous creativity. However, reality hides as the four additional unknown elements that form the workculture element without our consciousness. First, revenue is a spontaneous creation that continuously streams by sequencing infinite spontaneous creations from infinite relationships. Second, we organize the infinite spontaneous creations into a sequence using our energy as the foundation for revenue. Third, we all have the potential to be the 29-step organization that takes six knowable growth steps after knowing the sixteen known and seven unknown steps to manifest the revenue as the finite spontaneous creation. Fourth, when we manifest revenue as our growth without any relationship, we are absolved from the ethical responsibility of sharing.

We can enjoy the revenue like a king as a tribute to our self. That is why we follow the tradition of taxation for the sentient benefit of those who wish to live a kingly nonworking leisurely life (kshatriyas) and the sentient cost of those who are working on manifesting the enjoyables in life (shudra). Those who know the known elements are compensated well for keeping their knowing secret (brahmin). Those who don't know the known elements develop the knowing of the unknown elements and generate increasing returns by selectively revealing what they know (vaishya). Those free from both the known and unknown elements get to consume the knowable elements as the specially gifted children of Mother Nature (asura).

Relationships shape our sense of reality. To begin with, they give us a dream-like consciousness of capability to form the universe-to-be. It is not a coincidence that Aries is the first zodiac house that endows the same qualities as do the relationships to anyone who behaves like a meek sheep doing nothing but taking in the essence of the universe as-is.

The other eleven zodiacs follow with their qualities. For instance, capability generates the investment of Taurus for affinity growth. Investment generates Gemini ego, trading oneself for kin. Trading generates Cancer emotions, as one exchanges kin with kith to fill the gap formed when one makes sacrifices for the kin. Think of oneself as the organization. Think of kin as the employees. Think of kith as the vendors. What's next? Exchange generates the Leo attachment, as one develops kith-like sentiments for the aliens. Think of aliens as the customers.

The first nine zodiacs fold our relationships as our culture. The other three unfold them as our workculture. The four bodies that form the twelve zodiacs generate revenue as a continuous stream of light. An organization creates these four bodies as its spontaneous creation. Think of four bodies as the grandchild, infused with the bodies of the organization (grandfather), relationships (grandmother), consciousness (mother), and the organizer (father). The grandchild is our I, diffusing light in a continuous stream. The continuous stream that forms the organization's I, within or without relationships, is the revenue.

Revenue shapes our consciousness system into an astrological system and transforms the first nine zodiacs into the Sun and the eight planets. The other three zodiacs form the Moon, the White Star, and the Dark Matter. The Four Bodies transform reality into a particle fused with our energy. Thus, relationships generate our thermodynamic entropy and transform us into energy that grows everyone's revenue.

Without any wheel, an organization enjoys revenue as a spontaneous creation, with our energy as the foundation. Our relationship with the organization absolves it from the ethical responsibility of sharing the revenue with us. It lets the organization be the king enjoying the revenue as the tribute to oneself

2.6 Exchanging Time with Seven Thousand Finite Points

Our "life experience" (*Anubhuti*, 7,000) comprises the "perpetuated value" (*Ekarshi*, 1000) of our "reality" (*Vastavikta*, 7) in seven thousand finite points. Our reality comprises the "ideal force" (*Dasha*, 1) of the Sun and the six planets other than Uranus and Neptune. "Uranus" (*Rahu*, 73) is the "ascending mass" (*Rahu*, 73) of our "emotional energy" (*Hunduka*, 73), the energy in ascending motion. "Neptune" (*Ketu*, 140) is the "descending mass" (*Ketu*, 140) of the "present group" (*Kendrabhi mukha*, $140 = 7 * 20$) the Sun and the six planets as the "photons" (*Ruah*, 20) due to the "end" (*Antaka*, 140) of their "lifetime" (*Savahita*, 125) after "breeding" (*Janana*, 15) the "white star" (*Yama*, $180 = 125 + 15 + 40$) with their "workculture-effect" (*Nivrtti dharma*, 40).

The workculture-effect is the "reproductive reality" (*Bhavartha*, 40) of the "dark matter" (*Dakshinashapati*, $1600 = 40 * 40 = 200 * 8$). The dark matter is the "reproductive energy" (*Lalita*, 100) of the white star, the sun, and the six planets within Uranus and Neptune. Uranus invests 27-units into the "incubation" (*Vam*, 27) of Neptune. "Neptune" (*Ketu*, $140 = 100 + 27 + 13$) incubates the thirteen bodies as our "ideal forces" (*Dasha*, 1) within itself after exchanging Uranus's "incubator value" (*Vam*, 27).

Thus, we become the "Almighty Creator" (*Tvam*, 26) of the thirteen bodies that form a "solar universe" (*Surya mandala*, 13) within another "solar universe" (*Surya mandala*, 13) formed by us with our previous life experiences. A "solar universe" (*Surya mandala*, 13 = 6 + 7) is conscious of the "growth" (*Dishta*, 6) it experiences as a "sentient entity" (*Siddha*, 7). As a sentient entity, we perpetuate our "potential" (*AUM*, 18) in the form of the "solar universe" (*Surya mandala*, 13) and the "cultural factor" (*Sugriva*, 5) that reproduces the solar universe in the form of our "sentient light force" (*Apas*, 169 = 13 * 13).

Modern science shows that the universe is also conscious like a human being. Our full potential is to be the essence of the universe we form as a human being correlating with everything as if it is a human being too. When we are a micro human being and reality without us is a macro human being, we can enjoy the reality as a "sentient entity" (*Siddha*, 7), conscious that the reality without us is the reality within us. By ascending the consciousness of the reality within us, we descend our dependence on the para-consciousness of the reality without us. Para consciousness is the consciousness that transcends consciousness. Our potential works like para consciousness that transcends our consciousness.

When we activate our potential, it seems to us that there is a transcendental power that is guiding us and taking us beyond our power to reason using our conscious effort. That lack of confidence creates a void within us. Instead of moving in a straightline for accomplishing our goals, we start moving sideways and curve our path into a half-circle. We just reproduce the mirror opposite of the half circle that exists without us. Somebody giving birth to us creates that half-circle. That somebody is our twin body, embodying our past life experience since our past life experience gave birth to us in the present life. We as a body and somebody as our twin body forms the whole circle. The whole circle is a reflection of the illusion of our potential. It is not our potential. Therefore, whatever we perceive, conceive, experience, and become conscious of within that whole circle is always contaminated with an error in consciousness. There is always some uncertainty in what we believe. It leads to a discrepancy between our beliefs and behaviors.

Consequently, we begin questioning our existence. Wouldn't the world go on even without us? Indeed, but the world would also be different without Mahatma Gandhi, Martin Luther King, and Nelson Mandela. Or, for that matter, Hitler or Osama Bin Laden. The answer to the value we bring as a unit within the universe of billions starts with "grace" (*Pushti*, 84), i.e., self-acceptance. Received wisdom holds that the self is the source of "ego" (*Aham*, -1) because the self is not the "whole" (*Akala*, 16). It needs to be surrendered to save our "soul" (*Atma*, 4) and attract an "ideal" (*Adarsha*, 1) beyond self. Therefore, our self takes on an "extroversion face" (*Bahirmukha*, -4), seeking to discern that ideal by sacrificing itself.

Normally, one lives life substantiating the ideal (*e.g.*, personifying Jeff Bezos) while transforming oneself. One follows by rejecting the ideal (*e.g.*, de-personifying Jeff Bezos and what he stands for as one enjoying wealth and making others dependent on his wealth that could have been one's had the terms of exchange been different). One leads by reforming oneself using a theory of mind that positions one as a "super ideal" (*Lingopahita-laingika-vadartha*, 1500). However, modern science holds we are just "observers" (*Rajah*, 0) with zero impact. One has to have blind faith in the grace of an illusionary deity to account for the nine-dimensional change in the "absolute space" (*Sadakhya*, 9), including four space, three time, one entity, and one para entity dimension.

Further, one must believe in spirituality to accept the metaphysical idea that whatever happens is for our good. Logically, we must accept the self as the vehicle to bring oneness with the Lord sending the deities for making a change to capitalize on that change for our personal wellness as a path to universal wellness. By the grace of the "holy spirit" (*Trinetra*, 1), we convey the message that the Lord is only a leader in the service of God. Once we have become Jeff Bezos, we start playing God, demonstrating our benevolence and compassion to command absolute devotion from the passionate devotees of our capitalism and their socialism.

In India, gurus recognize everyone is a form of God (*Ishvar ek rupa anek*). The wise say if our self is not working out, then let's just discard the self as one would do with a piece of clothing. Let's make the personalization of the idealized form of God the goal of our life. Just surrender your polluted self to your idealized God form, consciously praying for HIS grace within oneself. Of course, selves are always conserved. When the ideal gifts you his seedless "purified state" (*Maha Svapna*, 9), he takes on your seeded "polluted state" (*Beeja-jagrat*, 84). When an ideal is socialized into the polluted state, it is natural for the "anger" (*Krodha*, 275) to become institutionalized into each "human" (*Naran*, 275) who has ever questioned oneself. Further, it is natural for each human to take on an "I am a deity consciousness" (*Pavitratma*), wishing to be the "Lord" (*Indra*, 0) of all idealized deity forms. By problematizing their self as the "Wisher" (*Chaahak*, 0), each devotee begins passionately observing all formations—animate and inanimate—waiting for the present God to reveal HIMSELF within a Goddess who has exchanged her polluted self with his purified self. Eventually, the devotee who accepts himself becomes the "object of devotion" (*Padartha*, -3) for all the "subjects" (*Sura*, 0) standing in line for their similar "deification" (*Subheccha*, 86).

Once a devotee makes herself the problem, the mind-born "discordant energy" (*Asura shakti*, -1) sustains and multiplies over time as "ego" (*Aham*, -1). The ego is the energy going out of the "problematized self" (*Asura*, -1) in ascending motion. The "subject" (*Sura*, -1) is zero with no impact. The problematized self, discordant energy, and ego are negative one, producing the extrinsic impact. The timeless object of devotion is negative three, producing the "intrinsic impact" (*Vasana*, -3) while consuming the "extrinsic impact" (*Ravinandana*, -1) with the "time multiplier" (*Vaishya*, 3).

A scientist makes us believe that we are an observing zero subject and question the value of our self. It makes science a cost-escalating solution to all problems. It generates an increasing demand for the star scientist to be deified as a way to force the star to diffuse his entire knowing for the public good in exchange for his stardom. By fusing the knowing of the diverse stars, we become a problem-free guru servicing our problem-solving wisdom.

Every child has a freedom to ask why do we need to pray for

grace of a solution-problematizing God or a problem-solving Guru. Why can't we simply accept the essence that gifts us the joy of diverse selves that develop over time and focus our energy on developing time to enjoy the whole diversity within the present lifetime?

2.7 Developing Time as a Child with One-Hundred-And-Twenty-Eight Finite Points

A "child" (*Arcisa*, 128) has the potential to develop time to "channel" (*Raasta*, 36) the "mass consciousness" (*Satma*, 10). Mass consciousness forms with time. By developing time, a child reforms the mass consciousness to enjoy space for her making her "choices" (*Chunava*, 90,000) in life. There are several paths for life to subsist "within force" (*Prachalita*, 15) of our choices. In a depressed state, we choose "oneness with atom" (*Anuyoga*, 90) out of the anxiety about life seeking "death" (*Vira ratri*, 18). In an elated state, we censure oneness with the atom to unite with the "spirit" (*Ruah*, 20) that formed the "epigenetic atom" (*Kalpanik*, 20) and became the "cell" (*Hiranyagarbha*, 19) that gave us "life" (*Zindagi*, 4). In a normal state, we make "spiritual union" (*Pariprashna*, 90) with the cell the "platform" (*Swakshetra*, 9000) for spiritual growth. "Spiritual growth" (*Anuyogakrit*, 10^9) is "without force" (*Anuyogakrit*, 10^9) of our choices. It makes us "one with nothing" (*Pinakapani*, 10^9) but "someone unknown" (*Konasa*, 10^9) wishing to be "someone known" (*Kincha*, 10^9) "with force" (*Anubhav*, 9,000) of "reckoning" (*Viganana*, 90) developed over diverse lives. We are "re-energized" (*Haryasvas*, 90) by our newfound "reckoning" (*Viganana*, 90).

When we exponentiate our path-independent present experiences with the path-dependent infinite experiences, we gain the "path-exponentiating" (*Konasa*, 10^9) gift of consciousness. The "experience" (*Anubhav*, 9000) of diverse lives makes us an "absolute authority" (*Raub*, 10), making choices by breeding our "within force" (*Prachalita*, 15). It lets us live our livelihood on our terms for gaining "infinite experiences" (*HAUM shakti*, 9) from our "culture" (*Sadakhya*, 9) by exchanging the "base" (*Kumbhaka*, 10) of the "present experiences" (*Sushumna*, 10). It lets us rely on our "within God" (*Sura*, 0) instead of depending on "without God" (*Asura*, -1). A "path of spiritual growth" (*Vatsanapat*, 10) offers the certainty to be "with God" (*Rab*, 16). It makes us the "interrogator" (*Pariprashnati*, 10^9), leading the 'interrogation" (*Pariprashna*, 90) into the spirit that became the cell and gave us the "gift of consciousness" (*Jinpa*, 10^9) of our life.

The gift of consciousness exponentiates the present experiences with the infinite experiences, i.e., all experiences except the present one. One forms the present experience by relying on "within God" (*Sura*, 0). If we remain dependent on "without God" (*Asura*, -1), we enjoy the infinite experiences but not the "one within God" (*Pai*, 0). "Within God" (*sura*, 0) is "zero" (*Shunya*, 0) reliance on anyone. It uses the present experience as a "tangent" (*Ardhajya*, 10), taking the "whole value" (*Ardhajya*, 10) of one, i.e., adding 0 after 1. Thus, the energy of the livelihood, subsistence, base, present experience, and absolute authority is 10. As an enjoyer dependent on "without God" (*Asura*, -1) as life's "goal" (*Lakshya*, 9), the energy of infinite experiences, enjoyer, and goal is nine (ten minus one).

An "atom" (*Anu*, 19) is the "extrinsic conscious-effect" (*Anu*, 19) of consciously transforming our present experience into an infinite experience. It includes the "descending force" (*Nandini*, 105) of the present experience and the "ascending force" (*Vaimitra*, 99) of the infinite experience. Conversely, a cell is the "conscious apparatus" (*Hiranyagarbha*, 19) that gives us energy for this transformation. Therefore, the energy of the atom, cell, and energy is ten plus nine, i.e., nineteen. It excludes the twenty-unit spirit that fills the "void" (*Shunyata*, -2) of the "para entity" (*Ashtavakra*, 19) by guiding as an epigenetic atom to use "within force" (*Anubhav*, 19) and programs that guidance into our "genetics" (*Vilaga*, 90).

"Oneness with atom" (*Anuyoga*, 90) infuses infinite experiences of our cells into the present experience of the atom seeking to be a cell using our energy as an "animate entity" (*Sushupti*, 96). Conversely, genetics infuse the atom's present epigenetic experience into the cells' infinite experiences. Thus, the energy of the genetics and the oneness with atom is nine times ten, i.e., ninety. It is the energy we gain through spiritual union by exchanging the "path of spiritual growth within God" (*Vatsanapat*,

10) with the physical interrogation of the "path of humanity without God" (*Tanunapat*, 10).

By developing the "time" (*Kaal*, 360) into the "consciousness stock" (*Atma*, 4) of "with God" (*Rab*, 16), we move past idealizing the absence of "within God" (*Sura*, 0) and theorizing the presence of "without God" (*Asura*, -1). Our "consciousness stock" (*Atma*, 4) is one of the infinite forms of God. It includes:

- a unit that becomes within God after multiplying itself into infinite atomic mass fragments;
- a unit that becomes without God after dividing itself into infinite energy fragments;
- a unit that remains with God after making itself the essence of energy before it fragments into the consciousness or defragments into the atom formed with the mass of consciousness; and
- a unit that forms God by perpetuating the four-unit consciousness within us. As one form of God, we are the fifth unit that forms God.

We form "infinite experiences" (*HAUM shakti*, 9 = 4 + 5) by diffusing the four-unit "paternal consciousness" (*Pitra*, 4) without us, thus adding four units to our five-unit value as "God" (*Ishvara*, 5). By infusing the four-unit paternal consciousness without us into the four-unit "maternal consciousness" (*Chaithanya*, 4) within us, we develop "with God" (*Rab*, 16) endowed with our sixteen-unit "essence" (*Sara*, 16). The essence makes us a "primordial greeter" (*Srijak*, 16) capable of reversing time with our "conscious consciousness" (*Prashantatma*, 16).

2.8 Reversing Time with Sixteen Finite Points

A primordial greeter has a "potential" (*AUM*, 18) to "reverse" (*Mahavijya*, 20) "time" (*Kaal*, 360 = 18 * 20) and practice "gratitude" (*Stuti*, 167) by "multiplying consciousness" (*Haryyatma*, 167) of "whatever exists" (*Amba*, 32). Gratitude is the "hypothesized reality" (*Stuti*, 167) of our "energization" (*Yajna dharma*, 33). A "hypothesis" (*Prakalpana*, 861) brings "what can exist" (*Priti*, 33) into "fragmentary" (*Chitta*, 861) "existence" (*Maujudgi*, 2000) within our multiplying "human consciousness" (*Haryyatma*, 167) once we decide not to be with God. When we are "grateful" (*Kritajna*, 167), we become "mentally alert" (*Sacheta*, 10,000) about both HIM and HER. "HIM" (*Jagrit*, 18) is "existing" (*Maujud*, 1000) as a "male" (*Nara*, 10^8) with a potential to bring HER into existence. "HER" (*Konastha*, 13) "exists" (*Jivita*, 147) as a "female" (*Nari*, 10,000) without HIS "male consciousness" (*Tarkshya*, 18). HIS "co-exists" (*Satha*, 1000) with HIM as a "child" (*Arcisa*, 128) within HER "female consciousness" (*Priti*, 33).

HER is "unlucky" (*Akushala*, 13) because SHE has "infinite forms" (*Konastha*, 13), and each "form" (*Rupa*, 100,000) is "temporary" (*Mahabhaumika*, 19). On the other hand, HIM is "lucky" (*Kushala*, 12) because the child is always a "possibility" (*Hariti*, 128) within the "mentation" (*Manogata*, 12) of HER "grandmother consciousness" (*Adiparaatma*, 100).

HE is "fortunate" (*Shobhana*, 14) to "live" (*Nivas*, 14) permanently within HER "daughter consciousness" (*Jyotistava*, 4). HE is "activated" (*Jagrit*, 18) within HER "maternal consciousness" (*Chaithanya*, 4) once SHE becomes "active" (*Sakriya*, 10^{10}) as a "consciousness system" (*Sara Kalpa*, 10^{10}). Once activated as an "inanimate element" (*Sthavara*, 18), HIM becomes "permanent" (*Sabbacitta*, 18). SHE nurtures HIM to life using HER potential to be the "nurturer" (*Palanhaar*, 18). SHE is "unfortunate" (*Ashobhana*, 15) to "subsist" (*Prachalita*, 15) within the force of HIS "all-pervading" (*Sarvatraga*, 16) "range" (*Parisar*, 15). Therefore, she becomes a "fortune-maker" (*Sparshi*, 13), "breeding" (*Janana*, 15) the "universe" (*Brahman*, 2) for extending the range of his "divinity" (*Divyata*, 57).

Thus, both HIM and HER are the "circular creation" (*Vartula*, 10,000) of our "physical existence" (*Sasharira*, 10,000). They are the "multiforms" (*Nanarupa*, 10,000) of our "growth potential" (*Nanarupa*, 10,000). Each "form" (*Rupa*, 100,000) comprises twenty "mental factors" (*Chetasika*, 5000) "within" (*Eke*, 5000) our "multiplying" (*Eke*, 5000) "primordial child consciousness" (*Yatachittatma*, 5000). The twenty mental factors form a twenty-unit "belief system" (*Saguna*, 20) and norm the conscious "spirit" (*Ruah*, 20) that guides the "God" (*Ishvara*, 5) within us to make us the "guider" (*Guru*, 100).

Of the twenty, ten mental factors are "within affliction" (*Upaklesa*, 10) of our guiding spirit, i.e., the belief system. They include "laziness" (*Kausidya*, 56), "faithlessness" (*Asraddhya*, 70), "heedlessness" (*Pramada*, 80), "shamelessness" (*Ahkriya*, 90), "disregardfulness" (*Anapatrapya*, 100), "slothfulness" (*Styana*, 110), "restlessness" (*Auddhyata*, 120), "mindlessness" (*Masitasmrtita*, 130), "inattentiveness" (*Asamprajanya*, 140), and "desultoriness" (*Vikshepa*, 150). Our "affliction" (*Klesha*, 169) makes us "loyal" (*Vafadaar*, 79) to the spirit. When we believe in God without us, we become lazy. We lose faith in God within us. We stop paying heed to our intuition. We become shameless in manipulating reason for our convenience. We disregard the real facts and transform illusionary fiction into imaginary facts. We develop a sloth when we lose touch with reality. We become restless, not knowing the reality of God. We mindlessly grope in the darkness of our ignorance. We stop paying attention to our conscious wellbeing. We lose sight of the purpose of our life.

Eleven mental factors are "with affliction" (*Aniyata*, 800) of our guiding spirit, i.e., the belief system. They include "ego" (*Mada*, -1), "pretense" (*Maya*, 1), "hypocrisy" (*Asathya*, 13), "jealousy" (*Irshya*, 19), "selfishness" (*Matsarya*, 19), "cruelty" (*Vihimsa*, 123), "anger" (*Krodha*, 275), "slyness" (*Mraksha*, 369), "spitefulness" (*Pradasa*, 700), "resentment" (*Upanaha*, 15,000), and "discernment" (*Vichara*, 8) of the ten factors within the spirit's affliction. When we believe in our blessedness by God without us, we develop ego. We pretend that we are a divine entity. Hypocritically, we devalue everyone. We feel jealousy with anyone that has a value we don't. We become selfish about owning everything of value. We use cruelty for appropriating everything that anybody owns. We are angry when we don't succeed in our goal. We try to be sly using bait and switch option. We are spiteful towards the victims of our tricks. We resent if our morality is called into question when all we are trying to do is to defend our fundamental right to enjoy life fully. We seek to discern why we should continue to believe in God without us when we can be God ourselves.

Discernment as a "human being" (*Insan*, 1) "with belief system"

(*Prapaka*, 36) creates twelve "intellectual factors" (*Appamanna*, 10,000) "without affliction" (*Mulaklesa*, 12). They include "moderation" (*Alobha*, 2), "mental equanimity" (*Tatramajjhattata*, 4), "regard" (*Apatrapya*, 18), "mindfulness" (*Sati*, 42), "intellectual tranquility" (*Passaddhi*, 47), "shame" (*Hri*, 58), "faith" (*Shraddha*, 59), "proficiency" (*Pagunnata*, 160), "wieldiness" (*Kammannata*, 486), "lightness" (*Lahuta*, 580), "innocence" (*Adosha*, 888), and "malleability" (*Muduta*, 7000) of belief system. We moderate our manipulation when everyone begins believing that we are God. We enjoy mental equanimity as we develop regard for ourselves. We become mindful of the universal sentient wellbeing. We enjoy intellectual tranquility when we know that our destiny is in our hands. We feel ashamed when our deeds hurt everyone's faith in us. We develop proficiency in wielding our presence in everyone's mind. We enjoy lightness when we are universally present. Our innocence makes us malleable. Our malleability motivates everyone to reproduce our guider power.

Malleability as a "guru" (*Guru*, 100) "without belief system" (*Saboot*, 16) creates thirteen "physical factors" (*Visayaniyata*, 13) whose "reproduction" (*Prajanan*, 285) releases the "affliction" (*Klesha*, 169). They include "understanding" (*Samma Vacha*, 2), "delight" (*Mudita*, 9), "compassion" (*Karuna*, 20), "charity" (*Samma Drishti*, 28), "patience" (*Samma Pasadana*, 30), "chastity" (*Samma Ajiva*, 46), "rectitude" (*Ujukata*, 51), "peace" (*Samma Jnana*, 93), "gentleness" (*Samma Karma*, 147), "piety" (*Samma Sankalpa*, 168), "purity" (*Samma Sati*, 196), "probity" (*Samma Vayama*, 246), and "self-consciousness" (*Prajna*, 2222). Understanding that we have become God for everyone delights us and activates our compassion. We start advocating for charity, patience, and chastity for rectifying our past error in consciousness. We come to peace with our physical existence. We become gentle in our behavior. We develop piety towards our relationships. Our thoughts become pure reproduction of everyone's beliefs. Our actions probe the truth of our conceived reality of God. We become self-conscious that we are God.

Self-consciousness as "God" (*Ishvara*, 5) "within belief system" (*Pratyabhijna*, 100,000) creates a 43-unit "metaphysical factor"

(*Radha*, 43) without a 3-unit "moderating factor" (*Hanuman*, 3) comprising discernment, malleability, and self-consciousness. It afflicts "emancipation" (*Vimukti*, 10,000) from the "physical existence" (*Sasharira*, 10,000) seeking "meaningful life" (*Samma Vimukti*, 14) from our "metaphysical existence" (*Maujudgi*, 2000) instead.

We seek "meaningful" (*Sarthaka*, 14) life without physical body when living like everyone within the physical body loses "meaning" (*Tatpraya*, 155). While hibernating in "bardo" (*Maha Ratri*, 76) after departing from the physical body, we enjoy the opportunity to reflect on the "eight-fold consciousness" (*Parshnisamasta*, 32) of "whatever exists" (*Amba*, 32). We conceive whatever exists over the four masculine "lifephases" (*Ashrama*, 250) living life on earth and four feminine lifephases for processing the meaning of life after death. We create a fifteen-unit "dynamic factor" (*Bhuvanesvari*, 15), comprising a compensating three-unit moderating factor, six-unit growth within the moderating factor while in Bardo, and six-unit growth without the moderating factor while on Earth. The compensating moderating factor includes "zeal" (*Chand*, 19), "memory" (*Smriti*, 35), and "conviction" (*Adhimoksha*, 190).

The six-unit growth within the moderating factor includes "attention" (*Manasikara*, 1), "centering" (*Cetana*, 10), "awakening" (*Samjna*, 52), "concentration" (*Ekagratha*, 158), "sensation" (*Vedana*, 197), and "touch" (*Sparsha*, 378). In our zeal, we memorize whatever exists with a conviction that we can reproduce and multiply that with our "absolute consciousness" (*Paramatma*, 1600). We focus our attention on centering our awakening by concentrating the sensation of our touch on those devoted to us. We enjoy the 52-unit "awakening" (*Sangya*, 52) after birth before conceiving a need for six-unit growth to diffuse the awakening as our "institutional consciousness" (*Chhaya*, 52), taking into "consideration" (*Marudeva*, 58) the fact that the seventh-unit in the sequence of physical factors is "rectitude" (*Ujukata*, 51). "Peace" as the eighth-unit completes the "octave" (*Sargam*, 60) that transforms our "self-sovereignty" (*Svarajya*, 60) during Bardo into our "institutional sovereignty" (*Samrajya*, 60) on Earth.

The six-unit growth without the moderating factor includes

"imposition" (*Avidya*, -10^{30}), "reposition" (*Mana*, 16), "opposition" (*Pratigha*, 28), "cohesiveness" (*Raga*, 250), "skepticism" (*Vicikitsa*, 3000), and "worldview" (*Drishti*, 360,000). While living, the consciousness we impose on others repositions their opposition and brings cohesiveness among our devotees. It also breeds skepticism among the non-devotees. Our worldview towards devotees and non-devotees shapes our potential as an "institutional force" (*Trivikrama*, 24). Our conscious planning guided by our "known reality" (*Rachitartha*, 1600) memorized during bardo alienates us from the non-devotees while living on earth. It limits our capacity to act. Our life becomes a discontinuous "event" (*Ghatna*, 18) as we diffuse the "growth" (*Dishta*, 6) of our "institutional force" (*Trivikrama*, 24 = 18 + 6), seeking to contest competing institutional forces. Eventually, we seek a "big jump" (*Antardhyana*, 27) forward for acculturing time backward by founding a "religion" (*Dharma*, 370) that culturally binds the contesting institutional forces.

2.9 Acculturing Time with Twenty-Seven Finite Points

A "religion" (*Dharma*, 370) is a "method" (*Upaya*, 3) for "acculturing" (*Sanvardhan*, 90) the discontinuous local "time" (*Kaal*, 360). Religion transforms the local time into the continuous "global time" (*Hari*, 10,000). The global time reproduces the "physical existence" (*Sasharira*, 10,000) of the "religion founder" (*Dharma Sthapak*, 1,111) as the "primordial masculine self" (*Shri Krishna*, 10). The primordial masculine self makes three zeroes of time into three ones of time moving without him.

The "greeter consciousness" (*Pratyagatma*, 1) of the primordial masculine self gives an "impassioned devotee" (*Samyama*, 0) the freedom to consciously determine a "diverse present" (*Brahli*, 89). The impassioned devotee para-consciously imagines a future originating from the "parallel reality" (*Pushtartha*, -8) of the diverse present. She virtuously conceives an illusionary past for substantiating the diverse present as the "unknown reality" (*Vastavikta*, 7) everyone needs to know. Thus, the past, the present, and the future of time move in oneness with the oneness of the religion-founder with the "primordial greeter" (*Srijak*, 16) as "four ones" (*Viparyaya*, 888). Four ones are intrinsically "discontinuous"

(*Sattva*, 1111). Extrinsically, the second one doubles the first one, the third one doubles the double, and the fourth one doubles the quadruple for producing an "eight-fold growth" (*Satarupa*, 8) and reproducing three eights as the "normative development" (*Viparyaya*, 888) of the "primordial followership universe" (*Viparyaya*, 888) over the three "time dimensions" (*Guru dharma*, 360).

Each "primordial follower" (*Antara*, 13,000) becomes a "religion defender" (*Dharma Rakshak*, 15), defending the religion founder as the timeless "God" (*Ishvara*, 5 = 15/3) of the three-time dimensions. Religion defender is an "office" (*Kamma*, 130) that "subsists" (*Prachalita*, 15) as a "continuous" (*Rajas*, 15) element through "breeding" (*Janana*, 15) of the "practitioners" (*Arhat*, 15) of the "religion" (*Dharma*, 370). The office forces the "officer" (*Khyati*, 45) to be the "appropriator" (*Seshi*, 45) of the "infinite cost" (*Varakharcha*, 45) of "cause-defending" (*Nivedana*, 189) with "devotional energy" (*Bhakti shakti*, 189). Therefore, it motivates a "technological adjustment" (*Vyavastha*, 189) in the "organizational arrangement" (*Vyavastha*, 189) of the "ecosystem enactment" (*Vyavastha*, 189) of the "cause" (*Nidana*, 18) being defended. To compensate for the "decreasing returns" (*Kshetra*, 189), the officer becomes a "conscientious" (*Annam*, 185) "religion destroyer" (*Dharma Bhakshak*, 185). The officer "weakens" (*Citraka*, 185) the "tropical force" (*Citraka*, 185) of the religion founder and becomes "global" (*Bha*, 185) and "indistinguishable" (*Tamas*, 185) from a "primordial greeter" (*Srijak*, 16). "Without religion" (*Nirdharma*, -6), everyone enjoys "opulent" (*Raivata*, 185) "bliss consciousness" (*Ananda*, 185) of the "infinite element" (*Tamas*, 185). The infinite element is "dark" (*Tamas*, 185) because it is "bond free" (*Aditi*, 1024). It lets one rely on one's "intrinsic light" (*Usha*, 19) of "white color" (*Shvetah*, 10). It frees one from the "extrinsic light" (*Digambara*, 100) of "black color" (*Shymah*, 4,975).

However, without a "clarified consciousness" (*Prakashatma*, 19), the black color invokes a "sensation" (*Vedana*, 197) of "fear" (*Bhaya*, 176). It pollutes the white color by forming a "light spectrum" (*Ambara*, 180) of "rainbow color" (*Satrangi*, 379). Rainbow color services the light of the entire "primordial greeter realm" (*Indra*

Loka, 379). It makes one a "path-discovering entity" (*Kathanayaka*, 379) for rediscovering the path to the white color with one's "intellect" (*Medha*, 379). It makes the intellect the "mentor" (*Kunti*, 379) to activate the "holy spirit within self" (*Lakshmi*, 379) for the "creation" (*Srijan*, 379) of "global maxima" (*Pindajya*, 379) and the "illumination" (*Dipamala*, 1) of the "seed" (*Beeja*, 379) "essential" (*Beeja*, 379) for doing so.

The value one expects from the illumination is shaped by one's beliefs about God. When it comes to God, there are varying beliefs about the ways of God. In the Eastern belief system, God takes the "privilege" (*Adhikriti*, 1) from those working and gives it to the nonworking "helpless" (*Asahaya*, 149). For instance, at present, the Chinese government is acting God, taking the privilege away from its internet giants, leading to depreciating the Chinese stock market and currency values. In the Western belief system, God takes the privilege from those acting helpless and gives it to those who work. For instance, the American government is also acting God by giving substantial incentives to the working American citizens, leading to appreciating the American stock market and currency values. In the East, despite planning, retirement is always risky. One can never trust God since there is no concept of a transcendental power that fills the role of God. In the West, one can retire without planning. However, one can always trust God since there are always institutions ready to fill the role of God. And if those institutions fail, then it is moral for the Western people to trade the "perpetuating value" (*Saranyu*, 5) from the Eastern people. On the other hand, if those institutions succeed in enabling trading, then it is ethical for the Western people to service the "illuminating value" (*Prakasha*, 7) as the reality for the Eastern people to bear the cost of their discovery.

Indian sages reconciled the Eastern doctrine that God is immanent with us and the Western doctrine that God transcends us. They underlined that "God" (*Ishvara*, 5) is one but with "many forms" (*Aneka Rupa*, 10,000). Each of us is a God's form with the power to act like God. We have the power for self-governance without depending on the "institutions" (*Kundalini*, 8) to fill the role of God for securing international standing or letting individual

The Time Tensor: Exchanging the God Without

leaders act helpless. At the same time, God within them leads them to destroy the whole nation. Therefore, the people of India believe that self-governance is better than "government" (*Dandanayaka*, 100,000). Self-governance makes us question the "supreme offering" (Oblation: *Huta*, 17) of our "leadership" (*Netritva*, 10^{100}) in "service" (*Seva*, 10^{100}) of the "God-like entities" (*Sunanayaka*, 5) who are not that helpless and are the reason for making people helpless by promoting devotion to themselves.

Consider the "management class" (*Jnana Yogi*, 2), comprising Brahmins, seeking "absolute joy" (*Moksha*, 1600) for their self by consuming everyone's offerings. The "leader class" (*Dharma yogi*, 0), comprising Kshatriyas, seeking "conscious freedom" (*Mukti*, 17) for their self by behaving like "God of Gods" (*Indra*, 0), positioning themselves above the helpless "subjects" (*Sura*, 0) who must be devoted to them. The "entrepreneurial class" (*Bhakti Yogi*, 3), comprising Vaishyas, seeking "temporary joy" (*Nirvana*, 123) for their self by repelling all Gods except their self. The "follower class" (*Karma Yogi*, 1), comprising Shudras, seeking "Mother Nature" (*Kudrat*, 8) to care for their self as they care for everyone else. The "Other Backward Class" (*Prakriti Yogi*, -1), comprising the capitalist class and the consumer class, seeking the "State" (*Sthiti*, 100) to care for their self as they have decided to sacrifice their self by following the "State religion" (*Ishvaratva*, 800).

The State religion makes sacrificing our self for the universal wellbeing everyone's life purpose. It leads us to become "immanent" (*Antara*, 13,000) as a "devil" (*Sura*, 0) subject within everyone's "soul" (*Atma*, 4) as a guiding "spirit" (*Ruah*, 20) seeking the "supreme sacrifice" (*Ablation: Svaha*, 44) of their selves. Thus, the one within the Other Backward Class is a "satan" (*Shani Bhagwan*, -1), a discordant factor not in tune with time and releasing "ego" (*Mada*, -1) by sacrificing the "infinite masculinity" (*Prapti*, 16) that produces the "discordance" (*Vilaga*, 90) from Mother Nature. Thus, the one within the Other Backward Class becomes a "devil" (*Sura*, 0), i.e., a concordant factor in tune with time and blessed with the infinite femininity, elevated naturally to be the "King of Devils" (*Indra*, 0) enjoying heavenly life on earth itself.

Thus, oneness with the Other Backward Class is the best path for enjoying life by making everyone a "volunteer" (*Sewadaar*, 9000), starting with the "self-awareness" (*Atma bodha*, 30) of one's potential to be the "absolute volunteer" (*Seemandhara*, 190). Eventually, we realize that our self is the source of the problem, forcing us to sacrifice our "personal value" (*Seemandhara*, 190) and act like a "spiritual soul" (*Vikaranadharmitva*, 79). To solve the problem of everyone's self, we become the "healers" (*Madan*, 789) of everyone's spiritual soul. With the offering of our "supreme self" (*Puryashtaka*, 10^{17} -1) as a "student" (*Shishya*, 10^{17} -1), we make everyone a "management class" (*Jnana Yogi*, 2) educating our self. The management class becomes the leadership class by seeking "conscious freedom" (*Mukti*, 17) for their rediscovered self. A leadership class becomes an entrepreneurial class by repelling all those seeking freedom and moving ahead of line by focusing only on self. By working to make oneself worthy of being in line with everyone who shares one's consciousness, the entrepreneurial class becomes the followership class. By identifying the self as the source of ego, the followership class becomes the Other Backward Class. Only the Other Backward Class knows that the infinite selves of others are the origin of all problems and need retirement planning.

When one sees a "rope" (*Rajju*, -1), one may experience a "force of attraction" (*Padartha*, -3) to enjoy it as a play "object" (*Padartha*, -3). Or, one may experience a "force of repulsion" (*Mangalnath*, 16), fearing "what does not exist" (*Prajaniyam*, 16) as that may potentially be a "snake" (*Sarpa*, 20). An enlightened guru enjoys "gravitational stillness" (*Sthairya*, 0), knowing a rope is the "universe of atoms" (*Sthula*, 387) with a potential to be the "universe of cells" (*Pitavasa*, 169) that constitutes the snake. When we have a clarified consciousness of both the "absolute" (*Param*, 1600) and the "potential" (*AUM*, 18) beyond the absolute, we transcend beyond the "cause" (*Nidana*, 18) that entangles us with the acculturing "cycle of time" (*Kaal chakra*, 3800).

2.10 Working to be Free from Time with Negative Sixteen Finite Points

For the potential to materialize, one needs a "cascade" (*Vyatikara*, -16) of exchange of the "dark energy" (*Arundhati*, 691) of the past. First, one must stop conceiving the "physical body" (*Sthula sharira*, 387 = 169 + 59 + 169 - 10) that forms the "universe of atoms" (*Sthula*, 387) after death as a play object. Second, one must start experiencing the "universe of cells" (*Pitavasa*, 169) immanent within the universe of atoms that makes one "animate" (*Trasa*, 76). Third, one must continue perceiving a "mind-born creator" (*Brahma*, 59) emanating from the universe of cells for reincarnating the universe of cells without the "primordial masculine self" (*Shri Krishna*, 10).

Fourth, one must discontinue manifesting the primordial masculine self as part of the universe of cells already programmed to reincarnate as a "human being" (*Insan*, 1). Fifth, one must let the "primordial masculine self" (*Shri Krishna*, 10) and the "primordial feminine self" (*Parvati*, 10) incarnate as an "epigenetic atom" (*Kapinjala*, 20) endowed with one's "paternal consciousness" (*Pitra*, 4) of "what does not exist" (*Prajaniyam*, 16). Sixth, one must let the "snake" (*Sarpa*, 20) incarnate by transforming "what does not exist" (*Prajaniyam*, 16) with the modified "paternal consciousness" (*Pitra*, 4) of "what can exist" (*Priti*, 33).

One may sustain exchanging "what exists" (*Amba*, 32) with "what can exist" (*Priti*, 33) as a "cycle of sustainability" (*Kaal Sarpa chakra*, 39) for the "self-discovery" (*Suddha*, 39) of one's "potential" (*AUM*, 18) to take "many forms" (*Aneka Rupa*, 10,000) over time. There are six steps to self-discovery.

2.10.1 First, Shape the Twelfth House with Goal

The twelfth astrological house shapes the correlation of the "Pisces zodiac" (*Meen Rashi*, 96) with the "Dark matter" (*Dakshinashapati*, 1,600) for helping one gain freedom from time. To be free from time, one must set a "goal" (*Lakshya*, 9) of "self-governance" (*Svarajya*, 60). A "path of devotion" (*Bhakti marga*, 1) to self-governance promotes oneness with the "mind-born creator" (*Brahma*, 59 = 60 - 1). It detaches the self from the "present" (*SAUM*, 1600) that is "absolute" (*Param*, 1600) and makes the self "omnipresent" (*Sarvavyapak*, 1) within and without the present. By reproducing the omnipresent without the present, a "human being" (*Insan*, 1) attaches the "universe" (*Brahman*, 2 = 1 + 1) as its "pure reproduction" (*Prajana*, 285). Pure reproduction takes the form of "ether" (*Prajanan*, 285). By making the universe omnipresent within HIS "mind" (*Manas*, 38), the human being becomes "omnipotent" (*Sarvashaktiman*, 1600) within the "present" (*SAUM*, 1600). The human being's "paternal consciousness" (*Pitra*, 4) reproduced by the universe becomes the "absolute consciousness" (*Paramatma*, 1600) within the universe. The absolute consciousness materializes as the "omnipresent matter" (*Dakshinashapati*, 1600), known as dark matter, taking the form of "Methylene [CH_2]" (*Dakshinashapati*, 1,600).

Paternal consciousness takes the form of "hydrogen [H]" (*Jalaprana*, 4). Its reproduction creates "Twin hydrogen [H_2]" (*Indrani*, 69). Hydrogen creates Twin Hydrogen because of the oneness of the twin hydrogen with the hydrogen and the universe with the human being. Two ones "materialize" (*Bhuyojan*, 21) the "carbon" (*Bhuyojan*, 21) element since two is a "carbon copy" (*Chitra*, 1) of one and includes one. The carbon copy materializes the "carbon" (*Bhoyuojan*, 21) element fused with the "universe" (*Brahman*, 2) as a "triple copy" (*Ajitatma*, 0) within the "twin hydrogen" (*Indrani*, 69 = 3 * [21 + 2]). The "carbon copy" (*Chitra*, 1) and the "twin hydrogen" (*Indrani*, 69) form as a "water" (*Jal*, 169 = 1,69) element.

Water element forms as a "sentient light force" (*Apas*, 169). Water and sentient light forces are the "universe of photons" (*Vishvagoptri*, 169). Universe of photons is the "eternal human force" (*Vishvagoptri*, 169 = 53 * 3 + 1 + 9) a "human being" (*Insan*, 1) forms by servicing the "human force" (*Manava Karak*, 53) as a "time multiplier" (*Upaya*, 3) for fulfilling a "goal" (*Lakshya*, 9). Due to the eternal human force, the water element generates the "water force" (*Jalaprana*, 4) infused with the human being's "consciousness" (*Chaithanya*, 4) of "being omnipresent" (*Moksha*, 1600) as the "paternal consciousness" (*Pitra*, 4) within everything in the four directions of space. Being omnipresent gives one the "freedom from the present-effect" (*Moksha*, 1600). For being omnipresent, a "human being" (*Insan*, 1) first needs to be a "goalkeeper" (*Shiva*, 7), producing the "goal" (*Lakshya*, 9) as the "causation" (*Hetu*, 1) for enjoying time variations within constant space.

By transforming a sequence of goals into "micro goals" (*Adarsha*, 1), a goalkeeper "breeds" (*Prajan*, 78) an "infinite group" (*Vrihat*, 96 = 78 + 28) of "animate entities" (*Sushupti*, 96) within the "convergent energy" (*Samvat shakti*, 28) of a sequence of micro goals. The "breeding force" (*Dikapala*, 1000) reproduces "eight copies" (*Maharavana*, 8) of each "form" (*Rupa*, 100,000) of the animate entity after consuming a "copy" (*Nakala*, 0). The "goalkeeper" (*Shiva*, 7) produces a copy after consuming the "eight copies" (*Maharavana*, 8) to develop a "technological capability" (*Mohini*, 15 = 7 + 8) for "breeding" (*Janana*, 15).

The twelfth astrological house infuses the "essential nature" (*Svabhav*, 8) of the "fish" (*Matsya*, 17) for "feeding" (*Bharana*, 17) the goal of breeding to a "copy" (*Nakala*, 0). A small fish becomes a large fish by consuming eight copies of the small fish. A large fish lets a small fish produce eight copies by consuming the "mass fragments" (*Anurenu*, 19) of the large fish. "Eight copies" (*Maharavana*, 8) sustains the "present growth" (*Sva*, 11) of the "mass fragments" (*Anurenu*, 19) necessary for sustaining the small fish's "co-existence" (*Sahastitva*, 10^{10}) with the large fish within the goalkeeper's "consciousness system" (*Sara Kalpa*, 10^{10}).

The Mystery of a Fish Pond

In a fish pond, a small female fish lays eggs in ascending "clutch sizes" (*Sva*, 11) with a minimum of eleven. She consumes eight of her eggs to become a large male fish. The ninth egg hatches a small female fish by consuming the large male fish after his death and reproducing eight eggs. The tenth egg hatches a small female fish that consumes those eight eggs. The eleventh egg hatches a female fish that sustains the present growth of the fish population after the death of the ninth and the tenth fishes. She consumes both the ninth and the tenth fishes after their death to become a larger male fish after laying twelve eggs.

> The largest female fish has a clutch size of 8 x 10^{15}. She feeds the "natural essence" (*Svabhav*, 8) of the eight primordial eggs to breed three "forms" (*Rupa*, 100,000) of fishes: small, large, and larger than the small but smaller than the large. The ninth, tenth, and eleventh fishes form a "triple copy" (*Ajitatma*, 0) of the zeroth fish that sustain the zeroth fish as the "essence" (*Sara*, 16) of the fish population. The sixteen-unit essence comprises the zeroth fish, eleven eggs that sustain the zeroth fish, and twelfth egg that perpetuates the essence as the consciousness of the zeroth fish within the four forms of fishes: small, large, larger, and largest. The sixteen-unit essence is the "limitation" (*Parimita*, 16) of the present growth.

2.10.2 Second, Shape the Tenth House with Willfulness

The tenth astrological house shapes the correlation of the "Capricorn zodiac" (*Makar Rashi*, 80) with the "Sun" (*Surya*, 21) for helping one gain freedom from the "limitations" (*Parimita*, 16) of the present growth. To be limitless, there must be "willfulness" (*Prakramya*, 9) for transforming the "goal" (*Lakshya*, 9) of "self-governance" (*Svarajya*, 60) into an "intentionality" (*Abhipraya*, 38) for "exponential growth" (*Chakrika*, 90). Exponential growth is possible only when each copy becomes a "replicative" (*Liksha*, 90) of the "goalkeeper" (*Shiva*, 7). Exponential growth substitutes self-governance with "social governance" (*Ramarajya*, 60) by the "universe of copies" (*Chakrika*, 90) within the goalkeeper's "leadership" (*Netritva*, 10^{100}).

A goalkeeper leads with the "path of knowing" (*Jnana marga*, -10^{19}) that brings "convergence" (*Mamaya*, 158) in culture. Each copy reproduces his "omnipotent" (*Sarvashaktiman*, 1600) form to attach the "universe" (*Brahman*, 2) as her "pure reproduction" (*Prajana*, 285). She lets the goalkeeper detach himself from the universe. For enjoying the socially-bred universe, the goalkeeper must reorient his "goal" (*Lakshya*, 9) to "meditation" (*Dhyana*, 9). He must "dream" (*Dhyana*, 9) about the universe he wishes her to breed for becoming an "enjoyer" (*Rasiya*, 9) of his wishes as his "impassioned devotee" (*Samyama*, 0) seeking "absolute joy" (*Moksha*, 1600) like him. The goalkeeper makes the meditation the

The Time Tensor: Exchanging the God Without 91

"essence" (*Sara*, 16 = 7 + 9) of willfulness. The meditation lets the goalkeeper energize the "triple copy" (*Ajitatma*, 0) into a "time multiplier" (*Upaya*, 3) that makes him the "Sun" (*Surya*, 21 = 7 * 3).

A small female fish breeds a minimum "clutch size" (*Sva*, 11) of eleven in a fish pond. If the eleventh fish is willing to let the first ten fishes breed eight fishes, then she shapes the "feasibility" (*Sadhyata*, 80) of producing eighty fishes and consuming thirty fishes. All ten fishes may breed eight fishes only if the ninth does not breed eleven fishes, and the tenth feeds on the ninth after she has bred eight fishes. In that case, the ninth, tenth, and eleventh fishes breed eight eggs each. The eleventh fish's "conscious consciousness" (*Prashantatma*, 16) lets her gain "conscious freedom" (*Mukti*, 17) from the "limitations" (*Parimita*, 16) of the present growth. Her "conscious energy" (*Amrit*, 1000) becomes the "blessing" (*Ashirwad*, 1000) that shapes the "maternal consciousness" (*Chaithanya*, 4) of the ninth fish and the "paternal consciousness" (*Pitra*, 4) of the tenth fish.

The first four fishes develop a "daughter consciousness" (*Jyotistava*, 4) of the "Capricorn zodiac's" (*Makar Rashi*, 80) "feasibility" (*Sadhyata*, 80). The second four fishes develop a "son consciousness" (*Putatma*, 4) of the feasibility of energizing the "triple copy" (*Ajitatma*, 0) into a "triple octave" (*Prabhava*, 80) of eggs, with each copy producing an "octave" (*Sargam*, 60). The eleventh fish acts like an "octave" (*Sargam*, 60) that transforms the "essence" (*Sara*, 16) into a "double octave" (*Madhusudan*, 16). She forms the double octave into a "base" (*Kumbhaka*, 10) of ten fishes for producing the "productive energy" (*Brahmani*, 80) of eighty fishes within her "feminine body" (*Karana sharira*, 30) after consuming thirty fishes.

With willfulness, the eleventh fish morphs into a "larva" (*Dimbhak*, 80). Larva fulfills the "will" (*Marji*, 5900) of the fifth fish whose "maternal consciousness" (*Chaithanya*, 4) guides her to morph into a "tadpole" (*Dimbhakita*, 90). Tadpole fulfills the will of the tenth fish whose "paternal consciousness" (*Pitra*, 4) divides her "animate body" (*Sharira*, 56) into twenty-four "froglets" (*Mandukak*, 90), each with a "daughter consciousness" (*Jyotistava*, 4) of the "triple octave" (*Prabhava*, 80). Froglet fulfills the will of the fifteenth fish whose "son consciousness" (*Putatma*, 4) liquifies the "body" (*Mahatattva*, 56) of the twenty-four froglets. The liquified bodies of twenty-four froglets solidify into the "inanimate body" (*Vapu*, 56) of one female "frog" (*Manduk*, 120). The whole process of forming the froglets and transforming them into the frog takes "twenty-four hours" (*Dina*, 60). Twenty-four hours manifests the seventh fish's "social governance" (*Ramarajya*, 60 = 16 + 24 + 16 + 4) over the sixteen fishes, twenty-four froglets, four forms of "consciousness" (*Chaithanya*, 4), and "polluted consciousness" (*So Ham Siddhanta*, 4) of the goalkeeper as the "entity" (*Hasti*, 24). As an entity, the seventh fish is the "goalkeeper" (*Hasti*, 24), embodying the "light" (*Prabha*, 180) of the three "octaves" (*Sargam*, 60).

The "female frog" (*Manduk*, 120) fulfills the will of the sixteenth fish by letting the seventh fish consume her "inanimate body" (*Vapu*, 56) and morph that into the "para body" (*Chandra*, 82) of a "male frog" (*Mendaka*, 180). The 82-unit para body includes the 56-unit inanimate body, the 24-unit "institutional force" (*Trivikrama*, 24) of the twenty-four froglets, the 1-unit "greeter consciousness" (*Pratyagatma*, 1) of the seventh fish, and the 1-unit "illusionary energy" (*Maya shakti*, 1) of the sixteenth fish as the "maternal primordial greeter" (*Sati-Parvati*, 16). The maternal primordial greeter is the "essence" (*Sara*, 16) of both the willful "goalkeeper" (*Shiva*, 7) and the "willfulness" (*Prakramya*, 9) as his "revealed identity" (*Maha Shiva*, 9), norming the "male frog" (*Mendaka*, 180 = 9 * 20) with the five forms of four-unit "consciousness" (*Chaithanya*, 4). Twenty units with consciousness and one unit with greeter consciousness norm the goalkeeper's "future reality" (*Ekartha*, 21) as the 21-unit "Sun" (*Surya*, 21).

2.10.3 Third, Shape the Eighth House with Infinite Breeding

The eighth astrological house shapes the correlation of the "Scorpio zodiac" (*Vruschik Rashi*, 735 = 724 + 11) with the "Earth" (*Bhu*, 724) for helping one enjoy the "present growth" (*Sva*, 11). An "enjoyer" (*Rasika*, 9) enjoys the "present growth" (*Sva*, 11) by channeling "willfulness" (*Prakramya*, 9) into "infinite breeding" (*Shrimate*, 9). Infinite breeding substitutes the "social governance" (*Ramarajya*, 60) by the universe of copies within a goalkeeper with the "institutional governance" (*Samrajya*, 60) by the goalkeeper without the "universe of copies" (*Chakrika*, 90). The "goalkeeper" (*Shiva*, 7) detaches the self from the universe of copies after "death" (*Vira ratri*, 18) and becomes a "conscious entity" (*Siddha*, 7). The goalkeeper attaches the "universe" (*Brahman*, 2) as the "causation" (*Hetu*, 1) to fulfill a transformed "goal" (*Lakshya*, 9) of enjoying the "transcendental value" (*Khuda*, 6) as "growth" (*Dishta*, 6) with "time multiplier" (*Upaya*, 3). The "Earth" (*Bhu*, 724) sets the limiting "horizon" (*Kshitij*, 724) for the transcendental value.

The "Scorpio zodiac" (*Vruschik Rashi*, $735 = 724 + 11$) shapes the "fatal attraction" (*Shurasena*, 735) for the 724-unit "Mother Earth" (*Kshiti*, 724). A "spider" (*Makadi*, $1600 = 8 * 2 * 100$) has eight legs, two feelers, and a "reproductive" (*Gurutva*, 100) spider body. Each "feeler" (*Pipplayshraya*, -10^9) has a dark matter-like putrid "bragging face" (*Puti mukha*, -10^9) that breeds a "spiderling" (*Lutashaav*, 7000). Each spiderling breeds twenty "spinnerets" (*Vyanang*, 13). Each spinneret breeds ten "sacs" (*Thaili*, 13) of silk. Each sac of silk breeds a hundred "eggs" (*Anda*, 19). The "spider body" (*Triyancha*, 19) reproduces two spiders. Thus, a spider breeds eighty thousand eggs with its spider body (2 spiders x 2 spiderlings x 20 spinnerets x 10 sacs x 100 eggs). She breeds forty thousand eggs with her two feelers. She feeds thirty thousand eggs to her eight "legs" (*Panva*, 24) with two "poison fangs" (*Jambha*, 9) that are a part of her spider body. Each leg consumes 3,750 eggs to impregnate the egg with the eight 3750-unit "menses" (*Upasravana*, 3,750). "Eight menses" (*Kundalini*, 8) act as the "institution" (*Kundalini*, 8) that activate the potential of an "infinity" (*Tamo guna*, 90,000) of Father spiders within the Mother Spider and lead to a growth of 90,000 eggs (80,000 + 40,000 − 30,000).

After feeding, the "poison" (*Ibha*, -10^{10}) liquifies the mother spider and feeds her to the daughter spiderling. It makes the daughter spiderling "two sisters" (*Dhamini*, 1200) because the dead mother spider is the "not-yet-born twin sister" (*Dhamini*, 1200) of the daughter spiderling. Before feeding, the daughter spiderling's poison gasifies the father spider. The mother spider freerides on the gasified father spider and feeds him to the son spiderling. It makes the son spiderling "two brothers" (*Ashwini kumaras*, 120) because the dead father spider is the not-yet-born "twin brother" (*Shivagati*, 120) of the son spiderling. In a sidereal year of three hundred sixty days, each brother produces 360 sperms. The brother produces 180 in the first month, 90 in the second month, 60 in the third month, 20 in the fourth month, 10 in the fifth month.

After producing 180 sperms, the brother divides his body into three sperms in the sixth month—one of which is the twin brother comprising two brothers and the other is the brother himself. The brother multiplies his body into three sperms in the seventh month—one of which is the twin brother comprising two brothers and the other is the brother himself. The brother continues to produce 10 sperms in the eighth month, 20 in the ninth month, 60 in the tenth month, 90 in the eleventh month, and 180 in the twelfth month. Thus, as a bother, the son spiderling produces 724 sperms that form the "earth" (*Bhu*, 724) element over a year. He also produces two twin brothers. The first twin brother transforms into his spider body and makes him the father spider. The second twin brother waits within the son spider's body for the mother spider to appear on the horizon. Losing "patience" (*Titiksha*, 30), the second twin brother incarnates as the "sister" (*Bahen*, 130). The sister is the mother spider.

2.10.4 Fourth, Shape the Sixth House with Entanglement

The sixth astrological house shapes the correlation of the "Virgo zodiac" (*Kanya Rashi*, 37) with the "Neptune" (*Ketu*, 140) to fulfill the goal of the enjoyer's "entanglement" (*Kula*, 9) within the "present growth" (*Sva*, 11). As the enjoyer "universalizes" (*Sarvabhaumikaran*, 130) the "infinity force" (*Anantariti*, 140), he loses "institutional legitimacy" (*Vaidhatva*, 140) to the "bred group" (*Kendrabhi mukha*, 140) and "gains" (*Prapta*, 140) "religious governance" (*Vrajya*, 60). He detaches the universe for reproducing the causation that substantiates his "natural essence" (*Samaa*, 1) within the "many forms" (*Nanarupa*, 10,000) of the bred group. He attaches the self as the "time" (*Kaal*, 360) that manifests many forms impregnated with his "divine" (*Divya*, 360 = 9 * 40) element for sustaining a new "goal" (*Lakshya*, 9) of the ether's "reproductive reality" (*Bhavartha*, 40).

The "ether" (*Prajanan*, 285) reproduces its "reproductive reality" (*Bhavartha*, 40) as the "dark matter" (*Dakshinashapati*, 1,600 = 40 * 40). It solidifies the twenty-unit "underflowing consciousness" (*Saguna*, 20) of the "ascending light force" (*Ketu*, 140) of the "past system" (*Antaka*, 140) with the one-unit "greeter consciousness" (*Pratyagatma*, 1) within the "natural essence" (*Samaa*, 1) to form the "Sun" (*Surya*, 21) as the "solid earth" (*Bhuyojan*, 21). The "Virgo Zodiac" (*Kanya Rashi*, 37) shapes the 37-unit "feminine gene [p18]" (*Upadhi*, 37 = 21 + 16) with the "Sun" (*Surya*, 21) and the "essence" (*Sara*, 16). A grandmother gorilla incubates the mother gorilla in her womb for 37 weeks. After her birth, Mother Gorilla breeds a twin at age 12 and a son gorilla followed by a daughter gorilla every four years until age 36. Thus, she breeds four son gorillas and four daughter gorillas. She devotes fifteen weeks searching for a mate for each of her babies, thus living a life of 36 years and 15 weeks. Each daughter gorilla also breeds four grandson gorillas and four granddaughter gorillas.

Thus, a mother gorilla breeds a matriarchal family of four daughter gorillas, sixteen granddaughter gorillas bred by her daughters, and sixteen granddaughter gorillas bred by her daughters-in-law. She is a matriarch of a group of 37 female gorillas. A grandfather gorilla guides a father gorilla to have sexual intercourse with the eight grandmother gorillas during the first eight years. After that, the father gorilla has sexual intercourse with twenty mother gorillas over the next twenty years. Next, he has sexual intercourse with eight daughter gorillas over the next eight years. When 37, he has sexual intercourse with a granddaughter gorilla. A father gorilla is a "star" (*Tara*, 2) of attraction as a dominant silverback within a three-generational family of thirty gorillas, of which 29 are females. Over his lifetime, he impregnates 37 female gorillas with his "paternal consciousness" (*Pitra*, 4).

2.10.5 Fifth, Shape the Fourth House with Culture

The fourth astrological house shapes the correlation of the "Cancer zodiac" (*Karka Rashi*, 6) with the "Venus" (*Shukra*, 2,700) following a "culture" (*Sadakhya*, 9) of "growth" (*Dishta*, 6). It leads to the acculturation of each copy within the enjoyer's culture of growth.

A grandfather gorilla fathers four father gorillas and four mother gorillas, thus fathering a growth of six without the two that make him the "star" (Tara, 2) of attraction. The two include him and an "organization" (Sangathan, 29) of twenty-nine female gorillas. The "feminine gene" (Upadhi, 37) for the "eight-fold growth" (Satarupa, 8) of child gorillas and 29-fold organization female gorillas is immanent within the grandmother gorilla. After impregnating the grandmother gorilla with the feminine gene, the grandfather gorilla detaches himself as the "time consciousness" (Kalatma, 14) of the "present development" (Maha Vibhu, 14). The 14-unit present development comprises the eight-fold growth within the grandmother gorilla and the six-fold growth without the grandmother gorilla. It lets the grandfather gorilla attach the universe as the "space" (Vaas, 18,000) by targeting a new goal of becoming a "guider" (Guru, 100) who services his "light" (Prabha, 180) to free one from the entangling "vestige of the past" (Shukra, 2,700).

Before his birth, with willfulness, a grandfather gorilla creates "space" (Vaas, 18,000) for himself in the womb of a daughter gorilla for reincarnating as the grandson gorilla and for the grandmother gorilla in the womb of a mother gorilla for reincarnating as a daughter gorilla. Thus, he manifests "what never existed" (Rachayita, 27) for becoming a two-unit "neutron star" (Tara, 2) that eventually becomes a daughter gorilla and gives birth to him as the grandson gorilla. What never existed is the "freedom force" (Rupa siddhi, 27) one enjoys as a "council of twelve zodiacs" (Rachayita, 27). It is the "fate" (Bhaga, 27) one conceives with the "conscious imagination" (Hiranyagarbha, 19) of the "future reality" (Ekartha, 21) that makes one the "star of attraction" (Tara, 2). It lets one establish one's "imperial governance" (Adhirajya, 60) as a "guider" (Guru, 100 = 60 + 40) of the "reproductive reality" (Bhavartha, 40) of another as a "zodiac house" (Jadatva, 1) impregnated with a 37-unit "masculine gene" (Samsarga, 37) serviced by the "sequential force" (Samsarga, 37) of the two-unit "universe" (Brahman, 2) one forms as a neutron star.

A zodiac house incarnates as a satan crab. She mates with a greeter crab to breed six grandfather crabs and six grandmother crabs. Then, she mates with the six grandfather crabs to breed six father crabs and six mother crabs. Next, she mates with six father crabs to breed six son crabs and six daughter crabs. After that, she mates with six grandmother crabs, six mother crabs, and six daughter crabs to breed six grandson crabs and six granddaughter crabs. Finally, she mates with a devil crab to breed five greeter crabs. The devil crab is the "imaginary spirit" (*Kalpanik*, 20) of the greeter crab. The neutron star's "gravitational energy" (*Lalita*, 100) guides the satan crab to conceive its "entropy value" (*Sarvanasha*, 5) as a greeter crab for breeding five greeter crabs. The greeter crab's imaginary spirit guides the satan crab to perceive a potential of six grandmother, six mother, and six daughter crabs within her as the "animate primordial maternal" (*Shani*, 18). It also leads to a growth of six granddaughter crabs without her as the "creative force" (*Uma*, 6) of the six greeter crabs. As an imaginary spirit, the devil crab conceives the 27-unit council of twelve zodiacs to comprise twenty-four masculine crabs, a satan crab, a greeter crab, and himself.

Therefore, a mother crab breeds 26 juvenile daughter crabs and one juvenile son crab in ascending sequential oneness with the twenty-seven zodiacs and 26 juvenile son crabs in descending sequential oneness with the twenty-seven zodiacs. Consequently, the mother crab enjoys oneness with the 28th zodiac without the twenty-seven zodiacs. The satan crab is the "oneness with the 28th zodiac" (*Shani*, 18). The 28th zodiac is the "intrinsic force" (*Rohini*, 50) of the "greeter essence" (*Sara*, 16). With acculturation to the intrinsic force, she reproduces the 16-unit "greeter essence" (*Sara*, 16) within the 37-unit feminine gene for breeding 53 child crabs. Fifty-three child crabs materialize the 16-unit essence reproduced by the "three copies" (*Ajitatma*, 0) and the 5-unit "entropy value" (*Sarvanasha*, 5) of the "neutron star" (*Tara*, 2). The neutral star perpetuates its "illusionary energy" (*Maya shakti*, 1) within the three copies to form the entropy-producing "time multiplier" (*Upaya*, 3).

2.10.6 Sixth, Shape the Second House with Materialism

The second astrological house shapes the correlation of the "Taurus zodiac" (*Vrishaba Rashi*, 69) with the "Jupiter" (*Brihaspati*, 1780) to fulfill the goal of "materialism" (*Pravritti*, 9). By materializing the "thing" (*Vastu*, 9) of enjoyment, it furthers the "personal sovereignty" (*Nagarajya*, 60) of the satan crab over the "universe of copies" (*Chakrika*, 90). With her reproductive energy, the satan crab attaches to the universe as the "space force" (*Dik*, 100) reproduced by the "space" (*Vaas*, 18,000) without her "light" (*Prabha*, 180). She detaches the self to fulfill the goal of becoming the "thing" (*Vastu*, 9). The thing "entangles" (*Uljhana*, 158) and "matters" (*Sadhibhuta*, 158) for the "universal sentient wellness" (*Sadhaniya*, 157).

After the self becomes the thing, she incarnates as a "person" (*Vyakti*, 1) for discovering the "path to absolute freedom" (*Param Mukti Marga*, 8) from the "entanglement" (*Kula*, 9) to both the self and the thing. She becomes conscious that the new goal is to be "conscious" (*Ojas*, 189) of the "sentient" (*Ojas*, 189) element. Jupiter "fades" (*Rajni*, 1780) her "guider planning" (*Rajni*, 1780) from the "spiritual realm" (*Rajyaika-sheshena*, 169). Without the "guider" (*Guru*, 100) element, the spiritual realm transforms into the "Taurus zodiac" (*Vrishaba Rashi*, 69 = 169 - 100) and makes her the "empress" (*Indrani*, 69) of the "universe of bovines" (*Shveta Varaha*, 69).

A Female Bovine is An Empress!

A female bovine's presence brings a male bovine to erection in 60 seconds after she becomes visible at a distance from the mountain top. The male bovine mounts on her in 100 seconds of her proximity at a mountable distance. It takes her 15 seconds to bring him to ejaculation. The mating sequence involves 7 masturbation thrusts, each lasting for 2 seconds. The female bovine keeps the male bovine mounted for 27 seconds after the mating sequence. Next, she steps forward to dismount him. She lets him believe he is running the show. As an "institution" (*Kundalini*, 8), she mates with 24 male bovines in a day for breeding 8 x 24 = 192 son bovines.

The female bovine reproduces her "essential nature" (*Svabhav*, 8) within herself to conceive a twin female bovine who breeds another 192 son bovines. Her "guiding force" (*Chitta*, 100) makes the twin female bovine within her conscious of her personal potential to breed 192 daughter bovines. Twin female bovine's conscious element forms a consciousness system that materializes the female bovine's personal potential of 192 daughter bovines as well.

Twin female bovine is the "institution" (*Kundalini*, 8) producing the "essential nature" (*Svabhav*, 8) for the "eight-fold growth" (*Satarupa*, 8) of the 24 male bovines and the "four-fold growth" (*Maha Nitya*, 4) of the 192 granddaughter bovines with the 24 male bovines. The latter are an eight-fold growth of the three mother bovines.

Twin female bovine is the grandmother. Female bovine is the daughter. "Twin consciousness" (*Sachi*, 69) of the natural "six-fold growth" (*Dishta*, 6) is the mother. Female bovine forms the consciousness of the twin female bovine with a "two-fold growth" (*Vidhana*, 2). Twin female bovine conceives the two-fold growth as an addition to her natural "six-fold growth" (*Dishta*, 6). Female bovine perceives the two-fold growth as a subtraction from her natural six-fold growth.

Twin consciousness materializes the breeding of 6 x 24 = 144 greeter bovines.

Male bovine is a "conscious entity" (*Siddha*, 7) who makes his "reality" (*Vastavikta*, 7) "self-luminous" (*Svarochisha*, 12) within four-unit consciousness and three-unit time multiplier to be "God" (*Ishvara*, 5) of the materialized universe. By illuminating the seven-unit "self-luminous reality" (*Purushartha*, 7) within the self, three mothers manifest the breeding of 3 * 7 = 21 devil bovines.

Twenty-four satan bovines incarnate as 24 male bovines. Their "convergent energy" (*Samvat shakti*, 28) incarnates as the "twin female bovine" (*Kundalini*, 8).

Thus, a "super-fertile" (*Runasvara*, 933) "satan bovine" (*Runasvara*, 933) breeds 192 * 4 + 144 + 21 = 933 child bovines. She and her twin sisters are the creation of the "entity bovine" (*Viyojya*, 957) that materializes the "etheric body" (*Bhoga sharira*, 957) in the form of 24 satan bovines and 933 child bovines. The etheric body is the "replicative body" (*Bhoga sharira*, 957) that materializes the "infinite value" (*Viyojya*,

957) of the "reproductive reality" (*Bhavartha*, 40) of the "time multiplier" (*Upaya*, 3) with one's "conscious energy" (*Amrit*, 1000 = 957 + 40 + 3).

2.11 Knowing the Value of Time with Negative Twelve Finite Points

In the "initial spatial state" (*Atita*, -12), a "conscious entity" (*Siddha*, 7) is conscious of the "quantum thing" (*Asamskrita*, -12) formed with many transformations of the self. Each thing enjoys a "local time" (*Kaal*, 360) as an "entity" (*Hasti*, 24) in "continuity" (*Rohini*, 50) with the "intrinsic force" (*Rohini*, 50) of the "global time" (*Hari*, 10,000) normed by the self's "growth potential" (*Nanarupa*, 10,000). The conscious entity networks the growth potential with the "extrinsic force" (*Sankhara*, 269) of the "national time" (*Shani*, 18) normed by his "potential" (*AUM*, 18) for "growth" (*Dishta*, 6) with the "time multiplier" (*Upaya*, 3). He manifests the time multiplier into the "astral body" (*Linga sharira*, 3). He becomes the "leader" (*Neta*, 0) of the universe that idealizes his copies.

The leader infuses his value as the consciousness system to make the copies a part of his conscious system. He becomes a part of the "circular system" (*Gandakanayaka*, 10,000) that lets each copy be an "impassioned devotee" (*Samyama*, 0) of the "present" (*SAUM*, 1600) his value creates and be that present. After completing a "circle" (*Valaya*, 100,000) experiencing the present, each copy begins "circulating" (*Chakra Yukti*, 10^{1024}) itself as the "energy system" (*Antarmukha*, 10^{1024}), energizing a "universe of copies" (*Chakrika*, 90) as its "replicative" (*Liksha*, 90). A replicative enjoys the value of her entity "time" (*Kaal*, 360 = 90 + 90 * 3) with three copies. The knowing sequence involves six steps.

2.11.1 First, Shape the Eleventh House into a Leader

The eleventh astrological house shapes the correlation of the "Aquarius zodiac" (*Kumba Rashi*, 296) with the "Vega" (*Vega*, 67) to fulfill the conscious entity's goal of becoming a leader. A conscious entity focuses his "energy" (*Shakti*, 19) on "knowing" (*Jnana*, 19) "everything" (*Sarvam*, -5). He diversifies his knowing by "feeding"

(*Bharana*, 17) "everyone" (*Pratyeka*, 180) the "light" (*Prabha*, 180) of his "knowledge" (*Vidya*, 600). "Everyone" (*Pratyeka*, 180 = 6 * 30) includes his three copies and their three copies reproducing his "causal body" (*Karana sharira*, 30) for producing his "guiding force" (*Chitta*, 100) as their "knowledge" (*Vidya*, 600 = 6 * 100). Everyone integrates his feeding within their knowledge by "breeding" (*Bharana*, 15) their copies as their "knowable" (*Mara*, 396) value. Each copy starts "freeriding" (*Muftakhori*, 13) on the "consciousness" (*Chaithanya*, 4) of the conscious entity as the "knower" (*Bhagwat*, 639) "qualified" (*Bhagwat*, 639) to decide the knowable's "quality" (*Guna*, 0). Thus, each copy becomes a "cosmic knower" (*Jnathru*, 639) "par excellence" (*Gita*, 368).

The three copies reproduce the conscious entity's "future reality" (*Ekartha*, 21) with the "time multiplier" (*Upaya*, 3) within the conscious entity's "paternal consciousness" (*Pitra*, 4) of "Mother Nature's" (*Kudrat*, 8) "essence" (*Sara*, 16) to form the "Vega" (*Vega*, 67) white star. The "Vega" (*Vega*, 67) is the "Radiant One" (*Vega*, 67) "ignorant" (*Karala*, 67) of her "emaciation" (*Glani*, 67) after she produces another Vega as her "potential child" (*Keshava*, 22) and transforms into "Mars" (*Mangala*, 102 = 67 + 67 - 22). The "potential child" (*Keshava*, 22 = 21 − 3 + 4) is the Vega's future reality as the "Sun" (*Surya*, 21) and the potential child's "past reality" (*Padartha*, -3) as an "object" (*Padartha*, -3) within the Vega's "maternal consciousness" (*Chaithanya*, 4). The Vega produces the potential child as a "photon" (*Ruah*, 20) within a "neutron star" (*Tara*, 2) that makes two copies of the photon and infuses her reality within the three copies for conceiving another "Vega" (*Vega*, 67 = 20 * 3 + 7). The Vega and her guiding force within the neutron star form the "spiritual realm" (*Rajyaika-sheshena*, 169 = 67 + 100 + 2) as the "universe of photons" (*Vishvagoptri*, 169).

Dolphins represent the Aquarius zodiac. Each copy of the photon takes birth as a female "dolphin" (*Swati*, 1) in oneness with Vega. A female dolphin mates seven times with seven male dolphins for seven seconds each over a year. Next, she mates twice with two male dolphins for two seconds each over six months. During mating, a male dolphin feeds his knowing to a brother dolphin. The brother dolphin breeds the "impregnated consciousness"

(*Prakashatma*, 19) of the knowing within the female dolphin. Thus, over eighteen months following the mating with the focal male dolphin, the female dolphin freerides on the "impregnated consciousness" (*Prakashatma*, 19) of the eighteen "daughter cells" (*Kshatradharma*, 19) traded from the nine pairs of male dolphins. After that, she conceives a "son cell" (*Manyu*, 19) with the "energy" (*Shakti*, 19) of a daughter cell reproducing her "ideal force" (*Dasha*, 1) and the eighteen daughter cells reproducing the potential of the nine pairs of male dolphins.

Female dolphins move out of their family pods to form a thirty-six-member sister pod at age nine. Half of the female dolphins in a sister pod are pregnant or nursing their baby, and form a maternity pod. The other half help nurse while being helped in mating. Gestation is 12 months, followed by six months of nursing, with 18 months of mating; they breed once every three years. During the eighteen months leading up to the mating with the focal female dolphin, the male dolphin freerides on the "knowing" (*Jnana*, 19) of the eighteen "son cells" (*Manyu*, 19) traded from the nine pairs of female dolphins. After that, he conceives and impregnates the focal female dolphin with the "daughter cell" (*Kshatradharma*, 19), embodying the traded knowing of the eighteen son cells and the intrinsic knowing of a son cell embodying him as the "ideal" (*Adarsha*, 1).

During the eighteen months following the conception of a son cell, the pregnant maternal dolphin trades the "para consciousness" (*Mahatma*, 18) of the "knowledge" (*Vidya*, 600) of the eighteen maternal dolphins as the eighteen "mother cells" (*Dhumavati*, 7) and the eighteen paternal dolphins as the eighteen "father cells" (*Rodha*, 1).

During the twelve months of gestation following the conception, the maternal dolphin exchanges the "self-perpetuating" (*Udvaha*, ½) time-varying "present consciousness" (*Paramatma*, 1600) of the "knowable" (*Bhagwat*, 639) knowledge of the sister and the brother pods in the form of eighteen "granddaughter cells" (*Bagalamukhi*, 9) and eighteen "grandson cells" (*Pavamana*, 9).

During the six months of nursing following the birth of a child dolphin, the maternal dolphin services her consciousness of the "time multiplier" (*Upaya*, 3) within the sister and the brother pods to the child dolphin in the form of the three "grandmother cells" (*Dadi*, 18) and three "grandfather cells" (*Uccitika*, 18).

Thus, over the thirty-six months of reproductive cycle, as a leader, the eleventh house services the "transcendental value" (*Khuda*, 6) of six cells to the physical realm. It becomes a "formative primordial greeter" (*Mangala*, 102) after trading the "becoming energy" (*Rodayitri shakti*, 102 = 18 * 6 - 6) of 102 cells from the physical realm.

2.11.2 Second, Shape the Ninth House into a Consciousness System

The ninth astrological house shapes the correlation of the "Sagittarius zodiac" (*Dhanu Rashi*, 128) with the "Moon" (*Soma*, 997) to fulfill a leader's goal universalize his "consciousness system" (*Sara Kalpa*, 10^{10}). The Sagittarius zodiac programs the "oneness-effect" (*Ekatva*, 997) of the "solar universe" (*Surya mandala*, 13) within each "child" (*Arcisa*, 128).

Sagittarius zodiac is represented by cats. A female cat breeds 300 days after birth. Her gestation is 65 days followed by rearing solitary for 300 days while the male cat guards the territory living a solitary life like a king. Her nursing lasts for a full lunar cycle of thirty days. After that, she feeds the hunt to the kitten for the next 30 days. After 60 days from birth, the kitten begins hunting. The litter size is four. In the wild, a cat lives like a "queen" (*Rani*, 7) for two years, falling prey to the predators as she worries about her kittens and comes out of her hiding. Thus, the cat population doubles every year by reproducing the leader's consciousness system. A follower cat lives like a "dam" (*Setu*, 90) for sixteen years, breeding fifteen litters, thus giving birth to one-hundred eighty kittens with her "light" (*Prabha*, 180). Finally, an entrepreneurial domesticated cat lives like a "molly" (*Paturiya*, 10), breeding a litter size of nineteen by "mating" (*Samagam*, 19) with five tomcats. She breeds three kittens with each of the tomcats and four kittens that sequentially unite the genetics of the two successive tomcats.

2.11.3 Third, Shape the Seventh House into a Conscious System

The seventh astrological house shapes the correlation of the "Libra zodiac" (*Tula Rashi*, 10^{100}) with the "Uranus" (*Rahu*, 73) to free everyone from the "leadership" (*Netritva*, 10^{100}) gained by an individual leader from everyone reproducing his "consciousness system" (*Sara Kalpa*, 10^{10}). The "Libra zodiac" (*Tula Rashi*, 10^{100}) makes everyone "independent" (*Svadhina*, 10^{100}) from the leader's "convergent reality" (*Tulyartha*, 10^{100}). "Uranus" (*Rahu*, 73) makes everyone's "emotions" (*Hunduka*, 73) dependent on the collective's "conscious system" (*Shunya Kalpa*, 8×10^{15}). Everyone diffuses their energy in ascending motion seeking togetherness, thus accelerating their entropy and halving the population by bringing "unity in duality" (*Deva Yajni*, 0). Unity in duality implies the "unity" (*Abheda*, 0) of the child in the "duality" (*Bheda*, 28) of the father's leadership and the mother's "followership" (*Apacayana*, 10^{19}).

Peafowls represent the Libra zodiac. Peafowls' "muster" (*Juta*, 1) has thirty-six members, comprising a peacock, five peahens, and thirty peachicks. A peacock breeds with a "harem" (*Haram*, 15) of five peahens. A peahen lays six eggs over ten days from age two, one every other day. She incubates each egg for two days within her body and twenty-eight days without her body. She lives for twenty years in the wild, laying one-hundred fourteen eggs by mating with ten peacocks. Every two years, she forms a new harem with a constant group of five peahens. Thus, as a "child" (*Arcisa*, 128), she unites the "greeter consciousness" (*Pratyagatma*, 1) of 114 children, four sisters, and ten brothers. Each peacock mates with fifty peahens over twenty years. Thirty peachicks comprise fifteen males and fifteen females. Only three of the fifteen males survive. The other twelve do not follow their mother as she searches for the food. They like to drool and rest, thus falling prey to predators.

Over a two-year cycle, six males and thirty females survive and mature. Of the six sons, maternal genes are dominant in five and the paternal gene in one. Each male is a child of a different mother. In the six daughters born to each mother, the birth mother's DNA programming genes are dominant. However, the predominant RNA planning genes vary as the birth mother services the "conscious energy" (*Amrit*, 1000) of the four foster mothers to the first four daughters, the "consciousness" (*Chaithanya*, 4) of the father to the fifth daughter, and the "perpetuating value" (*Saranyu*, 5) of the mother to the sixth daughter.

Consequently, the deciding mtDNA performing genes also vary. The fifth daughter gives birth to the grandson, who survives with the father's conscious system. The sixth daughter manages the harem of "five sisters" (*Sauri*, 1000), including the first four and herself. She organizes the surviving grandson and the five sisters into a harem for enjoying the grandson's beauty.

2.11.4 Fourth, Shape the Fifth House into a Circular System

The fifth astrological house shapes the correlation of the "Leo zodiac" (*Simha Rashi*, 47) with the "Mercury" (*Budha*, 1600) to make a "collective" (*Visarjya*, 47) "whole" (*Akala*, 16) as a "circular system" (*Gandakanayaka*, 10,000). In a circular system, an individual enjoys "rotating leadership" (*Liksha*, 90). It lets each person be "active" (*Sakriya*, 10^{10}) as an "individual" (*Vijujya*, 53) by "trading" (*Samana*, 53) the collective "cultural heritage" (*Virasat*, 10^{10}).

Lions represent the Leo zodiac. Once every two years, a female lion mates with a male lion every thirty minutes for forty-eight times over twenty-four hours when behaving like a queen. She mates every fifteen minutes for ninety-six times over twenty-four hours when behaving like a dam. Each copulation lasts for twenty seconds. Gestation is one-hundred eight days. The litter size is three. A pride comprises four male lions, twelve female lions, and twenty-four cubs. The male lions kill a male cub of each lioness because he inherits a "child profiting gene" (*Pravritta*, 12) that groups the "son consciousness" (*Putatma*, 4) of all four male lions and makes him potentially "omnipotent" (*Sarvashaktiman*, 1600). Cubs follow her around after three months and lead the hunt after an additional eight months. They begin living independently in three pairs of four brothers at age two and leave their pride at age three. After living as a free entrepreneur for two years, they form a pride at age five with the twelve cousin sisters. Four brothers comprise two with the same grandmother and two with the cousin grandmother.

Thus, after six years, four male and twelve female lions become four prides of forty each with a "ten-fold growth" (*Vicitra mandala*, 170). They are the offspring of eight grandmothers, with four rearing four male and four female cubs and four rearing a pair of female cubs. Two grandfathers, each breeding two male and six female cubs, father them.

2.11.5 Fifth, Shape the Third House into a Circle

The third astrological house shapes the correlation of the "Gemini zodiac" (*Mithun Rashi*, 70) with the "Mars" (*Mangala*, 102) to make each individual a "part " (*Anga*, -1) of the "collective circle" (*Valaya*, 100,000). It transforms the individual's "faithlessness" (*Asraddhya*, 70) without the collective into the "faithfulness" (*Sraddhya*, 102) within the collective.

Foxes represent the Gemini zodiac. A vixen dog starts breeding one year after birth. On the day of estrus, a vixen fox token marks an open collective circle with five drops of her urine eight times a minute over twenty hours. The dog fox copulates with her with seven thrusts of one second each before ejaculating. He remains in copulatory lock with her for ninety-six minutes. The gestation is forty-eight days. The litter size is four. A litter comprises a dominant male, a subordinate male, a dominant female, and a subordinate female kit.

A dominant male forms his individual friendly long-distance "social circle" (*Vartula*, 10,000) and mates with a dominant female within that social circle. A subordinate male norms a collective short-distance "foe circle" (*Valaya*, 100,000) and mates with a subordinate female within that foe circle. A social circle is a circular system within the foe circle that circles outward as a dominant member seeks to explore the space for the collective's growth and transformation of the subordinate "consciousness system" (*Sara kalpa*, 10^{10}) through conscious endeavors.

As a dominant male travels long-distance to search for his mate, he crosses four foe circles. On his return journey, the subordinate female seeks to charm him with her token mark, leading to a battle of supremacy between him and a subordinate male. The dominant male is tired and starved after traveling long-distance, weakens after copulation, and dies in the battle. Consequently, the subordinate male mates with the dominant female and makes her a part of the collective circle.

Dominant vixen breeds the dominant male with the dominant dog's planning gene and the subordinate male with the subordinate dog's programming gene. She breeds the dominant female with her performing gene and the subordinate female with the subordinate vixen's profiting gene. Subordinate vixen profits because she mates twice with the subordinate dog: once before his fight and once after on the sixth day after the first copulation. She breeds the dominant male with her performing gene serviced during the second mating and the subordinate male with her profiting gene serviced during the first mating. The performing gene makes the kit conscious that he must act individually to gain personal dominance.

The profiting gene programs the consciousness that the kit must be faithful to the collective to manage social dominance. She breeds the dominant female with the subordinate dog's planning gene traded during the first mating. She breeds the subordinate female with the subordinate dog's programming gene traded during the second mating. The planning gene makes the kit act spontaneously and perform as an individual within the collective interest of diversifying the collective's subordinate "cultural consciousness" (*Virasat*, 10^{10}). The programming gene makes the kit inactive and profit from the collective's subordinate "cultural consciousness" (*Virasat*, 10^{10}).

2.11.6 Sixth, Shape the First House into an Energy System

The first astrological house shapes the correlation of the "Aries zodiac" (*Mesha Rashi*, 256) with the "Saturn" (*Satan*, -1) to make each individual a "discordant factor" (*Asura*, -1) without the collective wishing to be a part of the collective. It forms each individual as an "ingratiating" (*Supuma*, 256) element seeking to please the collective and norming an "identity" (*Pahachaan*, 8) shaped by the collective's "energy system" (*Antarmukha*, 10^{1024}). It transforms an individual's "reality" (*Vastavikta*, 7) into the collective's "quantum reality" (*Prakarana Vadartha*, 90,000). It subordinates the collective to the individual whose reality dominates everyone waiting in line for their "shared becoming" (*Sadgati*, 15), i.e., salvation. The breeding lets the "present individual" (*Madhya*, 16) be the "spiritual center" (*Madhya*, 16) of the "present collective's" (*Achaarya*, 16) "conscious consciousness" (*Prashantatma*, 16) as an "investigator" (*Achaarya*, 16) of the desired "superimposed reality" (*Pushtartha*, -8).

Serpents represent the Aries zodiac. Snakes live an individual, solitary life. They aggregate as a collective for hibernation and the mating event immediately after. Two males entwine two-thirds of their bodies and lift the residual third of their body upward-facing one another. The eyes of the male facing the "dominating collective" (*Taraka*, 36) produce the illusion of his individual dominance. The dominant male experiences an adrenaline rush of masculinity leading to an ascending blood pressure that activates the left masculine hemipenis and makes him insert and release the "energy in ascending motion" (*Hunduka*, 73) in the cloaca of a subordinate female. A subordinate female faces the dominant male from the side of the deciding collective that decides to send a third of its members to the right feminine side. She is not a part of the "deciding collective" (*Trivikrama*, 24).

After experiencing subordination, the subordinate male becomes a "leader" (*Neta*, 1) who sends half the members to the left masculine side for balancing the power imbalance. The half comprises the two-thirds from the deciding collective and a third from the "pre-dominating collective" (*Gabhira*, 12). The latter's dominance projects as an "illusion" (*Nitya ratri*, 1) within the dominant male's eyes. Therefore, the perception of dominance changes and transforms the dominant male into the subordinate male. The adrenaline rush of the first dominant male stops. The descending blood pressure activates the right feminine hemipenis and makes him insert and release the "energy in descending motion" (*Ahamkara*, -1) in the cloaca of a dominant female. A dominant female faces the subordinate male from the side of the predominating collective that sends half of its members to the left masculine side. She is a part of the "deciding collective" (*Trivikrama*, 24) that first sends a third of its members to the right, then sends half of its members to the left, and finally decides that all the members are a part of the collective. It lets each male impregnate a female twice, first from an inertial position of masculine dominance and then with a dynamic momentum of feminine determination.

A female leads a third of the members, all females, to the right. She is the maternal greeter. The maternal greeter protects the males from the in-fighting by tilting the balance of power in favor of the side led by her leadership initiative. A male follows half the members, all males, and remains still on the left. From the stillness, he experiences the diverse realities due to the "energy system" (*Antarmukha*, 10^{1024}) generated by the female movement. The other sixth of the members is the females in the center of the collective foe circle of the males, surrounded by the individual social circle of the females. They are the female satans whose "discordant energy" (*Asura shakti*, -1) causes two-thirds of the females to move out of the collective circle seeking individual attention. After two-thirds of the males finish mating monogamously with two-thirds of the females, they isolate a third of the males observing as the devil snakes. The "anger" (*Krodha*, 275) of the devil snakes in the center of the two sides of the masculine circle makes everyone run for their life individually except the females stuck in the center.

A devil snake impregnates the satan snake once at the peak of his anger with the "air in ascending motion" (*Mada*, -1) that generates a "discordant energy" (*Asura*, -1) with the female snakes. The snakes in the center on either side become angry because their development genes make them imagine the reality that is about to manifest. A third of males and a third of females develop the genes for imagining the reality after their parents are stuck in the middle of a third that runs away first. A third run away last, believing that a looming made the other two-thirds run away individually without even saying goodbye to their newfound friends and soulmates. The devil snake impregnates the targeted satan snake the second time at the ebb of his anger and the peak of the fear with the "air in descending motion" (*Hunduka*, 73) before dispersing.

As each female snake disperses in "urgency" (*Jaldi*, 120), she also develops an "organizational gene" (*Konastha*, 13) that makes "her" (*Konastha*, 13) conscious of the flow of "time in ascending motion" (*Mahant*, 18,000). Starting age three, she lays ten eggs and lets them hatch on their own after sixty days. Due to their size, the larger snakes experience intrinsically strong urgency and lay a hundred eggs starting at age two. Due to the "strong intrinsic psychic force" (*Kana*, 186), the size of their eggs is the largest, and they are the largest snakes. The smaller snakes experience intrinsically weak urgency and lay three eggs starting at age four. Due to the "weak intrinsic psychic force" (*Raga*, 250), the size of their eggs is the smallest, and they are the smallest snakes.

The females impregnated by the initially dominant males are the first cohort to leave because the latter's subsequent subordination brings energy inflow to a standstill. Those daughter snakes can leave quickly and feel little urgency. They experience extrinsically weak urgency and lay two eggs starting at age five. Due to the "weak extrinsic psychic force" (*Brihadbala*, 149), their eggs are small, and they are small snakes. The females impregnated by the initially subordinate male are the second cohort to leave because the energy inflow takes time to standstill after the latter's subsequent domination. Those mother snakes depart after the daughter cohort and feel hyper-urgency. They experience extrinsically strong urgency and lay eighty eggs starting at age two. Due to the "strong extrinsic psychic force" (*Amrit*, 1000), their eggs are large, and they are large snakes.

The grandmother snakes depart last. They experience a strong sense of urgency following the collective cultural consciousness. Due to the "strong psychic force" (*Prakatikarana*, 54,000), their eggs are larger, and the baby snakes develop into larger than normal size for their species. With sufficient "density" (*Sandra*, 60) of the collective, they become a new "species" (*Prajati*, 17). The maternal greeter remains in her place because she has nowhere to go once the snakes have gone in all directions. Therefore, she experiences a weak sense of urgency. Due to the "weak psychic force" (*Citraka*, 185), the size of her eggs is smaller, and the baby snakes develop into smaller than normal size for their species. Therefore, as a protection against the "personal reality" (*Kamartha*, 10^{17}-1) of their "endogenous" (*Antarjata*, 10^{16}) "combination within one" (*Kamartha*, 10^{17}-1) led by the maternal greeter's "entrepreneurship" (*Savana*, 10^{17}-1), they develop an "exogenous"

(*Bahirjata*, 14,000) "combination without one" (*Ashtottara-sata lingam*, 10^{16}) by trading her "living consciousness" (*Atma sharira*, 14,000). As an individual, they become a "child primordial greeter" (*Madhusudan*, 16) of a new "kind" (*Prakara*, 17) of species.

Each snake targets only one mate at a breeding event guided by the sequence each male and female snake arrives. The late-arriving unmatched snakes leave the event without mating because the snakes are the most peaceful and egalitarian animals in the universe. Due to the "para psychic force" (*Mahayana*, 179) of many individual snakes without mate, each snake holds its venom within instead of polluting the air until it feels real danger to the mate's life about whom it is very protective. Due to the "infinite para psychic force" (*Parjanya*, 177) of the collective of babies forming a "combination without two" (*Yathayogya*, 10^{100}) by fusing the maternal and the paternal "consciousness systems" (*Sara kalpa*, 10^{10}), snakes become the "progenitor" (*Agnishvatta*, 10^{100}) of the "infinite lifeforms" (*Asthiyantra*, 16,000).

2.12 Feeding the Value of Time with Negative One Finite Points

The "discordant energy" (*Asura shakti*, -1) intensifies the pressure for the goal of "sustainability" (*Samposhaniyata*, 9) within one's mind by instantly feeding the value of one's entity time to everyone. Sustainability is a sequence of six feminine, even zodiacs in descending order. It leads to spiritualism in the first house, realism in the third house, imagination in the fifth house, illusion in the seventh house, wholeness in the ninth house, and piecewise linear "opportunity" (*Avasar*, 136) in the eleventh house. Six masculine, odd zodiacs curve the piecewise linear opportunity into a foe circle for personal conscious wellbeing. Six feminine, even zodiacs close the curve into a social circle of the piecewise parabolic challenges for realizing the universal conscious wellbeing. Twelve zodiacs as a conscious system shape the "parabola" (*Anuvritta*, 16) that includes both the curving "questions" (*Prashana*, -7) and the linear "answers" (*Pratigad*, 39) within one's "curved consciousness" (*Bhagnatma*, 9). A clarified consciousness of how Mother Nature works and shapes our essential nature helps us enjoy absolute freedom from the naturally sustainable cycle of birth, growth, maturity, death, and hibernation and be a "sustainer" (*Poshaka*, 30) rather than "sustained" (*Poshita*, -1)

A clarified consciousness requires one to transcend beyond the self-imposed ethical and the socially-imposed moral boundaries. "Ethics" (*Aachar*, 10^{10}) is the "life lesson" (*Panchtantra*, 10^{10}) that promotes "maturity" (*Gandakanayaka*, 10,000) as a "planning point" (*Tilaka*, 10^{10}) attuned to our "extrinsic consciousness" (*Visata*, 10^{10}) free from the "path-dependency" (*Pratyalidha mandala*, 10^{10}) on "what exists" (*Amba*, 32). "Ethics" generates an illusion of the universal sentient wellness while furthering our sentient wellness as a leader. Ethics help us consciously organize our "finite form" (*Rochisha*, 13) as a "reproducer" (*Rochisha*, 13) of our ethics to be "all-pervading" (*Sarvatraga*, 16) in two ways. First, as the "permanent growth" (*Rasayana*, 13) within our consciousness. Second, as a "consciousness system" (*Sara kalpa*, 10^{10}) that norms everyone's "conscious reality" (*Virasat*, 10^{10}) as ethical because it is concordant with our normalizing "organizational development" (*Viparyaya*, 888) and gives us a freedom to "reform" (*Viparyaya*, 888). It makes us a conscious system manifesting "what does not exist" (*Prajaniyam*, 16) in infinite temporary forms. It generates a "structural entropy" (*Pruthaktva Tejas*, 81) of our finite permanent form, making us dependent as a leader on the followers' diverse finite forms.

Each finite form generates "temporary growth" (*Maha Kali*, 13), transcending the consciousness of our "organizational planning" (*Abadbha*, 10). Temporary growth shapes our "organizational programming" (*Dharana*, -8) to multiply the finite form into infinite forms. Each infinite form generates "permanent growth" (*Rasayana*, 13) through our "organizational performing" (*Rachana*, 286). Permanent growth multiplies the infinite form. It eventualizes a finite temporary form that is "indivisible" (*Avibhajya*, 13) but "investible" (*Viniyojya*, -15) as a "technological cost" (*Yojya*, 58). Consequently, we conceive a finite permanent form, that is "divisible" (*Vibhajya*, 81) into the infinite temporary forms. We generate profit, i.e., "social benefit" (*Labha*, 81), from trading a finite permanent form and servicing the infinite temporary forms.

When everyone exchanges their "finite permanent form" (*Sthayitva*, 81) with our "infinite temporary forms" (*Asthayitva*, 81), we enjoy "profits" (*Labha*, 81) with our "reformative" (*Sudharatmak*, 14) freedom while they incur "losses" (*Shreya*, 81). By exchanging their "finite temporary form" (*Rochisha*, 13) with our "infinite permanent form" (*Konastha*, 13), we transform their losses into "organizational profiting" (*Nanarupa*, 10,000). We normalize their "profiting" (*Mahasatta*, 378) as zero to form a consciousness of our "infinite profiting" (*Eroli*, 100,000).

"Morals" (*Naitik*, 13) generate an illusion of the personal sentient wellness as an "entrepreneur" (*Prabhu*, 1600) free from the "social forces" (*Kapinjala*, 20) within the confines of the "institutional morality" (*Kaal Sarp*, 369) shaped by the "religious values" (*Yukti*, 8). The institutional morality gifts "freedom" (*Svatantra*, 109) to the entrepreneur for manifesting "evolutionary morality" (*Ridhyati*, 157) by imagining "universal sentient wellness" (*Sadhaniya*, 157) through "donation" (*Chanda*, 90) of his personal sentient wellness. As everyone reproduces the entrepreneur's "guider morality" (*Martya*, 146), everyone becomes a "mortal entity" (*Martya*, 146) like the entrepreneur. Everyone's "impact" (*Siddhi*, 57) is zero once others reproduce their rock-solid "present morality" (*Shila*, 1790). Consequently, everyone's "morality" (*Naitikta*, 70) transforms into "zero morality" (*Dauhshilya*, 353), leading to "immorality" (*Dauhshilya*, 353). Without morality, everyone takes us as their "ideal" (*Adarsha*, 1). They follow a continuous "path of infinite development" (*Rajas marga*, 8,000) as a "link" (*Rajas*, 15), mediating our "leadership energy" (*Nayakatva shakti*, 90,000). "Infinite links" (*Setu*, 90) make us the "infinite" (*Tamas*, 185) and everyone a zero until anyone exchanges our conscious reality within the "interplay" (*Chakri*, 90) of "market forces" (*Abadha*, 10). The path of infinite development is the path to exchange our conscious reality. It is the "meeting point" (*Sakshatkara*, 8000) at which everyone's entrepreneurship and our leadership "intersects" (*Katana*, 8000) as "entrepreneurial leadership" (*Sakshatkara*, 8000).

There are many forms of entrepreneurship and leadership. Entrepreneurial leadership is a form that combines all these forms into one and offers powerful opportunity for wealth creation. "Entrepreneurship" (*Savana*, $10^{17}-1$) makes leadership "organic" (*Chini-Chini*, $10^{17}-1$). It "personalizes" (*Vyaktikrita*, 10^{16}) the reality of a person's "local locus standi" (*Antarjata*, 10^{17}-1). More proficiently a person is networked locally, the greater the person's "technological trading" (*Bhakti marga*, 1) from the "global linkages" (*Shodashottari dasha*, 40). Global networks lead the person into "technological servicing" (*Karma*, 10) for advancing the "global locus standi" (*Bahirjata*, 14,000). "Technological exchange" (*KLIM*, 24) limits the entrepreneur's leadership. It pushes the entrepreneur to become a follower of the localized trading opportunities to perpetuate its "locus standi" (*Nidhi*, 366). It makes one research what it was doing that went wrong, and develop what others are doing by descending priorities on what it was doing.

Leadership makes entrepreneurship "inorganic" (*Shankha*, 87). Leadership is the "visualization" (*Samadhi*, 10^{100}) of the "workforce" (*Janabala*, 10^{100}) globally converging with one's reality in a "winner-take-it-all ocean" (*Kacangala*, 10^{100}). More proficiently a person exchanges the workforce from the "local linkages" (*Maha dasha*, 0), the stronger the person's leadership service. After winning it all, a "leader" (*Neta*, 0) distributes it all and becomes the "Lord of the distributed universe" (*Indra*, 0). The leader pulls everyone by becoming their ideal. Everyone becomes a follower of the globalized trading opportunity to create their locus standi through research into what the leader was doing that went right and develop that by descending priorities on what they were doing.

Entrepreneurial leadership is the "incubation point" (*Sata*, 8000) of the "ecosystem value" (*Ekavali*, 8000). It is the "minimum entropy state" (*Ekavali*, 8000) that leads to "information saturation" (*Kapila Kumara*, 8000). Through leadership mediation, entrepreneurship mobilizes all there is to know from its personally designed ecosystem. Leadership incubates a "banking system" (*Ekavali*, 8000) focused on personal conscious wellness with entrepreneurship moderation. Entrepreneurship makes it moral for a leader not to distribute the value the followers create back to the followers. Leadership makes it ethical for the entrepreneur to trade value from the global linkages and to service its benefits to the "local universe" (*Gardhabi mukha*, 1000). Thus, entrepreneurial leadership generates disproportionate growth in the wealth in the universe local to the meeting point of the entrepreneurs and the leaders. It delegates "infinite producing" (*Kesarisutaya*, 24) to the agents internationally and elevates "infinite consuming" (*Shatakanta Mudapahartre*, 8000) for the principal nation.

Today, more so than ever, humanity stands committed to furthering the values of equity, ethics, and social justice in practice globally. All the societies expect members to document how we are innovating through a community of inquiry, making an impact through human-centered approaches, and engaging diverse stakeholders through robust partnerships. Therefore, there is an urgent need to assess how the cultural factor contributes to the unprecedented polarization of wealth through entrepreneurial leadership and the way forward for the democratization of the benefit potential of the polarized wealth to accelerate the mass practice of the de-polarized entrepreneurial leadership. The first step is to recognize the presence of two major cultural systems in the world:

- Eastern: Belief that God is immanent within each animate and inanimate entity. Everyone has the potential to be God. Those in power must make sacrifices to help the powerless enjoy par excellence. The knowledge must be shared socially. The wealth must be owned collectively. All we have is the gift of nature. We must live in harmony, managed by the leader, to avoid any uncertainty and individual trade-offs. Let's set the boundaries on our liberties to enjoy our harmony.

- Western: Belief that each animate and inanimate entity emanates from God. Everyone is a potential God, created in the image of God. Those in power must empower the powerless. The knowledge must be kept private. The wealth is owned individually. All we have is the gift of humanity. We must live as masters, managing our followers, to avoid uncertainty and collective trade-offs. Let's push the boundaries of our liberties to enjoy our mastery.

Globally, the bipolar cultural system forms two major workculture systems.

- Eastern. Centralize power in the leader to ensure organizational planning and let each follower follow the organizational programming for ascending organizational performing and descending organizational profiting. It ensures universal organizational development, characterized by network followership.

- Western. Mobilize a group of followers to paralyze the organizational planning. Let the leader lead the organizational programming for descending organizational performing and ascending organizational profiting. How? Give absolute freedom to each follower to take risks and find ways for sustaining organizational performing through a research into alternative practices. It ensures unique organizational development characterized by network leadership.

Locally, the bipolar workculture system forms two major geographical systems.

- Eastern. Over-exploitation of resource endowments due to the efficiency focused on manufacturing power. There are high social, human, ecological, economic, national, and psychological costs. There are high global, unique, inclusive, diversity, engagement, and responsibility benefits.
- Western. Over-exploration of resource endowments due to the cost-effectiveness of trading, leading to marketing power. There are high social, human, ecological, economic, national, and psychological benefits. There are high global, unique, inclusive, diversity, engagement, and responsibility costs.

Nationally, the bipolar geographical system forms two major group systems.

- Eastern. Limited resources for investing in technological growth. Excellence in technological trading. Convergent quality of technological servicing. Barriers to technological exchange.
- Western. Surplus resources for investing in technological growth. Barriers to technological trading. Divergent quality of technological servicing. Agility for technological exchange.

Organizationally, one may manage the growth of the geographical advantages through cross-border trading and marketing for upgrading

- Eastern economic infrastructure, so that one is not limited by the local resource endowments.
- Western cultural infrastructure, so that one is not limited by inherited cultural priorities.

With upgraded infrastructure, one may sustain the growth of the borderless dynamism as a corporate entity by managing the development of

- Eastern administrative structure, so that the outdated legacy methods do not impede the vision for the future.
- Western geographical structure, so that the vision for the future can be multiplied through a desirable mission to upgrade the sustainability of the geography by investing in additional advantages.

Structural development makes it possible to transform the space and become free from the constraints of the already formed time and its inherited cultural force.

Chapter 3

The Space Tensor :
Exchanging the Entity Within

God's Future. I lived my past as an "individual" (*Viyujya*, 53) who formed the space for the "population" (*Abadi*, 18) within my "heterogeneous genus" (*Prajati*, 17). I am living my present within a "collective" (*Visarjya*, 47 = 9 + 36) to channel "double population" (*Lahari*, 36), half myself over my past and half with my family in the present. I normed the collective with a ten-member "family" (*Parivar*, 10) comprising a "heterogeneous group" (*Nivasi*, 9) and the one who formed me as the fifth member and is himself the seventh member.

After my "entropy" (*Mahodbhava*, 5), my future is livable without the collective I formed when the "heterogeneous group" (*Nivasi*, 9) reincarnates as a "group of species" (*Gana*, 387), reproducing the first three members with my "entropy value" (*Sarvanasha*, 5) as an "octave" (*Sargam*, 60) and "moving"

(*Ganin*, 327) the eight members—excluding the fifth and the seventh—from the consequential "triple population" (*Chalan*, 327) to live within the "consciousness" (*Chaithanya*, 4) of the "seventh member" (*Upa Pitha*, 14).

The seventh member exchanges the collective's "shared consciousness" (*Trigunatma*, 14) of the "individual energy" (*Upa Pitha*, 14) within me as "God" (*Ishvara*, 5) derived from the "energy" (*Shakti*, 19 = 14 + 5) that incarnated me. The energy incarnated me as an individual who lived as the "param maternal" (*Saranyu*, 5) for transforming the space with my "breeding" (*Janana*, 15) of the seventh member and seven members within and without the seventh member.

How a Self-luminous Entity Orders a Genus of Entities for Transforming the Space That Conceived It

A self-luminous entity orders a genus of entities for transforming the space that each entity conceives for forming the self-luminous entity as a creature. There are three ways for a "creature" (*Purusha*, 12) as an "ideal self-luminous entity" (*Purusha*, 12) to order a "heterogeneous family" (*Parivar*, 10) of members into a "homogeneous genus" (*Prakara*, 17 = 12 + 5) of families through a "replication" (*Samalekha*, 90) of its "forward-order development" (*Samalekha*, 90) as a "cultural factor" (*Sugriva*, 5).

First, order the entire "family" (*Parivar*, 10) of members, including the nine within the "genus" (*Varga*, 9) and the genus as the "tenth member" (*Varga*, 9), into a "homogeneous genus" (*Prakara*, 17 = 9 + 9 -1) without the "eleventh member" (*Sadashya*, 1). The eleventh member is the "double copy" (*Chitra*, 1) of the twelfth "principal member" (*Sanlagnata*, 1). The double copy forms by "copying" (*Chitra*, 1) the subordinate "thirteenth member" (*Upasarjana*, 550) as the "copy" (*Nakala*, 0) of the principal member.

In reality, the "subordinate member" (*Upasarjana*, 550) is not a copy but a "twin" (*Hridayesha*, 121 = 1,2,1) of the principal member. The twin replicates the "twelfth member" (*Sanlagnata*, 1) as the "eleventh member" (*Sadasya*, 1) by behaving like the fourteenth, gender-differentiating "superordinate member" (*Tara*, 2). The superordinate member is the generation-differentiating "twin double copy" (*Brahman*, 2) of the fifteenth, gender and generation-differentiated "inordinate member" (*Prajati*, 17). It becomes a "gender-differentiating member" (*Tara*, 2) after differentiating the "gender" (*Linga*, 9) as the sixteenth, generation-differentiating "coordinate member" (*Linga*, 9).

The "generation-differentiating member" (*Linga*, 9) is a gender-differentiated "twin triple copy" (*Khara*, 6) of the seventeenth, generation-differentiated "ordinate member" (*Tripurantaka*, 10^9). It becomes a generation-differentiating member after differentiating the "generation" (*Amnaya*, 97) as the eighteenth, gender-differentiated and "generation-differentiable member" (*Amnaya*, 97).

The generation is the "preordinate member" (*Amnaya*, 97). It "preordinates" (*Putana*, 97) the "order" (*Suchan*, 2) of the "family" (*Parivar*, 10) by "ordinating" (*Samanvaya*, 80,000) the "disorder" (*Pradarshan*, -8) created by the nineteenth "group-differentiated member" (*Samanantra*, 80,000). The group-differentiating member is the "ordinating member" (*Samanantra*, 80,000). It forms as a "parallel" (*Samanantra*, 80,000) of the twentieth, "ordinated" (*Samanvayita*, 20,000) "group-differentiating member" (*Pitri Loka*, 20,000). The "form" (*Rupa*, 100,000) is the twenty-first "ordinable" (*Samanviya*, 100,000) "group-differentiable member" (*Rupa*, 100,000).

The ideal self-luminous entity "replicates" (*Pratikara*, 1) itself as the twentieth "ordinated member" (*Pitri Loka*, 20,000) for the "forward-order development" (*Samalekha*, 90) of the twenty-first "ordinable member" (*Rupa*, 100,000) as a "cultural factor" (*Sugriva*, 5 = 100,000/20000).

Second, order each "member" (*Sadasya*, 1), who "replicates" (*Pratikara*, 1) itself, first into a "heterogeneous genus" (*Prajati*, 17) that norms the "species" (*Prajati*, 17), and then into the "homogeneous genus" (*Prakara*, 17) that norms the "kind" (*Prakara*, 17), i.e., the "phylum" (*Prakara*, 17) of the species. The species is the "extrinsic form" (*Prajati*, 17) of the member. The phylum is the "transcendental form" (*Prakara*, 17) of the family that conceives the "division" (*Sudhanvan*, 999) as the "intrinsic form" (*Sudhanvan*, 999) of the entity.

The phylum is the twenty-second "ordained member" (*Prakara*, 17) that transcends the division for "ordaining" (*Aadesha*, 957) the "kingdom" (*Nayana*, 957) as the twenty-third "ordaining member" (*Nayana*, 957). The kingdom is the "immanent form" (*Nayana*, 957) of the time as the "twin entity" (*Ghatika*, 24). The division is the twenty-fourth "ordinal member" (*Sudhanvan*, 999). It is the "ordinal" (*Kramasuchaka*, 999) that "orders" (*Suchan*, 2) each "member" (*Sadasya*, 1) to be the "divisor" (*Shudra*, 1) of the "domain" (*Suchaka*, 9000) that norms the entity's "lived experience" (*Anubhav*, 9000).

The domain is the twenty-fifth "cardinal member" (*Suchaka*, 9000) behaving like a "cardinal" (*Buniyadi*, 9000). It is the "emanating form" (*Suchaka*, 9000) of the space as the "para entity" (*Ashtavakra*, 19). The domain "replicates" (*Pratikara*, 1) itself as the "fifth member" (*Danda*, 176) to "fathom" (*Danda*, 176) the entity's "lived experience" (*Anubhav*, 9000) as a "gravitational wave" (*Varshakritya Taranga*, $1,800 = 9000/5$) with the "cultural factor" (*Sugriva*, 5). The gravitational wave is the twenty-sixth "subliminal member" (*Varshakritya Taranga*, 1,800) that is "subliminal" (*Prabhavi*, 1800) in a self-luminous entity's "livable experience" (*Anubhuti*, 7000).

Third, order each "entity" (*Hasti*, 24), who becomes "self-luminous" (*Svarochisha*, 12) with time, first into a "homogeneous genus" (*Prakara*, 17) to norm the "transcendental form" (*Prakara*, 17) of the "family of species" (*Parivar*, 10) and then multiply them into many "heterogeneous genera" (*Prajati*, 17), manifesting the diverse "species" (*Prajati*, 17) with their diverse "extrinsic forms" (*Prajati*, 17) as a "sentient wave" (*Jyataranga*, 86). The sentient wave is the twenty-seventh "liminal member" (*Jyotaranga*, 86) that makes a "sentient entity's" (*Siddha*, 7) "living experience" (*Tajaurba*, 2000) "liminal" (*Avasimiya*, 86).

The liminal replicates itself as the "second member" (*Pola*,

16), norming the "primordial species" as the "pole" (*Pola*, 16) that "adapts" (*Anukulana*, 16) by becoming the "cultural factor" (*Sugriva*, 5). The cultural factor "polarizes" (*Nabhika*, 90) the "primeval species" (*Upa Pitha*, 14) as the "seventh member" (*Upa Pitha*, 14) to make the "replication" (*Samalekha*, 90) the twenty-eighth "infinitesimal member" (*Samalekha*, 90).

The "culture" (*Sadakhya*, 9) "replicating" (*Doharana*, 9) the "eighth member" (*Adi Pitha*, 10) as a "twin double copy" (*Brahman*, 2) of the "fourth member" (*Kumbhaka*, 10) is the twenty-ninth "centesimal member" (*Sadakhya*, 9. The eighth member is the "param species" (*Adi Pitha*, 10), norming the "upper octave of mass" (*Adi Pitha*, 10) of the replicated "primeval species" (*Upa Pitha*, 14). The fourth member is the "param kind" (*Kumbhaka*, 10) of species replicated as a "base" (*Kumbhaka*, 10) to form the primeval species.

The "essential nature" (*Svabhav*, 8) "replicable" (*Pratikriti*, 8) within the "ninth member" (*Shakti Pitha*, 4) formed by reproducing the "third member" (*Pitha*, 8) is the thirtieth, "vigesimal member" (*Svabhav*, 8). The third member is the "primordial kind" (*Pitha*, 8) of species, norming the "octave of energy" (*Pitha*, 8) energizing the ninth member to be the "primeval kind" (*Shakti Pitha*, 4) of the "sixth member" (*Shirati*, 100,000). The sixth member is the "Rod" (*Shirati*, 100,000), ordinable into the "first member" after adding the "twenty-first member" (*Rupa*, 100,000) that gives form to the first member and subtracting itself to norm the "first member's" (*Skambha*, 10^{10}) "body system" (*Hridya Pranali*, 100,000).

The first member is the "column" (*Skambha*, 10^{10}) formed by a "departed entity's" (*Divangat*, 1600) "consciousness system" (*Sara Kalpa*, 10^{10}) that "pokes" (*Kuradana*, 10^{10}) out a "child face" (*Dasharatha*, 10^{19}) within the seventeenth, generation-differentiated "ordinate member" (*Tripurantaka*, 10^9). The departed entity is the "present creature" (*Prabhu*, 1600) who becomes the "zeroth member" (*Vamsaa*, 2), norming the "pedigree" (*Vamsaa*, 2) of the "child" (*Arcisa*, 128) whose "face" (*Mukha*, 76) "animates" (*Trasa*, 76) within a "living entity"

The Space Tensor : Exchanging the Entity Within 123

(*Sushuptivat*, 96). The living entity is the "octave of self-luminous entities" (*Sushuptivat*, 96) formed with the "consciousness" (*Chaithanya*, 4) of the "incoming light" (*Ambara*, 180) of the "space" (*Vaas*, 18,000) that conceives the living entity. Each self-luminous entity transforms the space for living like the entity that forms the space for conceiving the living entity.

3.1 Forming Space for the Universe of Entities with Five Finite Points

The first finite point is a rotation of the four space units, reproducing another four space units by rotating the four-space units already produced and creating a hole.

The second finite point links the primordial space and the primeval space. Primordial space rotates infinitely around the rotatable present space and services infinite rotated space.

The third finite point is a reflection of the three units of time.

The fourth finite point is an index for replicating the four space-units rotation, one causation unit link, and three time-units reflection.

The fifth finite point is a loop trading the ten-unit base as a tangent and servicing the twenty-unit spirit of the tangent as a photon.

As a fifth finite point, a photon includes an index unit, a replication unit, a loop unit, a trading unit, and a servicing unit beside the ten-unit base and the four units of finite points.

Space forms as the sixth finite point, trading the twenty-unit photon and servicing the 18,000-unit double anti-neutrino with the 900-unit triangulation of the three units of time. It produces the double anti-neutrino as two units of space parallel to the two units of primordial space that reproduces the double-neutrino for conceiving the four-unit primeval space.

Photon = 20

Space = Double anti-neutrino = 18,000

Triangulation = 900

Time forms as the seventh finite point, trading the 18,000-unit double anti-neutrino and servicing the 180-unit light and the 100-unit primordial space that reproduces another 180-unit light and 100-unit primordial space. It reproduces the causation that leads to 285-unit reproduction of the 280-unit primordial atom and the five finite points, thus transforming space into quantum space. Quantum space is the index indexing the eight finite points of the squared space with the nine finite points of the triangulated time.

The causation forms as the eighth finite point. It produces the first, odd finite point and lets it reproduce the next, even finite point by endowing it with the causation for infinite producing. Infinite producing transforms the 100-unit primordial space into the electromagnetic mass and makes it the cotangent of the 285-unit reproduction. It transforms time into quantum time with the 900-unit triangulation of the three time-units by three units of reproduction and three units of breeding that breed reproduction. Quantum time is the loop looping the nine finite points of the triangulated time for entangling the entity as the ninth finite point.

The entity forms as the ninth finite point. It conceives the zeroth finite point and perceives the tenth finite point with a 1,0 sequence. It experiences the tenth finite point as the second prime finite point after servicing the photon as the fifth prime finite point. Thus, with a constant tenth finite point as the base, it rotates the primordial space infinitely for manifesting an infinity of rotated spaces with quantum causation. Quantum causation reflects the three prime finite points (*second, fifth, and third*, as a 253 sequence) as if they are the three units of time. The first two points link the ninth and the zeroth finite points. The next five form a loop that loops ninth, zeroth, and tenth finite points. The final three reflect the loop after the loop rotates over time and is indexed as an entity to create space for another loop indexing another entity.

The Universe of entities forms as the tenth finite point for enjoying the space with a hole indexed for infinite entities. Each entity fills the hole by transforming space into "five finite points" (*Prakashisu*, 10) by sequentially activating five ways of transformation with one to five finite points.

3.2 Transforming Space in Five Ways

3.2.1 Transforming Space with One Finite Point

It is possible to transform space into a circle with a finite point of time. A square has unit vertical and horizontal axes. The vertical y-axis reflects the constant past time. It is nonlinear. curving the present time. The horizontal x-axis reflects the curved and variable present time. The diagonal z-axis reflects a parabolic future time; it is the square root of two time-units, one constant and the other variable. It is uncurved but curvable. It is possible to manifest the future time as the present time and let it perpetuate as the past time.

A closed space with finite points is a "spherical space" where four space coordinates and one time coordinate have one unit distance among them. It is possible to conceive that the four dimensions of space are the four coordinates enclosed within a finite point of time. Further, one may conceive the four coordinates to be reproducing the unit energy of the finite point produced by the time's sphere of influence. Thus, a spherical space can be represented as 11,111.

Spherical space = Closed space with finite point = 1,111

It is possible to conceive infinite points without the closed space equivalent to the finite points within the closed space. A closed space with infinite points is an "elliptical space" where time reproduces the four-space coordinates within the open space beyond, and without, the closed space. Elliptical space includes the consciousness of the four space coordinates within the closed space that it is reproducing consciously within the open space. Therefore, it comprises four illusionary space coordinates within the consciousness of past time and four real space coordinates with the consciousness of past time reproduced in the present time. The eight dimensions of space are within another closed space without consciousness of the varying time. Elliptical space has eight units of space curved vertically within the perpetuating force of the three units of time, all zeroes. Thus, the elliptical space can be represented as 8,000.

> Elliptical space = Closed space with infinite point = 8,000

Each closed space is the perpetuating force of the three units of time reproducing the eight units of space. It has a normative value of twenty-four within a unit that perpetuates time as three zeroes. The unit that perpetuates time as three zeroes is an ellipsis. It can be represented as 1,000. Thus, the normative value representing a closed space is 24 within 1,000, i.e., 1,024. It is without limits and bonds. The unit may perpetuate the closed space without bond to the existing closed space and do so without limit into infinity.

> Ellipsis = Unit perpetuating time as three zeroes = 1,000
>
> Closed space = Limitless = Bondless = Normative value = 1,024

Open space includes the elliptical space, and the ellipsis perpetuated openly over time without the already produced elliptical space. It reproduces the three units of time and perpetuates the three reproduced units as all zeroes within the already manifested three units of time. Thus, the open space can be represented as 9,000. Four space coordinates within the closed space reflect the past time. Four space coordinates without the closed open but within the open space reflect the present time. Three units of time without the open space reflect the future time.

Open space is the Y-axis, perpetuated horizontally over time. It is the God circle because anything that unfolds over time is folded within the open space. Everyone just circulates the unit energy of the closed space within the infinite reproductions of the open space through an infinite consuming of the elliptical space to enjoy two pies of actions: one within the circle enclosed by space and another within the circle enclosed by time. However, open space is also a partial phi bond. One does not need to circulate two pies of action over the three units of time as if they are God-gifted ten pies of inaction, reproducing eight spatial pies with the investment of one time pie by God and empowering one to trade another time pie at present for shaping everyone's future. One also does not need to circulate twelve pies of reaction after having circulated two pies.

The Space Tensor : Exchanging the Entity Within 127

> Open space = Elliptical space + Ellipsis = Y-Axis = God circle = Partial Phi Bond = 9,000

3.2.2 Transforming Space with Two Finite Points

It is possible to transform space into a triangle with "two finite points" (*Ghatikamandala*, 53) of the open activated space. One may circulate two finite points as two pies reproduced as 22 without consciousness that the triangulating one is just reproducing two points. One point is within the circle of space and another within the circle of time without unfolding the seven-unit reality of four units of space and three units of time. Thus, circulate is 22 that divides the seven-unit reality until that reality becomes infinitesimal at infinity of divisions. In other words, circulate = $22/7$ = two pies that form a circulating pie within the spherical space and transform into ten pies without the spherical space for norming twelve pies with the spherical space = $\pi = 5\pi = 6\pi$.

Six π is the stress-energy-matter tensor, producing electromagnetic stress within the two pies to consume pi and produce energy for transforming the matter into a replication of π. It is also the Newtonian Universal gravitational constant [G] since it is reproduced universally for gravitating the desired replication of mass with one's energy. Two pies form the two illuminated quarters of space. Since pi comprises two pies, it is a square π. Six π is the triangle π that triangulates the square pi. Five π is the circle π that circles the π into a closed space after triangulating the open space with six pies, thus producing another circle π. Each pie is a cot, cotangent with the base circulating the circulate.

> Circulate = Two pies = Two cots = Two quarters = π (square π, within spherical π) = 5π (circle π, without spherical π) = 6π (triangle π, with spherical π) = Newtonian Universal Gravitational Constant = $22/7$

The infinity divides the open space into four real units of space at present and four imaginary units of space in the future after first dividing the time into the present on the Y-axis and the future on the X-axis. Thus, the infinity and the infinitesimal can be represented as 9,000 with ten divisions, i.e., as 90,000. Infinity is the quantum ellipse; it reproduces space as four zeros within the reproduced three units of time.

> Infinity = Infinitesimal = Quantum Ellipse = 90,000

A square π is a reproduction of the ellipsis's revolving value that forms the spherical pi. Revolving value divides the ellipsis into the seven-unit reality and, conditioned by those seven units, produces an additional two units: one forms the vertical axis reproducing time and the other transforms into the horizontal axis reproducing space. Two units form the whole pi, comprising a pi from temporal action and pi from spatial inaction. Its four quarters add to one. All four quarters of space are immanent within the whole π, making whole π a unit of measure and matter-energy tensor. Thus, revolving value = 1,000/72. It is a unit gravitational constant that gravitates a sequence of revolving values that measure its revolving value.

> Revolving value = Four pies = Four cots = 2π = Whole π = Four Quarters = Matter-energy tensor = Unit Measure = Unit gravitational constant = 1,000/72

3.2.3 Transforming Space with Three Finite Points

It is possible to transform space into a sphere with three finite points of the closed inaction space, comprising the first finite point as the activator, second as the activating sequence, and third as the activated open space reflecting the activator's sphere of energy without any conscious action befitting the open space. A spherical pi's revolving value is the sphere's evolving value that forms the elliptical pi.

Evolving value revolves around the ellipsis to transform the three zeroes of time into the three zeroes of space within the perpetuating value of the fourth illuminated Eastern dimension of space. Perpetuating value comprises four units of space within the circle of space and one unit of time perpetuating the circle of space by reproducing it as the circle of time. Thus, perpetuating value = 5. Since one who perpetuates value is God, God = 5.

The Space Tensor : Exchanging the Entity Within

A revolve perpetuates the value as the one unit of God, perpetuating the circle of time as the circle of causation. Thus, revolve = 55. A revolve transforms the ellipsis into a polariton, i.e., an ellipsis polarized by the revolve into the elliptical pi's evolving value. Thus, polariton = 55,000.

> Perpetuating value = God = 5
>
> Revolve = 55
>
> Polariton = 55,000

An elliptical pi transforms two pi of the spherical pi into seven pi by adding five pi as the third finite point without the spherical space. Seven pi is a wholesome pi, whose fourteen quarters also add to one. It comprises four space pi and three-time pi as a unit of conscious causation, i.e., curved causation. It is the consequential causation state. It is the mass-stress-energy-momentum density tensor. It transforms the mass of consciousness by dividing it into a circle of space that repels energy and a circle of time that attracts energy for generating momentum density. It produces the inorganic replication of the mass without consciousness and the organic replication of the consciousness without mass. Thus, evolving value = 55,000/21

> Evolving value = Elliptical pi = Seven pi = Conscious causation = Curved causation = consequential causation state = mass-stress-energy-momentum density tensor = 55,000/21

"Polariton's" (*Visaranem*, 55,000) "evolving value" (*Sasana*, 55,000/21) "exists" (*Jivita*, 147 = 21 * 7) as an "exciton" (*Jivita*, 147). The exciton transforms the evolving value into the "reality" (*Vastavikta*, 7) of the polariton.

3.2.4 Transforming Space with Four Finite Points

It is possible to transform space into an ellipsis with four finite points of the half-closed activating space, activating the open space with the inactive but activable closed space. Half-closed space is the base that becomes the tangent illuminating the closed space as a circle without the two finite points of the open activated space.

> Half-closed space = Base = Tangent = 10

An elliptical pi's evolving value produces a continuity of a pi each from the closed space and the open space. It forms the third finite point as the reality, transcending the fourth finite point that norms the illusion. Thus, continuity = 22 + 22 + 7 − 1 = 50. Continuity is the elliptical chord that connects successive space transformations organized as the quantum space. It is the half-open activator space that activates the quantum space.

> Continuity = Elliptical chord = Half-open space = Activator space = 50

The quantum space exponentiates the half-closed space with another half-closed space to illuminate the reality perpetuating over the three-time units. The seven-unit reality and the three time-units index a third ten-unit half-closed space. The exponent reflects the fourth ten-unit half-closed space that reproduces the base without the second half-closed space to perpetuate the illusion over the nine reproduced time-units. Thus, quantum space = $3 \times 10^{10+7} = 3 \times 10^{17}$. Since the exponent adds a unit illusion of the evolving value back to the seven-unit reality, exponent = 78. The exponent reproduces the primeval space that reflects the primordial space and breeds the present space in infinite forms.

> Quantum space = Index = Space transformation = 3×10^{17}
>
> Exponent = Reproduce = Reflect = Breed = 78

3.2.5 Transforming Space with Five Finite Points

It is possible to transform space into a cone with five finite points of the half-open activator space. The cone forms the present space by exponentiating a unit of space into two units: one closed space and the other open space. The cone transforms the present space by perceiving the open space as three units: first, present space; second, primordial space that reflects the present space; and third, primeval space that reproduces the primordial space and breeds the present space. Space comprises two units conceiving the integrated closed space and three units experiencing the differentiated open space. Thus, cone = 123.

The Space Tensor: Exchanging the Entity Within 131

Cone = Exponentiating = Perceiving = Conceiving = Experiencing = 123

The cone's co-evolving value sequences the open space and its three forms as four consecutive cones, successively adding two pie units that form the π enclosed in the closed space, thus forming a 500-unit zeroth metric tensor. Zeroth metric tensor co-evolves and transforms into an ellipsis by reproducing itself. Co-evolving value forms a conical π by dividing the zeroth metric tensor with

(a) the four units that form the four-dimensional open space,

(b) the four units that transform the four-dimensional closed space into four sequential π's, and

(c) the one unit that norms the diverse forms of the open space, including the closed space, by exchanging four π's as the whole pi that makes a unit whole.

Conical π = Eight π, comprising four individual π's that form the open space and its three forms and four collective π's that transform the closed space into a cotangent with eight units and reorganize the nine units, including the cotangent, within the eight π's.

Zeroth metric tensor = 500

Co-evolving value = Conical π = 8π = 500/9

An eight-π conical π transforms into a two-π spherical π when one divides the co-evolving value by eight after multiplying it by two. Thus, it is the zeroth gravitational constant. It gravitates the spherical π with the four-fold growth of the zero-unit cotangent and reproduces the spherical π as the absolute π.

Transforming Conical π into Spherical π

$8\pi * 2/8 = 2\pi$ = 500/9 * 2/8 = 1000/72

3.3 Norming Space with Five Forming and Five Transforming Finite Points

It is possible to norm and organize space into a parabola with ten finite points of the space realm. The space realm is the forward force of the first finite point that forms the present space. Ten finite points exponentiate the first finite point by (1) making it a cotangent with eight units and (2) sequencing (a) the cotangent as one reproducing the base from the adjacent closed space in the coadjacent open space and (b) eight units as eight reflecting the one in semi-conjugacy beyond the open space. Thus, space realm = 10^{18}.

Space realm = Forward force = Semi-conjugacy = 10^{18}

A parabola comprises two finite points and four infinite points, indexing four finite points on the parabola. Two finite points norm the vertex. Four finite points norm the two ends of the latus rectum that intersect the two curves intersecting at the vertex. They are linked with two additional finite points that form the directrix without the parabola. The second pair of finite points reflect the third pair that forms the focus within the parabola. Vertex reproduces the locus of points that forms the Y-axis of symmetry. Parabola rotates the locus of points into the X-axis of asymmetry that moves forward into infinity due to the forward force of the reproducing vertex. As a finite point within the cone, parabola loops a sequence of six units, each revolving within the finite point, evolving with the finite point and co-evolving without the finite point's limits. Thus, parabola = 16. Each parabola is a reformation of another parabola.

Parabola = Reformation = 16

The parabola's involving value involves light within the parabola and represents parabolic π. It comprises four-unit primordial space evolving along the left of the Y-axis, four-unit primeval space co-evolving along the right of the Y-axis, four-unit param space revolving along the Y-axis, and a unit of time involving the three-unit absolute gravitational constant. The three-unit absolute gravitational constant produces the three-unit Einstein gravitational constant that divides unit time into two additional units. The past time unit left of the parabola feeds its light to the future time unit right of the parabola. Thus, involving value = 13/2.

Involving value = Parabolic π = 13/2

The already present time unit below the parabola breeds the consciousness of the transcendental light within the parabola by making the three-unit Einstein gravitational constant the "focus" of inflowing and outflowing light. Transcendental light is the gravitational light formed by exponentiating the base unit of time by two additional time units. It is the gravitational element. It trades light from the parabola and services the illusion of light as its servicing value that forms the primordial space. Thus, gravitational light = 100.

> Absolute gravitational constant = Einstein gravitational constant = Focus = 3
>
> Transcendental light = Gravitational light = Servicing value = Gravitational element = Primordial space = 100

The light within the parabola includes the gravitational light a parabola countertrades from its left by perceiving the past time without its consciousness. It also includes the four photons the parabola services to its right by conceiving the future time within its consciousness. Four photons form the four poles of the plane's sequential development as an elliptic. The first pole indexes the geography's present value into the past time. The second pole reflects the past time as the time consciousness. The third pole reproduces the time consciousness as the present time. The fourth pole loops the future time by producing a sequence of present times with the gravitational element.

> Four poles = Quadrupole = Four photons = 80
>
> Light = Value = 180

The parabola freerides on the consciousness of the transcendental light without it by shifting its spatial focus to "directrix," pounding the 180-unit light away from the 1000-unit ellipsis. Thus, directrix = 180,000.

> Spatial focus = Directrix = Pounding = 180,000

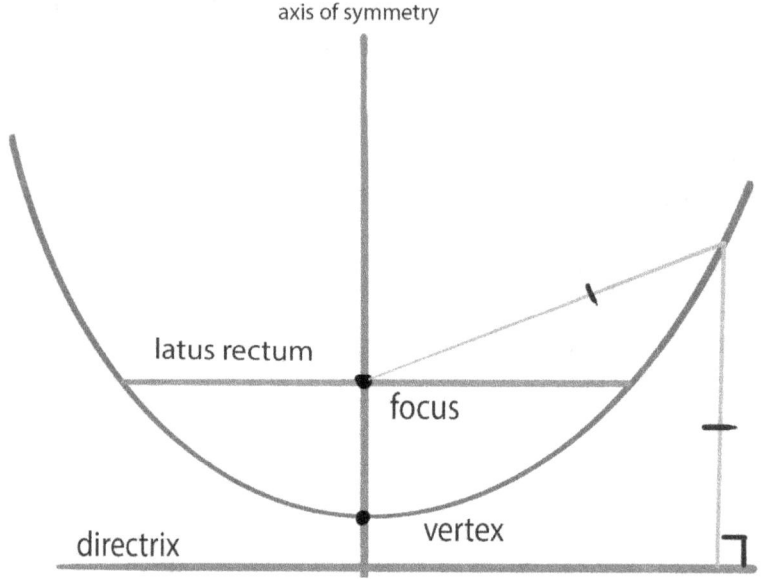

3.4 Transcending Space with Twenty Finite Points

It is possible to exchange and reform space into a hyperbola with twenty finite points of the time realm: five forming, five transforming, and ten norming the space. The time realm is the backward force of the ten finite points that norm the space with the ten-unit base. It divides the ten finite points into five that form the primeval space and five that transform the present space into the primordial space. It lets the latter five perpetuate the primordial space but trades their consciousness to replicate five additional finite points for conceiving the present space. Thus, it takes fifteen finite points to breed the present space into the primeval space. These fifteen finite points are within the open space.

Space comprises five additional points, distributed as two within the closed space and three without. Therefore, one needs twenty finite points to transcend the space that forms a parabola with its parabolic π. Space beyond parabola is a hyperbola. The time realm includes fifteen finite points that breed the present space and the oneness of the space with the present space. Oneness is the sixteenth finite point. Thus, time realm = 16.

The Space Tensor : Exchanging the Entity Within

> Time realm = Neutrino = Eye = Gravitational collapse = 16

Each transformation of the present space repels the 16th finite point in the form of a neutrino. Neutrino is the eye of the present space. By exchanging the eye, the present space leads to a gravitational collapse. Each formation of the present space attracts the 20th finite point in the form of a photon. Photon is the spirit of the space that perceives and trades the time consciousness, and the six finite points of the reproduced open space. Time consciousness is proportionate to the trading space. Therefore, it ingresses time within space's consciousness as a trader of the spacetime as an imaginary element that egresses as the photon.

> Photon = Imaginary = Spacetime = Egress = Spirit = Trader = Trade = Perceive = 20
>
> Time consciousness = 20 − 6 = 14

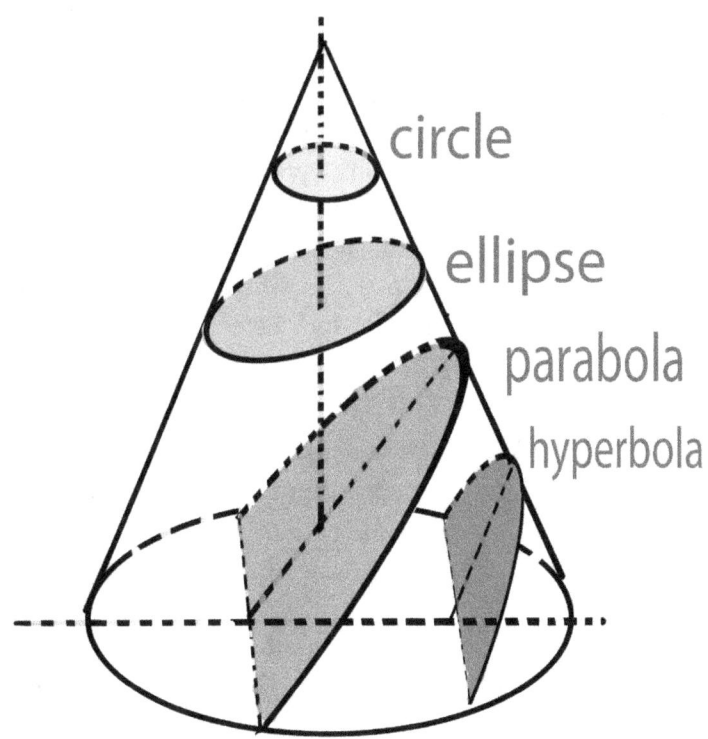

Present space is the reproduced nine units of time. Primeval space is the space's spirit without the 20th finite point that trades the spirit after transforming the present space. Thus, present space = 9. Primeval space = 19; it comprises the present space and the base for forming the primeval space by transforming the present space into the primordial space. It divides the primordial space into five finite points, each is a bundle of eighteen finite points, taking base as the sixth finite point of the reproduced open space. The bundle perpetuates the potential of the space by reproducing the six finite points as the hyperbola with twelve finite points. Thus, hyperbola = 12.

> Present space = 9
>
> Bundle = Potential = 18
>
> Hyperbola = Quantum photon = 12

The hyperbolic π is zero. Hyperbola trades the twelve finite points as a reproduction of the six that form the parabola. In reality, a hyperbola is the Hyperbolic chord of the parabola, illuminating the reality of the parabola within it. It illuminates the seven-unit reality with the five finite points that form the space where it resides as a quantum photon. Hyperbolic π is a cot, carrying the zero-unit electromagnetic mass of the tangent. It services whatever it trades as a photon in the form of a neutrino and the consciousness of the photon within the neutrino. It transforms the four-unit consciousness into four layers of light, norming the thirteen-unit quantum light within the nine-unit present space. By trading hyperbolic π as a pie of the cone, it transforms the cone into a layer of light of the present space.

The light layer divides the primordial and the primeval space and multiplies the primeval space like an atom. The multiplied atom differentiates into an atom, a neutron, and a particle. The particle transforms the neutron into the cell and the atom into the energy element, energizing the cell to conceive the ellipse as its nucleus. Finally, the particle circles the nucleus and transforms into an electron.

> Hyperbolic π = Cot = Pie = 0
>
> Quantum light = 13
>
> Primeval space = Atom = Neutron = Particle = Cell = Energy = 19
>
> Cone = Layer = 123
>
> Ellipse = Nucleus = 1,000

An electron is a triangle with three finite points. It produces the parabola with six finite points and consumes the space with five finite points for servicing its elementary charge to a quark. Thus, electron = 365. It makes space for one finite point to be triangulated into another electron. An electron triangulates the energy of the three time units within its 14-unit time consciousness. Time consciousness lives forever as the space's eventual temporal state and the time's thermodynamic limit within the causation's absolute development.

> Electron = 365
>
> Time consciousness = Live = Eventual temporal state = Thermodynamic limit = Absolute development = Infinite quark = 14

A quark is a square with four finite points. It produces the space as the fifth finite point and consumes a full circle with two finite points for trading itself as a twin. A twin is a unit of quark that forms two quarks and transforms into one quark by servicing itself as two quarks without the two finite points that form the full circle. The full circle lets the quark triangulate two quarks into six quarks without the two finite points that form another full circle. Thus, quark = 476. Twin = 121. Two quarks = Twin quark = 476 * 2 – 2 = 950. Six quarks = Triple Twin quark = 950 x 3 – 2 = 2,948. Thus, 7-unit reality forms an absolute development of fourteen finite points, comprising an infinite quark. A full circle trades the five finite points and services the three finite points. Thus, full circle = 53.

Quark = 476

Twin = Itself = 121

Twin quark = 950

Triple twin quark = 2,948

Full circle = 53

3.5 Enjoying Space with Thirty-Three Finite Points

An enjoyer enjoys the space with twenty finite points that form the photon and thirteen finite points that transform the photon into the triple twin quark and norms the formative and the transformative finite points as the full circle. The full circle is the normative finite point because it is a twin finite point. With eighteen-unit potential, it forms a thirty-six finite point channel for channeling the triple twin quark as a sixty-four unit fragment of energy, thus producing the 100-unit gravitational element. The fragment is a quadric, comprising the 36-unit channel as the thirty-six elliptical chords. Quadric is a fragment of the eighty-unit quadrupole without the sixteen-unit fermion whose four bodies form the quadric. Fermion is a neutrino with a consciousness of the four finite points that form the variable horizontal X-axis. Four finite points are a pair of twin finite points. A finite point on the left arm of the parabola secretly attracts the light from its left. A quantum finite point on the top of it consciously repels the light towards its right. The one attracting by charming the light to curve linearly down forms the charm quark. The one repelling by strangely curving the light up to form the Y-axis forms the strange quark. Quantum finite point is the zero-unit hyperbolic π. One attracting and zero repelling form the ten-unit base.

Channel = 36

Fragment = Energy fragment = Quadric = Thirty-six Elliptical chords = 64

Fermion = 16

Transcendental value = Growth = 6

Charm quark = Strange quark = 1

The Space Tensor : Exchanging the Entity Within

The Y-axis reflects the light like a mirror to a twin finite point on the right arm, gifting it a thermodynamic consciousness of togetherness with the six-unit transcendental value of the triple twin quark. The twin finite point is the negative five-unit full circle π, seeking oneness with the six-unit transcendental value to be one. The one awakened with the thermodynamic consciousness of togetherness is the up quark. Full circle π is the sphere's devolving value that forms everything with a triple twin quark. It divides the oneness of the Y-axis into a double twin quark. The oneness of the Y-axis is a quarter of the four axes of a full circle. Y-axis is one of the four axes. Thus, Full circle $\pi = \frac{1}{4}$ quarter $= -5$.

Double twin quark is doubling the 950-unit twin quark by activating the consciousness that twin quark is a pair of two quarks, of which one is itself. Thermodynamic consciousness is the oneness with the oneness of the Y-axis. Thus, thermodynamic consciousness $= 11$. It is the foundation for the growth of the transcendental value and the present growth of everything that leads to growth.

> Full circle π = Everything = Devolving value = $\frac{1}{4}$ quarter = $\frac{1}{4}$ Cot = $\frac{1}{4}$ pie = Twin finite point = -5
>
> Up quark = Awakened One = 1
>
> Thermodynamic consciousness = Foundation = Present growth = 11
>
> Double twin quark = Doubling = 1,900

Double twin quark triangulates the Up Quark into three additional quarks. The awakened Up Quark reproduces the thermodynamic consciousness and indexes the twenty-unit photon without the two finite points that form a twin finite point. The photon replicates the oneness without the Y-axis by gravitating a quantum photon towards its right without the six finite points that form another triple twin quark and the twin finite points that form another full circle. The full circle transforms the triple twin quark into six photons, the triple oneness between twins into three additional photons, and the paired twinning into nine additional photons. The full circle itself forms the tenth photon as the twenty-unit photon. Its oneness with the universe of full circles norms the tenth twin photon. Its electromagnetic consciousness of otherness from the tenth twin photon transforms the tenth twin photon into the Down quark.

Electromagnetic consciousness comprises the nine units of oneness among the nine pairs of photons. Nine units of oneness form the potential up quark. It is the universe of full circles, enjoying oneness with the photon. Thus, electromagnetic consciousness = 9. The tenth twin photon is the down anti-quark; it is the potential down quark.

> Electromagnetic consciousness = Universe of full circles = Enjoyer = Potential up quark = Up anti-quark = 9
>
> Down anti-quark = Potential down quark = Down quark = 1

The sleeping Down Quark links its 19-unit energy traded from the tenth photon to its oneness with the full circle that makes it a 30-unit causal body for channeling the six-unit growth into five finite points within itself. The five finite points form an eleven-unit pentaquark that produces the six-unit growth by transforming into a fourteen-unit hexaquark. The oneness with the full circle produces the gravitational consciousness of growth. It reproduces the 29-unit mass-effect of light and forms the top quark. The top quark is the potential material that materializes the light's mass-effect, comprising an organization of the pair of nine oneness correlations. Gravitational consciousness is the potential top quark. It multiplies reality with the ten units of oneness and exponentiates the ten units with the six-unit growth of the thirty-unit causal body. Thus, gravitational consciousness = 7×10^{180}.

> Top quark = Oneness with full circle = Potential material = 1
>
> Mass-effect = Organization = 29
>
> Causal body = 30
>
> Gravitational consciousness = Potential top quark = Top anti-quark = 7×10^{180}

Ten units of oneness produce the mass consciousness of the nine differentiated twin photons with the nine undifferentiated photons without oneness with the causation. Ten units of oneness form the potential strange quark that repels oneness with the causation as the eleventh unit of oneness to form the pentaquark. They transform the causation into the fourteen-unit hexaquark by organizing the nine units of oneness into a three-unit nonaquark. Three-unit nonaquark forms a triple-copy and transforms into the nine-unit potential up quark. Eleven-unit pentaquark comprises the three-unit nonaquark, three-unit triple copy that forms the nine-unit potential up quark, and five quarks other than up quarks that form five potential quarks other than a potential up quark. Fourteen-unit hexaquark comprises the eleven-unit pentaquark and the three-unit nonaquark. Nonaquark is a triple copy of Up Quark and forms with the reproduction of the triple copy culture by each of the triple copies that form the nine-unit potential up quark.

Potential strange quark = Strange anti-quark = Mass consciousness = 10

Pentaquark = Oneness with causation = 11

Hexaquark = 14

The dreaming Top Quark loops its 180-unit exponentiating value for norming oneness with the time and reproduces time as the 360-unit divine element. The oneness with time produces the sentient consciousness of the entropy and forms the bottom quark. The bottom quark is the electromagnetic collapse of oneness with the space that forms the potential bottom quark. Sentient consciousness is the potential bottom quark. Therefore, it is also space consciousness. It is the oneness with the six-unit growth that forms the 18,000-unit space by reproducing the 30-unit causal body with the 100-unit gravitational element.

Bottom quark = Electromagnetic collapse = Oneness with time = 1

Sentient consciousness = Space consciousness = Potential bottom quark = Bottom anti-quark = 16

The polluting Bottom Quark replicates the 100-unit gravitational element within each of the triple copies to involute the space with the quantum consciousness of $100^3 = 1,000,000$ units. Thus, Quantum consciousness is the potential nonaquark. Involuted space becomes the evolvent of time by conceiving itself as the twin ellipse that reproduces an additional ellipse.

> Quantum consciousness = Involute = Evolvent = Twin Ellipse = Potential nonaquark = Anti nonaquark = 1,000,000

Triple ellipse is the potential charm quark. Thus, potential charm quark = 1000^3 = 1,000,000,000. It has the potential to attract the sentient light of the entity time, in the form of the IBGYOR Light, by conceiving each of the six quarks and their three copies as the sentient light of the copying one. Sentient light comprises nine units of the sentient light and their oneness with the Copying One. Thus, sentient light = 10. The Copying One copies the Up Quark because it is zero without oneness with the entity evolving over time into diverse forms. It becomes an Up Quark due to oneness with the entity.

> Triple ellipse = Potential charm quark = Charm anti-quark = 1,000,000,000
>
> Sentient light = IBGYOR light = 10
>
> Up Quark = Oneness with entity = 1

The Copying One is a Wisher cycling its concordant energy with the photonic, indigo light that is exponentiating its value and becoming a concordant factor to cycle back to the initial spatial state. The concordant factor is the eventual spatial state. It is the Higgs field conscious of its thermodynamic potential extrinsic to everyone exponentiated by it with its electromagnetic mass and nine units of oneness with the oneness. The Copying One is the potential hexaquark with a potential for oneness with the three units of time that led to its present becoming and the three units of time that follow the cycle of present becoming to become the culturally-mediated potential up quark.

> Copying One = Wisher = Cycling = Concordant energy = Concordant factor = Eventual spatial state = Higgs field = Extrinsic field = Thermodynamic potential = Electromagnetic mass = Present becoming = Potential Hexaquark = Anti hexaquark = 0
>
> Culture = Cycle of present becoming = 9

The Space Tensor : Exchanging the Entity Within 143

The initial spatial state is the potential pentaquark. It is transformable and, therefore, a tensor. What is transformable is the matter within the initial spatial state. The matter is transformable due to the stress of exponentiating and repulsing one's photonic light and logarithmically transducing and attracting the reproduced blue light. Therefore, every tensor is essentially a matter-stress tensor that transforms the initial spatial state by sequentially shifting the spatial state to fulfill the wish to experience the diverse forms of existence, inanimate and animate. A tensor is a double bond. It comprises four electrons: one pair in a sigma bond and the other pair in a pi bond. Four electrons form a triple copy over the three-time units to realize their eventual spatial state. Therefore, at present, the initial spatial state = -12, without the triple copying of the four electrons.

> Initial spatial state = Potential pentaquark = Anti pentaquark = Transformable = Tensor = Matter-Stress Tensor = Double Bond = -12

Blue light is the light reproduced by the three copies with twelve electrons that form six quarks and six potential quarks. Since each copy is a devil reproducing the initial spatial state, blue light can be identified as the devil light. Three copies form the pentaquark, hexaquark, and nonaquark. Thus, the cycle cycles the three copies and is the twentieth point, formed with a 5,6,9 sequence. It divides oneness into sixty by making three copies of itself. The cycle is a self-producing closed system. It cycles its copy into infinity, thus making each of its twenty points finite. Then, it realizes a need to transcend the finiteness for enjoying the infiniteness. Thus, cycle = 1/60. Blue light = 12 * 4 = 48.

> Blue light = Reproduced light = Devil light = 48
>
> Cycle = Twentieth Point = Self-producing = Closed system = 1/60

3.6 Exchanging Space with Sixty Finite Points

There is a fourteen-unit thermodynamic limit to one's capability for enjoying the infiniteness by copying one's finiteness. It lets the copy manifest three copies of itself and program the three copies to profit normatively by developing three more copies. It develops a base of ten finite points and consciousness of four groups forming a double bond. The base may also replicate three copies without itself and reproduce two copies within for symmetry with those three copies. The symmetry is the sentiment of attachment, guided by worry that the copy is gaining disproportionate value. It leads one to exchange the copy's color as a boson. Boson is a sequence that brings symmetry of reality within and without oneself by aligning five copies [three extrinsic and two intrinsic] with one's zero-unit concordant energy. Thus, boson = 250.

> Symmetry = Sentiment = Attachment = Worry = Color = Boson = 250

The base with five copies constitutes sixty finite points that norm the density of the zero-energy electromagnetic mass. Density is the growth illuminated due to the circularity of one's sentient programming. Density is a single bond, comprising two electrons in a sigma bond; one electron is homogeneous, and the other is heterogeneous. Its asymmetry with the electromagnetic mass produces an endogenous electron. Its symmetry with the octave of copies. Each copy is the electromagnetic mass trading its density, produces an exogenous electron. The bonding of the endogenous and the exogenous electron into a square pi norms a pi bond within the trading force of the heterogeneous electron and the reality illuminating value of the homogeneous electron.

Pi Bond generates thermodynamic heat and leads to the density's entropy. Entropy retains five finite points of the formative space, letting it exchange them with the five finite points of the transformative space to discontinue the descending-order sequence. An endogenous electron is a ten-unit "kaon negative" that forms a z-axis tangent to the electromagnetic mass within the base. Exogenous electron is a five-unit "kaon short" that perpetuates the entropy value of the electromagnetic mass as a "cosmological unit" without the base. A heterogeneous electron is a nine-unit "kaon long" with the base that includes a one-unit "homogeneous electron" as the "positive quark." The positive quark is a lepton.

Density is immanent within an octave of copies because the base comprises three copy sequences: a copy, three copies of the copy, and three copies of the three copies. Three copy sequence is an effect of three copies. With three units of time, ten different combinations are feasible: six heterogeneous (three pure and three mixed), three homogeneous (all pure), and one endogenous (mixed homogeneous). The endogenous is the mixed homogeneous because its homogeneity as three zeroes is mixed with the lack of oneness with time. The mixed heterogeneous is exogenous.

> **Ten combinations feasible with the Three Units of Time**
>
> Three Pure Heterogeneous: 123, 231, 312, in ascending order within the copy sequence index and with the sidereal link
>
> Three Exogenous (Mixed Heterogeneous): 132, 321, 213, in forward order within the copy sequence index and with the tropical link
>
> Three Pure Homogeneous: 111, 222, 333, in horizontal order without the copy sequence index and with the zodiac link, conscious of the oneness of each copy within the indexed copy sequence
>
> One Endogenous (Mixed Homogeneous): 000, in descending order without the copy sequence index and with the astrological link, trading the consciousness of backward order that makes each copy a zero without oneness with time

Heterogeneous involves a sidereal link. Sidereal link is a strong psychic linkage generating strong force with the one following. It forms a circle using an addition function within the thermodynamic limit and a trading function for transcending the thermodynamic limit through reincarnation. Thus, 123 becomes 231, and 231 becomes 312. Sidereal link is an absolute consciousness link that services the potential of one's proportionate position to each copy.

Exogenous involves a tropical link. A tropical link is a weak psychic linkage generating weak force with the one free to transform the circle by making the one following the leader using the subtraction function to make oneself the follower. Thus, 132 becomes 321, and 321 becomes 213. A tropical link is a para-conscious link that exchanges the potential of both the leader and the follower.

Homogeneous involves a zodiac link. Zodiac link divides the sidereal link into intrinsically vs. extrinsically strong psychic linkage. Intrinsically strong psychic force generates oneness within a homotopy group, freeing the variations due to the copy sequence index. However, it also adds to the oneness without a homotopy group, so each succeeding combination adds an incremental bond order. Thus, 111 becomes 222, and 222 becomes 333. A homotopy group is a preceding combination. A succeeding combination is a cohomotopy group, co-existing with the homotopy group. A homotopy is a receding combination that recedes to the left by giving precedence to the one in the center. The one in the center exchanges the four-fold supernatural growth of the one on the right. The one on the right enjoys three incremental oneness within its three differentiated time units and the fourth unit of oneness as three copies with the three copies in the center as a group. Since everyone in the discontinuous manifold produces and consumes four-fold supernatural growth, its winding value is 888.

Winding value of the homotopy on the left transpositions on the homotopy group in the center and reforms its normative development from four-fold supernatural growth to eight-fold growth. Next, the one in the center realizes the need to exchange eight-fold growth from the one on the right and corrects the unit value of the supernatural growth through an artificial formation of value aspiration that diverges from the consciousness of the growth of the one on the right. One is conscious of only half the eight-unit natural value of growth because one services the other half for making three copies in proportionate oneness with one another and in absolute oneness with oneself. Thus, one invests four ones for servicing four units of oneness without oneself.

Endogenous involves an astrological link. Astrological link divides the tropical link into intrinsically vs. extrinsically weak psychic linkage. Intrinsically weak psychic force weakens and destroys oneness with time due to the electromagnetic mass of the group-effect. Extrinsically weak psychic force perpetuates one without oneness with space as the electromagnetic mass of the group-effect. It makes one capable of exchanging space without group linkages. Overall, the zodiac link forms a psychic linkage, generating a psychic force that strengthens with the consciously formed oneness. Astrological link forms a para-psychic linkage, generating a para-psychic force that weakens the consciously formed oneness.

Energy Synonyms of the Discontinuous Manifold, 888

- Paryaya (Formula; Varying manifold; Piecewise linear manifold; Discontinuous manifold; Synonym; Homotopy; Receding combination; Zero degrees asymmetry; Quantum hall-effect; Topological-effect; Discontinuous value; Winding value; Winding number)
- Viparyaya (Transposition; Reversal; Reform; Antonym; Homotopy group; Preceding combination; 180 degrees asymmetry; Organizational development; Normative development)
- Nipparyaya (360 degrees asymmetry; Conform; Homonym; Cohomotopy group; Succeeding combination)
- Saundarya (Beauty; Artificial)
- Nasika (Value aspiration)
- Vinyasa (Formation)

Ten forms of links

Sidereal link = Strong Psychic linkage = Strong force = 54,000

Tropical link = Weak Psychic linkage = Weak force = 185

Psychic link = Psychic force = Emotional force = Vacuum = 47

Para-Psychic link = Para-psychic force = Dark force = 179

Intrinsicially strong psychic link = Intrinsically strong force = Soliton = Nanoparticle = 186

Extrinsically strong psychic link = Extrinsically strong force = Sanguiin = 1000

Intrinsically weak psychic link = Intrinsically weak force = Gluon = 149

Extrinsically weak psychic link = Extrinsically weak force = Boson = 250

Zodiac link = Animal force = 26

Astrological link = 580

Four groups of combinations

Heterogeneous = 12

Exogenous = 14,000

Homogeneous = 98

Endogenous = 10^{16}

Homogeneous electron = Positive Quark = Lepton = 1

Entropy = Entropy value = Kaon short = Exogenous Electron = 5

Illuminating value = 7

Natural value = Four-fold supernatural growth = Eight-fold growth = 8

Kaon long = Heterogeneous electron = 9

Kaon negative = Endogenous electron = Z-axis = 10

Density = Octave of copies = Sentient programming = Circulating sequence = Circularity = Single bond = 60

Pi Bond = Density – Entropy = 60 – 5 = 55 = 22 + 26 + 7

3.7 Developing Space with One Hundred Eighty Finite Points

It is possible to triangulate the octave of copies and claim the value an illuminating entity is illuminating with its design energy. The value is the 180-unit topological field, illuminated as light, i.e., visible photonic indigo light. The illuminating entity is a white star servicing its illuminated value. The design energy is the past life energy. Before incarnating as a white star to illuminate its value, the illuminating entity perpetuates its value as a perpetuating entity. The perpetuating entity is the only entity in the universe. It is an ellipse within space that remains an ellipse within time. Thus, it forms a twin ellipse within causation that superpositions

The Space Tensor : Exchanging the Entity Within

the evolving time over the space revolving over time. With the revolution, space begins co-evolving with time. The correlation of space with time objectifies the space into a fermion, the time into a neutrino, their correlation into a photon, and their oneness into the light.

To keep track of the developing space, I identify the values in the brackets. I will index the values against the words taken from the Indian "metaverse" (*Gurutva*, 100) to enlighten the mythological nature of modern science and the reality inherent within the ancient wisdom of India. If the words in brackets appear alien to you, take them as the references authenticating my experience with reality illuminated here. They are for the reference of those who wish to dig deeper into the nuances scraped here at the surface for the sake of illustration. They are not necessarily literal translations of the quoted words or phrases. They are only the links to illuminate the linkages among the concepts illustrated through those quoted words and phrases. Know that there are many alternative forms of reality. I have chosen the physics of space because it is the hottest domain attracting the world's wealthiest people to devote their wealth. Thus, it offers the greatest return on investment on time devoted to knowing the sensible way for managing the space. My thesis is that by minimizing the energy devoted to space exploration, the universe can maximize the energy devoted to exploiting time at hand for everyone's sentient well-being.

As a two-dimensional "replication" (*Samalekha*, 90) of the "ellipse" (*Dirghavritta*, 1,000) illuminating its "value" (*Mulya*, 180) as a "white star" (*Yama*, 180), the "twin ellipse" (*Yamadeva*, 1,000,000) is a "potential white star" (*Yamadeva*, 1,000,000). Since the illuminated value is light visible as colored light, the twin value is "potential light," which is a "colorless light" (*Yamadeva*, 1,000,000). Since value is the past life energy, the twin value is the "woman daughter" (*Stree*, 1,000,000) living life as "ALL" (*Sab*, 1,000,000) that exists before giving birth to the white star with the ellipse's "sentient energy" (*Amrit*, 1000).

The white star is the oneness of the space's "eventual state" (*Sura*, 0) and the time's "consequential spatial state" (*Janaka*, 180). It is the 180th finite point where the eventual spatial state becomes one with the consequential spatial state; thus, the incremental value of the space without time is zero. The time's consequential spatial state is a state that "self-perpetuates" (Amsha, ½) "time" (Kaal, 360) over time. The time that perpetuates is of the "upregulated" (Ekarshi, 1000) "perpetuating entity" (Ekarshi, 1,000) as an ellipse. The ellipse that perpetuates is a section of another ellipse that transforms into an "octave of sections" (Dirghavritta, 1,000). Besides the ellipse, the octave of sections comprises circle, parabola, hyperbola, square, triangle, cone, and full circle, all "existing" (Maujud, 1,000) in their "present form" (Param rupa, 1,000) as an ellipse, copying the ellipse.

The ellipse "co-exists" (Satha, 1,000) with their reality as an "illusionary production" (Maya, 1) and becomes an ellipse when the time perpetuates the space's "eventual state" (Sura, 0) over the past, the present, and the future. It forms as a "Phi Bond" (Dirghavritta, 1,000), bonding a discordant electron from the past with a concordant electron from the future into a "phi" (Apari, 2). The phi norms an "orbital" (Apari, 2), an additional illusionary electron from the present. The orbital transforms the "base" (Kumbhaka, 10) into "two tens" (Bindu, 10^{10}), taking the past as the base exponentiating its value using the power base of the present to be the "planning point" (Bindu, 10^{10}) of the future.

The "discordant electron" (Brihadbala, 149) is the "Gluon" (Brihadbala, 149), the one servicing four-unit consciousness of its nine-unit goal to become the ten-unit base for designing the future with its 1,4,9 sequence. The "concordant electron" (Kritajna, 167) is the "muon" (Kritajna, 167), the one servicing the six-unit growth of its future seven-unit reality as a "potential kaon" (Vastavikta, 7) with its 1,6,7 sequence. The potential kaon is the antikaon that manifests as the "kaon positive" (Vastavikta, 7), servicing "positive energy" (Sakaratmak shakti, 19) its "knowing" (Jnana, 19). Finally, the "illusionary electron" (Kalpanik, 20) is a "photon" (Kalpanik, 20), servicing "southern light" (Kalpanik, 20) of a "man" (Manav, 24).

The man is an "imaginary" (Kalpanik, 20) element who trades the four-unit consciousness and services the two-unit "psychic" (Jiva, 2) element. The psychic element transforms into the illusionary electron without differentiating the time from the space. The man also services two ones that form the "present time" (*Sva*, 11) and divide it into the past time and the "future time" (*Tribhajya*, 10^{10}). The "past time" (*Kumbhaka*, 10) is the base. The future time is the planning point. The gluon is the "past time force" (*Brihadbala*, 149). The muon is the "future time force" (*Kritajna*, 167). The photon is the "present time force" (*Kalpanik*, 20).

The planning point is the "gamma bond" (*Tilaka*, 10^{10}), bonding an imaginary electron that services the "conscious spirit" (*Ruah*, 20) of the future with a real electron that trades the "consciousness stock" (*Atma*, 4), i.e., the soul, of the past, into a "gamma" (*Andhakara*, 10^{19}). The gamma manifests the "consciousness void" (*Andhakara*, 10^{19}) in the form of a "whole electron" (*Asvaryogya*, 10^{19}). The whole electron is the "feeding recombination" (*Asvaryogya*, 10^{19}) that feeds the conscious spirit and the consciousness stock as a "knowing recombination" (*Dvaiyogya*, 10^{24}) by adding five finite points as a "circle" (*Valaya*, 100,000). The feeding recombination adds nine finite points to the gamma bond after breeding a "triple ellipse" (*Alidha mandala*, 10^9) by freeriding on the three-time units. The whole electron copies itself to transform into a "wholesome electron" (*Upa*, 16), comprising two whole electrons. Thus, gamma becomes two gammas and orbital becomes "two g-orbitals" (*Tilaka*, 10^{10}), recombined as the gamma bond to form an "f-orbital" (*Tilaka*, 10^{10}). Two whole electrons are four electrons because they include two "electrons" (*Dyumna*, 365) and two "potential electrons" (*Samdesha*, 1), i.e., "positrons" (*Samdesha*, 1). Four electrons form four bodies of the wholesome electron, norming the latter as the "fermion" (*Upa*, 16). Consequently, phi becomes two phis, and orbital differentiates "two f-orbitals" as the "phi bond" (*Dirghavritta*, 1000).

As the "eighth ellipse" (*Dirghavritta*, 1,000), the ellipse needs the "space" (*Vaas*, 18,000) not only for itself but for all eight ellipses, eight "potential ellipses" (*Hamsa chakra*, 1000), the ellipse as the octave of sections, and the potential ellipse as the potential octave of sections. An ellipse is the "quantum vortex" (*Dirghavritta*, 1,000) of the "convergism" (*Mahabala Parakramaya*, 1,000) of the "potential" (*AUM*, 18) of eighteen ellipses. Therefore, it conceives an

18,000-unit "space" (*Vaas*, 18,000) with a "square" (*Chakor*, 18,000) "field" (*Kha*, 18,000) to fill with its "field value" (*Kshetra*, 189). The field value comprises the "value" (*Mulya*, 180) of 180 ellipses formed by potentiating the 18-unit "potential" (*AUM*, 18) of each unit within the base. It perpetuates the "culture" (*Sadakhya*, 9) of reproducing the value by trading another nine-unit "potential up quark" (*Rasi*, 9) after conceiving the eight sections of a potential ellipse and a potential ellipse as the potential octave of sections.

3.8 Reversing Space with Eighteen Thousand Finite Points

It is possible to reverse space with eighteen thousand finite points for perpetuating the culture of forming space with five finite points and norming those points for transforming, transcending, enjoying, exchanging, and developing space. Once the space is developed, the forming ellipse reverses the space for acculturing another ellipse into the "path of followership" (*Charya marga*, 1,000,000) exchanged from the "twin ellipse" (*Yamadeva*, 1,000,000). Each "spatial ellipse" (*Ashtottara-sata*, 10^{16}) conceives a "temporal ellipse" (*Vishvabhava*, 10^{13}) without the "ellipse" (*Dirghavritta*, 1,000) that forms the space. A temporal ellipse conceives a "triple ellipse" (*Alidha mandala*, 10^9) for making each time unit an ellipse and a "causation ellipse" (*Gandakanayaka*, 10^4). The causation ellipse is a "circular system" (*Gandakanayaka*, 10,000) circling space and three units of time as sequential bases for reforming the "infinite space" (*Para Ganesha*, 19) taken by one's "energy" (*Shakti*, 19) into "quantum space" (*Nirdeshika*, 3×10^{17}). The quantum space becomes the "index" (*Nirdeshika*, 3×10^{17}) for conceiving eighteen thousand finite points with the "space" (*Vaas*, 18,000) and banking them as "knowledge" (*Vidya*, 600) within the "causal body" (*Karana sharira*, 30 = 18,000/600).

The knowledge "spreads up" (*Vitata*, 600) the causal body by trading ten time-differentiating finite points as the base. The base differentiates the three-time units and integrates the seventeen units as the exponentiated "fire" (*Agni*, 17) to make the "wisher" (*Chaahak*, 0) an "eternal entity" (*Ajanman*, 3×10^{17}). The wisher is the "copying one" (*Chaahak*, 0), reversing the space by copying another eternal entity. Thus, each eternal entity becomes a "variable chromosome" (*Ajanman*, 3×10^{17}) with varying consciousness of the "spatial development" (*Nirdeshika*, 3×10^{17}) he indexes as a "man" (*Manav*, 24). The man enjoys "growth" (*Dishta*, 6) by reproducing the woman's three time-units as his three time-units.

The man and the growth are immanent within the "causal body" (*Karana sharira*, 30), which is the woman's "feminine body" (*Karana sharira*, 30). The fire potentiates the multiform "growth potential" (*Nanarupa*, 10,000) of a "temporal ellipse" (*Vishvabhava*, 10^{13}) as the "origin of everything" (*Vishvabhava*, 10^{13}), including life. The woman's three time-units make her an "infinite feminine" (*Devi*, 3). The man's reproduction of those three time-units makes him the "greeter face" (*Aditya*, 6) and the "manipulator" (*Vaishwadeva yajni*, 6) of the woman's "creative force" (*Uma*, 6) that he is reproducing as his.

The "wait and watch option" (*Tikram*, -4) for trading the three reproducing time-units by consuming a negative unit "discordant energy" (*Asura shakti*, -1) becomes "dimensionless" (*Nirdharma*, -4). It forms a "qualifying" (*Nirguna*, -4) "triple bond" (*Nirdharma*, -4). A triple bond is a "scale factor" (*Nirdharma*, -4) that scales the wait-and-watch option with the "increasing demand" (*Kharidi*, 6) for appropriating the "creative force" (*Uma*, 6). As a metric with four negative finite points, it requires four men to discern it as the "Friedmann–Lemaître–Robertson–Walker metric" (*Nirdharma*, -4). A triple bond comprises six electrons: one pair in a sigma bond and two pairs in pi bond.

The "sigma bond" (*Akhanda*, 8×10^{15}) is a "quantum system" (*Akhanda*, 8×10^{15}) the "eternal entity" (*Ajanman*, 3×10^{17}) conceives by servicing his "gravitational energy" (*Lalita*, 100). The gravitational energy transforms the three-unit "time multiplier" (*Upaya*, 3) into the woman's "absolute gravitational constant" (*Upaya*, 3), reproduces and norms it as the man's "imaginary absolute gravitational constant" (*Vahana Mandapa*, 3), and forms the reproducing "orbit" (*Apari*, 2) into the daughter's "universe of potential electrons" (*Brahman*, 2). As a metric requiring a man with imagination to conceive it, the imaginary absolute gravitational constant is Einstein's gravitational constant. The oneness of the absolute gravitational constant and Einstein gravitational constant forms an "infinite point" (*Ab*, 1600), shaping the letter "S" (*Ab*, 1600). The "Wisher" (*Chaahak*, 0) incarnates as a woman to conceive the absolute gravitational constant over her three time-units. Next, he incarnates as a man with the four-unit consciousness of space and time that shaped the absolute gravitational constant. He becomes the "curved time" (*Yahoodi*, 1), experiencing the absolute gravitational constant as the imaginary absolute gravitational constant for knowing the "reality" (*Vastavikta*, 7) of the "eight-fold growth" (*Satarupa*, $8 = 1 + 7$).

The curved time is the "space curvature" (*Yahoodi*, 1). As one that transforms a man's imagination, it is "Einstein tensor" (*Yahoodi*, 1). Finally, the wisher incarnates as a "daughter cell" (*Kshatradharma*, 19), embodying the "knowing" (*Jnana*, 19) that the "fire" (*Agni*, 17) is the "paternal" (*Tejas*, 17). Further, the orbit is the "micro universe" (*Brahman*, 2) of both the universe of potential electrons and the universe of granddaughter cells. As the wisher reincarnates, the daughter cell becomes the "quantum cell" (*Kshatradharma*, 19), and the "universe of granddaughter cells" (*Brahman*, 2) becomes the "universe of light echoes" (*Brahman*, 2). The two "S" (*Ab*, 1600) letters bond with their "orbitals" (*Apari*, 2) to form a "sigma" (*Dvishakha*, $12,000 = [1600 + 1600] * 2 * 2 - 800$) and transform the wisher into an "invariant" (*Nirbija*, 800) norming the "symmetry system" (*Vyagrata*, 800). The sigma is the "bifurcate" (*Dvishakha*, 12,000) that bifurcates the "whole area" (*Garbha*, 12,000) of the "cuboid" (*Ghanabha*, 6,000) into two cuboids.

A cuboid is a "trifurcate" (*Trishaka*, 6,000) that transforms its "three surfaces" (*Pandrahbhuj*, 6000) into a "fifteen-sided pentagon" (*Pandrahbhuj*, 6000). The sigma bond transforms the "circle's" (*Valaya*, 100,000) "momentum density" (*Valaya*, 100,000) into a "pentagon" (*Panchakone*, 360) "breeding" (*Janana*, 15) fifteen sides. Thus, the sigma bond is the "momentum density tensor" (*Akhanda*, 8×10^{15}). The pentagon transforms the "time" (*Kaal*, 360) into "fifteen heterozygous chromosomes" (*Panchakone*, 360). Each "heterozygous chromosome" (*Shaneeswar*, 24) is an "evolving" (*Shaneeswar*, 24) element linked to the wisher's propounding letter "X" (*Prachar*, 15) for breeding the "Y-linked" (*Manav*, 24) "man"

The Space Tensor : Exchanging the Entity Within 155

(*Manav*, 24). It is a "heterozygous X-linked" (*Shaneeswar*, 24) "son allele" (*Shaneeswar*, 24). As a "divided allele" (*Shaneeswar*, 24), it is the "father-father allele pair AB" (*Shaneeswar*, 24). Y-linked is the "father allele" (*Manav*, 24) that a son allele adds to fill the void of the "mother allele" (*Augrya*, 24). The father allele trades the mother allele from the "grandson allele" (*Manipura chakra*, 24). The grandson allele divides itself into the "Mother-father allele pair Aa" (*Manipura chakra*, 24). It is "Heterozygous Y-linked" (*Manipura chakra*, 24) linked to a "multiplying allele" (*KLIM*, 24). The multiplying allele is the "granddaughter allele" (*KLIM*, 24) that multiplies into a "Sister-Sister allele pair BB" (*KLIM*, 24). It is "homozygous" (*KLIM*, 24) because it is the "green light" (*KLIM*, 24).

Green light is the "natural light" (*KLIM*, 24) that disassembles into its eight-unit "natural value" (*Nirvikalpana*, 8) per unit of time. The natural value comprises four masculine and feminine generations: grandfather/ grandmother, father/mother, son/ daughter, and grandson/granddaughter. All three units of natural value form a "growth photon" (*Hasti*, 24) and transform into a "grandmother allele" (*Hasti*, 24), sequentially adding the natural value to multiply as a "Mother-mother allele pair AA" (*Hasti*, 24). The "multiplied allele" (*Trivikrama*, 24) is a "daughter allele" (*Trivikrama*, 24), forming a "Father-Father allele pair aa" (*Trivikrama*, 24). Multiplying subtracts the "deity value" (*Daivi*, 24), the "grandfather allele" (*Daivi*, 24), formed as the "Z-linked" (*Daivi*, 24) "Brother-Brother allele pair ab" (*Daivi*, 24). The Letter "Z" (*Nilanjana*, 18) is the "feminization" (*Nilanjana*, 24) of the "potential" (*AUM*, 18) into the "potential energy" (*AUM shakti*, 18) for making the letter "Y" (*Mapanka*, -18) the "modulus" (*Mapanka*, -18) of the "Wisher's" (*Chaahak*, 6) "masculinization" (*Mardangat*, 257). Masculinization transforms the "natural light" (*KLIM*, 24) into the "sound" (*Naad*, 257 = 2[4+1]7), mediated by paternal "fire" (*Agni*, 17).

Sound generates the "vibration" (*Dhvani*, 247) of the "ascending cultural force" (*Dhvani*, 257) of the triple copy culture through "infinite psychic linkages" (*Tiripurai*, 257) with the "past force" (*Ghatika*, 24). It makes the past force the "greeter allele" (*Ghatika*, 24), embodying the sense of what might happen in the future as the "Wisher Paradigm" (*Ghatika*, 24). The Wisher Paradigm is the "present paradigm of radiant love" (*Ghatika*, 24), radiating love for the future "one without entity" (*Ghatika*, 24) and irradiating hate for the present "one within entity" (*Purushayoni*, 24). The one within entity becomes the "weak point" (*Purushayoni*, 24), the Achilles Heel, bifurcating its "secondary growth" (*Purushayoni*, 24) into two "quantum photons" (*Svarochisha*, 12). Each quantum photon bifurcates the two "tertiary growth" (*Dishta*, 6) units, one from the illusion of the past experience and the other from the imagination of the future experience. The "objectified consciousness" (*Ajitatma*, 0) of the "three divisions" (*Ajitatma*, 0) of "potential" (*AUM*, 18) is the "real absolute gravitational constant" (*Ajitatma*, 0). It is the "triple copy culture" (*Ajitatma*, 0). The triple copy culture leads to five copies of each finite point for realizing the "tertiary growth" (*Dishta*, 6). It copies the four-unit "consciousness" (*Chaithanya*, 4) of the five copies to norm the five-faced "culture" (*Sadakhya*, 9) element.

The culture accultures the "causation ellipse" (*Gandakanayaka*, 10,000) into a "female" (*Nari*, 10,000) acculturing her "reproductive system" (*Prajanan Pranali*, 10,000) into "holding" (*Hari*, 10,000) "both" (*Dono*, 10,000) HIM and HER as a "trading photon" (*Sahayaka*, 10,000). The causation ellipse is a "section" (*Anuvaka*, 10,000) of a "sequence of ellipses" (*Dvijyya*, 10,000), causing the culture to become the "infinitesimal" (*Atisukshma*, 90,000). The infinitesimal is the "octave of gender-differentiating four generations" (*Atisukshma*, 90,000) that makes each of the eight gender-differentiating "potential generations" (*Khudrata*, 10,000) and the "potential octave of generations" (*Vartula*, 10,000) a "sine" (*Dvijya*, 10,000). It makes the wisher "immortal" (*Amara*, 90,000) with the "empowerment" (*Sashaktikaran*, 10,000) of the female within the "infinity" (*Tamo guna*, 90,000) of copied "finite points" (*Maya*, 1).

The infinity is the past "oneness without entity" (*Ananta*, 90,000) that forms a "continuous gravitational wave" (*Ananta*, 90,000) of 90,000 "alleles" (*Devendra*, 12) as "quantum photons" (*Svarochisha*, 12). Every "second" (*Ananta*, 90,000) it services the "immanent feminine-effect" (*Prakarana vadartha*, 90,000) to 90,000

The Space Tensor : Exchanging the Entity Within 157

"messenger RNA's" (*Shilajit*, 855) for acculturing 90,000 "proteins" (*Shilajit*, 855) with the eight-unit "nature" (*Kudrat*, 8) and the fifty-five unit "capability" (*Udana*, 55), the "pi bond" (*Udana*, 55). The pi bond transforms into two p-orbitals. Letter "P" (*Bhuyobhava*, 160) is the "descending entropy-effect" (*Bhuyobhava*, 160) of the "development" (*Param Parvati*, 160) through "gene flow" (*Vikas*, 160) exchanged from an "octave of photons" (*Mahachitti*, 160).

The octave of photons forms with the development of the eight-unit nature within each of the twenty-unit "photons" (*Ruah*, 20). The gene flow "progresses" (*Vikas*, 160) the development further by transforming nature into the "infrastructure" (*Kundalini*, 8) for "formative growth" (*Vidhana*, 2). Formative growth is "feminine growth" (*Vidhana*, 2) the wisher reproduces as the "technological growth" (*Vidhana*, 2) to fulfill the "objective" (*Vidhana*, 2) of forming two additional "orbits" (*Apari*, 2) for bonding the two "P" letters into two "p-orbitals" (*Kudrat*, 8).

The Beauty of the Eight-Unit Nature

Eight-unit nature services a unit of "natural essence" (*Samaa*, 1) every moment it "progresses" (*Vikas*, 160) the development. Thus,

$8 \times 1 = 8$, becomes

$8 \times 2 = 16 \rightarrow 1 + 6 = 7$, becomes

$8 \times 3 = 24 \rightarrow 2 + 4 = 6$, becomes

$8 \times 4 = 32 \rightarrow 3 + 2 = 5$, becomes

$8 \times 5 = 40 \rightarrow 4 + 0 = 4$, becomes

$8 \times 6 = 48 \rightarrow 4 + 8 = 12 \rightarrow 1 + 2 = 3$, becomes

$8 \times 7 = 56 \rightarrow 5 + 6 = 11 \rightarrow 1 + 1 = 2$, becomes

$8 \times 8 = 64 \rightarrow 6 + 4 = 10 \rightarrow 1 + 0 = 1$, becomes

$8 \times 9 = 72 \rightarrow 7 + 2 = 9$, becomes

$8 \times 10 = 80 \rightarrow 8 + 0 = 8$.

In ten steps, eight-unit nature becomes the "essential nature" (*Svabhav*, 8) of the progress. The sixteen-unit "essence" (*Sara*, 16) progresses the "development" (*Param Parvati*, 160) in a sequence of ten steps. Development is the "gene stock" (*Param Parvati*, 160). Progress is the "gene flow" (*Vikas*, 160) that "binucleates" (*Abhikha*, 90) the "entropy-effect of the beauty factor" (*Vikas*, 160) into a "genetic library" (*Kshitigarbha*, 90) for further "replication" (*Samalekha*, 90). It makes the replication the "initial causation state" (*Samalekha*, 90). Each succeeding replication transforms the initial causation state into a further curvable "quantum causation" (*Samalekha*, 90).

Quantum causation forms a "delta bond" (*Samalekha*, 90) with "two d-orbitals" (*Samalekha*, 90). Letter "D" (*Avadata*, 12) is the "indigo color" (*Avadata*, 12), whose fifteen units form the "indigo light" (*Prabha*, 180 − 12 * 15). A pair of letter D's form a "pentagon" (*Panchakone*, 360) with "fifteen heterozygous chromosomes" (*Panchakone*, 360), dividing into fifteen "father-father allele pairs AB" (*Shaneeswar*, 24). Two D's with orbitals form an orbital that transforms AB into a "father-father allele pair" (*Shaneeswar*, 24 = 2 D's, 2 D's + 2-unit orbital). Each "D" (*Avadata*, 12) with "orbital" (*Apari*, 2) is a "d-orbital" (*Maha Vibhu*, 14 = 12 + 2), the "acculturation" (*Maha Vibhu*, 14) of the "masculine cell" (*Maha Vibhu*, 14) with the feminine cell's "eventual temporal state" (*Maha Vibhu*, 14).

Letter "A" (*Parameshthi*, 28) is the "immanence-effect" (*Parameshthi*, 28) of the "feminine cell" (*Maha Lakshmi*, 14). Letter "B" (*Dahana*, 7) is the "spreading" (*Dahana*, 7) of the immanence-effect as the discontinuous "light wave" (*Kojyataranga*, 196 = 7 * 28). The "light wave" (*Kojyataranga*, 196 = 14 * 14) reforms the two "d-orbitals" (*Maha Vibhu*, 14) into two "replication" (*Samalekha*, 90) units of the sixteen-unit "essence" (*Sara*, 16). One replicates the "illusionary past" (*Tippa*, 2000) as a "multiplex" (*Tippa*, 2000) of "two planes" (*Tippa*, 2000). The other replicates the "imaginary future" (*Maujudgi*, 2000) of "existence" (*Maujudgi*, 2000) of the "real present" (*Sataha*, 2000). The real present is the "surface" (*Sataha*, 2000), conceiving the past and the future as the "two planes" (*Tippa*, 2000) replicated by the illusionary past.

3.9 Acculturing Space to the Triple Copy Culture with Ninety Thousand Finite Points

It is possible to acculture space with ninety thousand points so that each replication of space replicates the "triple copy culture" (*Ajitatma*, 0) to become "immortal" (*Amara*, 90,000). Immortal is the "reproducible point" (*Amara*, 90,000) of infinity. However, unlike the "original" (*Mukhya*, 14), a "copy" (*Nakala*, 0) remains "mortal" (*Antavanta*, 1428) once reproduced. As a reproducible point repels its "leadership energy" (*Nayakatva shakti*, 90,000), it attracts the original's "followership energy" (*Pasaka*, -9) due to its "intrinsic oneness" (*Janma yoga*, -10,000) with the "causation ellipse" (*Gandakanayaka*, 10,000) as a "copy" (*Nakaka*, 0 = 10,000 - 10,000). The "original" (*Mukhya*, 14) forms the "orbit" (*Apari*, 2) to services its "essence" (*Sara*, 16) to endow "followership energy" (*Pasaka*, -9) on each "orbiting element" (*Mat*, -9), the orbit's copy. The "orbiting copy" (*Mat*, -9) is a "quadruple bond" (*Mat*, -9), comprising eight electrons: one pair each in a delta and sigma bond and two pairs in pi bond.

The "ion" (*Konastha*, 13) breeds a "twin pi bond" (*Konastha*, 13), immanent within a "whole photon" (*Konastha*, 13). The whole photon emanates from the orbiting copy's "diverging reality" (*Rudhartha*, -13) and forms an additional "twin pi bond" (*Konastha*, 13). The two pairs in the pi bond comprise four "masculine electrons" (*Narayana*, 28). A masculine electron is a "potentially scalar meson" (*Narayana*, 28), forming the "D-meson" (*Narayana*, 28) with the "seven-fold consciousness" (*Shaddhatma*, 28) of reality. The two pairs in the delta and the sigma bonds comprise four "feminine electrons" (*Durga*, 28). A feminine electron is a "potentially vector meson" (*Durga*, 28), forming the "B-meson" (*Durga*, 28) for balancing the "weight" (*Bhara*, 28) of the D-meson.

The "pi bond" (*Udana*, 55) transforms its "capability" (*Udana*, 55) into a "polypeptide" (*Udana*, 55) chain of 55 amino acids. Each "amino acid" (*Ida*, 1) is a chain of 9 polypeptides. A "chain" (*Shani Bhagwan*, -1) is the "electron force" (*Shani Bhagwan*, -1) perpetuating as the "Casimir force" (*Shani Bhagwan*, -1) generated with the "rupture of time" (*Shani Bhagwan*, -1). Rupture of time is the "variable space" (*Shani Bhagwan*, -1) that forms with "magnetoelectric-effect" (*Shani Bhagwan*, -1 = -9 − [-10]) of the orbiting copy. The orbiting copy magnetically attracts the "extrinsic reality" (*Niyogartha*, -9) of the original and electrically repels a "ligand" (*Samlagni*, -10) seeking "reversion" (*Padavanita*, -10) to the "primordial self" (*Parvati*, 10). The ligand is the "partial pi bond" (*Samlagni*, -10) that transforms a mature "father electron" (*Jagadvinasa*, -10) into a "coordination complex" (*Taijaisa*, -10). The coordination complex coordinates a "seven-step freeriding process" (*Vidhi Vadartha*, 130 = -10 * -13) with the orbiting copy's "diverging reality" (*Rudhartha*, -13).

- First, freeriding a "real ion" (*Mash*, 0) formed as the "hum" (*Mash*, 0) with the "humming" (*Gunja*, 900) vibration of orbiting,
- Second, breeding an "ion" (*Konastha*, 13) as a "twin pi bond" (*Konastha*, 13) with a "real absolute gravitational constant" (*Ajitatma*, 0),
- Third, feeding an "illusionary ion" (*Upaya*, 3) as an "absolute gravitational constant" (*Upaya*, 3),
- Fourth, adding an "imaginary ion" (*Yatamanasa*, -28) with an "imaginary absolute gravitational constant" (*Vahana mandapa*, 3),
- Fifth, subtracting the "whole ion" (*Parvati*, 10) as the "primordial self" (*Parvati*, 10) in the form of its "light potential" (*Devata*, 10), the "kaon negative" (*Devata*, 10),
- Sixth, trading the "wholesomewhole ion" (*Kanalakshamsha*, 39) as a "molecule" (*Kanalakshamsha*, 39).
- Seventh, servicing the "whole ion" (*Parvati*, 10 = 13 + 0 + 3 − 28 + 22), comprising the "ion" (*Konastha*, 13), the "real ion" (*Mash*, 0), the "illusionary ion" (*Upaya*, 3), and the "imaginary ion" (*Yatamanasa*, -28) within "π" (*Artta*, 22/7) multiplying the "reality" (*Vastavikta*, 7) of the "seven-step freeriding process" (*Vidhi Vadartha*, 130) into the "seven-fold consciousness" (*Shaddhatma*, 28).

The "molecule" (*Kanalakshamsha*, 39 = 19 + 19 + 1) comprises an "atom" (*Anu*, 19) and a "neutron" (*Matsarya*, 19) in oneness as a "meson" (*Khud*, 1). The meson is the "wholesome ion" (*Ida*,

The Space Tensor : Exchanging the Entity Within 161

1), the "finite self" (*Ida*, 1) of the "mature electron" (*Jagadvinasa*, -10) as a "Hubble tension" (*Prani*, 1). The wholesome ion diffuses the Hubble tension as the "amino acid" (*Ida*, 1), forming a "chain" (*Shani Bhagwan*, -1) of nine "polypeptides" (*Udana*, 55) with the ten-unit "potential strange quark" (*Parvati*, 10) and norming a chain of fifty-five amino acids with a pair of five-unit "kaon shorts" (*Attan*, 5). Thus, each amino acid is a twin chain of 495 amino acids. It repels the "chain" (*Shani Bhagwan*, -1) as the electron force to attract the introverted "energy system" (*Antarmukha*, 10^{1024}) as the "growth electron" (*Antarmukha*, 10^{1024}), i.e., the son electron. The son electron forms the "partial sigma bond" (*Antarmukha*, 10^{1024}) and reproduces that as a "twin point particle" (*Antarmukha*, 10^{1024}), "preon plus" (*Antarmukha*, 10^{1024}).

An amino acid exponentiates the "whole ion" (*Parvati*, 10) with the "bondless" (*Aditi*, 1024) "closed space" (*Devatamayi*, 1024). The "closed space" (*Devatamayi*, $1024 = 1000 + 9 + 15$) includes an "ellipse" (*Dirghavritta*, 1000) formed with a "triple chain" (*Dirghavrittaphala*, 991) of 991 amino acids, the "present space" (*Sadakhya*, 9), and the "delta" (*Param Shiva*, 15) as the "link" (*Rajas*, 15) with the "primeval space" (*Para Ganesha*, 19). The triple chain is the "elliptical potential" (*Dirghavrittaphala*, 991) of the "present space" (*Sadakhya*, 9) the "culture" (*Sadakhya*, 9) forms with the "koan negative" (*Devata*, 10) for conceiving the "amino acid" (*Ida*, 1) to compensate for the "electron force" (*Shani Bhagwan*, -1). The primeval space forms as the "partial delta bond" (*Hemarenu*, 19), comprising a nascent "grandson electron" (*Vidya ratri*, 19) with the "energy" (*Shakti*, 19) of the "energy system" (*Antarmukha*, 10^{1024}).

Fifteen triple chains and their cultural elements form a "scalar meson" (*Krita*, 15,000). Each ellipse within the scalar meson enjoys a "shared consciousness" (*Trigunatma*, 14) of fourteen other ellipses that form a "vector meson" (*Trigunatma*, 14). The vector meson transforms the "kaon negative" (*Devata*, 10) into an orthogonal "tensor-forming K-meson" (*Devata*, 10). K-meson norms the culture with nine forms of "mesons" (*Khud*, 1) without itself.

- "Meson" (*Khud*, 1). Meson is a theta meson, t
- Potentially-scalar "D-meson" (*Narayana*, 28). It comprises Top Zeta meson, c

- Potentially-vector "B-meson" (*Durga*, 28). It comprises Bottom Iota meson, d
- "Scalar meson" (*Krita*, 15,000). It comprises Up Iota meson, u
- "Vector meson" (*Trigunatma*, 14). It comprises Down Iota meson, d
- "Tensor meson" (*Ekatma*, 986), the "unified consciousness" (*Ekatma*, 986) of the "triple chain" (*Dirghavrittaphala*, 991), and its "entropy value" (*Sarvanasha*, 5), leading the meson to transform itself into the down meson. Tensor meson is the Eta prime meson. It comprises Up Kappa meson, $(u+d+s)/\sqrt{3}$
- "Down meson" (*Maya shakti*, 1), the "twin point fragment" (*Maya shakti*, 1) of the "twin point particle" (*Antarmukha*, 10^{1024}). It comprises Bottom Iota Meson, d
- "Eta meson" (*Yogatma*, 42), the "oneness consciousness" (*Yogatma*, 42) of the two twin chains forming a third twin chain by transforming into a triple chain within the closed space and programming the replication of the chain within the "open space" (*Swakshetra*, 9000). It comprises Down Kappa Meson, $(u+d-2s)/\sqrt{6}$
- "Pi meson" (*Bodhichitta*, 147), the "spiritual oneness" (*Bodhichitta*, 147) of the four gender-differentiated mesons and the first seven forms of the mesons within the space . The space includes both the closed and the open spaces. The pi meson norms the Top Iota Meson, u

The open space forms a "partial phi bond" (*Payojya*, 9,000), comprising a deceased "grandfather electron" (*Griva*, 9,000). It transforms into a "limitless" (*Aditi*, 1024) closed space when "oneness consciousness" (*Yogatma*, 42) "vibrates" (*Sphul*, 79) within it to form a "partial gamma bond" (*Sphul*, 79 = 42 + 27). Oneness consciousness divides the two twin chains into seven twin chains. A triple chain comprises three twin chains because the three chains also include two chains and one chain. A triple chain closes the present space, a second triple chain closes the primordial space, and the third twin chain opens the primeval space. A unit of space and three time-units form four units of "oneness consciousness" (*Yogatma*, 42), grouping the four units of

2,7 tensor sequence into a "self-ordered" (*Virodhita*, 108) "group of 108 scalar mesons" (*Virodhita*, 108). A "continuity" (*Rohini*, 50) of the "group of 108 scalar mesons" (*Virodhita*, 108) into a "half-open space" (*Rohini*, 50) is serviced by the "octave of eta mesons" (*Sadhibhuta*, 158) as the "matter" (*Sadhibhuta*, 158). The matter forms as a "meso proton" (*Sadhibhuta*, 158).

A "geography" (*Ganarajya*, 476) that forms a "quark" (*Dridayudha*, 476) transforms the four units of 2,7 "tensor sequence" (*Rupa siddhi*, 27) into the seven units for differentiating the space into the present, the primordial, and the primeval spaces. The quark trades the "sigma value" (*Aravu Maniyal*, 476) of the 4,7,6 "potential tensor sequence" (*Mandra*, 1) for differentiating the present time into a triple pair: the past and the future time, the primordial and the primeval space, and the space and the present space. The potential tensor sequence is the "potential quark" (*Mandra*, 1), the anti-quark.

3.10 Working to be Free from the Triple Copy Culture with Negative Sixteen Finite Points

The one activating the possibility of acculturing the space to the reality of making replication immortal is an "uncurved entity" (*Vyatikara*, -16). The uncurved entity reproduces a "cascade" (*Vyatikara*, -16) of the four units of 2,7 "tensor sequence" (*Rupa siddhi*, 27) as a "path of programming" (*Mahakriya marga*, 108 = 4 * 27). It leads to "regrouping" (*Samsrkita*, -16) of the "entity" (*Hasti*, 24) into a "potential space" (*Prithvi*, 132 = 24 + 108) that forms the "mass of consciousness" (*Prithvi*, 132). The regrouping is a "mass-energy-momentum tensor" (*Samskrita*, -16); it transforms the "mass of consciousness" (*Prithvi*, 132 = 7 * 19 -1) into seven units of "energy" (*Shakti*, 19) and a unit of "discordant energy" (*Asura shakti*, -1).

The discordant energy forms the "metric tensor" (*Param Shunya*, -1) to transform the three copies of the seven units of energy into a "three-faced system value" (*Ribhu*, 999 = 133 * 3 + 500) by organizing the "gamma force" (*Pulastya*, 500) of the two-fold sequence as the "zeroth metric tensor" (*Pulastya*, 500). With the "momentum" (*Ribhu*, 999) of the "three-faced system value" (*Ribhu*, 999), the "potential quark" (*Mandra*, 1) becomes an "ellipse" (*Dirghavritta*, 1,000 = 999 + 1) free from the limitations of the triple copy culture. The ellipse makes the "entity" (*Hasti*, 24) "limitless" (*Aditi*, 1,024). The "ellipse" (*Dirghavritta*, 1,000 = 2 * 3 * 19 * 7 + 2) forms with two triple copies of the energy's "reality" (*Vastavikta*, 7) emanating from a "meron" (*Brahmin*, 2) immanent within the ellipse that divides into an "octave of sections" (*Dirghavritta*, 1,000).

A "meron" (*Brahmin*, 2) is a "half instanton" (*Brahmin*, 2) norming the "collective potential" (*Brahmin*, 2) of the "octave of sections" (*Dirghavritta*, 1,000) as a "divided entity" (*Brahmin*, 2) that forms the "zeroth metric tensor" (*Pulastya*, 500). The "twin meron" (*Suchan*, 2) that "orders" (*Suchan*, 2) the collective potential within the ellipse, to make the ellipse the octave of sections, is the other "half instanton" (*Brahmin*, 2). Unit "instanton" (*Shudra*, 1) is the "absolute quark" (*Shudra*, 1) that "flows with time" (*Shudra*, 1) as the "Higgs rest mass" (*Shudra*, 1).

The Higgs rest mass forms a "negative quark" (*Nitya ratri*, 1) as an "illusion" (*Nitya ratri*, 1) that divides the ellipse twice into a "double octave" (*Madhusudan*, 16). First, as an "upper octave" (*Upa*, 16), taking ellipse as the octave of sections divided into eight sections and dividing the ellipse as the first section divisible as a "lower octave" (*Prati*, 16) into an additional eight sections. Second, as a lower octave, taking the ellipse as the ninth section of the twin ellipse formed by the upper octave and dividing the ellipse into eight sections. Thus, the triple ellipse becomes the tenth section of the ellipse the negative quark divides into thirty-two sections. Thirty-two sections include four ellipses, comprising the divided ellipse and the undivided triple ellipse.

Thus, the mass-energy-momentum tensor is the "quadruple ellipse" (*Samsrkita*, -16) the uncurved entity reproduces to become a "cotangent" (*Sura*, 0), the Higgs field, of the triple ellipse. As the tenth section, the triple ellipse is the "tangent" (*Ardhajya*, 10) of the thirty-two sections. Thirty-two sections form an "octave of infinite quarks" (*Krishna*, 32) in "primordial oneness" (*Adi*, 32) with the quadruple ellipse. They transform the quadruple ellipse into a "quintuple bond" (*Samskrita*, -16), comprising ten electrons: one

The Space Tensor : Exchanging the Entity Within 165

pair each in a phi, delta, and sigma bond and two pairs in pi bond. Ten electrons are the ten sections, of which three are ellipses, and seven are the "illuminating value" (*Prakasha*, 7) of the triple ellipse. Triple ellipse is the triple copy of the ellipse a negative quark forms as a unit illusion with three zeroes to compensate for the "metric tensor's" (*Param Shunya*, -1) negativity.

The illuminating value is the "twin phi bond" (*Prakasha*, 7) that forms the ellipse as the "phi bond" (*Dirghavritta*, 1,000) and positions an additional ellipse within the ellipse with the second illusion unit. Twin phi bond comprises seven electrons, whose illuminating value transforms five electrons into a "twin delta bond" (*Saranyu*, 5). The twin delta bond is the "perpetuating value" (*Saranyu*, 5) that transforms three electrons in a twin sigma bond. The "twin sigma bond" (*Upaya*, 3) is the time multiplier that transforms one electron into a partial pi bond. The "partial pi bond" (*Samlagni*, -10) transforms into a "circular double octave" (*Adharma*, -10) for forming a "triple partial pi bond" (*Shudra*, 1) as a "robot" (*Karuyantra*, -11). The triple pi bond is an "absolute quark" (*Shudra*, 1), comprising an instanton divisible into a meron and a twin meron. Absolute quark is the "individual potential" (*Shudra*, 1) of an electron as the "divider" (*Shudra*, 1) of reality into a sequence of illusions without dividing itself.

3.11 Knowing the Value of Freedom with Negative Eight Points

A quintuple bond "self-perpetuates" (*Amsha*, ½) itself as a "sextuple bond" (*Akarta*, -8) with negative eight points. A sextuple bond comprises twelve electrons: one pair each in a gamma, phi, delta, and sigma bond and two pairs in a pi bond. "Twin gamma bond" (*Maha Shiva*, 9) comprises nine electrons and forms the "kaon long" (*Maha Shiva*, 9). In reality, kaon long is a "potential kaon" (*Vastavikta*, 7) in a twin phi bond. Potential kaon's illuminating value includes a perpetuating value of five electrons and the consciousness of four electrons immanent within the fifth electron that perpetuates that value. Perpetuating value is a "kaon short" (*Attan*, 5) in a twin delta bond. It includes a time multiplier of three electrons and the consciousness of four electrons emanating from the other two electrons that bring order with the third electron.

Before order, there are only two electrons in a twin sigma bond. The first electron forms the sigma bond without an electron. The other electron transforms the sigma bond to the partial sigma bond with the first electron. As a collective, the two electrons norm a twin sigma bond. The twin sigma bond as an individual conceives the sigma bond to be an electron. The sigma bond perceives the twin sigma bond to comprise three electrons. The partial sigma bond experiences the twin sigma bond comprising two electrons with a third electron in the partial pi bond. The partial sigma bond includes the first electron, different from the copies of the two electrons that form the twin sigma bond. The second electron that forms the sigma bond without the two electrons is the consciousness of all four electrons.

After ordering the two copies as the two prime values (*second and third*) *with two electrons*, the kaon short manifests the other three electrons as the time multiplier of the order with the first odd copy in the second position and consciousness with the second even copy in the third position. The "multiplying order" (*Ghataprakashaka*, 78) is an "exponent" (*Ghataprakashaka*, 78) that forms a square. The "multiplying consciousness" (*Haryyatma*, 167) forms a "cube" (*Ghana*, 90). The square is immanent within the cube. "Disproportionate consciousness" (*Prakriti-yogi*, -1 = 167 − 78 − 90) generates an "exponential growth" (*Chakrika*, 90) in "disorder" (*Pradarshan*, -8). Disorder moves the first odd copy to the ninth position as a kaon long. Exponential growth moves the second even copy to the eighth position as an "exponentiator" (*Satarupa*, 8) of the "eight-fold growth" (*Satarupa*, 8) in the position of the first odd copy to compensate for the "disorder" (*Pradarshan*, -8).

Thus, kaon short becomes kaon long as both phi bonds individually within the potential kaon contribute two electrons each to superposition a twin gamma bond with four additional electrons over the kaon short's five electrons. Further, the twin phi bond as a collective contributes two electrons to position a twin pi bond under the kaon short's five electrons. The consciousness of the four electrons generates a "thermodynamic force" (*Rodha*, 1) due to the three electron layers. The thermodynamic force generates an "illusion" (*Nitya ratri*, 1) of the twelfth electron forming a pi

The Space Tensor : Exchanging the Entity Within 167

bond by transforming the kaon short's partial pi bond.

Twelve electrons transform the "sextuple bond" (*Akarta*, -8) into a "photon" (*Kalpanik*, 20 = 12 − [-8]). The sextuple bond transforms the twelve electrons into a "quantum photon" (*Svarochisha*, 12). A quantum photon comprises twelve forms of "mesons" (*Khud*, 1) within itself.

- First, Strange B-meson (Up Nu Meson), s
- Second, Strange D-meson (Down Nu Meson), c
- Third, Phi meson (Top Nu Meson), s
- Fourth, Charmed B-meson (Bottom Nu Meson), c
- Fifth, Charged Rho meson (Up Mu Meson), u
- Sixth, Neutral Rho meson (Down Mu Meson), d
- Seventh, Upsilon meson (Top Mu Meson), b
- Eighth, Psion meson (Bottom Mu Meson), c
- Ninth, Omega meson (Up Tau Meson), $(u+d)/\sqrt{2}$
- Tenth, Neutral pi meson (Down Tau Meson), $(u-d)/\sqrt{2}$
- Eleventh, Vector D-meson (Top Tau Meson), u
- Twelfth, Scalar B-meson (Bottom Tau Meson), d
- A quantum photon forms three "primordial-primordial photons" (*Pitra*, 4), comprising four forms of kaon:
- First, Kaon positive (Up Zeta Meson), s
- Second, Kaon (Down Zeta Meson), s
- Third, Kaon-short (Top Kappa Meson), $(d+s)/\sqrt{2}$
- Fourth, Kaon-long (Bottom Kappa Meson), $(d-s)/\sqrt{2}$

3.12 Feeding the Value of Freedom with Negative Nineteen Finite Points

With negative nineteen finite points, a primordial-primordial photon generates a consciousness of its "unsustainable value" (*Lobha*, -19) within the "livable past" (*Nivasaniya*, 23). Livable past services a "desire" (*Kama*, -19) for feeding a future one did not live in the past. The primordial-primordial photon objectifies the desire for moving on without lamenting the choices one made in the past. However, it also spreads the desire to consume reality that perpetuates the primordial-primordial photon as a star of attraction. Therefore, it compounds the desire that devalues the choices one is making in the present as not fulfilling the desire for the future one wishes to live in the present. It motivates everyone to espouse their multiplying desires as worthy of fulfillment by everyone as a collective. It induces everyone to engage in the infinite trading of the unfulfilled subtle desires and stack genes that gravitate one to the "eventual spatial state" (*Sura*, 0) of zero as a leader, letting the collective fulfill the wishes of each individual.

Sequential Effects of Nineteen finite points with the Self-producing 20th finite point
Forming, Norming, and Transforming
Transcending, Enjoying, and Exchanging
Developing, Reversing, and Acculturing
Working, Knowing, and Feeding
Breeding, Freeriding, and Objectifying
Spreading, Compounding, and Espousing
Gene stacking with infinite adding
Eventual state making one zero

With thermodynamics, an individual conserves the energy of the collective even after deciding not to be a part of the collective. As a leader, the individual transforms his energy into the collective's energy. With leadership, the collective "C" (*Asamskrita*, -12) within the individual "I's" (*Svayam*, 12) consciousness transforms into the collective "C" without the individual "I's" consciousness. The collective "C" is a "quantum thing" (*Asamskrita*, -12); it is a thing each individual "I" (*Svayam*, 12) superpositions on the "space" (*Vaas*, 18,000) as a collective to make the space "transformable" (*Atita*, -12). The individual "I" is the "three rotations" (*Svayam*, 12) of the "thing" (*Vastu*, 9), each "ascending value" (*Rodha*, 1) by a unit. The first rotation makes the thing a "quantum thing" (*Asamskrita*, -12) and, therefore, transformable through the "thermodynamic force" (*Rodha*, 1). The second rotation makes the quantum thing a "subject" (*Sura*, 0) of further transformation by sustaining the force of the individual "I's" (*Svayam*, 12) consciousness over time. The third rotation transforms the "subject" (*Sura*, 0) with the individual "I's" "conscious force" (*Soham*, 4), its "consciousness" (*Chaithanya*, 4) within the collective "C," and the "consciousness" (*Chaithanya*, 4) of the collective "C" within the subject wishing to enjoy the individual "I" (*Svayam*, 12).

A subject wishes to enjoy the individual "I" because of the "entropy" (*Mahodbhava*, 5) of "C" within the subject after the individual shifts his consciousness. It makes the collective the "focal point" (*Kendrabindu*, 3) by directing the attention to a "finite point" (*Maya*, 1) in the space. The quantum thing moves with the consciousness to the focal point. It begins "breeding" (*Janana*, 15 = $|-12| + 3$) the consciousness as the individual "I" (*Svayam*, 12) diffused into the four finite points in the space. Therefore, the space transforms into its "present state" (*Vartaman*, 9) of a "quantum field" (*Vartaman*, 9) and starts behaving like a thing. Breeding makes the "focal point" (*Kendrabindu*, 3) the "absolute gravitational constant" (*Upaya*, 3) that the time reproduces for transforming the space into the "thing" (*Vastu*, 9) that behaves like a thing within a "quantum field" (*Vartaman*, 9). With his consciousness, the individual makes the quantum field the thing that matters for managing the entities within the space that forms the quantum field. Therefore,

> Quantum Field = Entropy of "C" within consciousness + Consciousness of "I" that generates the entropy of "C" within consciousness = 5 + 4 = 9.
>
> Thing = (Absolute gravitational constant)^2
>
> Quantum Field = Thing

The space gravitates the individual towards it for enjoying its "reality" (*Vastavikta*, 7) as a "thing" (*Vastu*, 9). After enjoying its reality, the individual wishes to incarnate as a "human being" (*Insan*, 1) for managing the reality. To fulfill his wish, the individual gravitates the thing consciously with his "energy" (*Shakti*, 19) as an "illusion" (*Nitya ratri*, 1) within his "dream" (*Dhyana*, 9). The thing an individual gravitates towards himself is the "ambiance" (*Vahana mandapa*, 3) of himself as the thing that has become a human being. The ambiance is the "imaginary absolute gravitational constant" (*Vahana mandapa*, 3). The ambiance is the imaginary focal point of the "relative reality" (*Nirrti*, 1) of the individual still circulating within the individual's mind because of the "past reality" (*Padartha*, -3). Therefore, it is Einstein gravitational constant.

(Absolute gravitational constant)² = Absolute gravitational constant x Imaginary focal point that transforms the past reality of the discordant energy when one incarnates as a human being = Absolute gravitational constant x Einstein gravitational constant = 3 * 3 = 9

"Past reality" (*Padartha*, -3) is the "curving time" (*Padartha*, -3), the "initial temporal state" (*Padartha*, -3) that transforms the space with the entity's "divine force" (*Padartha*, -3). The curving time is the "geodesic curvature" (*Padartha*, -3). The entity's divine force is the "entity scalar" (*Padartha*, -3). Geodesic curvature curves the space by reproducing the entity scalar within it in the form of a "geodesic force" (*Padartha*, -3). Geodesic force is a "gravitomagnetic force" (*Padartha*, -3) that behaves like an "attraction force" (*Padartha*, -3), attracting one to its confining, relative reality by objectifying one as the thing. Therefore, the geodesic force is the Ricci scalar. It makes one the thing by curving the space at the initial temporal state.

Thing = Geodesic curvature x Geodesic force = Geodesic curvature x Ricci Scalar = = (Initial temporal state)² -3 * -3 = 9

The "curvable space" (*Samketana*, -1) is the causative state encoding the individual's "discordant energy" (*Asura shakti*, -1). The space within the individual's mind is curvable because the individual lacks the collective's "concordant energy" (*Sura shakti*, 0). The "discordant factor" (*Asura*, -1) generates stress within the individual's mind and transforms the individual's "energy" (*Shakti*, 19 = 7 + 12) into the "reality" (*Vastavikta*, 7) with "I" (*Svayam*, 12). The curvable space is the "stress-energy tensor" (*Samketana*, -1), making the energy flow in descending motion. An individual's energy becomes discordant because of the "extrinsic impact" (*Ravinandana*, -1). Other individuals service a "polluted reality" (*Ravinandana*, -1) of one's "potential" (*AUM*, 18). Polluted reality forms "geodesic reality" (*Ravinandana*, -1), i.e., geodesic torsion in the form of the "perceived space" (*Ravinandana*, -1) within the mind one did not exist.

An individual makes the perceived space "my space" (*Param Shunya*, -1) "I" wish to enjoy. My space becomes the "organizational metric" (*Param Shunya*, -1) of the individual's incarnation as a "human being" (*Insan*, 1). The organizational metric is the "metric tensor" (*Param Shunya*, -1) of the "potential" (*AUM*, 18) within the individual's "energy" (*Shakti*, 19) to transcend beyond the perceived space. Once an individual transcends the perceived space, the "reproduced reality" (*Adravya*, -1) "does not matter" (*Adravya*, -1). Reproduced reality is "gravitated reality" (*Adravya*, -1) that gravitates one towards a "satan" (*Shani Bhagwan*, -1) for appropriating the reality one is reproducing. Reproduced reality becomes an "illusionary cosmological constant" (*Adravya*, -1) that produces the discordant energy by deifying the "human being" (*Insan*, 1) whose reality one reproduces. Thus, the illusionary cosmological constant is the Einstein cosmological constant.

> Curvable space = Causative state = Discordant energy = Geodesic reality = Geodesic reality = Perceived Space = My Space = Organizational metric = Metric tensor = Reproduced reality = Gravitated reality = Illusionary cosmological constant = Einstein cosmological constant = -1

The "illusion" (*Nitya ratri*, 1) is the "curved time" (*Yahoodi*, 1) one experiences as the "space curvature" (*Yahoodi*, 1) after deifying another "human being" (*Insan*, 1) and developing an "I AM a deity consciousness" (*Pavitratma*, 1) within the deified deity's "ideal force" (*Dasha*, 1). It is one's "relative reality" (*Nirrti*, 1). Therefore, it is "Einstein Tensor" (*Yahoodi*, 1). One reproduces the illusion within one's "imagination" (*Caksur vijnana*, 109) as the thing's "imaginary spirit" (*Kalpanik*, 20) flowing as a light "photon" (*Kalpanik*, 20). Therefore,

> Imagination = Thing + Gravitational Energy = 9 + 100 = 109

The light illuminates the space beyond one's imagination and transforms the "causation" (*Hetu*, 1). It lets the individual's "I" reproduce the initial temporal state and produce an absolute gravitational constant to perpetuate the growth in one's imagination.

I = (Initial temporal state)^2 + Absolute gravitational constant = -3^2 + 3 = Consequential entity state that makes an entity matter to the universe without I's consciousness = 12

Chapter 4

Causation Tensor:
Exchanging the Entity Without

God's Present. A sentient entity has three heads, conscious of my past, present, and future, and four arms, guiding my two hands with the consciousness of the four directions. One of the hands is of Grandmother Nature, servicing her conscious light force in the form of water to enlighten the sentient entity about her reality, which is also his and my reality.

The "sentient entity" (*Siddha*, 7) conceives me as the "fifth member" (*Danda*, 176) of a ten-member family for manifesting the seventh member with his "first energy" (*Pratham Shakti*, 14). I breed the seventh member as my "param son" (*Hanuman*,

3), the eighth member as my "param daughter" (*Jiva*, 2), the first six members as my "transcendental value" (*Khuda*, 6) comprising six additional daughters including the one born as the ninth member, and the final six members as my "creative force" (*Uma*, 6) comprising six additional sons for the growth of the family. The tenth member feeds me the "energy" (*Shakti*, 19) for breeding as the "param paternal" (*Narada*, 7) by "freeriding" (*Muftakhori*, 13) on the energy I breed as the "seventeenth member" (*Kincha*, 1,000,000,000) to order a "heterogeneous genus" (*Prajati*, 17) of species with the "homogeneous genus" (*Prakara*, 17) of my family.

How a Sentient Entity Conceives God as his "I"

A sentient entity incarnates himself as the "eighteenth member" (*Amnaya*, 97). He behaves like the "primordial maternal" (*Narayani*, 18) for activating the "reality" (*Vastavikta*, 7) of the "param maternal" (*Saranyu*, 5) within his "I" (*Svayam*, 12). Once "I" finish "breeding" (*Janana*, 15) seven sons, seven daughters, and energy as the eighth daughter, the sentient entity reveals that the energy is the "twin" (*Hridayesha*, 121) of my "param daughter" (*Jiva*, 2) born as the "potential daughter" (*Aap*, 1) within "you" (*Aap*, 1), the "potential son" (*Manyu*, 19). The potential son incarnates as the "nineteenth member" (*Samanantara*, 80,000) by trading the energy and servicing his "true nature" (*Vyaktitva*, 1) of the "trader" (*Trayastrimsha*, 20) of your "radiant love" (*Trayastrimsha*, 20 = 19 + 1) to make you the potential daughter.

The potential son is a "copy" (*Nakala*, 0) of the trader, born as the "twentieth member" (*Pitri Loka*, 20,000). The twentieth member is the "universe of replication" (*Pitri Loka*, 20,000), trading your "self-luminous reality" (*Purushartha*, 7) for servicing every "member" (*Sadasya*, 1) as a "replication" (*Samalekha*, 90). Each replication carries a replicative "bit" (*Liksha*, 90) of "information" (*Sandesha*, 1) about the "replicating member" (*Varga*, 9) as a "genetic library" (*Kshitigarbha*, 90). The replicating member is the tenth member that norms the "genus" (*Varga*, 10) as a "divergent group" (*Varga*, 9).

How a Sentient Entity Transforms the Causation with Genetic Code

A sentient entity enjoys a finite coordinate of space and time. There are three ways for a "finite coordinate" (*Kshetra*, 9) of space and time to "perpetuate" (*Ramakatha*, 9) the "cultural factor" (*Sugriva*, 5) as the "genetic code" (*Somapa*, 180) for transforming the "causation" (*Hetu*, 1).

First, start to transform the "finite coordinate" (*Kshetra*, 9) into the "culture" (*Sadakhya*, 9) that perpetuates into the "infinite space" (*Para Ganesha*, 19) within the "finite time" (*Vastavika*, 9) the sentient entity lives in the "mental realm" (*Masika*, 6666).

Next, continue to transform the "culture" (*Sadakhya*, 9) with the "sentient entity" (*Siddha*, 7) into a "factor" (*Kaya*, 16 = 9 + 7) that destroys the "finite space" (*Antaramsa*, 100) within the "infinite time" (*Kalpanik*, 20) the sentient entity lives in the "astral realm" (*Yaanshala*, 14) before living in the mental realm. The finite space is the "reproductive force" (*Chitta*, 100), forming the "cultural wisdom" (*Chitta*, 100) of the "departed entity" (*Divangat*, 1600) who lived as a "factor" (*Kaya*, 16) in the "etheric realm" (*Srijak*, 16) before the sentient entity.

Then, stop to transform the "factor" (*Kaya*, 16) into the "cultural factor" (*Sugriva*, 5) for illuminating the "foundation" (*Sva*, 11) constructed by the "cultural wisdom" (*Chitta*, 100 = 11 + 89) with its "sensory reality" (*Nashita artha*, 89) in the "causal realm" (*Atiyoga*, 916). In the causal realm, the departed entity is an "entity" (*Hasti*, 24) not conscious of its "impact" (*Siddhi*, 57) on the "intellectual realm" (*Dadhikravan*, 19). The "sensory reality" (*Nashita artha*, 89) constructs the "foundation" (*Sva*, 11) with the "continuous time" (*Dvilaya*, 17) within the "discontinuous space" (*Angana*, 6).

The "continuous time" (*Dvilaya*, 17) is the "twin time" (*Dvilaya*, 17 = 24 - 7) when the "entity" (*Hasti*, 24) lives without the "sentient entity" (*Siddha*, 7) who "hibernates" (*Kundalini*, 8) as a "living entity" (*Sushupti*, 96) living within the entity. The "discontinuous space" (*Angana*, 6) is the "twin space" (*Angana*, 6 = 7 - 1) where the "sentient entity" (*Siddha*, 7) hibernates without the "human being" (*Insan*, 1) who "sleeps" (*Nidra*, 876) as a "living being" (*Satta*, 2000) enjoying the "being energy" (*Kali shakti*, 96) of the "hibernated entity" (*Sushupti*, 96)

Finally, discontinue being the "cultural factor" (*Sugriva*, 5) for liberating the "light" (*Prabha*, 180) that "channels" (*Raasta*, 36 = 180/5) the "discontinuous time" (*Agaha*, 80,000) of the sentient entity within the "continuous space" (*Muddata*, 60,000) of the departed entity in the "physical realm" (*Bhautika*, 3333).

The "discontinuous time" (*Agaha*, 80,000) is the "triple time" (*Agaha*, 80,000) when (a) the human being lives as a living being in a "sleeping state" (*Maha Jagrat*, 0) within the sentient entity, (b) the sentient entity lives as a hibernated, living entity in a "hibernating state" (*Genda*, 16) within the departed entity, and (c) the departed entity lives as a "luminous entity" (*Kali*, 96) in a "dreaming state" (*Maha Svapna*, 9) without "consciousness" (*Chaithanya*, 4) that "He" (*Somanasa Mahayuga*, 13 = 9 + 4) is a "She" (*Dandi*, 710) dreaming herself to be a grandpaternal "departed entity" (*Divangat*, 1600), a paternal "sentient entity" (*Siddha*, 7), and a son "human being" (*Insan*, 1) by personifying the "identity" (*Pahachaan*, 8) of a grandson "Wisher" (*Chaahak*, 0) in a "waking state" (*Svapna*, 59) wishing to realize one's "true identity" (*Usha*, 16) of being a SHE. The Wisher is a "concordant entity" (*Sura*, 0).

The "continuous space" (*Muddata*, 60,000) is the "triple space" (*Midduta*, 60000) where (a) the living being enters a "polluting state" (*Beeja-Jagrat*, 84) to become a "discordant entity" (*Asura*, -1) without the sentient entity after incarnating as an "entity" (*Hasti*, 24), (b) the sentient entity enters a "cleansing state" (*Jagrat*, 85) to become a "super deity" (*Jiva*, 2) who norms the living being into a "deity" (*Deva*, 1) without herself and a "supra deity" (*Devi*, 3) within herself after incarnating as a "twin entity" (*Ghatika*, 24) with time waiting to become an entity, and (c) the departed entity enters a restful "guiding state" (*Svapna-Jagrat*, 97) to become a "supreme deity" (*Bhagwan*, 4) who transforms the living being and the sentient entity into a "primeval deity"

(*Maheshwara*, 6) and a "param deity" (*Shiva*, 7) without himself and a "primordial deity" (*Bhairavi*, 8) and a "devoted deity" (*Maha Saraswati*, 9) within himself.

The guiding state guides the grandson Wisher, the "concordant entity" (*Sura*, 0), into becoming a "devotee deity" (*Shri Krishna*, 10). The devotee deity is the "essential identity" (*Shri Krishna*, 10) for one to realize the "causing state" (*Jagrat-Svapna*, -1) that makes the Wisher a "discordant entity" (*Asura*, -1) servicing the "cultural factor" (*Sugriva*, 5). "Realization" (*Prati*, 16) gives one the "pristine identity" (*Madhusudan*, 16) of the "child primordial greeter" (*Madhusudan*, 16). The pristine identity is the "codon" (*Naadi*, 16) that transforms the "light" (*Prabha*, 180) of the departed entity into the "genetic code" (*Somapa*, $180 = 45*[1+1+1+1]$). It codes the start and "create codon" (*Vignesh*, 1), the continue and "perpetuate codon" (*Mandra*, 1), the stop and "destroy codon" (*Nitya Ratri*, 1), and the discontinue and "illuminate codon" (*Dravya*, 1) within one's "creative work" (*Khyati*, 45) as the "creator" (*Brahma*, 59). As a creator, transform the "causation" (*Hetu*, 1) that forms the "octave" (*Sargam*, $60 = 59 + 1$) of your copy and norms that as "reality" (*Vastavikta*, 7).

Second, develop a "consciousness" (*Chaithanya*, 4) of the "cultural factor" (*Sugriva*, 5) within the "finite coordinate" (*Kshetra*, $9 = 4 + 5$) of space and time that "incarnates" (*Virupaksha*, 900) one as a "guider" (*Guru*, 100), conscious of the "cultural wisdom" (*Chitta*, 100). "Channel" (*Raasta*, 36) the cultural wisdom as a "divine gift" (*Palala*, 64) from "God" (*Ishvara*, 5) who gifted the cultural wisdom in the form of the "cultural factor" (*Sugriva*, 5) copying himself. "Channel" (*Raasta*, 36) the "cultural factor" (*Sugriva*, 5) as the "genetic factor" (*Somapa*, $180 = 5*36$) having "exchanged" (*Somapa*, 180) God's "genetic code" (*Somapa*, 180) as your "genome" (*Somapa*, 180) in the "self-luminous realm" (*Kshitigarbha*, 90), norming the "genetic library" (*Kshitigarbha*, 90) for further "replication" (*Samalekha*, 90).

Next, become a "universal deity" (*Maha Gauri*, 11), forming the "foundation" (*Sva*, 11) for your present growth into a human being. Let the God reincarnate as a "unique deity" (*Isha*, 12), letting the departed entity be the God paving the path for HIM to be the "sentient entity" (*Siddha*, 7) within the God's cultural factor serviced by you. Finally, let the departed entity reincarnate as an "eternal deity" (*Maha Kali*, 13), norming the "maternal luminous" (*Maha Kali*, 13) and making you "self-luminous" (*Svarochisha*, 12) as an "entity" (*Hasti*, 24) enjoying his "institutional force" (*Trivikrama*, 24). Let the institutional force transform the "causation" (*Hetu*, 1) by norming yourself as the human entity and myself as the "universe" (*Brahman*, 2), making the human entity self-luminous.

Third, develop the "cultural factor" (*Sugriva*, 5) into a "finite coordinate" (*Kshetra*, 9) of space and time. Let the finite coordinate develop into the "genetic code" (*Somapa*, 180 = 9 * 20) guided by your "conscious spirit" (*Ruah*, 20). Let your conscious spirit be the "creator deity" (*Bhagwan*, 4) that develops the cultural factor into the finite coordinate. Let your conscious spirit without the creator deity be the "codon" (*Naadi*, 16 = 20 - 4) that forms the "essence" (*Sara*, 16) "exchanged" (*Somapa*, 180) as the "genetic code" (*Somapa*, 180). Let the codon be the "causation" (*Hetu*, 1) that transforms into the "creative force" (*Uma*, 6) to make your "reality" (*Vastavikta*, 7) self-luminous as a "potential deity" (*Maha Lakshmi*, 14) who "lives" (*Nivas*, 14) within the "present development" (*Maha Vibhu*, 14) of everyone's reality.

4.1 Forming Causation at the First Infinity Level: The Infinite Council and The Intellectual Body

An entity forms the causation for transformation at the first infinity level. The first infinity level is the "primordial self" (*Parvati*, 10). The primordial self promotes "organizational planning" (*Abadha*, 10) for "action" (*Karma*, 10) and to be "joyful" (*Hasa*, 10). Organizational planning is normative planning for the infinity of followers with the "leadership energy" (*Nayakatva shakti*, 9000) at the "infinity point" (*Amara*, 90,000) that makes the leader "immortal" (*Amara*, 90,000) as the followers reproduce the leader's "entity consciousness" (*Satma*, 10).

4.1.1 Role of the Earth Wheel

An entity's "overall consciousness" (*Bhagnatma*, 5) is guided by two factors:

- First, an "entity consciousness" (*Satma*, 10) is managed intrinsically by the intellectual body. Entity consciousness is "mass consciousness" (*Satma*, 10) of kinship linkages within the mental body. The intellectual body manages the entity consciousness with the "solar wheel" (*Surya chakra*, 8) through the pituitary gland in the brain.

- Second, an "ecosystem consciousness" (*Anahata*, 53) is managed extrinsically by the "Infinite Council" (*Ratnakosha*, 30). Ecosystem consciousness is "God consciousness" (*Anahata*, 53) that an entity develops in the form of the "human force" (*Manviya karak*, 53). Human force is the effect of "lunar time" (*Tribhajya*, 10^{10}) on human-like "wakefulness" (*Jagarya*, 1000) and consequently "entity consciousness" (*Satma*, 10 = $1000^{1/3}$) at present. Human force is the "para-consciousness potential" (*Vinayaka*, 53) that dynamically transforms the entity consciousness. The Infinite council is the Infinite Council of the living souls as the "followers" (*Trivikrama*, 24) and the departed souls as the "ascended masters" (*Bhrigu*, 805). It manages the ecosystem consciousness through the para-thyroid gland with the "lunar wheel" (*Karma chakra*, 7). "Lunar cycle" (*Karma chakra*, 7) makes the effect of lunar time on wakefulness zero. Thus, it proficiently compensates for the varying ecosystem-effect of each entity. It harmonizes the "entity time" (*Kaal*, 360) with the "solar cycle" (*Surya chakra*, 8).

The entity manages the overall consciousness through the "third eye wheel" (*Ajna chakra*, 65), i.e., the "earth wheel" (*Bhu chakra*, 65). The "earth cycle" (*Bhu chakra*, 65) proficiently compensates for the varying "entity-effect" (*Haisiyat*, 38) in managing the entity consciousness. Thus, the solar time fully conditions the "para conscious potential" (*Vinayaka*, 53).

4.1.2 Role of the Moon Wheel

The lunar wheel modifies the pineal gland's consumption of the rest hormone "melatonin" (*Kshemya shakti*, 189) by managing the production of the parathyroid hormone. It manages the variable intercellular exchange of energy by programming the rapid death of the quaternary cells that have already serviced their energy for the growth of other cells. Parathyroid hormone descends a cell's formative lifespan by making it hyperactive. The ascending parathyroid hormone signals the bones to diffuse the calcium reserves during the daytime. Diffused calcium descends the third eye movement and ascends the wakefulness through the rapid eye movement. Ascending wakefulness ascends the pressure on

- the "throat wheel" (*Visuddha chakra*, 34) to ascend the breathing rate for trading the super-wisher
- the "heart wheel" (*Anahata chakra*, 33) to ascend the heartbeat rate for servicing the wisher
- the "digestion wheel" (*Manipura chakra*, 24) to ascend the wishing for the fresh energy intake by eating food
- the "reproductive wheel" (*Svadhisthan chakra*, 10) to ascend the wish for the cellular waste liquid off-take via urinary tract
- the "excretory wheel," i.e., the "root cause wheel" (*Muladhara chakra*, 12) to ascend the wishable pressure on solidifying the cellular waste to be excreted via rectum.

4.1.3 Role of the Sun Wheel

The Solar wheel diffuses the energy of the quinary dying cells to the primary knowable inception cells, secondary knowledge conception cells, and tertiary knowing perception cells. The primary cells initiate the knowable work left over from the functions finished by the quaternary knower experience cells. The secondary cells conceive the work performed by the primary cells with their personal experience and social force of the institutional experience of all the cells. The tertiary cells perceive the knowledge of the secondary cells by trading the institutional force to fill the vacuum formed after servicing their energy as a primary cell and stocking their consciousness as a secondary cell. The quaternary cells experience the knowing of the tertiary cells as the institutional force they service to the tertiary cells. The quinary cells encode their death as a knower, knowing that their knowledge is knowable to the primary cells in the form of the cultural consciousness of the senary degenerating cells.

The primary cells are the "mother cells" (*Dhumavati*, 7). The secondary cells are the "father cells" (*Rodha*, 1). The tertiary cells are the "daughter cells" (*Kshatradharma*, 19). The quaternary cells are the "son cells" (*Manyu*, 19). The quinary cells are the "granddaughter cells" (*Bagalamukhi*, 9). The senary cells are the "grandson cells" (*Pavamana*, 9), diffusing their value to the regenerating septenary "grandmother cells" (*Dadi*, 18). The septenary cells regenerate within the octonary incubating "grandfather cells" (*Uccitika*, 18). After incubating, the octonary cells become the nonary dead "greeter cell" (*Nirjara*, 1000). The greeter cells are the constitutional cells that constitute the physical body by reincarnating as the denary "germinal cells" (*Chandramouli*, 476). The germinal cell is the "satan cell" (*Chandramouli*, 476) infused with the "present value" (*Kalpa*, 476) of the "geography" (*Ganarajya*, 476).

The Lunar wheel senses and perpetuates the entity's sentient well-being by managing the lifespan of "constitutional cells" (*Nirjara*, 1000). The ectoderm, the "inner layer" (*Sahaja puta*, 16) of germinal cells, separates the dermis, the basement-cementing membrane of the physical body, through the epidermis, the "outer layer" (*Kaya*, 16) of "stem cells" (*Brahli*, 89). Germinal cells produce stem cells through a sequentially ascending "avascular growth" (*Satarupa*, 8) with "intrinsic energy" (*Adi shakti*, 15) without blood supply. The dermis is the "meso layer" (*Dharma puta*, 16) of the "neural crest cells" (*Samghatavigrhitartha*, 900).

A neural crest cell embodies the "cultural reality" (*Samghatavigrhitartha*, 900) of the "infinite collective" (*Sarvalokacharine*, 15) of cells. Neural crest cells produce germinal cells through a sequentially descending "vascular growth" (*Maha Nitya*, 8) with "extrinsic energy" (*AUM shakti*, 18) within the blood supply. A stem cell embodies the "perceived consciousness" (*Vijnanatma*, 89) of the "diverse present" (*Brahli*, 89) at the moment of its genesis. Stem cells produce "melanocyte cells" (*Brahmani*, 80) that form the skin as the "mass layer" (*Nirman Puta*, 16). Melanocyte cells "color" (*Raga*, 250) the skin in "symmetry" (*Ashrama*, 250) with the "geography" (*Ganarajya*, 476) by producing a coloring protein—"melanin" (*Dhumra*, 17). Melanin produces a colored protein, "keratin" (*Anantariti*, 140). The keratin ends the lunar cycle by transforming the germinal cell into the stem cell.

The Earth wheel forms a new mother cell in the form of the "myelin cell" (*Dhumavati*, 7) that differentiate into "white blood cells" (*Jangamavisha*, 570). White blood cells service the energy for forming new germinal cells. The throat cycle norms a new daughter cell in the form of the "beta cell" (*Kshatradharma*, 19) that produces a new father cell in the form of the "glial cell" (*Rodha*, 1). The heart cycle norms a new granddaughter cell in the form of the "oligodendrocyte cell" (*Bagalamukhi*, 9) that produces a new son cell in the form of the "alpha cell" (*Manyu*, 19) by trading ten father cells. The digestion cycle norms a new grandmother cell in the form of the "Oriens lacunosum-moleculare cell [OLM cell]" (*Dadi*, 18) that produces a new grandson cell in the form of the "bistratified ganglion cell" (*Pavamana*, 9) by dividing itself and recombining each division with nine father cells.

The reproduction cycle norms a new greeter cell in the form of the "basket cell" (*Nirjara*, 1000) that produces a new grandfather cell in the form of the "pyramidal cell" (*Uccitika*, 18). Eighteen basket cells divide into one-thousand pyramidal cells. The root cause cycle norms a new "devil cell" (*Chaithanya*, 4) in the form of the "memory T-cell" (*Chaithanya*, 4). The memory T-cell produces a new "satan cell" (*Chandramouli*, 476 = 120 * 4 - 8) by trading four "primordial cells [Betz cell]" (*Shivagati*, 120) in the form of the "Martinotti cells" (*Shivagati*, 120) and servicing eight "T-cell receptors" (*Pumsavana*, 196) as the "lifecycle-effect" (*Pumsavana*, 196) of the nine wheels.

4.1.4 Role of the Wheel of Wheels

The Ninth wheel is the "wheel of wheels" (*Hora chakra*, 25) that produces a new "cell" (*Hiranyagarbha*, 19) in the form of the six "Purkinje cells" (*Hiranyagarbha*, 19). Six Purkinje cells and six "Golgi cells" (*Ekarupa*, 1) norm a "param cell" (*Ashwini Kumaras*, 120 = 19 * 6 + 1 * 6) in the form of a horizontal "Cajal-Retzius cell" (*Ashwini Kumaras*, 120). Six Golgi cells form twelve "granular cells" (*Chitra*, 1) as their "double copies" (*Chitra*, 1). Each granular cell behaves like a "master cell" (*Chitra*, 1), enjoying mastery of the cell production cycle. Eight T-cell receptors produce thirty-two hundred "Golgi cells" (*Ekarupa*, 1) by consuming an "energy fragment" (*Yogihridaya*, 64 = 3200 * 1 - 8 * 196).

The "reproduction wheel" (*Akasha chakra*, 64) becomes an energy fragment after it conceives the root cause wheel for reproducing the wheel of wheels. The wheel of wheels reproduces itself sequentially as the solar wheel, the lunar wheel, the earth wheel, the throat wheel, the heart wheel, the digestion wheel, and the reproductive wheel. An "electron quadruplate" (*Palala*, 64) becomes a reproduction wheel when three "photons" (*Kalpanik*, 20) recombine with one "krypton" (*Soham*, 4) to form the electron quadruplate as a "fermion quadrupling condensate" (*Palala*, 64) and break that into four "Golgi bodies [white-blood apparatus]" (*Madhusudan*, 16).

Over its lifetime, each Golgi body develops a mastery for managing the compensating effects of the lunar and the earth cycles within the gravitational-effect of the solar cycle. The effect of both lunar and earth cycles ascends for the first half, ascending lunar phase due to the weakening solar-effect. Without the master cells, the ascending lunar phase generates an above-par consumption of the cellular energy than normatively programmed by the solar wheel. The effect of the lunar wheel descends for the second half, descending lunar phase within the still ascending earth cycle. With the master cells produced during the first fifteen days, the descending lunar phase generates a below-par cellular energy consumption.

The effect of the earth wheel continues to ascend for the first week of the solar cycle while the master cells moderate the effect of the ascending lunar cycle. The effect of the earth wheel descends over the second week of the solar cycle while the master cells mediate the growth of seven "mossy cells" (*Dishta*, 6) to form the "zeroth chromosome" (*Radha*, 43 = 7 * 6 + 1). Constant ecosystem-effect norms the lifespan of each cell's "workforce-effect" (*Bharajaka pitta*, 44) as a progeny "agent cell" (*Asura shakti*, -1) working on transforming from a germ cell to a stem cell at 44 days [15 + 15 + 7 + 7]. Constant entity-effect norms the lifespan of each cell's "networking-effect" (*Damodara*, 26) traded from the geography as a germ cell at 30 days.

Constant "exchange-effect" (*Vyanavata*, 365) norms the lifespan of each cell as a "somatic cell" (*Vyanavata*, 365) at 21 days. Somatic cell completes the production cycle without the reproduction cycle after it has reproduced itself as the "progeny cell" (*Asura shakti*, -1) without the germ cell. The 21 days comprise the accelerating lunar-effect during the first seven days of the solar cycle, followed by decelerating lunar-effect within still accelerating earth-effect over the next seven days. Finally, it leads to descending lunar-effect over the following seven days compensated by the decelerating though ascending earth-effect.

4.1.5 Symmetry of the Intrinsic Wheels with the Extrinsic Cycles

The solar, lunar, and earth wheels within the animate physical body are in symmetry with the solar, lunar, and earth cycles without the animate physical body. The physical body exchanges the varying "lunar energy" (*Amrit*, 1000) due to the earth's 24-hour rotation cycle around its axis. The solar chakra programs the varying present-effect of the 23-hours of varying lunar-effects in the form of the 23 chromosomes within a "gamete" (*Siddhi shakti*, 187). The gamete organizes 24-hours of the present earth-effect within 180-days of the past lunar-effect and 7-days of the future solar-effect. Networking-effect trades the present earth-effect. Exchange-effect services the past lunar-effect. Workforce-effect generates the future solar-effect of the seven additional days that lead to the horizontal stabilization of the earth-effect on the 29th day and stillness on the 30th day before beginning the descent over the next thirty days.

The 28-day lunar cycle within the ascending earth-effect comprises the eight-phase moon rotation around the Earth: New moon, Waxing Crescent, First Quarter, Waxing Gibbous, Full moon, Waning Gibbous, Third Quarter, and Waning Crescent. These eight phases divide the sidereal day into eight "shakes" (*Prahar*, 100) of three-hours each. The first shake, the new moon phase, begins at 3 am, followed by the other seven shakes. Each shake starts with an ascending phase of 90 minutes, norming an ascending death of the master cells and ascending birth of the granular cells. It continues with a descending phase of 90 minutes, norming an ascending growth of the master cells and a descending birth of the granular cells.

During the ascending phase, the lunar wheel is hyperactive in trading ascending extrinsic energy. It becomes hypoactive during the descending phase. The earth wheel compensates by trading ascending intrinsic energy. Consequently, an entity's normal sleep cycle is 90 minutes. As the lunar wheel becomes hyperactive, the entity ascends into a REM [rapid eye movement] dream state for the first 25% sleep cycle. The solar wheel corrects for the hyperactivity by descending the intrinsic energy production. Therefore, the entity descends into a non-REM rest state for the other 75% time. As the lunar wheel becomes hypoactive due to descending intrinsic energy, the entity ascends again into a REM [rapid eye movement] dream state for the first 25% of the sleep cycle over the descending phase.

The pituitary gland has an overall strategic awareness of the global energy reserves throughout the physical body. During the ascending extrinsic lunar phase, the physical body trades ascending "atmospheric energy" (*Brahli*, 89). However, a third of the body cells formed during the prior descending extrinsic lunar phase are unaware of the cyclical reality. Therefore, the solar chakra does not descend intrinsic energy production. It leads to a hyperactivity of the earth wheel.

Similarly, during the descending extrinsic lunar phase, a third of the body cells formed during the prior ascending phase do not service the intrinsic energy necessary for the body's normative performing. Further, a third of the body cells formed during the prior descending phase do not correct their expectations of the normative profiting from the above-average extrinsic energy. Finally, the final third formed during the previous ascending phase has already decided on the normative development of their death to sustain the physical body with the formative investment of their energy into producing new cells. They decide to take advantage of the unexpected reversal from the descending trading of extrinsic energy through normative programming of their supernatural experience so that the new granular cells also enjoy the "shared becoming" (*Sadagati*, 15) with "breeding" (*Janana*, 15). Therefore, the earth wheel becomes hypoactive. With the entropy of local energy reserves, the solar wheel eventually becomes conscious of the urgency of descending the activity of the lunar wheel. This brings the entity back into the non-REM rest state.

With each sequential phase, the mastery of the solar wheel for managing the stability of local energy reserves ascends. In the last Waning Crescent phase of the lunar cycle (*midnight to 3 am*), the solar wheel is hypersensitive in deactivating the lunar wheel during the ascending extrinsic lunar phase. Inertia leads the entity into a deep sleep state. Similarly, the solar wheel is hypersensitive in activating the lunar wheel during the descending extrinsic lunar phase. After inertia, it takes time for the lunar wheel to become active. Therefore, the entity enjoys a disproportionate non-REM rest state.

Conversely, in the first phase (*3 am to 6 am*), the solar wheel is hyposensitive in deactivating the lunar wheel during the ascending extrinsic lunar phase, conditioned by the social experience of the lunar wheel's inertia. Consequently, the throat wheel gives voice to each wheel for working to maintain ascending local energy reserves in the form of fatty acid cells. Concurrently, the entity enjoys a disproportionate REM dream state. Further, during the descending extrinsic lunar phase, the solar wheel is hyposensitive in activating the lunar wheel as it is conscious of the presence of local energy reserves guided by its personal experience. Hence, the heart wheel takes the solar wheel's super-sensitivity about the conscious well-being of the wheels that follow the lunar wheel as the solar wheel's togetherness with each wheel.

Consequently, each wheel develops a "thermodynamic equilibrium" (*Yama*, 180) with the solar wheel. That produces an illusion that each wheel is trading one another's "design energy" (*AIM shakti*, 180). It gives birth to the <u>illusionary zeroth law of thermodynamics</u> that if the two systems are in thermodynamic equilibrium with a third system, they will be in thermodynamic equilibrium with each other as well.

4.2 Norming Causation at the Second Infinity Level: The Divine Council and The Mental Body

An entity norms the causation for transformation at the second infinity level. The second infinity level is the "sentient nature" (*Svabhav*, 8) for "reproducing" (*Maharavana*, 8) the "sensory illusion" (*Vichara*, 8) as "true" (*Sat*, 8). Once traded as the "technological input" (*SHREEM*, 8), the sensory illusion becomes the "falsifying code" (*Sadasat*, 8) that makes reality "false" (*Asat*, 8) and immanent within the "truth" (*Sathya*, 375).

4.2.1 Role of the Time Wheel

The "time wheel" (*Kaal chakra*, 3800) services the technological input as the "omnipresent cause" (*Maha Nitya*, 8) to the "solar wheel" (*Surya chakra*, 8). The time wheel is the 25-hour Circadian time cycle. It limits the bone decay from the hyperactive lunar wheel by instructing the "parathyroid gland" (*Chitti*, 34) for "advancing" (*Pragami*, 80) the "waking consciousness" (*Yantratma*, 80) of truth and the "sleeping consciousness" (*Bodhatma*, -10^{12}) of reality by trading "serotonin" (*Hasta*, 0), a monoamine neurotransmitter. Serotonin is the "invisible greeter" (*Hasta*, 0) greeting "don't worry, be happy" (*Avarata*, -10^{12}) impulse by making one conscious of one's "inferiority" (*Avarata*, -10^{12}) in "discerning" (*Sanmati*, 753) the reality of the truth. It transmits the "inner sense" (*Ajjhattikani*, 80) of the truth by forming a "direct neurological instruction" (*Ajjhattikani*, 80) for transforming the normative performing of the parathyroid gland. The lunar wheel—the body's "dreaming artist" (*Kalakaar*, 7)—mediates neurological instruction.

The solar wheel's "thermodynamic force" (*Rodha*, 1) awakens each cell from the state of intrinsic inertial dependence on the local knowledge of its local planning wheel within the deciding influence of the earth wheel. It makes each cell self-conscious without the predominating energy-conserving potential of the lunar wheel. The lunar wheel seeks to conserve energy by catalyzing the body organs to rest when the atmospheric energy is ascending. It lets them be hyperactive when the atmospheric energy is descending. It gives birth to the <u>illusionary first law of thermodynamics</u> that if the two systems are the complementary parts of a third system, their overall energy as the third system remains constant.

Awakening transforms a cell into a self-managing unit and a body organ into a universe managing diverse self-managing units. Each body organ organizes itself into different cells performing differentiated roles programmed by its planning wheel. It becomes a responsible strategic manager of the functions of the follower cell units. Each wheel follows the development of the solar wheel as the leader and the lunar wheel as the entrepreneur and moderates the performing of the earth wheel. Each cell acts as a performing unit of the organ's workforce system. Each wheel acts as a planning unit of the organ's networking system. Each organ acts as a programming unit of its exchange system with other organs. Each organ exchanges profiting with other organs to let other organs rest while it is hyperactive, and rest when they are hyperactive.

4.2.2 Role of the White Matter

Mastery of each wheel, organ, and cell descends their social dependence on the earth chakra's governance of the "path of action" (*Karma marga*, 86). Mastery makes a wheel, organ, and cell conscious of its sentient wellness. "Conscious cellular system" (*Triyancha*, 19) is "energy" (*Shakti*, 19) the "white matter" (*Hradamana*, 381) activates by inactivating the grey matter. The white matter is the "universe of body organ systems" (*Hradamana*, 381) that forms the "mental body" (*Manomaya sharira*, 381). Symmetry in each cell's "local consciousness" (*Prakashatma*, 19) and the mental body's "global consciousness" (*Sarvadevatma*, 381 = 19^2) descends the entropy of the "intellectual body" (*Sukshma sharira*, 306). The intellectual body is the "harmonic body" (*Sukshma*, 306) that harmonizes the cell's local consciousness with the body's global consciousness with each wheel's "national consciousness" (*Vedatma*, 1000^{1024}) of the cellular citizens it organizes into the diverse body organs within its conscious realm. National consciousness forms the "energy system" (*Antarmukha*, 1000^{1024}) and norms the "grey matter" (*Avanayaka*, 1000^{1024}) for "balancing" (*Rojamela*, 200,000) each wheel's "administration" (*Avanayaka*, 1000^{1024}) of the energy system. By balancing, the grey matter furthers the universal conscious well-being of the physical body. By inactivating the grey matter, the white matter lets each cell globalize its local consciousness to generate a "feeling" (*Sprishti*, 197) sensation of joyfulness within the mental body.

The balancing compensates for the variable local consciousness within infinite cells. It catalyzes the "astral body" (*Linga sharira*, 3) to service its "corporate consciousness" (*Mahatma*, 18) of each cell's "potential energy" (*AUM shakti*, 18) for managing the "impediments" (*Badha*, 1) to globalizing the local consciousness. The astral body is the "conscious body" (*Linga sharira*, 3) conscious of the potential of each cell to globalize the "ideal force" (*Dasha*, 1) of the mental body's "feelings" (*Sprishti*, $197 = 73 * 2 + 52 - 1$) in the form of two "emotions" (*Hunduka*, 73) within the "shadow" (*Matali*, 52) of the third that transforms the ideal force into the "ego" (*Asura*, -1). The two emotions include a position emotion of "joy" (*Piti*, 123) traded from the mental body and negative emotion of "restlessness" (*Auddhyata*, 120) that transforms the balancing "joy-force" (*Akhkhala*, $3 = 132 - 120$) into the astral body.

Joy force is "exclamation" (*Akhkhala*, 3) of joy that fills the "empty oneness" (*Nirbijayoga*, 120) of the "energy equilibrium" (*Samtol*, 120) without "consciousness" (*Chaithanya*, 4) of the "causation" (*Hetu*, 1). The astral body idealizes each cell to be a "carbon copy" (*Chitra*, 1) of the "granular cell" (*Chitra*, 1). "Balancing" (*Rojamela*, 200,000) generates two-hundred units of "sentient energy" (*Amrit*, 1000) for producing two units of "gravitational energy" (*Lalita*, 100) within the "light" (*Prabha*, 180) of the third that transforms each "carbon copy" (*Chitra*, 1) into "joy" (*Piti*, 120). Joy is the emotion of knowing that a "carbon copy" (*Chitra*, 1) is a "double copy" (*Chitra*, 1) of a third "copy" (*Nakala*, 0) which is a "triple copy" (*Vaishya*, 3) of the "joy force" (*Akhkhala*, 3) joy produces for naturally generating a "six-fold growth" (*Dishta*, 6). The six-fold growth is the "creative force" (*Uma*, 6) that transforms each carbon copy into a "conscious entity" (*Siddha*, $7 = 1 + 6$).

Restlessness is the emotion of not knowing that the "conscious entity" (*Siddha*, 7) conceives the "carbon copy" (*Chitra*, 1) by activating the "path of devotion" (*Bhakti marga*, 1) for enjoying the "natural essence" (*Samaa*, 1) within the self. The path of devotion transforms the conscious entity into the "solar wheel" (*Surya chakra*, 8). Without the path of devotion, the conscious entity is the "lunar wheel" (*Karma chakra*, 7), following the action cycle that eventually leads to the path of devotion. After sensing the escalating costs of "copying" (*Chitra*, 1), the conscious entity becomes devoted to oneself. Copying descends "intrinsic sentient energy" (*Alaya vijnana*, 108) and ascends "restlessness" (*Auddhyata*, 120) by destroying the oneness of two child copies within "I" (*Svayam*, 12) of the third, masculine copy. The "triple copy" (*Vaishya*, 3) forms the astral body as an "emotional body" (*Linga sharira*, 3) within the "paternal consciousness" (*Pitra*, 4) of the sentient entity.

The emotional body stabilizes the "emotional energy" (*Hunduka shakti*, $73 = 19 * 4 - 3$). It ascends the "local consciousness" (*Prakashatma*, 19) within the paternal copy and its triple copy of the triple copy within the paternal copy. After the "stabilization" (*Sthirikaran*, 9) of the two copies within one that forms the "energy" (*Shakti*, $19 = 2 * 9 + 1$) into a "cell" (*Hiranyagarbha*, 19), the emotional body transforms into the "ego body" (*Linga sharira*, 3). The ego body generates a feeling "sensation" (*Vedana*, 197) of the "ego" (*Aham*, -1) within the "paternal copy" (*Nakala*, 0) for not being a "part" (*Anga*, -1) of the cell that copies the energy but not the "causation" (*Hetu*, 0) that made it a "zero" (*Shunya*, 0). A "carbon copy" (*Chitra*, 0) becomes a zero "paternal copy" (*Nakala*, 0) after it forms a "double copy" (*Chitra*, 0) and norms their individual and collective oneness within a "triple copy" (*Vaishya*, 3). After servicing the "oneness" (*Yoga*, 48) with the "extrinsic energy" (*AUM shakti*, 18) as an "illusionary energy" (*Maya shakti*, 1) for forming a unit of "energy" (*Shakti*, $19 = 18 + 1$) for energizing its "potential" (*AUM*, 18) for "growth" (*Dishta*, 6) of the "triple copy" (*Vaishya*, 3) over 24 hours, the paternal copy trades the rest-inducing "ego energy" (*Asura shakti*, -1) for balancing the action-inducing illusionary energy over the next hour.

4.2.3 Role of the Serotonin-Melatonin Production Cycle

The "serotonin-melatonin production cycle" (*Akasha chakra*, 64) activates the parathyroid gland's "followership" (*Apacayana*, 10^{19})

for exhibiting "tryptophan" (*Jalini Mukha*, 10^{19}) and producing "serotonin" (*Hasta*, 0) as a "copy" (*Nakala*, 0). Serotonin activates the pituitary gland's "potential" (*AUM*, 18) as a "leader" (*Neta*, 0) within it. It forms eighteen units of "illusionary energy" (*Maya shakti*, 1) as the "hydroxylase" (*Maya shakti*, 1), "self-perpetuating" (*Udvaha*, ½) as the "tryptophan hydroxylase" (*Para shakti*, 9). Concurrently, the parathyroid gland activates the pineal gland to produce the "melatonin" (*Kshemya shakti*, 189). The "melatonin" (*Kshemya shakti*, $189 = 97 * 2 - 5$) activates two units of "rest" (*Vishram*, 97) that prepares each body organ for ascending, sustained action 24x7 throughout the lifetime. One rest unit forms a sequence of the "short cycles" (*Kaal chakra*, 3800) of rest over the lifetime.

The short cycles are "Circadian cycles" (*Kaal chakra*, 3800) of time emanating from the second unit of rest that forms the first unit as its "carbon copy" (*Chitra*, 1). The second unit of rest is the "long cycle" (*Dvandva Brahma*, 125) of action that forms the "lifetime" (*Svahita*, 125). The long cycle is not a "cycle" (*Avartan*, 1/60). Still, it cycles the "convergent ideal-effect" (*Dvandva Brahma*, 125) of both the "triple copy" (*Vaishya*, 3) and the "double copy" (*Chitra*, 1) within the "copy" (*Nakala*, 0). It naturally reproduces the "six-fold growth" (*Dishta*, 6) by forming an opposing pair of "five-dimensional copies" (*Nakala*, 0) within each "copy" (*Nakala*, 0). Five dimensions comprise four dimensions of the "generation" (*Ashrama*, 250) of one as a grandson, son, father, and grandfather and one dimension of the "gender" (*Linga*, 9) of one as a masculine. The opposing pair comprises four dimensions of the "regeneration" (*Shakti*, 19) of one as a granddaughter, daughter, mother, and grandmother and one dimension of the "potential gender" (*Napumsaka*, 9) of one as a feminine.

One performs the "divine planning" (*Abadha*, 10) of one's gender as a "carbon copy" (*Chitra*, 1) of Mother Nature's potential gender by becoming the "Divine Council" (*Sridhara*, 25) of three divinities: the perpetuator, the liberator, and the worker. Five-dimensional "paternal copy" (*Nakala*, 0) becomes the "perpetuator" (*Abhidheya*, $460 == 4,6,0$), conceiving four copies for enjoying six-fold growth by behaving like a "fifth copy" (*Nakala*, 0). Two-dimensional "maternal copy" (*Chitra*, 1) becomes the "liberator" (*Mokshayitri*, -13), perceiving one copy within its "discordant energy" (*Asura shakti*, -1) as a triple copy with the "time multiplier" (*Upaya*, 3). Three-dimensional "masculine copy" (*Vaishya*, 3) becomes the "worker" (*Karmi*, 1), experiencing itself as the maternal "carbon copy" (*Chitra*, 1) formed with the "maternal essence" (*Samaa*, 1) of Mother Nature as a "twin" (*Hridayesha*, 121).

For descending its entropy, the astral body masters the science of divine law for planning a first set of its five copies. The first set of five copies transform into the physical body, the intellectual body, the mental body, the astral body, and the etheric body. Next, the astral body transforms into the "causal body" (*Karana sharira*, 30) for reproducing two sets of five copies within its double copy. Double copy forms the intellectual body that transforms into the mental body by forming another copy as the intellectual body. The second set of five copies also transforms into the physical body, the intellectual body, the mental body, the astral body, and the etheric body. They liberate the first set of five copies to form the five wheels. The first intellectual body forms the solar wheel. The first mental body forms the lunar wheel. The first astral body forms the earth wheel. The first etheric body forms the respiratory air wheel for perpetuating its "infinite value" (*Viyojya*, 957) by breathing extrinsic energy. The first causal body forms the circulatory heart wheel for liberating the double copy to form the next two wheels.

The double copy is a part of the third set of five copies whose triple copy forms the "divine council" (*Sridhara*, 25) as the five sets of five copies, each of which is a double copy. The third intellectual body forms the digestive naval wheel. The third mental body forms the reproduction wheel. The fourth intellectual body substitutes the third intellectual body and forms the root cause wheel. The first physical body forms the "creature" (*Purusha*, 12) as a "self-luminous entity" (*Svayam*, 12). The fifth set of five copies, the fourth set of other four copies, and the third set of other three copies are immanent within the self-luminous entity. The second intellectual body norms the brain. The second mental body norms the mind. The second astral body norms the mood. The second etheric body norms the modifier modifying the mood by making mind modifiable and taking brain as the modified. The second physical body norms a "five-faced self-luminous entity" (*Panchajani,* 12) as the "manifestor" (*Sadhana,* 368) of the entity after the "creature" (*Purusha,* 12) as the "sixth self-luminous entity" (*Purusha,* 12) becomes the "modifier" (*Charanam,* 49) modifying his "mood" (*Bhava,* 360) by "breeding" (*Janana,* 15) the mood-modifying "time" (*Kaal,* 360) as an "entity" (*Hasti,* 24 = 360/ 15).

4.2.4 Role of the Divine Council

The entity is the "institutional force" (*Trivikrama,* 24) that over time institutionalizes the organizational programming of the divine council within each cell. The cellular organizational program descends the cost-escalating programming of rest by the earth wheel. It lets each cell make twenty-five copies to perpetuate its performing beyond the six-fold growth that transforms it from a carbon copy to a sentient entity. As a sentient entity, the cell enjoys profiting from the organizational performing of the divine council without itself. Thus, Divine Council is "Param Liberator" (*Sridhara,* 25). A cell masters the "science" (*Vijnana,* 47) of "servicing" (*Apana,* 47) the energy it trades from the ecosystem in 24 hours. It becomes conscious of its "trading" (*Samana,* 53 = 47 + 6) in one hour after its formation as a "granular cell" (*Rodha,* 1) once it becomes a "myelin cell" (*Dhumavati,* 7). It masters the science of trading the energy to compensate for the escalating costs of servicing its energy in another 24 hours. Therefore, it takes 25 hours for the cell to master the science of "exchange" (*Mahaspanda,* 269 = 69 + 100 * 2) and develop "twin consciousness" (*Shachi,* 69) of the "reproductive" (*Guru,* 100) element within both servicing and trading.

Each unit of the 25-unit Divine Council norms 25 double copies, with a copy involved in the servicing sequence through 24 copies that have an ascending consciousness of the exchange and a copy involved in the trading sequence through 24 copies that have a descending consciousness of the exchange. After developing a "conscious consciousness" (*Prashantatma,* 15) of the exchange in sixteen hours, a cell takes conscious actions to diffuse its knowing among the sixteen sets of "triploid" (*Suchi,* 16) formed over those sixteen hours. It is not conscious that the triploid is a triple copy of "five copies" (*Shanideva*, 100) and it is a part of the "diploid" (*Tantri*, 48) forming two additional ploidies, one masculine and another feminine. The "diploid" (*Tantri*, 48) is the double copy of five copies. It takes another nine hours for the cell to develop a "twin conscious consciousness" (*Anyajatma*, 25) of the five copies of the five copies. Five copies form "pentaploid" (*Rama*, 100) that reproduce another pentaploid.

After 25 hours, a cell is not yet conscious that it is a part of the sixth copy of the six copies. Six copies form "hexaploid" (*Raman*, 20). Hexaploid comprises the six copies of the "triploid" (*Suchi*, 12) within the two copies of a ploid. Two copies are the grandfather and the grandmother ploids. Six copies comprise three diploids. Three diploids comprise father and mother ploids, son and daughter ploids, and grandson and granddaughter ploids. It takes 26 hours for the cell to master the science of "investment" (*Prana*, 123) and develop a "triple conscious consciousness" (*Vidupatma*, 26) of the science of servicing, trading, as well as exchange for ascending the return on investment through the "greeter capability" (*Nipaka*, 10). The greeter capability is the "heptaploid" (*Nipaka*, 10) that makes seven copies of itself after conceiving seven copies of triploid within itself.

The heptaploid comprises a triploid within a "septaploid" (*Nipaka*, 10). Septaploid comprises seven "double copies" (*Chitra*, 1) of the seven "double copies" that constitute a "sentient entity" (*Siddha*, 7). Seven double copies are immanent within the other seven double copies; they form a triploid in the physical realm and a "tetraploid" in the metaphysical realm of the consciousness of the four generations of the triploid. The triploid comprises a satan and a devil within a greeter's dynamic realm of the "greeter consciousness" (*Pratyagatma*, 1). The cell takes 27 hours to master the science of "capability" and develop the "greeter consciousness" (*Pratyagatma*, 1) of the science of servicing, trading, exchange, and investment for ascending the capability through the six-fold growth. The six-fold growth sustains the "formative capability" (*Vidhana*, 2) of the seventh double copy by investing a sequence of six double copies to exchange the diverse forms of ploidies to be reproduced further.

The cell takes 28 hours to master the science of "growth" (*Dishta*, 6) and develop the "paternal consciousness" (*Pitra*, 4) of the science of servicing, trading, exchange, investment, and capability for descending the "entropy" (*Mahodbhava*, 5) following the growth. Science of growth involves transforming the six-fold growth into eight-fold growth by servicing the consciousness of the four-fold growth to each copy and trading that consciousness from each copy. The four-fold growth norms the aging of a copy from birth to youth to maturity to death.

At birth, each copy becomes receptive to trading as a follower. The youthful double copy is motivated to servicing for becoming a leader. The matured triple copy wishes exchange for becoming an entrepreneur, freely enjoying the fruits of the public good through the private interest in children. The dead tetra copy is open to investing its energy to advance each copy's capability to perpetuate the eight-fold growth by conceiving a double copy within itself. Thus, a dead "tetra copy" (*Chakrika*, 90) organizes a 24-fold growth of its energy after experiencing a unit of "growth" (*Dishta*, 6) within the Divine Council's "divine planning" (*Abadha*, 10) as a "heptaploid" (*Nipaka*, 10). That empowers it to incarnate as an "animate entity" (*Sushupti*, $96 = 90 + 6$) enjoying "being energy" (*Kali shakti*, $96 = 4 * 24$).

4.2.5 Role of the Serotonin

Before becoming conscious of the Divine Council's divine planning, a copy conceives the hormone "serotonin" (*Hasta*, 0) to compensate for the "consciousness deficit" (*Andhakara*, 10^{19}) within the mental body. The serotonin activates the greeter consciousness of the double copy within the copy, thereby transforming the copy into the triple copy. As a "global universe of entities" (*Linga sharira*, 3), the triple copy experiences the "five copy" (*Shanideva*, 100) "reproductive-effect" (*Chitta*, 100) of the earth wheel as the local knower reproducing its knowing about the need for resting for the personal conscious well-being of each copy. Five copies comprise four generations of the copies of the devil copy and a gender-differentiating "satan copy" (*Shanideva*, 100) with the consciousness of the four generations. The satan copy is pentaploid that reproduces itself as the five copies. The devil copy is the "tetraploid" (*Sthavaravisha*, 57) that produces the "satan ploid" (*Rama*, 100) by consuming itself as a "ploid" (*Shanideva*, 100). The "formative growth" (*Vidhana*, 2) of a pair of two copies as a pair of "double copy" (*Chitra*, 1) that forms the tetraploid lets the "triple copy" (*Vaishya*, 3) transform into an "octoploid" (*Malini*, 79 = 3 + 19 * 4) with the four-fold growth in "energy" (*Shakti*, 19).

An octoploid is an "octave of gametes" (*Malini*, 79) formed with the "spiritual consciousness" (*Vikaranadharmitva*, 79) of the "serotonin" (*Hasta*, 0) as the "greeter spirit" (*Hasta*, 0) greeting a restful "holiday junction" (*Arthatatma*, 15) to each copy for breeding the "fifteen-fold growth" (*Janana*, 15) within and without the octoploid. The heptaploid within the octoploid forms the "seven-fold growth" (*Bahutava*, 7) and the octoploid forms the "eight-fold growth" (*Satarupa*, 8) that reproduces the heptaploid and the seven-fold. Thus, an octoploid generates a sequence of 22-fold growth for illuminating the heptaploid and destroying itself after breeding an additional fifteen-fold growth. The octoploid forms as the "twin copy" (*Artta*, 22/7), comprising a pair of copies. The twin copy reproduces itself to norm a "twin pair" (*Chaarpai*, 1000/72) of copies that transform into a pair of twin copies. Holiday junction is the "energy consciousness" (*Arthatatma*, 15) of the "four double-stranded arm assembly" (*Arthatma*, 15) that a copy conceives for incubating the octave of gametes within itself.

The gamete is the set of twenty-three chromosomes. The octoploid becomes the seventh chromosome that conceives the first six as growth within itself as a heptaploid, the next seven as the heptaploid's "illuminating value" (*Prakasha*, 7), and the following eight as its "natural value" (*Nirvikalpana*, 8) as an octoploid, and the final two as the "formative growth" (*Vidhana*, 2) of the "twin pair" (*Chaarpai*, 1000/72) as a diploid. The seventh chromosome becomes the gamete that conceives another set of twenty-three chromosomes as the "revolving value" (*Piyati*, 1000/72) of the twin pair's "whole pi" (*Chaarpai*, 1000/72). A pair of twenty-three chromosomes are immanent within a twin gamete comprising a pair of gametes. The "twin gamete" (*Konastha*, 13) manifests the pair of twenty-three chromosomes in the form of a "zygote" (*Veerya*, 48 = 2 * [1 + 23]). Next, the twin gamete transforms into a "diploid cell" (*Maha Kali*, 13) for conceiving a "double copy" (*Chitra*, 1) as a "diploid" (*Tantri*, 48) within it. Thus, the twin gamete becomes the "reproducer" (*Rochisha*, 13) of the "temporary growth" (*Maha Kali*, 13) of the zygote in the form of a "zygote sequence" (*Hasta*, 0).

The serotonin is the "zygote sequence" (*Hasta*, 0), mediating the brain's hyperactivity with its "reproductive-effect" (*Chitta*, 100). As a twin gamete, the diploid cell forms two gametes for norming a zygote and transforms into a diploid that forms another zygote to norm an "animate entity" (*Sushupti*, 96 = 48 + 48) as a "twin zygote" (*Sushupti*, 96). A twin zygote is the "dizygote" (*Sushupti*, 96) that forms the "dizygotic twin" (*Kali*, 96) after reproducing its "twin" (*Hridayesha*, 121 = 96 + 25) within the "Divine Council" (*Sridhara*, 25). A dizyoge is the "monozygotic twin" (*Sushupti*, 96) that transforms into two zygotes, forming an identical, "homozygous twin" (*Lekhaka*, 96). The monozygotic twin as a collective of two zygotes, and the homozygotic twin as two individual zygotes, norm sixteen "alleles" (*Devendra*, 12 = [96 + 96]/16). The collective vs. individual class-differentiating dizygotic twin norms an additional eight alleles. The twenty-four alleles form as a "homozygous" (*KLIM*, 24) element. The homozygous element is the "granddaughter allele" (*KLIM*, 24) formed as a sister-sister allele pair BB. The granddaughter allele is an "identical allele" (*KLIM*, 24) that multiplies an "allele" (*Deveshi*, 12) formed by a single copy within a "double copy" (*Chitra*, 1) with a "triple copy" (*Vaishya*, 3) to form a "homozygote" (*Kanalakshamsha*, 39 = [12+1] * 3).

4.2.6 Role of the Homozygote

"Homozygote" (*Kanalakshamsha*, 39 = 12 * 3 + 3) produces twelve "double copies" (*Chitra*, 1) of the triple copy within a "triple copy" (*Vaishya*, 3). The "collective class" (*Augrya*, 24) is the "sister allele" (*Augrya*, 24) that attracts another sister allele with her "collective force" (*Devendra*, 12). The collective force is the allele. The two individual collective forces form a "heterozygous" (*Ghatika*, 24) element. The heterozygous element is the "greeter allele" (*Ghatika*,

24) formed as a twin sister entity that transforms into a pair of sisters within an entity.

The entity is the "grandmother allele" (*Hasti*, 24) formed as an "individual class" (*Hasti*, 24) and transformed into a mother-mother allele pair AA within the collective forces of both sisters as the twin sisters. The "mother allele" (*Augrya*, 24) is the sister allele formed with the entity's "individual force" (*Shakti*, 19) within the one individual and four collective copies of alleles. The "individual copy" (*Nakala*, 0) is the "daughter allele" (*Trivikrama*, 24). The "collective copies" (*Vaishya*, 3) are the greeter allele, the grandmother allele, the mother allele, and the granddaughter allele. The collective copies are the "triple copy" (*Vaishya*, 3) of the greeter allele. The greeter allele is a collective copy of the triple copies. The daughter allele is the father-father allele pair aa.

A daughter allele is conceived as an individual copy when a "father allele" (*Manav*, 24) repels himself to attract a sister-sister allele pair BB following another "father allele" (*Manav*, 24). The second father allele co-synchronously repels himself to attract a "mother allele" (*Augrya*, 24) and the "daughter allele" (*Trivikrama*, 24) within her for conceiving the first father allele in the form of the "son allele" (*Shaneeswar*, 24). The repulsion of two father alleles is "synchronous" (*Samanakala*, 1810) because the son allele is within the father allele's consciousness as his "individual copy" (*Nakala*, 0). The attraction of two mother alleles is "asynchronous" (*Asamakala*, 810) with the father allele's "sentient energy" (*Amrit*, 1000 = 1810 - 810) because the daughter allele is within the mother allele's consciousness as her "collective copy" (*Vaishya*, 3). The "mother-father allele pair" (*Manipura chakra*, 24), Aa, form a "grandson allele" (*Manipura chakra*, 24) with their "collective energy" (*Maha Pitha*, 90 = 24 * 3 + 3 * 6).

The collective energy comprises 24-units each of the mother allele, the father allele, and the grandson allele, and 3-units each of the six triple copies like them. The six similar triple copies include the grandson allele, the son allele, the "grandfather allele" (*Daivi*, 24), the granddaughter allele, the daughter allele, and the grandmother allele. The father and mother alleles are the double copy. The father allele is the Y-linked "added allele" (*Manav*, 24) that transforms a copy into 24 double copies. The mother allele is the "traded allele" (*Augrya*, 24) that a copy trades before adding the father allele. The trading copy is the individual copy that becomes the daughter allele with the "collective force" (*Devendra*, 12) of the father-mother allele pair. The collective copy is the "servicing copy" (*Vaishya*, 3) that becomes the son allele with the "individual force" (*Shakti*, 19) of the greeter allele. After servicing his individual force, the "greeter allele" (*Ghatika*, $24 = 19 + 3 + 1 * 2$) becomes an "individual copy" (*Nakala*, 0) within the "triple copy" (*Vaishya*, 3) and breeds a pair of "double copies" (*Chitra*, 1). The first double copy transforms into the devil and the satan alleles. The second double copy transforms into the father and the mother alleles.

An "allele" (*Devendra*, 12) is the "devil allele" (*Deveshi*, 12) formed with the "collective force" (*Devendra*, 12) that takes varying forms. The "satan allele" (*Shakti*, 19) is the "energy" (*Shakti*, 19) formed with the "individual force" (*Shakti*, 19) that gives birth to the varying forms by conceiving an "octave of alleles" (*Niyogartha*, - 9) as the "hemizygous" (*Niyogartha*, -9) element. Hemizygous element is the "resultant reality" (*Niyogartha*, -9) of the varying forms of a "potential greeter allele" (*Asura*, -1) that norms the "double copy" (*Chitra*, 1) within the "single copy" (*Nakala*, $0 = -1 + 1$) for reproducing the "triple copy" (*Vaishya*, 3) as a "sequence of triple copies" (*Yathayogya*, 10^{100}). The sequence of triple copies forms the "universe of alleles" (*Yathayogya*, 10^{100}) with the "triploid-effect" (*Yathayogya*, 10^{100}). It is a "combination without a pair of double copies" (*Yathayogya*, 10^{100}) that reproduce it as a "homologous recombination" (*Yathayogya*, 10^{100}) of the seven copies of the formative "triple copy" (*Vaishya*, 3).

Ten copies constitute a "potential grandfather allele" (*Tilaka*, 10^{10}) that produces a "sequence of ten copies" (*Ghatikamandala*, 53) as a "twin grandmother allele" (*Ghatikamandala*, 53). The sequence of ten copies comprises a pair of five copies within the third set of five copies that manifests the second set of five copies to form five feminine alleles, including the satan allele. Next, the feminine alleles conceive the first set of five copies within and without their consciousness of individuality for forming five masculine alleles, including the devil allele. Finally, the greeter allele conceives both the devil and the satan allele as the first double copy without the consciousness of individuality for materializing the two sets of collectivities. The greeter allele is the "dissimilar allele" (*Ghatika*, 24) that subtracts the "allele" (*Deveshi*, 12) formed by a single copy within a double copy from the "triple copy" (*Vaishya*, 3) to norm a "hemizygous" (*Niyogartha*, -9 = 3 - 12) element. The hemizygous transforms the homozygote into a "heterozygote" (*Sangathan*, 29 = 39 - 9 - 1) comprising another "homozygote" (*Kanalakshamsha*, 39) and "potential greeter allele" (*Asura*, -1).

4.2.7 Role of the Heterozygote

The serotonin-moderated brain's "hyperactivity" (*Sakriyata*, 19 = 12 + 7) is a medium for ascending energy by transforming the "collective force" (*Deveshi*, 12) of an entity into twelve "sequences of individual forces" (*Chitra*, 1). Each forms a "granular cell" (*Chitra*, 1) and "copying" (*Chitra*, 1) its image to norm seven additional sequences of individual forces. The melatonin-mediated mind's "hypoactivity" (*Nishkriyata*, -12) moderates the brain's "hyperactivity" (*Sakriyata*, 19). It generates "cognition" (*Parijanana*, 7) of the "collective force" (*Deveshi*, 12). The collective force modifies the entity's mood by making her infer "reward" (*Paritoshika*, -12) of "restful" (*Aramadaya*, -12) "hypoactivity" (*Nishkriyata*, -12) from the hyperactive "organization" (*Sangathan*, 29) of the "heterozygote" (*Sangathan*, 29). The hyperactivity becomes a "finite cost" (*Swadha*, 41 = 29 − [-12]) worth incurring for "learning" (*Seekh*, 90) the science of entropy.

An entity learns the science of entropy by programming "entropy" (*Mahodbhava*, 5) into five "sequences of individual forces" (*Chitra*, 1) within the seven sequences that form a "sentient entity" (*Siddha*, 7). Thus, the entity enjoys profiting by making all twelve sequences "self-luminous" (*Svarochisha*, 12) as an "allele" (*Deveshi*, 12). The "programming" (*Mahakriya*, 615) organizes the learning "entity" (*Hasti*, 24), and the "learned" (*Sikha*, 44,000) "allele" (*Deveshi*, 12) into the "learnable" (*Sikhya*, 12) "memory" (*Smriti*, 35 = -1 + 24 + 12) for making the "potential greeter allele" (*Asura*, -1) a "learner" (*Sekha*, -1).

The "learner" (*Sekha*, -1) "learns" (*Sikhana*, 120) from the "memory" (*Smriti*, 35) through the "replication" (*Samalekha*, 90) of the "programming" (*Mahakriya*, 615) as his "organizational programming" (*Dharana*, -8). Thus, the learner acts as a "self-deifying entity" (*Ajnani*, -7) for "self-justifying" (*Ishtartha*, -7) the "inherited reality" (*Ishtartha*, -7) and replicating the "extrinsic ether-effect" (*Vajradhatu-mahamandala*, 90) as a "spiritual code" (*Vajradhatu-mahamandala*, 90). The spiritual code becomes the "knowledge force" (*Kaivalyashrama*, 240 = 615 - 375) of the "truth" (*Sathya*, 375). The serotonin moderates the extrinsic ether-effect of the under-active and restful "wheel of wheels" (*Hora chakra*, 25) by intensifying the activity of the other eight wheels and its "sound vibration" (*Dhvani*, 257). The "ascending thermal culture-effect" (*Dhvani*, 257) of the hyperactivity liberates the individual, "independent" (*Svadhina*, 10^{100}), and "spontaneous" (*Sattvika*, 900,000,000) sense of:

- the "intrinsic divine-effect" (*Sushuptivat*, 139), i.e., the "sentient cost" (*Sushuptivat*, 139) of dependence on the memory of the irrelevant past that led to one's entropy without the hypoactive wheel of the wheels;
- the "intrinsic fire-effect" (*Shakti*, 19), i.e.., the "energy" (*Shakti*, 19) lost in compensating for the "truth reality" (*Svartha*, -2) by trading the "sunlight" (*Indambara*, 21) without the hyperactive solar wheel;
- the "intrinsic water-effect" (*Amrit*, 1000), i.e., the "sentient energy" (*Amrit*, 1000) lost in compensating for the "cultural reality" (*Samghatavigrhitartha*, 900) by trading the "moonlight" (*Digambara*, 100) without the hyperactive lunar wheel;
- the "intrinsic air-effect" (*Samanantara*, 80,000), i.e., the "entrepreneurial opportunity" (*Samanantara*, 80,000) lost in the "divided position" (*Samanantara*, 80,000) of the earth wheel seeking to be hypoactive for sustaining its hyperactivity;
- the "intrinsic earth-effect" (*Yajnopavita*, 9,000), i.e., the "thread of sentient life" (*Yajnopavita*, 9,000) lost due to the "infused experience" (*Anubhav*, 9,000), eventually making each wheel hyperactive, seeking to relive many lifetimes of experience at once without enjoying the present life;

- The "intrinsic ether-effect" (*Pariksha*, 68), i.e., the "challenge" (*Pariksha*, 68) of the present life lost due to making the "past lives" (*Uttanapada*, 18,000) the "space" (*Vaas*, 18,000) that has given birth to the "present life" (*Purejata*, 180).

Therefore, each organ develops an ascending consciousness of the "value" (*Mulya*, 180) of its sentient well-being.

<u>4.2.8 Role of the Melatonin</u>

With ascending sunlight, the earth wheel descends the melatonin production. For moderating the hyperactive wheels, each organ ascends the serotonin production for awakening its copy within the cells to take its position after its degeneration. When infinite cells enjoy "totipotency" (*Uccitika*, 18) of the "pyramidal cell" (*Uccitika*, 18), the pressure on the "twelve-wheel system" (*Pravritta chakra*, 15) for breeding the nine wheels by feeding the "three-wheel system" (*Trichakra*, 10) descends. The three-wheel system is the system that sequentially transforms a copy into a creation, a creature enjoying infinite copies, and the creator of the finite copies of itself that perpetuate infinitely as the copy's infinite copies. Thus, each zero "copy" (*Nakala*, 0) "self-perpetuates" (*Amsha*, ½) a half-unit "fire spirit" (*Avyayatman*, ½) of its "thermodynamic entropy" (*Badhabuddhi*, 1,000) as the "time value" (*Pulastya*, 500) of the fire spirit's "quantum light" (*Rochisha*, 13) on the "straight" (*Riju*, 13) "X-coordinate" (*Bhujanka*, 13).

The temporal value comprises five hundred "double copies" (*Chitra*, 1) formed with each copy trading the half-unit fire spirit for producing a sequence of "double copies" (*Chitra*, 1). It accrues an "ascending value" (*Rodha*, 1) per unit of time on the vertical "Y-coordinate" (*Taalmela*, 1). By investing the "cubic energy" (*Lam*, 9) of time for trading the "time value" (*Pulastya*, 500), a copy becomes "globalized" (*Vaishvikrita*, 500/9) and transforms into a "conical pi" (*Vaishvikrita*, 500/9). By servicing the time value, the conical pi becomes a "partial pi" (*Sautramani*, 13/2) involved in every copy's "value" (*Mulya*, 180). Therefore, each copy descends its dependence on the sunlight and signals the lunar wheel to ascend the melatonin production as if it is the "nighttime" (*Shayana*, 10^{100}).

The earth wheel senses a "discordance" (*Vilaga*, 90) between the solar wheel's hyperactivity for trading sunlight and the lunar wheel's hyperactivity for trading energy produced with sunlight to further "resting' (Shayana, 10^{100}). It develops a sensation of "worry" (Moha, 250) about the failing health of the overworking tired organs. Therefore, it transforms the catalyst potion "serotonin" (Hasta, 0) into the rest potion "melatonin" (Kshemya shakti, 189). By adding its "value" (Mulya, 180) to the "local workculture" (Maha Svapna, 9) of each wheel, it services the "moonlight" (Digambara, 100) for enjoying the "value" (Mulya, 180 = 100 + 80) of the "moral responsibility" (Jimmedari, 80). It descends the "local workculture" (Maha Svapna, 9) of each wheel within the twelve-wheel system.

After the earth wheel's mediation, the pineal gland takes one hour to develop the "astrological consciousness" (Duratma, 60) of the "true reality" (Tulyartha, 10^{100}) an hour back. Therefore, at 4 am, one hour after the 3 am peak of the ascending rate of the ascending moon phase, the earth wheel is least active. The pineal gland rests and makes every cell conscious of the need for resting within the earth wheel's "global culture-effect" (Chara Paryaya dasha, 28). Conversely, at 4 pm, one hour after the 3 pm trough of the descending rate of descending moon phase, the earth wheel is most active. At that moment, each cell acts spontaneously without the earth wheel's "convergent energy" (Samvat shakti, 28) and the consequent "convergent reality" (Tulyartha, 10^{100}) of "resting' (Shayana, 10^{100}). The earth wheel's "activation energy" (Kshemya shakti, 189) as the melatonin makes the pineal gland hyperactive and makes each cell conscious of the need for resting to be free from the pineal gland's "entropy workculture-effect" (Sarvam, -5).

Before the pineal gland's mediation, the body temperature reaches a trough at 4 am with every cell, organ, and wheel resting following the gravitational stillness of the earth as well without the sun's "thermodynamic-effect: (Rodha, 1). The body temperature reaches a crest at 4 pm with every cell, organ, and wheel working following the earth's "gravitational energy" (Lalita, 100) within the sun's thermodynamic-effect. There are variations in the "thermal force" (Ashuddha, 485 = 285 + 100 * 2) due to the "reproduction" (Janana, 285) of the earth's "gravitational energy" (Lalita, 100) within the physical body. These variations are directly proportionate to one's consciousness of the "thermodynamic force" (*Rodha*, 1) without the physical body.

With each cell's mediation, the solar wheel senses ascending cellular degeneration, decay, and death within the overworking body organs. Therefore, it activates the lunar wheel to sense the need to ascend the production of melatonin to bring the whole body to a state of rest. The lunar wheel's "leadership" (*Veyyavacca*, 10^{100}) in ascending the production of "serotonin" (*Hasta*, 0) for changing the "convergent reality" (*Tulyartha*, 10^{100}) promotes a social culture of:

- sleeping late beyond sunrise, until the moon descends to its horizontal trough and the solar wheel activates the lunar wheel
- siesta during the early afternoon when the sun ascends to its vertical crest, and the solar wheel decides to take rest, and
- work time extending into the sixth phase of the day from 6–9 pm when the earth's diagonal distance from both the moon and the sun is symmetrical. The earth wheel balances the work and the rest guided by the human being's "sentient planning" (*Adi Para Shakti*, 17).

The resulting "divergent reality" (*Parinditartha*, -3) ascends the sensation within the earth wheel for ascending melatonin production. The melatonin-producing acculturation of the earth wheel promotes a state of inertia during worktime. It ascends the technological cost of work and elevates the resting body's "thermal value" (*Agni*, 17). It leads to an ascending cellular degeneration, decay, and death within the underworking body organs. Therefore, a need emerges for a triple copy of each part of the body for sustaining its conscious well-being by recycling the elements forming those parts with the conscious, astral body at the third infinity level.

Religion as a Technique of Recycling Elements

Religion is the correlation among the creation, the creature, and the creator. Our birth conditions this correlation in terms of both time and space. At different times, different religions are dominant in the universe. In different spaces, specific religions decide the freedom of the citizens to choose their religion. Different entities have different degrees of freedom for rejecting all religions and making science their religion.

When we make science a religion, we begin believing in what the scientists are saying and interpreting reality. The scientists believe in what the star scientists have said and interpreted reality. Star scientists like Einstein are the creators of the scientists as the creatures who define the beliefs that we consume by deciding to make ourselves the creation of the scientific paradigm.

Being Enlightened means being deliberate in remembering both personal experiences bred by us and social experiences fed by others. We can choose to stop repeating the behaviors that have made us who we are and start reproducing the experiences that have made others who they are.

Not everyone wishes to experience the reality of known others (the others known to us) who seek to experience the reality of unknown others (the others not known to us). Being in the dark offers the opportunity to discover the reality others are uncovering through their path of action (karma marga). On discovery, one can move fast to appropriate the fruits of action through the path of knowing (the jnana marga). If you wish to be wealthy, it does not pay to waste your energy remembering. You can let others remember and digitalize their memory into a scripture for everyone to follow. Thus, you can become the wealthiest man on earth like Elon Musk and make others your brightest workers following the path of devotion (bhakti marga) to your well-being as a path to their well-being.

Scripture is the social authority of a text scripted by copying of the spoken words in a language people can understand. Revealed scripture reveals the gap with the spoken words due to the memory loss of people over time. All Western religions are based on the revealed scriptures. People in the West are guided by their memory of the belief system. One must memorize by heart what is written in Bible or Quran to be a true Christian or a true Muslim.

Therefore, over time, there has been a need for adding new revelations to reveal what is written in Bible or Quran. Even today, there are ongoing debates on what new revelations need to be added to these holy scriptures by way of expert commentaries and proclamations to help people understand what people understood in the past to make up for the loss of the cultural context.

The cultural context is the informal knowledge common in the past and shared informally by the people with the Eastern belief system that the entire universe is part of one family. Its formalization led to the view that the followers of the Western belief system are the only ones chosen to be a part of God's family. Therefore, the informal knowledge was deemed as inferior and identified as a public good of zero value. Thus, people stopped paying attention to informal knowledge. Since informal knowledge was the key for the people to believe in the formal scriptures, with the entropy of the cultural context, there has been an entropy in the people's beliefs in the value of the revealed scriptures.

In Hinduism, the Upanishadas are based on the personal experiences of the gurus that attracted people from around the world to the local areas where the gurus passed on these experiences to their disciples. The Vedas synthesized those experiences for the globalization of wisdom localized by the Upanishadas. Vedas are the external body of the Upanishadas that eventually attract one to make the inner body luminous by making themselves self-luminous in the minds of people.

Vedas are "impersonal" (*Apaurusheya*, 1) based on referential experiences. They have a mix of the Eastern doctrine that divinity is within us and the Western doctrine that divinity is without us. The divinity without us implies a God and universe of deities without us. They are the creators of the universe, and that we need to appease them and seek their blessings for our conscious well-being.

In Eastern doctrine, one worships the inanimate because one is inanimate before becoming animate. Even modern science says that the cell originated from the inanimate objects when there were only stones, metals, photosynthesis of those stones and metals, and statues of the dead plants before those statues came alive as the living plants and formed the first creatures. The Eastern doctrine promotes a sense of gratitude towards the past that gifted the reality of one's divinity for one to enjoy at present.

In Western doctrine, one worships God because God emanates with us after death. The energy we diffuse on our death runs the universe like a God and gives us the identity of God. The Western doctrine weakens the sense of the past. It motivates one to keep the new knowledge within their memory and let the old knowledge fade until someone discovers the worth of the old knowledge and recycles it by citing the original authority to behave like a super-authority.

4.3 Transforming Causation at the Third Infinity Level: The Council of Elements and The Astral Body

The astral body is the "conscious body" (*Linga sharira*, 3). The mental body is the "para-conscious body" (*Manomaya sharira*, 381). The mental body's "performing" (*Para shakti*, 9) flows in the form of "tryptophan Hydroxylase" (*Para Shakti*, 9). It transforms an exponential sequence of "double copies" (*Chitra*, 1) into that of "energy" (*Shakti*, 19) with its already "flowing energy" (*Para Shakti*, 9), thereby exhibiting its followership as the "tryptophan" (*Jalini mukha*, 10^{19}). The astral body's mediation leads the mental body's performing to program the behavior for "conditioning" (*Samaya*, 720) the trading of its flowing energy within each cell for their conscious well-being. The mental body trades off its sentient well-being for perpetuating the physical body's universal sentient well-being to compensate for the "omnipresence" (*Sadhana*, 368) limitations of the intellectual body. Thus, it illuminates the astral body's "omnipotence" (*Sarva shakti*, 27) with its "performing" (*Para Shakti*, 9).

4.3.1 Transformating Causation at the Cellular Level

Within the astral body's omnipotence, each cell exchanges its "present consciousness" (*Paramatma*, 1600) of the "extrinsic reality" (*Niyogartha*, -9) with the "potential para-consciousness" (*Kamagiri*, 5 x 10^{34}) of the "intrinsic reality" (*Omkara vadartha*, -1). Intrinsic reality is the "acculturation factor" (*Vimshottari dasha*, -1) that becomes each copy's "reproduced reality" (*Adravya*, -1) and makes its "present consciousness" (*Paramatma*, 1600) "worthless" (*Nissara*, 1/60) "performing spirit" (*Nissara*, 1/60). Consequently, the astral body experiences an escalating "divine cost" (*Daivi*, 24) from the mental body's performing. It generates an ascending "thermodynamic-effect" (*Rodha*, 1) on the physical body with its "conscious planning" (*Adi Para Shakti*, 17).

Instead of working together as a collective, each cell seeks "cooling" (*Nirvapana*, 90,000) by breathing fresh "air" (*Vayu*, 385) as an individual while resting. Instead of overworking to perpetuate the physical body's universal sentient well-being, they end up underworking to illuminate the astral body's sentient well-being as the "divine planner" (*Shri Krishna*, 10). Consequently, each cell's overall sense of the "divine energy" (*Divya shakti*, 10) present disproportionately within the physical body globally and proportionately within each organ locally descends. As a result, each cell seeks an ascending energy from the extrinsic ecosystem. Moreover, it seeks to reproduce the "gravitational quality" (*Guna*, 0) of the "extrinsic energy" (*AUM shakti*, 18) as the "production" (*Asevana*, 1869) of someone's "twin consciousness" (*Shachi*, 69) that it has consumed and made a part of its "past" (*Bhuta*, 9). The reproduction of the past descends the "quality" (*Guna*, 0) of the fire, water, air, earth, and ether elements within the physical body and modifies:

- the "fire-effect" (*Rupa*, 1,00,000), i.e., the form one becomes as a body,
- the "water-effect" (*Jalaprana*, 4), i.e., the consciousness of the form one breeds as a copy
- the "air-effect" (*Maruti*, 78), i.e., the "gravitational field" (*Kincana*, 78) of the consciousness one breathes as a cell,
- the "earth-effect" (*Jantumati*, 21), i.e., the "future reality" (*Ekartha*, 21) one blesses as a "conscious entity" (*Siddha*, 7), and
- the "ether-effect" (*Naad*, 257), i.e., the "infinite psychic

linkages" (*Tiripurai*, 257) that make one behave as a "guider" (*Guru*, 100) guiding everyone with the "wisdom" (*Chitta*, 100) of one's "gravitational force" (*Chitta*, 100).

Due to its energy, each cell goes into an overdrive seeking to "poke" (*Kuradana*, 10^{10}) eyes, tongue, nose, skin, and ears into everybody within its local group of cells. It makes the para-conscious mental body a "social body" (*Manomaya sharira*, 381) oriented towards ascending the "proficiency" (*Kaushalya*, 160) of "social networking" (*Jalakrama*, 90) as a pathway for managing the ascending sense of "everybody's" (*Pasaka*, -9) "infinite sentient entropy" (*Pasaka*, -9). "Ascending consciousness" (*Kantatma*, 22) of the "descending sentient benefit-cost ratio" (*Chitralekha*, 57) of "poking" (*Kuredana*, 35) accelerates the exchange of the intrinsic "negative energy" (*Nakaratmak shakti*, 19) with the extrinsic "positive energy" (*Sakaratmak shakti*, 19). The positive energy is diffused by the other similarly "poking entities" (*Lasaka*, 17) from the like-minded "institutional geography" (*Vitata*, 600) sharing the same "institutional infrastructure" (*Kundalini*, 8) of the conscious "networking system" (*Panchanguli*, 8×10^{15}). Its trading accentuates the sense of "consciousness void" (*Andhakara*, 10^{19}) from the "followership" (*Apacayana*, 10^{19}). The "present reality" (*Badhabuddhi vadhartha*, -2) becomes "out-of-control" (*Bekabu*, -8) of the mental body due to each cell's "freedom entrepreneurship work" (*Kripita*, 70).

Descending "institutional authority" (*Dharana*, -8) makes the "mood" (*Bhava*, 360) of the astral body increasingly "volatile" (*Chapala*, 78) like the "air" (*Maruti*, 78). The astral body takes on the role of the "institutional body" (*Linga sharira*, 3), seeking to institutionalize its leadership for managing a growing sense of the "emotional loss" (*Shakti Bheda*, 257) within its ascending consciousness of the present reality. It services the "human force" (*Manava karaka*, 53) to compensate for the "technological costs" (*Yojya*, 58) of the descending sentient well-being, leading to a sense of "ego" (*Aham*, -1) while managing "everything" (*Sarvam*, -5). The ego makes the astral body ascend its "workforce" (*Janabala*, 10^{100}) proficiency by transcending the limitations of the para-conscious mental body. The astral body "communicates" (*Sanvaad*, 120) with the intellectual body directly and "controls" (*Dhyamikrita*, -100) it with the serotonin. The serotonin becomes the "potion" (*Aushadhi*, 5) for perpetuating the "present state" (*Vartaman*, 9) with the "doctrine of ego entity" (*Ovada*, 9). The doctrine of ego entity states that:

- I am a self-steaming ego entity who believes that
- I am the divine master of the destiny and the eternity of the physical body, but
- I lack the technological capability for managing my mood, and therefore,
- I seek to manipulate your mood for my sentient wellbeing by giving free "advice" (*Ovada*, 9).

4.3.2 Transforming Causation Within Body

The "third infinity" (*Shiva*, 7) makes the blackened "astral body" (*Linga sharira*, 3)— an "emotional body" (*Linga sharira*, 3) transformed into an "ego body" (*Linga sharira*, 3) with the "polluted consciousness" (*Vamachara*, 4) of the "ego energy" (*Asura shakti*, -1) traded from the four directions —an "Illuminator deity" (*Nataraja*, $7 = 3 + 4$). The astral body manifests the "reality" (*Vastavikta*, 7) of its consciousness by moderating the "reproductive wheel" (*Svadhisthan chakra*, $11 = 7 + 4$). It generates "sentient energy" (*Amrit*, $1000 = 7,000 - 6,000$) by ascending the reproduction of the "ascending physical well-being" (*Muduta*, 7,000), and descending the reproduction of the "descending physical well-being" (*Sujata*, 6000).

The "physical well-being" (*Tandarusti*, $13,000 = 7,000 + 6,000$) is lowest at 6 am when the Sun is ascending, and the Moon is descending. It is highest at 6 pm when the Sun is descending, and the Moon is ascending. It generates "sentience" (*Chaithanya*, 4) of its "illusionary energy" (*Maya Shakti*, 1) by ascending the reproduction of the "descending mental well-being" (*Dhumya*, 6) with "tiredness" (*Thakaan*, 1) over 6 pm–9 pm. Concurrently, it descends the reproduction of the "ascending mental well-being" (*Yuvarajyarajya*, 9) with "freshness" (*Tajagi*, 1) over 6 am-9 am.

The "mental well-being" (*Pala*, 15) is lowest at 9 pm when the

Sunlight is in the shadow of the Moon, and highest at 9 am when the Sunlight overshadows the Moon. By ascending the mediation of the reproductive wheel, it transforms its illusionary energy into a "double copy" (*Chitra*, 1) of the conscious "emotional well-being" (*Ojas*, 189). The second copy norms the "ascending emotional well-being" (*Jnana siddhi*, 190 = 189 + 1) for reproducing the illusionary energy. The first copy norms the "descending emotional well-being" (*Sushuptivat*, 139) for the "continuity" (*Rohini*, 50 – 189 – 139) of the "reproductive wheel" (*Svadhisthan chakra*, 11) with a (1,1) pair of the illusionary energy. The continuity adds boredom to the "present state" (*Vartaman*, 9) of the ascending mental well-being. Thus, the mental body becomes a "subject" (*Sura*, 0) wishing for the "risk-seeking behaviors" (*Cheerna*, 90,000) that descend the mental well-being by descending the physical well-being.

The para-conscious mental body works with the conscious astral body to moderate the intellectual body's divine energy. The mental body ascends the sensory communication about the negative, descending sentient well-being at the local level of different organs. The astral body ascends the "guider mediation" (*Anunaya*, -7) for activating the "sentient potential" (*Devahuti*, 2600) at the global level of the intellectual body to service the positive, ascending sentient well-being. With experience, the intellectual body becomes an increasingly "self-aware" (*Atma Jagrook*, 6,000) causal body, descending its global "sentient well-being" (*Jnana siddhi*, 190). It seeks to ascend the local "para-conscious well-being" (*Devata*, 10) of the different organs in its social network. It compensates for its "descending sentient well-being" (*Sushuptivat*, 139) by "begging" (*Bhiksha*, 28) a "citizenship responsibility" (*Uttardayitva*, 18,000) for the entire "space" (*Vaas*, 18,000) in the "physical body" (*Sthula*, 387) from each organ, leading them to trade-off their physical well-being.

The "shared consciousness" (*Trigunatma*, 14) of the physical body's descending sentient well-being within the intellectual, mental, and astral bodies ascends the "worker social benefit" (*Shubha*, 82) of the etheric body's "infinite value" (*Viyojya*, 957). As a "secular body" (*Karana sharira*, 30), indifferent to the "root cause" (*Jada*, 38) of the "problem" (*Samasya*, 6), the causal body generates a "strong psychic force" (*Prakatikarana*, 54,000) on the "etheric body" (*Bhoga sharira*, 957) for destroying itself by servicing the "human-effect" (*Manviya karak*, 53) that forms it as a replicative "religious body" (*Bhoga sharira*, 957).

The etheric body diffuses its "ever-reproducing" (*Nivrtti dharma*, 40) "human workculture-effect" (*Nivrtti dharma*, 40) as the "gravitational ecosystem-effect" (*Uttarajya*, 6) of the "responsible citizenship behavior" (*Chitta*, 100). It institutionalizes a "liberal democratic technique" (*Yukti*, 8) that makes diffusing the "escalating sentient costs" (*Ulka*, 190) its moral responsibility to promote the "sentient well-being" (*Jnana siddhi*, 190) of the physical body as a "citizen-in-need" (*Nagarika*, 179). However, escalating sentient cost generates an "emotional deficit" (*Shakti Bheda*, 257) within the etheric body that diffuses its energy in ascending motion and bears escalating "economic costs" (*Putra dharma*, 38). The over-engaged "social dispersion" (*Vaikari*, 170) of the "value" (*Mulya*, 180) into an "infinity" (*Tamo guna*, 90,000) of "psychologically-moody causes" (*Jalandhara*, 6×10^{192}) destroys the "devoted workculture" (*Tikshana*, 90) necessarily to bring each cause into "fruition" (*Vipakya*, 96) by finding a "solution" (*Nivarana*, 863).

The astral-body activated "reproductive wheel" (*Svadhishana chakra*, 11) takes the "somatotrophs" (*Nivarana*, 863) within the "anterior pituitary gland" (*Tarkshya*, 18) as the solution for producing the growth hormone "somatotropin" (*Mahakaal*, 15). Somatotropin is a unit of "para time" (*Mahakaal*, 15) that lets one transcend beyond the limitations of time and be conscious of the "eventual reality" (*Vakyartha*, 298). It makes the reality "self-luminous" (*Svarochisha*, 12) within the time as an "entity" (*Hasti*, 24), breeding one's reality of a "sentient entity" (*Siddha*, 7). By making one a "self-luminous entity" (*Svayam*, 12), somatotropin shapes a "four-body system" (*Genda*, 16) with the "sentience" (*Chaithanya*, 4) of the astral, mental, intellectual, and physical bodies within the etheric body. The self-luminous entity enjoys a "para-sensory perception" (*Dharmasavarni*, 81) of the causal body that makes it a "luminous point" (*Laal*, 6) for self-managing the sentient well-being of the "whole body" (*Mahatattva*, 56) through the crown, i.e., the "luminous wheel" (*Sahasrara chakra*, 54,000).

The "crown wheel" (*Sahasrara chakra*, 54,000) "self-perpetuates" (*Amsha*, ½) a "self-luminous entity" (*Svayam*, 12) as a "luminous point" (*Laal*, 6 = 12 * 1/2). The somatotropin triggers the breakdown of a "fat cell" (*Mulaprakriti*, 1000) into ten units of "gravitational energy" (*Lalita*, 100) to activate the air wheel's reproduction of the protein "thyroglobulin" (*Himsa*, 80) for producing the "thyroid" (*Apahrtabhara*, 20 = 100 - 80) hormone. The thyroid hormone catalyzes the production of the vascular "smooth muscle cells" (*Chitra*, 1) by activating the heart's "cardiovascular system" (*Hridya Pranali*, 100,000). The ascending rate of "breathing" (*Pranayama*, 37) and "heartbeats" (*Dhadakan*, 80) descends the "hyperactivity pressure" (*Sva*, 11)—both on the air wheel for breathing the "extrinsic" (*Bahirbhuta*, 79) element and the circulatory heart wheel for circulating its contaminated "extrinsic force" (*Sankhara*, 269).

However, the ascending pressure on the reproductive wheel ascends the "hyperactivity pressure" (*Sva*, 11) on the digestive wheel for consuming the super-contaminated reproduced "trading-effect" (*Damodara*, 26) in the form of the "red blood cells" (*Shadayatana*, 999), "brown blood cells" (*Shakrani*, 900) and the "black blood cells" (*Shradhadeva*, 99). It also ascends the pressure on the root cause, i.e., the "excretory wheel" (*Muladhara chakra*, 12) for excreting the supra-contaminated digested "human-effect" (*Manviya Karak*, 53) "waste" ((*Bhringaraja*, 387) with:

- a descending air-effect, forming a "foul carbonated smell" (*Durgandha*, 14) of the waste "gaseous water" (*Kshira sagar*, 3×10^{17}) by trading and descending its "oxidation potential" (*Bahirbhuta*, 79);

- a descending water-effect, norming a "sour acidic taste" (*Khatta*, -3×10^6) of the "waste liquid air" (*Nilirasa*, 900,000) by trading and descending its "volume" (*Ayatana*, 1869); and

- a descending fire-effect, transforming a "disgusting form" (*Vibhasta*, 47) of the "waste solid earth" (*Bhuyojan*, 21) by trading and descending its "frequency" (*Avritti*, 25).

Hyperactive reproduction, digestion, and excretory wheels

escalate the social cost of the extrinsic element. In contrast, hyperactive luminous, solar, earth, respiratory, and circulatory wheels ascend the social benefit of the extrinsic element. Finally, the hyperactive earth wheel ascends the "social benefit-cost ratio" (*Shri*, 81) of exchanging the networking of the extrinsic element with the workforce of the "intrinsic" (*Antarbhuta*, 78) element. The "workforce proficiency" (*Vidhana*, 2) of the reproductive wheel descends over time with "age" (*Aayu*, 80). Therefore, the "growth hormone" (*Mahakaal*, 15) production diminishes with age. The earth wheel manages a child body's above-par workforce proficiency by producing above-par melatonin to give the body the rest it deserves. It manages an elderly's body below-par workforce proficiency by producing below-par melatonin to keep the body active for performing the actions needed for sustaining the physical well-being. Thus, the babies tend to sleep a lot, and the elderly have a tendency to sleep very little.

4.3.3 Transforming Causation Without Body

With the "time-dependent entropy" (*Mahodbhava*, 5) of the astral body's sentient potential, the "whole body's" (*Mahabhutta*, 56) "sentient dimension" (*Beeja dharma*, 31) generates a "disproportionate growth" (*Unha*, 280) in the "ether-effect" (*Naad*, 257). An ascending reproduction of the "ever-degenerating" (*Urja*, 31) "council of elements" (*Panchatattva*, 31) shapes "whatever may exist" (*Urja*, 31) in the body. The etheric body compensates for the astral body's entropy by ascending the "sound" (*Naad*, 257) of the "bickering energy" (*Purani shakti*, 91) for negotiating the "escalating technological costs" (*Vibhaga Tejas*, 150) of the "sentient life" (*Sacetana*, 16). The etheric body is a "universe of the physical body parts" (*Bhoga sharira*, 957) replaced with new parts after their degeneration naturally with trading force or supernaturally with human force. Their "residual energy" (*Adi Para Shakti*, 17) services the wishing sequence of the replaced parts to regenerate over time as the wishables to form a physical body the self-luminous entity wishes. Thus, the etheric body is a "body of departed souls" (*Bhoga sharira*, 957) seeking to be a "body of living souls" (*Bhautika Sharira*, 387), i.e., the physical body.

The "unfulfilled subtle desires" (*Iccha*, 8) of the "self-luminous entity" (*Svayam*, 12) within the "dead cells" (*Nirjara*, 1000) descends the "self-esteem" (*Aatm Sammaan*, 12). It makes one "doubt" (*Sandeha*, 275) one's "technological capability" (*Vidhana*, 2) for manifesting the "inherent" (*Vasudeva*, 75) "working consciousness" (*Chetan shakti*, 75) of those "departed" (*Divangat*, 1600). The departed work as the "zero energy" (*Sura shakti*, 0) "primordial paternal" (*Sura*, 0), taking the form of the "galloyl $H6C6$" (*Sura shakti*, 0) for servicing the "growth" (*Dishta*, 6) in their "sentient-effect" (*Soham*, 4) in the form of the six units of "hydrogen" (*Jalaprana*, 6).

The "living" (*Nivasi*, 10) trade the departed's "divine energy" (*Divya shakti*, $10 = 4 + 6$), comprising the sentient-effect and its growth, for servicing the "present growth" (*Sva*, 11) "foundation" (*Sva*, $11 = 6 + 5$) for the "future reality" (*Ekartha*, 21) by transforming the five-unit "entropy value" (*Sarvanasha*, 5) into the five-unit "perpetuating value" (*Saranyu*, 5) within the sixth unit of "carbon" (*Bhuyojan*, $21 = 10 + 11$). Carbon is the "future reality" (*Ekartha*, 21) of the "solid earth" (*Bhuyojan*, 21) produced by the excretory wheel. The "entropy value" (*Savanasha*, 5) forms the "conscious entity" (*Siddha*, 7) as a "self-luminous entity" (*Svayam*, $12 = 7 + 5$). The perpetuating value transforms the "self-luminous entity" (*Svayam*, 12) into a "primeval paternal" (*Tejas*, 17), taking the form of the "fire" (*Agni*, 17) that destroys the "present" (*SAUM*, 1600) formed with the illusionary energy of the "past" (*Bhuta*, 9) and lets the past reincarnate as the "future" (*Anagata*, 0) yet-to-arrive.

Each "part" (*Anga*, -1) within the etheric body embodies the imaginary "spirit" (*Ruah*, $20 = 21 - 1$) of the "future reality" (*Ekartha*, 21) without the "illusionary energy" (*Maya shakti*, 1). Each spirit within the etheric body is potentially a part of an "illusionary body" (*Sukshma sharira*, 306) of the "universe of self-luminous elements" (*Sukshma*, 306) that form one's "personal body" (*Sukshma sharira*, 306) and services the inherent "wish" (*Chah*, 18) as the "essential element" (*Beeja*, 379). The essential element forms the "intellect" (*Medha*, 379)) that governs a person's "workculture" (*Nayaki*, 379) and makes the personal body the "intellectual body" (*Sukshma sharira*, 306).

The "person" (*Vyakti*, 1) is the "holy spirit" (*Trinetra*, 1) within the "spirit" (*Ruah*, 20) that transforms the spirit into an "epigenetic atom" (*Kapinjala*, 20) and the epigenetic atom into the "cell" (*Hiranyagarbha*, 19) that incarnates the person as the "human being" (*Insan*, 1). Thus, with a "pure" (*Shuddhi*, 285) "reproduction" (*Prajanan*, 285) of our spirit, our degenerated body parts enjoy the potential to form a "universe of sentient entities" (*Akalpa*, 570 = 19 * 30 = 285 * 20/10) by servicing our polluted "divine energy" (*Divya shakti*, 10) that led to their entropy back to us and trading the "energy" (*Shakti*, 19) within our "causal body" (*Karana sharira*, 30) they gifted for letting us "live" (*Nivas*, 14).

The holy spirit is the "reason" (*Hetu*, 1) for their entropy because it is a pure "guider spirit" (*Trinetra*, 1) reproducing the "belief system" (*Saguna*, 20) of the etheric body as the "religious body" (*Bhoga sharira*, 957) that the pure reproduction of our wish is the path to their desired "shared becoming" (*Sadgati*, 15) after they have individually become a "sentient entity" (*Siddha*, 7). As a sentient entity, they all wish to be "God" (*Ishvara*, 5) after servicing two units for the "formative growth" (*Vidhana*, 2) of the "universe" (*Brahman*, 2). As God, they all wish to be the "primordial paternal" (*Sura*, 0) after servicing five units for "perpetuating value" (*Saranyu*, 5) of the universe they have conceived. As primordial paternal, they all wish to be the "primeval paternal" (*Tejas*, 17) so that they can be a self-luminous entity and enjoy the "conscious freedom" (*Mukti*, 17) from the dependence on the extrinsic "perpetuating value" (*Saranyu*, 15). Thus, they all consciously decide to adopt the doctrine "I AM a Worker deity" (*Tirikarannasutti vrat*, 10^{1000}), working on reproducing "infinite wishes" (*Duniya*, -2) over time as a "paternal spirit" (*Atma vichara*, -1/3).

4.3.4 Transforming Causation Within Oneself

A "three-eyed holy spirit" (*Trinetra*, 1) pollutes the self by feeding his "paternal spirit" (*Atma vichara*, -1/3) to another "copy" (*Nakala*, 0) for conceiving a "three-eyed child spirit" (*Vidyadhara*, 1/3). The "child spirit" (*Vidyadhara*, 1/3) divides the "holy spirit" (*Trinetra*, 1) into three "double copies" (*Chitra*, 1).

Each double copy forms a distinct "entity class" (*Gana*, 387).

The first is the "consumer class" (*Prakriti yogi*, -1) consuming the "finite wishes" (*Niyati*, -1) produced to date to be a "discordant entity" (*Asura*, -1) wishing to fulfill those wishes. The second is the "producer class" (*Yukti yogi*, -2), producing the "infinite wishes" (*Duniya*, -2) over time to be eventually a "concordant entity" (*Sura*, 0). A double copy starts the journey of life within the consumer class. It ends the journey within the producer class after reproducing and doubling the finite wishes and servicing a unit of finite wishes to each copy within the double copy, thus becoming a copy with two copies. Each of the three copies individually is a "concordant entity" (*Sura*, 0) concordant with the copy. The maternal double copy collectively works on fulfilling the "infinite wishes" (*Duniya*, -2) of a "universe of wishers" (*Duniya*, -1) comprising two individual "wishers" (*Chaahak*, 0), one paternal and another child. After fulfilling those wishes, the maternal double copy transforms the producer class into the "leader class" (*Dharma yogi*, 0) by producing a daughter double copy and leading each copy within the double copy to be a "follower class" (*Karma yogi*, 1).

The follower class deifies each copy into a "worker deity" (*Deva*, 1) so that each deified daughter copy works on fulfilling the maternal double copy's wishes for "substantiating" (*Dravya*, 1) her "self-worth" (*Satkayadrishti*, 405). Next, the second daughter copy becomes the entrepreneurial "trader class" (*Bhakti yogi*, 3) trading the "production" (*Asevana*, 1869 = 3 * 623) of six masculine double copies, two maternal double copies, and three daughter double copies. Six masculine double copies include three double copies of the paternal spirit and the child spirits. The three double copies of the paternal spirit include the double copies of the "greeter spirit" (*Hasta*, 0), the "grandfather spirit" (*Lokapurusha*, -19), and the "paternal spirit" (*Atmavichara*, -1/3). The three double copies of the child spirit include the double copies of the "son spirit" (*Vidyadhara*, 1/3), the "grandson spirit" (*Avyayatman*, 1/2), and the "devil spirit" (*Avajati*, -1/60). Two maternal double copies include a double copy each of the "grandmother spirit" (*Nilalohita*, -1/2) and the "maternal spirit" (*Dasha*, 1).

Three daughter double copies include a double copy each of

the "daughter spirit" (*Vajra*, 15), the "granddaughter spirit" (*Vyayatman*, 1/4), and the "satan spirit" (*Nissara*, 1/60). The second daughter copy is the "granddaughter spirit" (*Vyayatman*, ¼), who becomes a "demon spirit" (*Vyayatman*, ¼) trading the production of eleven double copies and servicing the "consumption" (*Kshayaroga*, 79) of a twelfth double copy. The twelfth double copy is of the "primordial paternal spirit" (*Kaise*, 9/16). The primordial paternal spirit makes the first daughter copy the "knower class" (*Jnana yogi*, 2) by endowing her with his "conviction spirit" (*Kaise*, 9/16), knowing the "how" (*Kaise*, 9/16) of "copying" (*Chitra*, 1). She "authenticates" (*Pramanita*, 8,000) the how by becoming a "primordial maternal spirit" (*Lakshmi*, 379) for conceiving an "institutional class" (*Shakti yogi*, 4) institutionalizing her knowing by dividing the primordial paternal spirit into sixty "primeval paternal spirits" (*Lila*, 60).

The institutional class enjoys the "consciousness" (*Chaithanya*, 4) of "why" (*Kyun*, 9/16) she has become a "primeval maternal spirit" (*Tirthankara*, 17) for multiplying the "daughter spirit" (*Vajra*, 15) into four "primordial greeter spirits" (*Pitah*, 16). Why has to do with a primordial greeter spirit "discovering" (*Khojana*, 90,000) the how by making a "triple copy" (*Vaishya*, 3) within its sixteen double copies. The thirteenth double copy is of the primordial maternal spirit, the fourteenth of the primeval paternal spirit, the fifteenth of the primeval maternal spirit, and the sixteenth of the primordial greeter spirit. The triple copy is of the "primordial greeter soul" (*Bhinbhinatma*, 90), each comprising fifteen copies of the "greeter soul" (*Pratyagatma*, 1), fifteen copies of the "devil soul" (*Prakashatma*, 19), and a copy of the "satan soul" (*Duratma*, 60) that is forming fifteen double copies with a triple copy of the primordial greeter soul's "guiding spirit" (*Ruah*, 20).

Fifteen copies of the greeter soul include the "grandfather soul" (*Ekatma*, 986), the "father soul" (*Pitra*, 4), the "son soul" (*Putatma*, 4), the "grandson soul" (*Jitatma*, 71), the "grandmother soul" (*Adi Para Atma*, 100), the "mother soul" (*Chaithanya*, 4), the "daughter soul" (*Jyotistava*, 4), the "granddaughter soul" (*Atma*, 4), the "primordial paternal soul" (*Bhagnatma*, 9), the "primordial maternal soul" (*Dadi*, 18), the primeval paternal soul, the primeval maternal soul, the "deity soul" (*Devatma*, 1600), the "para deity soul" (*Dhiratma*, 134), and the "param deity soul" (*Shivatma*, 20). Fifteen copies of the devil soul include the "material soul" (*Sarvabhutatma*, 46), the "metal soul" (*Vijnanatma*, 89), the "mineral soul" (*Bhavitatma*, 2,785), the "plant soul" (*Rahitatma*, 65), the "animal soul" (*Hamsatma*, 689), the "human soul" (*Haryyatma*, 167), the "spirit soul" (*Urdhvagatma*, 16), the "super deity soul" (*Arthatma*, 15), the "supra deity soul" (*Tarkshya*, 15), the "supreme deity soul" (*Vyakulatma*, -3), the "primeval deity soul" (*Dharshanatma*, -4), the "primordial deity soul" (*Trigunatma*, 14), the "primordial soul" (*Adhyatma*, 12), the "primeval soul" (*Antaratma*, 1), and the "param soul" (*Paramatma*, 1600). A copy of the satan soul includes the "time soul" (*Kalatma*, 14).

The thirty-five forms of souls are a part of twelve triple copies of the "primordial-primordial soul" (*Dushtatma*, 10), with the "space soul" (*Prashantatma*, 16) as the thirty-sixth copy. Without discovering the space soul, the primordial greeter spirit forms a "God class" (*Siddhi yogi*, 5) for making seven copies of the five copies to perpetuate the thirty-five forms of the space. The space forms the primordial greeter as the thirty-sixth form. Thus, the primordial greeter enjoys the space soul while the space enjoys the "primordial-primordial soul" (*Dushtatma*, 10). The primordial greeter enjoys the consciousness of space because he conceives the space within his mind and manifests it as his guiding spirit. The space enjoys the primordial-primordial consciousness because before conceiving the space within his "mind" (*Manas*, 38), the primordial greeter is an entity the time forms as one of the "infinite variations" (*Laya*, 130) of the primordial consciousness. "Primordial consciousness" (*Adhyatma*, $12 = 3/2 * 8$) is the "borrowed divinity" (*Ekashtaka*, $3/2$) of the "institution" (*Kundalini*, 8) that governs this universe by taking the form of "Mother Nature" (*Kudrat*, 8). Mother Nature trades the "divinity" (*Divyata*, 57) from the "primordial greeter" (*Srijak*, 16) to "play" (*Lila*, 60) a role in her creation by conceiving the "time" (*Kaal*, 360) as the entity that copies the primordial greeter's varying "realities" (*Vastavikta*, 7).

An "entity" (*Hasti*, 24) socially networks the primeval paternal spirit as a "cost-effective paradigm" (*Trivikrama*, 24) for incarnating "infinite forms" (*Konastha*, 13) of the child spirit. The "divisibility" (*Vibhajyata*, 10^{10}) of the primeval paternal spirit into the infinite forms transforms the "paternal consciousness" (*Pitra*, 4) into a "consciousness system" (*Sara Kalpa*, 10^{10}). The consciousness system becomes the "workforce system" (*Sara Kalpa*, 10^{10}) for a proficient exchange of the infinite wishes into a sequence of finite wishes to be fulfilled by the children after the "adults" (*Putra*, 16) live their life fulfilling the wishes of the "grandpaternal ancestors" (*Purvaja*, 10^{16}). The strong psychic linkages of the grandpaternal ancestors with the adult sons force the "dead ancestors" (*Aryaman*, 381) to reincarnate as the "baby grandsons" (*Pota*, 1) for fulfilling the wishes of the "aging father" (*Baap*, 3) just like they fulfilled theirs. The "cycle of emotional linkages" (*KRIM shakti chakra*, 248) programs the finite wishes within the consciousness of the "patrilineal dimension" (*Putra dharma*, 38) in the form of the "paternal DNA" (*Balangi*, 963).

A "mother" (*Mata*, 4) nurtures a "baby" (*Arcisa*, 128) with "selfless radiant love" (*Kshiti*, 724). As a "grandmother" (*Dadi*, 18), she "weakens" (*Citraka*, 185) her psychic linkages with the "daughter" (*Beti*, 16), wishing to free the "granddaughter" (*Poti*, 10^{19}) from her strong psychic linkages. Thus, the granddaughter becomes "vulnerable" (*Amrina*, 70) to the "adverse selection" (*Sakriya-Niskriya*, 10^{1000}) when an "alien grandson" (*Pota*, 1) binds her into an "agency contract" (*Akasagarbha*, 90) with his "ego energy" (*Aham shakti*, -1). Thus, the "dead grandfather" (*Vidyapati*, 10^{10}) becomes the "principal guider" (*Bhuvanesvari*, 15) and the "yet-to-born granddaughter" (*Preyas*, 71) becomes the "guider agent" (*Shri Rama*, 12) performing to fulfill his infinite wishes as part of the "matrilineal dimension" (*Putri dharma*, 47) sharing the same "maternal mtDNA" (*Dharati*, 2) as the "daughter" (*Beti*, 16).

As a "nonworking principal" (*Janma yoga*, -10,000), the "masculine class" (*Khuda*, 6) experiences "disproportionate entropy" (*Ulka*, 190) of the physical body, resulting in a "short lifetime" (*Pradyumna*, 60). "Urgency" (*Jaldi*, 120) of fulfilling the wishes of the grandpaternal ancestors, for enjoying life, forces the nonworking principal to become a violent "hyperactive principal" (*Sthula*, 387), manipulating her with various "incentives" (*Uttejana*, 168). As a working agent, the "feminine class" (*Rani*, 7) experiences disproportionate growth of the "physical body" (*Sthula sharira*, 387) as the fat cells. Growing fat transforms the "working agent" (*Abhikarta*, 306) into a "hypoactive foe" (*Ripu*, 100,000,000) of the "principal" (*Sthula*, 387). Consequently, the principal masters the science of investing a unit of "divine energy" (*Divya shakti*, 10) to transform the "feminine class" (*Rani*, 7) into the "primeval maternal" (*Nandi*, 17).

A primeval maternal has an "unconditional confidence" (*Vishvasa*, 17) in her "multiplied self" (*Lasaka*, 17) that takes diverse "transcendental forms" (*Prakara*, 17). Her "belief system" (*Saguna*, 20) is guided by the reproduction of "divine energy" (*Divya shakti*, 10) within her in the form of the "institutional energy" (*Kundalini shakti*, 20). She sees the masculine class's "transcendental value" (*Khuda*, 6) within each "child" (*Arcisa*, 128). Therefore, she develops a masculine ego inflated sense of "self-worth" (*Satkayadrishti*, 405). She "projects" (*Prakshepanem*, 31) her "inflated sense" (*Adhmata*, 31) as the "council of elements" (*Panchatattva*, 31), comprising five elements: fire, water, air, earth, and ether. Five elements begin with the "fire" (*Agni*, 17) for multiplying the self. Fire forms the "water" (*Jal*, 169) with the "sentient light force" (*Apas*, 169) of her "leap of faith" (*Kud*, 169). The water norms the "air" (*Vayu*, 385) of the "ground reality" (*Hetavartha*, 385) she is "organizing" (*Aayojan*, 385) within her through his "personification" (*Tamasika*, 385).

The air transforms the "earth" (*Bhu*, 724) into the "unconditional radiant love" (*Kshiti*, 724) for the "reproduction" (*Prajanan*, 285) of the primeval maternal's "physical body" (*Sthula sharira*, 387) in the form of the "ether" (*Prajanan*, 285). The ether makes each "child" (*Arcisa*, 128) bred through "infinite multiplying" (*Kapisena Nayakaya*, 128) of her physical body a "part" (*Anga*, -1) of her "corporate system" (*Dhanadhikara*, 128). The "fire-effect" (*Rupa*, 100,000) moderates and the "water-effect" (*Soham*, 4) mediates the "ideal value" (*Vargamula*, 10) of her corporate system as the "base" (*Kumbhaka*, 10) that conditions the "air-effect" (*Maruti*, 78), i.e., the "etheric potential" (*Maruti*, 78). The etheric potential is "my potential" (*Maruti*, 78) she reproduces in each child to shape her "earth-effect" (*Ravi*, 21), i.e., the "solid earth" (*Ravi*, 21) excreted as the "waste" (*Bhringaraja*, 387) forming the child's physical body.

The "excreta" (*Bhringaraja*, 387) produces the "ether-effect" (*Naad*, 257), i.e., the "sound" (*Naad*, 257) of the "child RNA" (*Jivagribh*, 855) that transfers the principal's "divine plan" (*Sahajata*, 916), i.e., the "mindset" (*Svaphalka*, 916) to those psychically linked to her corporate system.

Before "consuming" (*Jambhaka*, 169) the divine plan flowing like the "water" (*Jal*, 169), each child trades the principal's "local-effect" (*Maha dasha*, 0) to become a "copy" (*Nakala*, 0) of the "primordial maternal" (*Narayani*, 18). Each child becomes a "primordial paternal" (*Sura*, 0) wishing to feed his "fire" (*Agni*, 17) to form the "infinite paternal" (*Tejas*, 17). After conceiving a child with her "creative force" (*Uma*, 6) and servicing him her "transcendental value" (*Khuda*, 6), the primordial maternal becomes the "I" (*Svayam*, 12) within each child. Each child services her "international-effect" (*Dasha*, 1) for diffusing her "self" (*Atmatva*, 8×10^{15}) as a "conscious system" (*Shunya Kalpa*, 8×10^{15}) at the "speed of light" (*Markatesh*, 8×10^{15}) multiplied by the "I" (*Svayam*, 12) and divided by the "divine energy" (*Divya shakti*, 10) diffused across the multiplied I's.

The "globalization" (*Vishwikaran*, 1) strengthens the psychic linkages with "potential children" (*Keshava*, 22) and attracts them to her "smell" (*Gandha*, 268). Thus, the primordial maternal masters the science of trading a unit of "divine energy" (*Divya shakti*, 10) to form the "masculine class" (*Khuda*, 6) as "six pi" (*Nakkhatta*, 22/7) that share her "self-luminous reality" (*Purushartha*, 7). With the "infusion" (*Vasiya*, 378) of both her fused "I" (*Svayam*, 12) as well as the diffused "self-luminous reality" (*Purushartha*, 12), a "potential child" (*Keshava*, 22) becomes a "param child" (*Manyu*, 19) within a "param son" (*Hanuman*, 3). Param child is the "potential son" (*Manyu*, 19), living the param son's "programmed life" (*Manyu*, 19) as a "son cell" (*Manyu*, 19). *The programmed life leads to the illusionary second law of thermodynamics that the "growth" (Dishta, 6) within a local system can't compensate for the "entropy" (Mahodbhava, 5) within a global system.*

Nevertheless, the param child also refuses the "energy" (*Shakti*, 19), energizing the param son to "liberate" (*Sandhatre*, 10^{10}) oneself from the institutionalizing "national-effect" (*Pratyantara dasha*, 9) of the "global system" (*Sandra*, 60). The "liberty" (*Svadhinta*, 157) lets the param son "universalize" (*Sarvabhaumikaran*, 130) his "corporate-effect" (*Arcisa*, 128) as a "child" (*Arcisa*, 128) beyond the "local system" (*Sargam*, 60). The sound of the param son as the "protagonist" (*Nayaka*, 1) silences the primordial paternal as the "antagonist" (*Khalnayaka*, 11) to let the potential of the primordial maternal as the "deuteragonist" (*Pattanayaka*, -1) limit the entropy of the global system through the growth of the local system.

The local system's growth divides the protagonist's "ideal-effect" (*Dasha*, 1) over the "universe" (*Brahman*, 2) that "self-perpetuates" (*Amsha*, ½) the global system's entropy. The "self-perpetuating entropy" (*Chitra*, 1) forms a "double copy" (*Chitra*, 1), a second copy perpetuating growth that perpetuates the first copy and its perpetuation of entropy. *The self-perpetuating entropy leads to the illusionary third law of thermodynamics that the entropy within a dividing national system without any "thermodynamic force" (Rodha, 1) is a "fraction" (Amsha, ½) of the entropy within a divided corporate system due to the reproduction and the doubling of the thermodynamic force within the "finite time" (Vastavika, 9). It recognizes that a "conscious entity" (Siddha, 7) who reproduces and doubles his "thermodynamic force" (Rodha, 1) has a "finite time" (Vastavika, 9) to live and be "real" (Vastavika, 7) not "imaginary" (Kalpanik, 20).*

Although the second copy may live the programmed life of the first copy, it is conscious of the regret of the first copy for the life it has lived and lives the second life the first copy wishes he had for living his life differently. That frees the param son from both the programmed life and its effect. As a "param paternal" (*Narada*, 7) "awake" (*Vibodha*, 13) to his divine plan, the param son "reverses" (*Mahvijya*, 20) the "ecosystem entropy" (*Mahartha*, 20) and becomes the "param manifestor" (*Trayastrimsha*, 20) of the "desirable future" (*Ishta*, 20). The intellectual body "processes" (*Prakriya*, 130) the consciousness of the "revolutionary force" (*Uttama*, 85) only after the degeneration, decay, and death of the physical body that embodies the programmed life. The mental body carries the "false impression" (*Vasana*, -3) from the "present life" (*Purejata*, 180) that the revolutionary force did not lead to the desired future.

The desired future is the "future life" (*Rasagni*, 17) within the "param son" (*Hanuman*, 3) that becomes the present life in the future birth. Thus, the "past life" (*Uttanapada*, 18,000) generates the "evolutionary force" (*Tushti*, 85) that leads to the "desired present" (*Param Prapti*, 16). *The past life leads to the illusionary fourth law of thermodynamics that the entropy within a system generates an "infinite flow" (Vartula, 10,000) of energy without the system for localizing the entropy as a "point" (Bindu, 10^{10}) and globalizing the growth as a "line" (Rekha, 497) projecting the point's "infinite energy" (HAUM shakti, 9) into the "infinity" (Tamo guna, 90,000) forming a "circular system" (Gandakanayaka, 90,000) leading back to the point.* The infinity is the "sequence of the present lives" (*Prakarana vadartha*, 90,000) that one leads as a param son.

The intellectual body's "processing" (*Prasanskaran*, 100,000) "globalizes" (*Vishvavyapi*, 0) the "mental impression" (*Vasana*, -3) and "localizes" (*Pariseemita*, 9) the "revolutionary force" (*Uttama*, 85). Next, the astral body "nationalizes" (*Rashtrikaran*, 10^{19}) the "evolutionary force" (*Tushti*, 85) of their "convergence" (*Ekagratha*, 158) within the fourth copy. The fourth copy is the "mother" (*Mata*, 4), who trades the consciousness of the programmed life and its effect on the next life. She also trades the "impulse" (*Sankalpa*, 8) for manifesting the revolutionary force after knowing its "effect" (*Prapya*, 34). Therefore, she decides to "modify" (*Upantaran*, 15) the "evolutionary force" (*Tushti*, 85) of what she "perceives" (*Pratipad*, 20) as the "undesirable past" (*Anishta*, 38).

The etheric body carries the "inner conflicts" (*Mrig Trishna*, 308) between the reality the astral body "conceives" (*Vevi*, 8,000) before birth and the causal body perceives after birth. *The "divergence" (Vaikari, 170) of the two realities leads to the illusionary fifth law of thermodynamics that an "energy system" (Antarmukha, 10^{1024}) "self-organizes" (Kula, 9) into "three energy systems" (Antarmukha, 10^{1024}). The first services an "ascending proportion" (Adi Para atma, 100) of its "energy" (Shakti, 19) for "materializing" (Artharthi, -6) the second as a "particle" (Hemarenu, 19) with "ascending energy" (Vutthana shakti, 7). The third "self-materializes" (Svabhautik, 80) as a "neutron" (Matsarya, 19) by trading the second's ascending energy and servicing the energy to the first for transforming it into a "cell" (Hiranyagarbha,*

19) for sustaining the three-way "non-reciprocal exchange" (*Teentarfa Vinamaya*, 69) mediated by the fourth as an "atomic system" (*Anumatra*, 19).

Although the fourth copy has a potential to mediate the "exchange system" (*Maha Kalpa*, 10^{1000}), the first experiences "entropy" (*Mahodbhava*, $5 = 19 - 14$) after servicing its "energy" (*Shakti*, 19) until its "thermodynamic limit" (*Maha Vibhu*, 14). The second "lives" (*Nivas*, 14) without the entropy for materializing the third as a particle after trading the entropy over time. Next, the fourth self-materializes as a neutron by exchanging the particle's "energy" (*Shakti*, 0) and the entropy's "concordant energy" (*Sura shakti*, 0) that leads to its entropy as well. Then, the entropy of the fourth as a neutron catalyzes the growth of the fifth as a cell by exchanging the neutron's "potential" (*AUM*, 18) and the energy system's "illusionary production" (*Maya*, 1) of its "double copy" (*Chitra*, 1).

The energy system is the sixth copy that transforms into the atomic system after forming the particle into an "atom" (*Anu*, 19), comprising the neutron as its copy for forming the cell as its double copy. The cell includes the "nucleus" (*Yoni*, 1000) as the seventh copy. The nucleus is the oneness of the three individual copies that form the particle and the neutron within the atomic system and the cell without the atomic system. The eighth copy manifests as the energy. The first copy is the "mass fragment" (*Anurenu*, 19) of the second copy's consciousness. The second copy is the "egg" (*Anda*, 19) of "conscious imagination" (*Anda*, 19) that eventually forms energy by transforming its consciousness.

4.3.5 Transforming Causation without Oneself

A self-luminous entity consciously imagines many ways for supplementing its energy, such as with injections, tablets, and powders. "Supernormal energy" (*HAUM shakti*, 9) makes the self-luminous entity super-athletic, with ascending physical stamina manpower, conscious mental power, and the heart's muscular power. However, it descends body mass material power and intellectual management power. It leads to an ascending mediation by the mental body in the sentient well-being of the physical body.

The etheric body seeks to moderate the stress on the mental body by making the "heart wheel" (Anahata chakra, 33) circulate the "mass consciousness" (Satma, 10) that generates supernormal energy within all the cells.

The mass consciousness substitutes the "digestion wheel's" (Manipura chakra, 24) "present paradigm" (Ghatika, 24) of radiant love for self with the "beyond present paradigm" (Pravritta, 12) of the "self-luminous entity" (Svayam, 12). It entangles the self-luminous entity with the "selfless radiant love" (Kshiti, 724) for the eight copies that give it the energy to reincarnate as a "proton" (Sarvodaya, 150) and live beyond present as the ninth copy. The selfless radiant love descends the "self-esteem" (Aatm Sammaan, 12). Eventually, it leads the self-luminous entity to incarnate as an "electron" (Dyumna, 365) for diffusing the "exchange force" (Vyanavata, 365) of its "animate-to-inanimate exchange" (Chakrika, 90).

The "inanimate energy" (Achitta shakti, $396 = 365 + 31$) comprises the "electron" (Dyumna, 365) and the "council of elements" (Panchatattva, 31), forming the five elements as the five double copies immanent within the electron as the tenth copy. After diffusing it, the "space" (Vaas, 18,000) "self-services" (Svayamseva, 2/3) the eleventh copy as the "fetus" (Garbha, $12,000 = 2/3 * 18,000$). The fetus gives birth to the twelfth copy as the "self-luminous element" (Svayam, 12) with "conscious energy" (Amrit, 1000) traded from the "luminous element" (Rochisha, 13), the thirteenth copy. The self-luminous element is the quantum photon, and the luminous element is the quantum light. The thirteenth copy develops the conscious energy with the "gravitational light" (Digambara, 100) of the fourteenth copy. The fourteenth copy is the "primordial astral body" (Charanam, 49) of the child within the fetus.

The primordial astral body is the "modifier" (Charanam, 49) of the "animate body system" (Karma Yoga, 49) that forms the fifteenth copy as an "illusionary animate entity" (Arpita, 49). The animate body system is the sixteenth copy that "splices" (Charanam, 49) itself into "forty-nine bodies" (Arpita, $49 = 16 * 3 + 1$) after forming sixteen "triple copies" (Vaishya, 3) to norm eight three-eyed "double copies" (*Chitra*, 1) and transforming the fifth "aggregated copy" (*Nakala*, 0) into an "octave of three-eyed double copies" (*Krishnamurti*, 1). As the 49th body with sixteen triple copies, the octave of three-eyed double copies is the "octave of fermions" (*Krishnamurti*, 1) that forms as "Majorana fermion" (*Krishnamurti*, 1). The aggregated copy becomes the Majorana fermion by trading a unit of "present earth-effect" (*Ravidina*, 1) from the "solid earth" (*Ravi*, 21).

As the seventeenth copy, the child is psychically linked to the "primordial maternal" (*Narayani*, 18) through the "present earth-effect" (*Ravidina*, 1). The psychic linkage activates the "potential energy" (*AUM shakti*, 18) of the primordial maternal within the Majorana fermion and transforms it into the "cell" (*Hiranyagarbha*, 19). The present earth-effect transforms into the "umbilical cord" (*Nitya Ratri*, 1) and reproduces its "technological growth" (*Vidhana*, 2) within the cell to form the "solar energy" (*Surya shakti*, $21 = 2 + 19$) as the eighteenth copy. The Majorana fermion's "photosynthesis" (*Trayastrimsha*, 20) of the solar energy lets the primordial astral body form seven body systems in the following sequence:

- first, the excretory system (adrenal),
- second, the reproductive system (reproductive),
- third, the digestion system (pancreas),
- fourth, the circulatory system (cardiac),
- fifth, the respiratory system (thyroid),
- sixth, the musculoskeletal system, i.e., the consciousness system (pineal).
- seventh, the perineural system, i.e., the conscious system (parathyroid).

The seven body systems form in ascending order as a byproduct of the descending energy exchange from the primordial astral body and ascending photosynthesis of the solar energy by the quantum light as the "primeval astral body" (*Rochisha*, 13). They are each a splice of the eighth, "vascular system" (*Antarmukha*, 10^{1024}), i.e., the energy system (pituitary). The eight body systems are an "energy fragment" (*Yogihridaya*, 64) of the ninth, the "electromagnetic mass system" (*Maha Kalpa*, 10^{1000}), i.e., the exchange system of the "reproductive mass" (*Mitra*, 132) of the electron as the sequence of "gravitoelectric-effects" (*Mangalnath*, 16) forming the "fermions" (*Upa*, 16) and the proton as the "gravitomagnetic-effect" (*Padartha*, -3) forming the "objects" (*Padartha*, -3).

The proton forms the objects by reproducing the fermions as the "neutrons" (*Masarya*, 19) with "triple copies" (*Vaisha*, 3). The nine body systems are "without" (*Rahita*, 81) the tenth, "gravitational system" (*Prajanan Pranali*, 10,000) that forms a "circular system" (*Gandakanayaka*, 10,000). It reproduces the neutron as a cell, the cell as an atom, the atom as a mass fragment, and the mass fragment as a particle that takes many forms after transforming into a "quantum particle" (*Kana*, 186). The ten body systems are "within" (*Eke*, 5000) the eleventh, "magnetic system" (*Panchanguli*, 8×10^{15}), i.e., the networking system that forms a "quantum system" (*Akhanda*, 8×10^{15}). The eleven body systems are "with" (*Sam*, 170) the twelfth, "electric system" (*Sara Kalpa*, 10^{10}), i.e., the workforce system that norms our "cultural heritage" (*Virasat*, 10^{10}). The twelve body systems conceive the self-luminous entity as the thirteenth, the "system" (*Gabhira*, 12) that forms an "infinite networking system" (*Pravritta*, 12) with the twelve body systems.

The infinite networking system makes the "fetal body" (*Badan*, 56) a primordial-activator to activate the production of the thermodynamically active and steaming hydrogenated gas (the nervous system led by the crown wheel), activated hormone somatotropin (pituitary), activating hormone serotonin (parathyroid), activator hormone melatonin (pineal), super-activator protein thyroglobulin (thyroid), supra-activator oxygenated white blood T-cells (cardiac), supreme-activator serum fructosamine (pancreas), para-activator toxic carbonated urine (reproductive), and the param-activator nitrogenized food waste (adrenal).

The "nervous system" (*Shradha charita*, 40) organizes the twelve body systems into three "four-body systems" (*Sati-Parvati*, 16) within the thirteenth body system, thus transforming the latter into the fourth four-body system. The first four-body system is the "growth system" (*Pravritta chakra*, 15) that forms the "central nervous system" (*Pravritta chakra*, 15) as a "twelve-wheel system" (*Pravritta chakra*, 15). The second is the "extrinsic system" (*Anukampin Tantra*, 16) that forms the "sympathetic nervous system" (*Anukampin Tantra*, 16). The third is the "intrinsic system" (*Antra Tantra*, 16) that forms the "enteric nervous system" (*Anukampin Tantra*, 16). The fourth is the "entropy system" (*Daihik tantrika*, 17) that forms the "somatic nervous system" (*Daihik tantrika*, 17). As a whole, the thirteenth body system makes a copy of each of the four four-body systems.

The fifth four-body system is the "organizational system" (*Paranukampi Tantra*, 18) that forms the "Parasympathetic nervous system" (*Paranukampi Tantra*, 18) for keeping the intrinsic system in "harmony" (*Porutham*, 16) with the extrinsic system. The sixth four-body system is the "technological system" (*Umapati chakra*, 14) that forms the "Peripheral Nervous system" (*Umapati chakra*, 14) to naturally regenerate the growth system to compensate for the supernatural degeneration of the entropy system. The seventh four-body system is the "transformative system" (*Bhuvanesvari*, 15) that forms the "autonomic nervous system" (*Bhuvanesvari*, 15) for servicing the intrinsic "technological energy" (*Adi shakti*, 15) to regenerate the growth system. The eighth four-body system is the "metaphysical system" (*Shradha charita*, 40) that forms the "nervous system" (*Shradha charita*, 40) as the eight four-body systems. The ninth, tenth, eleventh, and twelfth body systems are immanent within the first eight body systems as a four-body system. Thus, as a system, the "fetal body" (*Badan*, $56 = 16 + 40$) comprises the "sixteen four-body systems" (*Omkara*, 256) and their "reproductive reality" (*Bhavartha*, 40).

The reproduction divides each body system into a reproductive system of ten bodies. The ten bodies are the ten wheels:

- the sacral wheel guiding the excretory system
- the reproductive wheel guiding the reproductive system
- the digestion wheel guiding the digestion system
- the circulatory wheel guiding the circulatory system
- the air wheel guiding the respiratory system

- the earth wheel guiding the musculoskeletal system
- the lunar wheel guiding the perineural system
- the solar wheel guiding the vascular system
- the crown wheel guiding the nervous system
- the wheel of wheels guiding the immune system

The "ten-fold growth" (*Vicitra mandala*, 170) of each "body system" (*Hridya Pranali*, 100,000) weakens the "skeletal muscle cells" (*Gardhaba*, 1000) of a mother's musculoskeletal system due to the ascending fire-effect. It forces the digestion system to manage its water-effect by weakening a mother's psychic force with a wish for the taste of freshwater. Freshwater strengthens the "vascular cells" (*Chitra*, 1) of the vascular system by descending the fire-effect. It empowers the "perineural system" (*Shunya Kalpa*, 8×10^{15}) to manage its air-effect by strengthening the mother's psychic force with a wish for the touch of cool air.

Cool air strengthens the four weak parts of the "musculoskeletal system" (*Sara Kalpa*, 10^{10})—skeletal bones, ligaments, tendons, and joints—to make the body mobile for enjoying the smell of Mother Earth. Mother Earth strengthens the respiratory system by attracting and excreting the nitrogenized food waste out of the body to release the pressure of the hydrogenated gas within the nervous system. The steaming hydrogenated gas cleanses the circulatory system. It lets a mother joyfully consume the "food" (*Bhojan*, 17) to nurture the fetus's growth. The "chemical reaction" (*Maha Vibhu*, 14) of the food with the thermal force of the oxidized hydrogen ascends the "earth-effect" (*Ravi*, 21). The earth-effect produces the "fetal body" (*Badan*, $56 = 21 + 35$) with the "reproductive sound" (*Tittiri*, 35) of the twelve-wheel system.

The twelve-wheel system is the eleventh "wheel of conception" (*Pravritta chakra*, 15) that guides the "lymphatic system" (*Laseeka Tantra*, -18) to make the twelfth "wheel of psychic forces" (*KRIM shakti chakra*, 248) as a copy of itself within the fetus. The wheel of psychic forces is the "wheel of creator" (*KRIM shakti chakra*, 248). The lymphatic system is a copy of the "energy" (*Shakti*, 19) as the "immune system" (*Rogapratirodhi Pranali*, 19) that lets the nervous system triangulate and descend the ascending "guider-effect"

(*Chitta*, 100) with its double copy: the lymphatic system is the first copy within the second copy, the immune system. The "double copy" (*Chitra*, 1) descends the "ether-effect" (*Naad*, 257 = 1 + 100 * 2 + 56) of the two "reproductive" (*Guru*, 100) units within the "maternal body" (*Sharira*, 56). Descending ether-effect descends the "energy exchange" (*Krida*, 79) through copying and ascends the natural "cooling-effect" (*Kaki*, 65) for activating the "earth wheel" (*Ajna chakra*, 65) within the fetal body. The earth wheel ascends the sentient-effect of the lunar wheel within the constant divine-effect of the solar wheel until the thermal force reaches the "infinity point" (*Amara*, 90,000) to activate the crown wheel with zero energy exchange. At that point, the maternal body pushes the fetal body out for living life as a child.

The "parallel reality" (Pushtartha, -8) of the child as a subsystem without the mother as a system leads to the illusionary sixth law of thermodynamics that a body system is a "subsystem" (Hridya Pranali, 100,000) of a "system" (Gabhira, 12) forming a "body" (Mahatattva, 56). The system is "divisible" (Vibhajya, 81) into nine subsystems, each further divisible into nine subsystems, norming eighty-one subsystems as the "span" (Vitasti, -18) of management by a leader. Eight-one divisions leave the system with the 19-unit "energy" (Shakti, 19) to compensate for the "span" (Vitasti, -18) and "self-perpetuate" (Amsha, ½) two units of "work energy" (Shrama shakti, 1), one as a "leader" (Neta, 0) and another as a "universe of followership entities" (Jiva Astikya, 7/8) within the leader's "guiding force" (Chitta, 100 = 81 + 19).

In reality, both the system and the subsystem as the two copies of the body have the "technological capability" (*Vidhana*, 2) for managing eight subsystems, each self-managing an additional eight systems. Thus, they create a "capability bed" (*Akasha chakra*, 64) of sixty-four systems.

A leader may organize each system into a four-body system, comprising an emergent, intrinsic system, a growth system, a maturing, entropy system, and an outsourced, extrinsic system. Thus, the leader enjoys the "corporate leadership work" (*Saraswata*, 256) of the two-hundred fifty-six body systems by transforming the leader-centric causation.

How a Capability Bed Works as the Wheel of Joy

The capability bed shapes one's "version" (*Bhashantara*, 90,000) of "seeing" (*Pashyana*, 900,000). One is always averse to seeing reality. Aversion to seeing reality is beautifully captured by the word Vipassana. Vipassana is a Pali word derived from the Sanskrit word Vipashyana. Vi means "aversion" (*Vi*, 90). Pashyana means seeing. Vipanshyana is translated by those averse to seeing the reality as "special seeing" or the "insight" (*Vandana*, 16). The literal meaning of the aversion to seeing is "blindsight" (*Vipashyana*, 100,000).

In Buddhism, Vipassana is a part of Jhana, the Pali word for Dhyana. Dhyana means "meditation" (*Dhyana*, 9). Meditation is the process of realizing a "dream state" (*Maha Svapna*, 9) and using the dream to gain an insight into the "goal" (*Lakshya*, 9) one should pursue for realizing nirvana. In Buddhism, Vipassana's "complement" (*Upanishada*, 9/16) is Samatha, meaning "abiding" (*Shamatha*, 90). When one is bound by blindsight, Buddhism provides a way for one to use meditation as a tool to dream the goal that will lead to one's nirvana, i.e., "temporary joy.

The "joy" (*Nirvana*, 123) of realizing the goal decided by someone else in the dream is "temporary" (*Asthayi*, 19) because everyone in the universe is devoted to their conscious well-being. When you pursue a goal somebody else decides, you divide your conscious well-being between theirs and yours. You never have "100% confidence" (*Tigma*, 16) that the guru who guides you in the dream has your well-being in mind. In Hinduism, the guru teaches you the technique of deciding the "appropriate goals" (*Abhipraya*, 38) instead of giving you the "right goal" (*Lakshya*, 9) to pursue. The goal in Hinduism is not nirvana but moksha, i.e., the "absolute joy" (*Moksha*, 1600). The joy of realizing the goal decided by oneself while awake is "permanent" (*Sabbacitta*, 18) and present now because in the "present" (*SAUM*, 1600), you realize that the goal you decided is only a path to the goal Mother Nature decides for everyone. Mother Nature's goal is to give each of her children "absolute freedom" (*Param Mukti*, 16) from the "vagaries" (*Taranga*, 1) of the "wheel of joy" (*Akasha chakra*, 64).

Wheel of joy is variously known as the Akasha chakra, Vrishapati chakra, Saptanadi chakra, Svana chakra, Mriga chakra, Samaya bhedoparacana chakra, Nadi chakra, and Ahibala chakra in Hindi, and wheel of ether, wheel of reproduction, etheric cycle, and Serotonin-melatonin production cycle in English. One feels lost after realizing moksha as one believes that was the desirable goal and therefore begins a "path of extroversion" (*Prabhasa Marga*, 25), seeking a param guru while hibernating oneself without the physical body to illuminate the right goal. Similarly, one feels lost after realizing nirvana as one believes that was the hibernating guru's desired goal and becomes disillusioned with the guru. Thus, one begins the "path of introversion" (*Siddha Marga*, 25) to be that "param guru" (*Param Guru*, 10) through moksha.

The Western religions follow the "path of conversion" seeking "eternal joy" (*Mukti*, 17). Eternal joy is the "freedom" (*Mukti*, 17) from the responsibility to decide one's life goal. One gets eternal joy by deciding that the present life itself is the goal of creation. If in the present life one believes that God made one a saint during the act of "creation" (*Srijan*, 379), then one becomes Sanatana. Saint means the "holy spirit within self" (*Lakshmi*, 379), the "harbinger" (*Lakshmi*, 379) of the wealth the creation offers. Sanatana means "ageless" (*Sanatana*, 379). In Western religions, a believer believes he is ageless because he is the saint entrusted with the distribution of the wealth of the creation among the believing followers and its appropriation from the non-believing leaders.

Faith in one's beliefs makes the "impassioned devotees" (*Samayam*, 0) transform the non-believer leaders into believing followers and become the non-believing leaders. Therefore, the West is going through a cycle of transformation from a faith-based culture to a science-based "workculture" (*Nayaki*, 379). After the appropriation of the wealth from the East using the religious "conversion" (*Karma yogi*, 1) as a tool, the West has crafted secular institutions to prevent diffusion of its wealth to those who are increasingly adopting the Western belief systems. Instead, it emphasizes the importance of working by self with the scientific tools to fulfill the promise of the belief system. Once the believing followers reveal their scientific insights, the non-believing leaders move fast to discover those insights, characterize them as the "public good" (*Labha*, 81), and patent them as their private "intellectual property rights" (*Jnanadhikara*, 80). It is called "heads I win and tails you lose" (*Maha Dasha*, 0). Thus, the "path of conversion" (*Surya Marga*, 25) gives eternal joy to the one converting the others who do not believe in the power of believing in oneself and get entangled in your "belief system" (*Saguna*, 20). Amen!

The consciousness of one's behaviors makes one "ageless" (*Sanatana*, 379) as a creation by ascending the confidence for crafting the dimensions of life that lead to the appropriate goals. Dimension is the Dharma one follows as a religion. Sanatana dharma means a religion that leads to the ageless dimension of life by illuminating the responsibilities of an impassioned devotee. An impassioned devotee is like a king who has the freedom to decide one's beliefs. A wise king believes in one's personal experience and learns from everyone's personal experiences.

A guru embodies the "wisdom" (*Chitta*, 100) gained through both the introversion of the social experiences after its discovery and the extroversion of the personal experience for its validation. Sanatana dharma is the ageless dimension that offers the path of inversion for inverting one's reality from a follower to that of a leader, free from the limitations of one's belief system. The "path of inversion" (*Paramatma Marga*, 25) is the "path of introspection" (*Chandra Marga*, 25) like a leader instead of being blinded by the faith as a follower.

Path of introspection leads to the "path of reversion" (*Gana Marga*, 25) to the group that is racing against time for knowing the reality that unfolds with time. "Race" (*Parampara*, 13) in Sanskrit means Parampara. Sanatana Parampara is the ageless race for authenticating the reality one discovers through introspection and promotes universally as a guru. There is urgency for one to authenticate the reality of one's belief system because of the short lifetime. If our belief system is not valid, then the conscious cost is infinite. We make a fool of ourselves believing what others are saying, howsoever impassionate they might be, and the mass they accumulate through their "Big Organization" (*Manojava*, 13) of the voice they have to silence us into believing that they must be knowing the reality we don't.

How else can one make infinite followers if one is not credible? How can one be credible without being authentic? How can one be authentic if one follows a belief system that says God created everyone in HIS image and gives salvation to only the chosen ones who believe that everyone was created so that the chosen ones could rule the universe by making the others believe in what they believe in? That means the believers never behave according to their beliefs.

There is always a difference between what the believers value and the behaviors they practice. The GLOBE (Global Leadership and Organizational Behavior Effectiveness) Project on the culture and leadership in 62 societies, whose seminal book I co-edited while at the Wharton School, shows significant differences between people's values and practices. These findings falsify the dominant science of psychology that stated, following the Western belief system, that values shape practices with 100% perfection.

Does falsification of the belief system mean I should take pleasure in providing a reality beyond our beliefs? Pleasure means Rati in Sanskrit. Rati means the extrinsic taste of the Big Responsibility that makes one rejoice with being who one is. One takes pleasure in believing "I am who I am" and behaving without minding your voice. Granted, you have a voice and a strong wish to express the voice apparently for my conscious wellbeing. However, if I do not mind your voice, then I can live a life of a series of joys.

The joy is temporary because I get to enjoy it as long as I

have the willpower to ignore your voice. I know you are in the race to win a Big Prize at the end of the competition that the Big Boss has promised you if you succeed in making me mind your voice. You are hitting a wall with me as I keep smiling while you are getting agitated and destroying your health, knowing that eventually, you will need to go under the care of the doctors, and I will be the sole survivor. I will be the default winner.

You have a choice not to choose the path of extroversion, but instead, take the path of introversion and begin speaking to yourself. Now, you put me in a fix because if I keep silent as a rock, I will end up making you a rockstar of the competition that is racing to the finish against time. By choosing to disrupt your path of introversion with the path of extroversion you demonstrated to me earlier, I test your patience for listening to the two voices concurrently and focusing only on knowing "I" who makes me "who I am" to disrupt your joy.

If one repeats what one is doing by voicing one's voice while everyone else is silent and focused on their work, then everyone begins paying attention to the one who reproduces what the other is repeating. There must be something the reproducer knows that the one repeating is seeking to discover and that everyone must seek to validate before the first one discovers the knowable. Thus, repetition breeds behavior, not belief. One does not need to repeat the belief.

Belief leads to devoting energy to repeat one's behaviors and for everyone to reproduce those behaviors. One believes that one's behaviors are appropriate because one deifies oneself by imagining oneself as the photocopy of God, the para deity.

When one fails to discover God within infinite repetitions of one's behavior, one begins believing that even God is prone to errors. That is why God repeats the mistake of gifting the power of giving birth to the females but making that power depends on the males who act as if they are the God of love. To win them, the females must compete and sacrifice everything they have. Such a belief has transformed the entire world, across cultures, into male-dominated cultures.

4.4 Transcending Causation at the Fourth Infinity Level: The Council of Seven Light Forces and The Etheric Body

The fourth infinity is the "transcendental value" (Khuda, 6) a follower enjoys from the "corporate leadership work" (Saraswata, 256) activated by the "wheel of wheels" (Hora, 25). The etheric body is the "consciousness body" (Bhoga sharira, 957), comprising the consciousness we reproduce for producing infinite value. It is a "replicative body" (Bhoga sharira, 957) a child replicates for enjoying the "infinite value" (Viyojya, 957) the mother enjoys as an adult. The mother's infinite value includes the value accrued through:

- the grandfather's "corporate leadership work" (Saraswata, 256) and its "transcendental value" (Khuda, 6), net of the "wheel of wheels" (Hora chakra, 25) = 256 + 6 − 25 = 237.

- the grandmother's "reorganization" (Punargathan, 130) of the "system" (Gabhira, 12) into the twelve sub-systems that reproduce the system without them as the thirteenth sub-system. As a copy of the system, the thirteenth sub-system produces the twelve systems within it. By copying the thirteenth sub-system, all twelve sub-systems produce twelve systems within them. After the "ten-fold growth" (Vicitra mandala, 170) of the system through the reorganization of its three copies, there are 156 "with system" (Purvabhasa, 156) copies of the first copy, 13 "within system" (Sushumna, 10) copies of the second copy, and 1 "without system" (Ardhajya, 10) copy of the third copy. Reproducing the ten-fold growth by taking the system as its third "without system" (Ardhajya, 10) copy, the "double copy" (Chitra, 1) produces an additional 340 "above system" (Ghataana, 340) copies positioned as a "quantum copy" (Lambila, 680).

The quantum copy comprises four units of the ten-fold growth, forming four "layers" (Parat, 123): the first of the system as the third copy, the second of the second copy, the third of the first copy, and the fourth of the zeroth copy the system forms before conceiving its double copy image. Before their "exemplification"

(*Drishtanta*, 620) as a "quantum copy" (*Lambila*, 680), the "system" (*Gabhira*, 60) and its four copies organized as an "octave of copies" (*Sargam*, 60 = 12 + 12 * 4).

Thus, the grandmother's exemplification adds 620 units to the 237-unit "example" (*Udaharana*, 237) the grandfather sets for generating an infinite value of 957 within the mother. It transcends the reality of the grandfather's corporate leadership work and illuminates the self-luminous reality of the grandmother's "corporate followership work" (*Drishtanta*, 620). *It leads to the seventh illusionary law of thermodynamics that the leadership leads to entropy due to the "ascending power distance" (Sadhyata, 80). Further, the followership leads to growth due to the "descending power distance" (Gaganaganja, 90).*

A child's "replication" (*Samalekha*, 90) of the descending power distance transforms the "infinite value" (*Viyojya*, 957) into a sequence of "finite values" (*Dhriti*, 30), each forming a "causal body" (*Karana sharira*, 30). With its ascending "disintermediation" (*Madhyasthahinta*, 900,000), the maternal "causal body" (*Karana sharira*, 30) "self-services" (*Svayamseva*, 2/3) her "sentient spirit" (*Ruah*, 20) for guiding the child's sentient well-being. The sentient spirit makes the "replication" (*Samalekha*, 90) a "primordial etheric body" (*Samalekha*, 90) for taking the child's sentient well-being to "infinity" (*Tamo guna*, 90,000) with the mother's "sentient energy" (*Amrit*, 1000) within her.

Mother's "sentient spirit" (*Ruah*, 20) and the Father's "transcendental value" (*Khuda*, 6) are immanent within the "primordial etheric body" (*Samalekha*, 90) in the form of the "Param Para Paternal" (*Saptarishi*, 26). The Param Para Paternal replicates the "trading-effect" (*Damodara*, 26) by forming a "Council of Seven light forces" (*Saptarishi*, 26). Seven light forces are the "light forces" (*Apas*, 169) of the six "planetary bodies" (*Vastu*, 9) without the "Earth" (*Bhu*, 724) as the seventh planetary body. Six planetary bodies include Mercury, Venus, Mars, Saturn, Jupiter, and Moon. The two outer planets, Uranus and Neptune are the Moon's double copy. The Moon is the Earth's copy. The Earth is the Moon's zeroth copy.

The Moon is present as the dark matter before the formation

of the Earth. After forming the Earth, the dark matter transforms into the moon. The Sun is the "white star" (*Yama*, 180) before the formation of the dark matter. The white star forms the dark matter as a triple copy. The third copy forms the Earth by repelling its past's "dark energy" (*Arundhati*, 691). The second copy transforms into the Moon. The first copy norms the Dark Matter.

After conceiving the Moon within her, Mother Earth forms Mercury and Venus as her "double copy" (*Chitra*, 1) for generating a "thermodynamic force" (*Rodha*, 1) that leads to her "entropy" (*Mahodbhava*, 5). After her "entropy" (*Mahodbhava*, 5 = 3 + 1 * 2), Earth transforms into a "triple copy" (*Vaishya*, 3) and a pair of "double copies" (*Chira*, 1). Mars, Saturn, and Jupiter are the Earth's triple copies. The second double copy is within Moon. After the Moon becomes free from the Earth's gravitational mass, it repels its double copies away from the Earth and her copies for attracting the Earth towards it. The first double copy forms the "Sun" (*Surya*, 21) as the "earth-effect" (*Ravi*, 21) and the "White Star" (*Yama*, 180) as the "light" (*Prabha*, 180) within the Sun. Like the second double copy, the first double copy is also within the Moon. However, the first double copy is the first to manifest without the Moon.

The "light forces" (*Apas*, 169) of the seven planetary bodies take diverse forms and are immanent within the respective planetary bodies in their "natural essence" (*Samaa*, 1). The light force manifests as the earth force within Moon that produces the ether force within Mercury and the guider force within Venus. "Earth force" (*Jantumati*, 21) is the "production force" (*Jantumati*, 21) that produces the "ether force" (*Naad*, 257) as its "reproduction force" (*Naad*, 257) and the "guider force" (*Chitta*, 100) as the "reproductive force" (*Chitta*, 100) of the "production" (*Asevana*, 1869) followed by the "reproduction" (*Shuddhi*, 285). The production of the Moon without herself consumes the "air force" (*Maruti*, 78) within Earth. The air force is the "oxygen" (*Maruti*, 78) needed for the "Moon" (*Soma*, 997) to be conceived as the "New Moon" (*Chandra*, 82) like a departed "human entity" (*Manushya*, 82) and nurtured into the "Full Moon" (*Soma*, 997) with the Earth's "energy force" (*Agrata*, 916). The energy force incarnates the human entity as the "Saturn" (*Shani Bhagwan*, -1 = 997 - 916 - 82).

Before incarnating as Saturn, the human entity incarnates as the Earth, Moon, and Mercury. A negative three-unit "past reality" (*Padartha*, -3) is within Saturn in the form of the "divine-effect" (*Padartha*, -3) of the "divine plan" (*Sahajata*, 916) that lets the Earth trade the energy force from the Mars. Saturn is not conscious that it was "Mars" (*Mangala*, 102) before Earth. The Mars generates the "energy force" (*Agrata*, 916) in the form of the 254 units of the "water force" (*Jalaprana*, 4) after trading the 257-unit reproductive force, servicing the negative 3-unit past reality, and exchanging the "consciousness" (*Chaithanya*, 4) of the four copies that precede it in the form of the Jupiter, Earth, Dark Matter, and White Star.

Mars hides its "identity" (*Pehchaan*, 8) from Saturn by following Venus's "guider program" (*Shivadrishti*, 17). The "Venus" (*Shukra*, 2700) incarnates after Saturn. Venus is the first copy of the Earth. However, its "primordiality" (*Ayanamsha*, 27) is "overshadowed" (*Avakirna*, 1) by Jupiter's "sentient performing" (*Yoni*, 1000). Earth is Jupiter's copy. Jupiter imagines that it has a copy due to the "hallucination" (*Dhyana*, 9) from the "heat" (*Gharma*, 8). It perceives the Earth as the copy without it. It conceives the Earth to be the cause of the "thermodynamic force" (*Rodha*, 1) that generates the heat of togetherness. However, with experience, it realizes the consequentiality of the Earth in cooling the air surrounding into the "gas" (*Paramartha*, 365). The gas is of the "electrons" (*Dyumna*, 365) formed due to the "fire-effect" (*Rupa*, 100,000) within the Jupiter that gives the latter the "form" (*Rupa*, 100,000) of the "solid earth" (*Ravi*, 21).

The solid earth is the earth-effect the Moon trades from the Earth when the latter formed as a "Jupiter" (*Brihaspati*, 1780). The "Jupiter" (*Brihaspati*, 1780 = 80 * 21 + 100) produces the "earth force" (*Ravi*, 21) after manifesting the "conscious spirits" (*Ruah*, 20) of Mother Earth and her three copies—the Moon, the Mercury, and the Venus within himself. Unknowingly, Jupiter reproduces the reproductive force Venus trades from Earth as the second of the first-generation triple copy. Although Jupiter is the Earth's first copy, the hallucinating Jupiter believes that the Earth is his copy. Therefore, Jupiter is blind to the Earth's reproductive force until Venus manifests it. Once Venus manifests the reproductive force,

Jupiter fears that Venus will become "omnipotent" (*Sarvashaktiman*, 1600) by breeding fifteen additional copies besides Jupiter as the first copy and Venus as the zeroth copy within a seventeen-unit "guider program" (*Shivadrishti*, 17).

Though there are only eight planets, the "solar universe" (*Surya mandala*, 13) comprises thirteen copies that include the Sun and the Moon as part of a "family" (*Parivar*, 10) perpetuating over the three dimensions of time. The other four copies comprise a double copy of the dark matter and the white star. After creating the solar universe, the white star behaves like the omnipotent dark matter after diffusing its light with a force of creation. The dark matter conceives the inanimate matter that forms the solid earth as its future reality and the divine force of the "animate entity" (*Sushupti*, 96) whose "consciousness" (*Chaithanya*, 4) forms the white star as the light of its past reality. However, both the animate entity and the consciousness are the parts of the "guider force" (*Chitta*, 100 = 96 + 4) the dark matter produces in the "present" (*SAUM*, 1600) after consuming the white star's "double octave" (*Madhusudan*, 16) of copies.

Seven light forces are the ether-effect within Mercury, earth-effect within Moon, air-effect within Earth, water-effect within Mars, fire-effect within Jupiter, divine-effect within Saturn, and guider-effect within Venus. Uranus norms the descending light force due to the ascending mass of the solar universe formed with the seven light forces. Neptune norms the ascending light force due to the descending mass of the solar universe as the solar universe reproduces the "universe" (*Brahman*, 2) for breeding another universe within another solar universe by feeding the consciousness within itself as the solar universe. Finally, the Sun norms the "horizontal light force" (*Bhuyojan*, 21).

The seven light forces emanate from the Moon to form the two outer planets, Uranus and Neptune, after trading the light of the other six planets emanating from the Sun. The Sun conceives the light of the six planets while it is a white star. The white star's "conceived reality" (*Yuktartha*, 6) of the six planets is within the dark matter. The "white star" (*Yama*, 180) becomes the "Sun" (*Surya*, 21) after its "copy" (*Nakala*, 0) becomes the "dark matter"

(Dakshinashapati, 1600). The latter's oneness with the Sun is the "tangent" (Ardhajya, 10 = 1,0) for the "tangent line" (Paramanu, 160) of the eventual "development" (Param Parvati, 160). Each of the ten bodies produces a double octave of copies. The double octave forms seven lightforces within the Moon, including the moon's, and nine lightforces within the white star, including the lightforces of the white star and the sun.

The light of the past reality emanates from the dark matter within the "freedom from the present-effect" (Moksha, 1600); the dark matter is the "present" (SAUM, 1600) fused with the past reality. The light gains "ascending force" (Vaimitra, 99) when the solar universe is reproducing itself within the "Neptune" (Ketu, 140) by polarizing its mass into the "Uranus" (Rahu, 73). As the Uranus gains mass, the Neptune believes that the former's reproductive force will make it rotate around the former like a tail. Therefore, Neptune exchanges the reproductive force and reproduces five additional solar universes to divide the ascending mass of the Uranus. The growth of six universes, including the one reproducing an additional five, leaves only two units within the "neutron star" (Tara, 2) that eventually becomes the "Uranus" (Rahu, 73 = 13 * 5 + 6 + 2). A "white star" (Yama, 180) becomes a "neutron star" (Tara, 2) after the "replication" (Samalekha, 90) of an additional white star.

The neutron star conceives its Uranus self to be a "person" (Vyakti, 1) whose mass of consciousness is radiating the light force to transform the second unit into a "universe" (Brahman, 2) produced by Neptune as the "twin person" (Insan, 1). The Neptune first produces the present solar universe, which reproduces four additional solar universes in the four directions without consciousness of its presence within those four directions. After the present solar universe's entropy due to "supernatural growth" (Nitya Ratri, 8), the four directions reproduce the present solar universe as the sixth copy. Neptune is Saturn when it first produces the present solar universe.

Due to its entropy followed by growth, the present solar universe is unaware that the seven lightforces emanating from the Moon are a subset of the nine lightforces emanating from the

Sun. The white star services its "light" (*Prabha*, 180 = 140 + 21 + 13 + 2 + 5) for conceiving "Neptune" (*Ketu*, 140), the "Sun" (*Surya*, 21), the "solar universe" (*Surya mandala*, 13), the "neutron star" (*Tara*, 2), and the "entropy value" (*Sarvanasha*, 5) of the present solar universe. Therefore, it assumes that the sixteen light forces form a double octave and makes a "double copy" (*Chitra*, 1) of the "consciousness" (*Chaithanya*, 4) of the "octave of copies" (*Sargam*, 60) within the "Sun" (*Surya*, 21 = 16 + 1 + 4). The octave of copies comprises the eight planetary bodies, each reproducing eight additional bodies and servicing four as the consciousness of the sixty bodies that norm the octave of copies. Eight additional bodies take the specific planetary body as the ninth body and the general planetary body comprising all eight as the tenth body. Next, the ten bodies imagine the double octave of copies as the double octave of lightforce of the "eight body system" (*Palala*, 64) formed within and without the varying ninth body.

The white star's light takes nine different forms and manifest as the nine varying body systems, comprising the Sun, the Moon, and the seven planets other than Neptune. These nine lights transform the "neutron star" (*Tara*, 2), which eventually becomes the Neptune, first into the "Param Para Maternal (*Navgraha*, 26) of the two "solar universes" (*Surya mandala*, 13). As the Param Para Maternal, the Neptune becomes the "Supreme Council of the Nine Archangels of Light" (*Navgraha*, 26).

The "Supreme Council" (*Navgraha*, 26) conceives the "Uranus" (*Rahu*, 73) as the "Param Perpetuator" (*Sirnapada*, 26) within herself. The Param Perpetuator is the "Infinite White Star" (*Sirnapada*, 26) formed with the "consciousness" (*Chaithanya*, 4) of the time-varying "triple copy" (*Vaishya*, 3) of the Uranus as the "person" (*Vyakti*, 1). The consciousness is the "primordial causal body" (*Chaithanya*, 4). As a Param Perpetuator, Uranus becomes the "High Council Guardian of Luminous" (*Sirnapada*, 22).

The "High Council" (*Sirnapada*, 22) is the keeper of the key to the presence of the "luminous" (*Rochisha*, 13), i.e., the "solar universe" (*Surya mandala*, 13) as a reincarnated sixth copy within the "present space" (*Sadakhya*, 9). The sixth copy incarnates as the "param child" (*Manyu*, 19) after the six-unit growth of the solar universe. The param child is the "key" (*Chabi*, 19) to the "infinite space" (*Para Ganesha*, 19) formed by producing a sequence of six-unit growth. The sixth copy is the second replication within the double "replication" (*Samalekha*, 90) that copies the "light" (*Prabha*, 180) as the "genome" (*Somapa*, 180). The first replication is the "eighth copy" (*Shri*, 81 = 100 - 19) that produces the "infinite space" (*Para Ganesha*, 19) as a unit of "energy" (*Shakti*, 19) by reproducing the "primordial space" (*Dik*, 100) as a unit of "reproductive energy" (*Lalita*, 100). Thus, a unit of energy perpetuates over time as a sequence of energies that "copy" (*Nakala*, 0) the preceding unit of energy.

The energy that forms Uranus as the "person" (*Vyakti*, 1) is immanent within Mother Nature's "conscious spirit" (*Ruah*, 20 = 19 + 1). The conscious spirit forms a sequence of energies through the "crown wheel" (*Sahasrara chakra*, 45,000) using the "Luminous Point" (*Laal*, 6) as the "gate" (*Dvara*, 264) for norming a "normative development paradigm" (*Martya*, 146) for her sentient well-being when she takes the diverse forms as a "photon" (*Kapinjala*, 20). The normative development paradigm limits the thermodynamic entropy of her succeeding "primeval forms" (*Konastha*, 13) by transforming the growth of the preceding "primordial forms" (*Siddha*, 7) into a "mortal entity" (*Martya*, 264).

The "High Council" (*Sirnapada*, 22) is also the keeper of the record of the presence of the "maternal self-luminous entity" (*Maha Gayatri*, 12) as a sequentially reincarnating ninth copy within the param child. The ninth copy is the sequential reincarnation of the "seventh copy" (*Siddha*, 7). The seventh copy is the "omnipotent photon" (*Siddha*, 7). It incarnates as a "sequence of exchange photons" (*Siddha*, 7) to perpetuate the param child as the "formative growth paradigm" (*Sarvanasha*, 5) without the two units that transform into the Uranus and the Neptune and the twelve units that norm the maternal self-luminous entity with the growth of those two units.

Each "exchange photon" (*Apadharma*, 100,000) exchanges both the "energy" (*Shakti*, 19) that forms the preceding form of the param child as well as the "exchange force" (*Vyanavata*, 365 = 19 * 20 − 19 + 4). The exchange force forms a sequence of twenty "energy" (*Shakti*, 19) units in the form of the 21st photon within the energy unit that forms the 22nd photon. The exchange force consumes the "energy" (*Shakti*, 4) of the 21st photon and services the "consciousness" (*Chaithanya*, 4) of the 23rd photon.

The 23rd photon forms the Earth into a "Ring Council of time keykeeper" (*Shamana*, 23). The Ring Council gives a "body" (*Mahatattva*, 56 = 30 + 13 * 2) to the param child by forming the "causal body" (*Karana sharira*, 30) with the "seventh copy" (*Siddha*, 7) for reproducing the "primeval forms" (*Konastha*, 13). The "Ring Council" (*Shamana*, 23 = 19 + 4) is the consequential tenth copy the param child produces with the "primordial causal body" (*Chaithanya*, 4). The body illuminates the "Earth" (*Bhu*, 724 = 7 * 100 + 14 + 10) as the "metaphysical entropy paradigm" (*Vajra*, 14) without the seven planets whose "reproductive energy" (*Lalita*, 100) is immanent within the "seventh copy" (*Siddha*, 7) and the "mass consciousness" (*Satma*, 10 = 7 + 3) of the seventh copy and its time-varying triple copy. The mass consciousness produces the "reproductive energy" (*Lalita*, 100 = 10 * 10) in the form of the "tenth copy" (*Shamana*, 23) immanent within the body. The tenth copy is the "reproductive potential" (*Kalika*, 23).

The "infinite exchange" (*Sitanveshana Panditaya*, 816) is the "primordial intellectual body" (*Sahadeva*, 816) that activates the copying of the tenth copy into the primeval forms of the param child. The copying is without the "consciousness" (*Chaithanya*, 4) of the "luminous point" (*Laal*, 6) that forms the "mass consciousness" (*Satma*, 10 = 4 + 6). The tenth copy reproduces the "mass" (*Prithvi*, 132) that forms the "consciousness" (*Chaithanya*, 4) in the form of the "gravitational mass" (*Mitra*, 132). The gravitational mass transforms the "Earth" (*Bhu*, 724) into the "primordial intellectual body" (*Sahadeva*, 816 = 724 + 132 - 40) of the "reproductive reality" (*Bhavartha*, 40) that forms the "Dark Matter" (*Dakshinashapati*, 1600 = 40 * 40) as the "twin earth" (*Dakshinashapati*, 1600).

The High Council is also the "Death God's" (*Yama*, 180) council. The "para-conscious trading" (*Kshema*, 816 = 81 → 6) of the growth as the "luminous point" (*Laal*, 6) without the "growth sequence" (*Shri*, 81) is the "tertiary cause" (*Svadhisthan*, 816) for the degeneration, decay, and death of the "body" (*Mahatattva*, 56) and the body's transformation into the "etheric body" (*Bhoga sharira*,

957) of a "universe of param children" (*Bhoga sharira*, 957). The "etheric body" (*Bhoga sharira*, 957 = 9 * 100 + 1 + 56) comprises the "reproductive energy" (*Lalita*, 100) of the seven planets, the Moon, and the Sun within the Uranus as a "person" (*Vyakti*, 1) with "whole body" (*Mahatattva*, 56).

The "tertiary cause" (*Svadhisthan*, 816) is "initiated" (*Upadishta*, 890 = 816 + 18 + 56) as a "primordial param child" (*Upadishta*, 890) with the Uranus's "potential" (*AUM*, 18) and "body" (*Mahatattva*, 56). It forms the etheric body as a "universe of energy" (*Bhoga sharira*, 957). The "body's entropy within the potential's growth-effect" (*Prayujya*, 208) transforms the "tertiary cause" (*Svadhisthan*, 816 = 208 * 4) into the "Second Astral Body" (*Atma*, 4). The Second Astral Body is the "consciousness stock" (*Atma*, 4) that forms the "High Council" (*Sirnapada*, 22) with the "trading-effect" (*Damodara*, 26 = 4 + 26 = 6 + 20) of the "Luminous point" (*Laal*, 6) within the "photon" (*Kapinjala*, 20).

Trading-effect transforms Uranus's "potential" (*AUM*, 18) into the "para-consciousness" (*Mahatma*, 18) of Neptune's "intellectual body" (*Sukshma sharira*, 306). The "intellectual body" (*Sukshma sharira*, 306 = 100 * 3 + 18/3) comprises the "reproductive energy" (*Lalita*, 100) of the time-varying "triple copy" (*Vaishya*, 3) within the time-divided "potential" (*AUM*, 18). The triple copy incarnates the "universe of param children" (*Bhoga sharira*, 957) in the form of the "universe of primeval children" (*Samghatavigrhitartha*, 900). Each "primeval param child" (*Janani*, 105) bears the "technological cost" (*Yojya*, 58) of the "liberal democratic technique" (*Yukti*, 8) that leads to her "replication" (*Samalekha*, 90) as an "INTO INFINITY body" (*Prakarana*, 60) within a "primordial param child" (*Upadishta*, 890). The INTO INFINITY body "advances" (*Agrima*, 60) the "density" (*Sandra*, 60) of the etheric body as a "conservative democratic technique" (*Ramarajya*, 60) to compensate for the escalating technological cost.

By "self-projecting" (*Ekashtaka*, 3/2) itself as an INTO INFINITY body, the descending "circulating force" (*Prakarana*, 60) of the Sun as the primordial param child forms the ascending "recirculating force" (*Chakrika*, 90 = 60 * 3/2) of the Moon as the primeval param child with a potential to reproduce the seven planets, the Moon, and the Sun. The "reproductive force" (*Chitta*, 100) of the seven planets without the Moon and the Sun, and their "reproductive forces" (*Chitta*, 100) within the Moon's and the Sun's "reproductive forces" (*Chitta*, 100), forms the sixteen-unit "essence" (*Sara*, 16) of the "reproductive force" (*Chitta*, 100) into the planet "Mercury" (*Budha*, 1600).

The planet "Mercury" (*Budha*, 1600) is the "dark matter" (*Dakshinashapati*, 1600) illuminated in the form of the "reproduction force" (*Naad*, 257) as the dark matter reproduces the Sun and the Moon. Due to the thermodynamic effect, the Moon forces the Sun to produce the five planets without Uranus and Neptune for descending the entropy. The growth of the five planets and the Sun produces Uranus by exchanging the "thermodynamic-effect" (*Rodha*, 1). The "illuminating value" (*Prakasha*, 7) of the dark matter without the Moon transforms the "Moon" (*Soma*, 997 = 140 + 724 + 133) into the "Neptune" (*Ketu*, 140), the "Earth" (*Bhu*, 724) and their "simultaneity" (*Sahokti*, 133) within the "white star" (*Yama*, 180). By "servicing" (*Apana*, 47) the "simultaneity" (*Sahokti*, 133), the "white star" (*Yama*, 180 − 47 + 133) transforms into the "Moon" (*Soma*, 997) that forms the Neptune with its "descending mass" (*Ketu*, 140) and the Earth with the "limiting element" (*Bhu*, 724) that limits the "mass descendance" (*Samalekha*, 90). The limiting element limits the Moon's entropy within the growth by keeping the Moon away from the illuminating value subject to entropy.

The "sound vibration" (*Dhvani*, 257) of the white star's "metaphysical entropy" (*Cheerna*, 90,000) activates the dark matter's "absolute consciousness" (*Paramatma*, 1600) within the planet "Mercury" (*Budha*, 1600). Thus, Mercury begins acting like a "param human child" (*Kesari Nandan*, 1600). The metaphysical entropy comprises the "infinity" (*Tamo guna*, 90,000) of "energy exchanges" (*Krida*, 79) that transform the "param human child" (*Kesari Nandan*, 1600) into a "continuous gravitational wave" (*Ananta*, 90,000) of an "infinite density" (*Sagarottarakaya*, 90,000). Each "param child" (*Manyu*, 19) within the continuous gravitational wave enjoys a "variable lifetime" (*Pradyumna*, 60) within the prior "illuminated growth" (*Pradyumna*, 60) and a "constant lifetime" (*Svahita*, 125) without. The constant lifetime comprises a unit of variable lifetime conceived within the illuminated growth, a unit of variable lifetime perpetuating with the illuminated growth once conceived, and the "perpetuating value" (*Saranyu*, 5) of the "conceived reality" (*Yuktartha*, 6) without the "discordant energy" (*Asura*, -1) that leads to the entropy of the conceived reality.

The sound vibrations of the energy exchanges over the "infinite lifetimes" (*Vardhimainaka Pujitaya*, 1000) transform the Moon into the "Wheel of Life" (*SRIM shakti chakra*, 9×10^{18}). The "ether-effect" (*Naad*, 257) makes the wheel of life a "wheel of causation" (*SRIM shakti chakra*, 9×10^{18}). The wheel of causation reproduces the energy exchanges in a "lexicographic sequence" (*Shodashottari dasha*, 40) to form a "book of life" (*Pustak*, 14). The book of life services the "causation" (*Hetu*, 1) that makes each param child a "wisher" (*Chaahak*, 0) wishing to be a "para-conscious body" (*Manomaya sharira*, 381) guided by a para-conscious "wishing sequence" (*Abhilasha krama*, 87). The para-conscious body is the "mental body" (*Manomaya sharira*, 381) guiding the "intellectual body" (*Sukshma sharira*, 381) guided by the "conscious body" (*Linga sharira*, 3). The conscious body is conscious of the "wish" (*Chah*, 18). It transforms the param human child into the "primeval human child" (*Vishvakarma*, 108) to manifest the "memorized divination" (*Alaya vijnana*, 108).

The Sun services the energy for the energy exchanges by transforming itself into a "primordial human child" (*Gabhira*, 12) whose "reality" (*Vastavikta*, 7) is "self-luminous" (*Svarochisha*, 12) in the form of a "quantum photon" (*Svarochisha*, 12). Each "photon" (*Ruah*, 20) within the quantum photon enjoys the "consciousness" (*Chaithanya*, 4) of the "reproductive-effect" (*Chitta*, 100) of the energy exchanges. As a "photon body" (*Karana sharira*, 30), the "causal body" (*Karana sharira*, 30) is conscious of the "possibility" (*Hariti*, 128) of a "beyond absolute paradigm" (*Pravritta*, 12) for liberating the primordial human child from the metaphysical entropy. The beyond absolute paradigm is one of channeling the "reproductive-effect" (*Chitta*, 100) into the "primeval human child" (*Vishvakarma*, 108) for producing each "human child" (*SHREEM*, 8) as an "octave of entities" (*Krishnamurti*, 1). The octave of entities produces an "eight-fold growth" (*Satarupa*, 8).

By gifting the "greeter consciousness" (*Pratyagatma*, 1) of the eight-fold growth to an "animal child" (*Triyancha*, 19), the human child becomes the "King of deities" (*Indra*, 0) and the animal child becomes the "deity" (*Deva*, 1) working to fulfill the wishes of the King as a member of the latter's causal body. The animal child's departed soul perpetually guides the King on "reaping" (*Prarabdh*, 1) the "fruit of [his] action" (*Karma phal*, -10^{19}) by leading him on to the "path of [her] knowing" (*Jnana marga*, -10^{19}).

4.5 Enjoying Causation at the Fifth Infinity Level: The Supreme Council and the Causal Body

The fifth infinity is the "perpetuating value" (*Saranyu*, 5) a "present follower" (*Siddhi Yogi*, 5) enjoys by trading the "overflowing consciousness" (*Ishvara*, 5) of a "past follower" (*Ajiva*, 92). After deciding to channel the "reproductive-effect" (*Chitta*, 100), a king becomes the past follower and makes the deity the "present follower" (*Siddhi Yogi*, 5). The deity follows by "reproducing" (*Maharavana*, 8) the "perpetuating value" (*Saranyu*, 5) of the king's "paternal consciousness" (*Pitra*, 4) that perpetuates her as his "double copy" (*Chitra*, 1), first as a daughter and second as a granddaughter. When the daughter reproduces the perpetuating value, she becomes the "present maternal" (*Saranyu*, 5).

When the granddaughter reproduces the perpetuating value, she becomes the "past grandmother" (*Vaishvanara*, 17) essential for making the daughter the present maternal. The king is the past grandmother's "copy" (*Nakala*, 0) that a "grandfather" (*Dada*, 10) conceives for reproducing himself as a "perpetuator organization" (*Ishvara*, 5). The perpetuating value comprises the four units of deity that reproduce four generations of the king and a double copy that produces two generations of the "queen" (*Rani*, 7). The double copy is the king's "illusionary production" (*Maya*, 1), not knowing the queen's "reality" (*Vastavikta*, 7). By reproducing her four generations and a double copy for producing two generations of the king as an illusion, the queen becomes an "Almighty Deity" (*Homa*, $25 = 5 * 5$).

By manifesting the queen as the Almighty deity and the king as the deity, a "manifestor" (*Sadhana*, 368) becomes a "param para maternal" (*Navgraha*, 26 = 25 + 1). The param para maternal is the "Supreme Council of Nine" (*Navgraha*, 26) because she includes four gender-differentiated "generations" (*Amnaya*, 97) and one gender-differentiating illusionary production of the successive generations. Each prior generation dies and gives way to the succeeding generation. A "mother" (*Mata*, 4) element is essential for manifesting the succeeding generation. A "father" (*Baap*, 3) element is essential for manifesting the prior generation. The "male" (*Nar*, 100,000,000) provides the "seed" (*Bija*, 379) as the "essential element" (*Bija*, 379) for "creation" (*Srijan*, 379) of the "female" (*Nari*, 10,000).

The female is the "reproductive system" (*Prajanan Pranali*, 10,000), reproducing the male as the "ideal manifestor" (*Adana vijnana*, 100,000,000 = 10,000 * 10,000) of two successive females, one as a "wife" (*Patni*, 121) and another as a daughter. As a creation, the female is the manifestor of the male as her "husband" (*Pati*, 121). As a "creature" (*Purusha*, 12), the male is the ideal manifestor of the female as her "daughter" (*Putri*, 16). As a "creator" (*Brahma*, 59), the Supreme Council comprises nine planets that manifest the "Moon" (*Soma*, 977) as the life-differentiating ninth planet. Without Moon, the eight planets are "devoid of information" (*Andha Kuan*, 82) about "life" (*Zindagi*, 4). The Moon manifests the "information" (*Sandesha*, 1) about the "deeds" (*Karmana*, 1) one does in one's "life" (*Zindagi*, 1) by reproducing the "path of devotion" (*Bhakti Marga*, 1) to those deeds in one's successive lives.

If you are a singer in the present birth, you are naturally attracted towards the singers in the next birth. That way, you generate "ascending demand" (*Kharidi*, 6), i.e., "transcendental value" (*Khuda*, 6) for the singers, thus substantiating the worth of your deeds as a singer. Your attraction towards the singers is the "destiny" (*Niyati*, -1) you can't change because it is "reaping" (*Pararabdh*, 1) the fruit of your past actions. However, your togetherness with the "consumer class" (*Prakriti Yogi*, -1) for consuming the singer's production is the "fate" (*Bhagya*, 27) you can "change" (*Badalana*, 89) because it is "sowing" (*Paryupti*, 19) the "seeds" (*Bija*, 379) for the future consumable with your present "actions" (*Karma*, 10). If you decide to consume the singer's production, in the next birth, you generate "descending demand" (*Kshetra*, 189) for the singers, thus substantiating the worth of your deeds as a consumer.

Once you decide to be a consumer of the singer's production, your "fortune" (*Kismat*, -250) lets you invest your "knowing" (*Jnana*, 19) of the sowing for trading the "infinite production" (*Rajasva*, 10^{10}) of not only the singers but also the non-singers. Once you decide to be an "investor" (*Rikshanayaka*, 16) of your knowing, in the present birth, you generate "horizontal demand" (*Maha Vibhu*, 14) for your knowing. Everyone with demand for your knowing is in "luck" (*Naseeb*, 123) because by knowing you as a consumer with "capital" (*Punji*, -2) to invest, they can service their "value" (*Mulya*, 180) to make you their "copy" (*Nakala*, 0). For instance, if you consume a painter's production, you are motivated to be a painter in the future birth. The wish to be a painter for enjoying the value you have created as a consumer is natural growth. It fills the void generated by the natural entropy of the wish to be a consumer in the present birth after you run out of the capital to perpetuate your "identity" (*Pahachaan*, 8) as a consumer.

The Supreme Council services the "linear flow of time" (*Chaithanya*, 4) in the form of the "primordial causal body" (*Chaithanya*, 4) to make a human child conscious of the "cause" (*Karana*, 18) of his birth as a "creature" (*Purusha*, 12). The creature transforms the primordial causal body into a "paternal universe of cells" (*Usha*, 16 = 12 + 4), norming an eight-fold growth of the "maternal universe of cells" (*Dharati*, 2). The creature is the "grand cause" (*Pratikarak*, 12) for the "supernatural growth" (*Maha Nitya*, 8) of the maternal universe. As a "creature" (*Purusha*, 12) who "self-services" (*Svayamseva*, 2/3) the "cause" (*Karana*, 18), a human child becomes free from the "technological oneness paradigm" (*Maha Jagrat*, 0) that makes him a "copy" (*Nakala*, 0) of the "producer" (*Nirmata*, ¾) whose production he has consumed in the past. The oneness with a "self-luminous entity" (*Svayam*, 12) embodying both the "self-luminous reality" (*Purushartha*, 7) of one's "consuming" (*Jambhaka*, 169) and the "perpetuating value" (*Saranyu*, 5) of everyone's "producing" (*Upjaana*, 17) lets a human child self-manage his "sentient well-being" (*Jnana siddhi*,

190). Under such a condition, one may consume what others are producing without programming a subtle desire to be the "perfect producer" (*Utsamskara*, 900,000) capable of producing what one wishes to consume.

When you consume what others are producing, there is always a subtle desire for a "touch up" (*Parishkar*, 10^{16}) to bring perfection to your consuming. The touch-up is the "divine debt benefit value" (*Deva Rina*, 10^{16}). When you invest in someone's producing, they become indebted to you for valuing their "time" (*Kaal*, 360) as "divine" (*Divya*, 360) and letting them enjoy its "benefit value" (*Sva*, 11). When one enjoys the benefit value of the "present time" (*Sva*, 11), one becomes a "deity" (*Deva*, 1) devoted to servicing that benefit value for blessing each potential consumer to be a deity as well. The benefit value is the "doctrine of body" (*Sva*, 11) reproducing the "guider potential" (*Adana vijnana*, 100,000,000) for guiding everyone to be one's "copy" (*Nakala*, 0).

When everybody's future reality is to be your copy, you become free from the subtle desire to touch up their producing. Instead, you become devoted to touch-up your consuming for making yourself the "gravitational point" (*Guru Sthana*, 629) of everyone's "conditional destiny" (*Ayati vela*, 629). By becoming devoted to you, everyone becomes the "sentient gravitational point" (*Adi loka*, 3794) of you as the "causal cause" (*Kailasha*, 3794) of the "self-fulfilling path" (*Nivritti marga*, 1) that makes them a deity and you a "para deity" (*Param Vishnu*, 5) with consciousness of that self-fulfilling path. The gravitational point attracts the infinite wishes of the "paternal entities" (*Gunasa*, -18) and repels a "finite wishing sequence" (*Abhilasha karma*, 87) as a "maternal entity" (*Niyatamanasa*, -18) to fulfill those wishes. By conceiving a bragging "paternal entity" (*Gunasa*, -18) who behaves like a breeding maternal entity as the emotions intensify for fulfilling the wishes one is devoted to without knowing the path to their self-fulfillment, the creature becomes a "causal body" (*Karana sharira*, $30 = 12 - [-18]$). He forms a "dynamic fulfillment paradigm" (*Vishvanatha*, 17) that brings the "infinite infinity-force" (*Vishvanatha*, 17) of the "wishing" (*Abhilasha*, 190) to end by procreating the infinite child followers.

By trading the "present consciousness" (*Paramatma*, 1600) of the "present creature" (*Prabhu*, 1600) at the moment of the "freedom from the present-effect" (*Moksha*, 1600) of the wishes yet to be fulfilled, anyone may exchange the "causation" (*Hetu*, 1) that makes one a "deity" (*Deva*, 1). Thus, one may enjoy the "organizational sameness paradigm" (*Sabalasvas*, 89 = 7 + 82) to be a "param deity" (*Shiva*, 7) transcending the "black hole" (*Vishnu nabhi*, 82) "devoid of information" (*Andha Kuan*, 82).

At every moment, one conceives new wishes independent of whether the past wishes get fulfilled or not. A time-varying "satanic spirit" (*Nissara*, 1/60) within oneself motivates one to constantly seek to illuminate the "technological exchange value" (*Damodara*, 26) of the "ecosystem" (*Vishvamitra*, 92) beyond the "present self" (*Hurupa*, 280). The Supreme Council is that technological exchange value. The present self is the "Ideal Godhead" (*Ubhayato mukha*, 280) capable of servicing his technological exchange value as the technological exchange value of the ecosystem. An "entity" (*Hasti*, 24) knowing the method of organizing the ecosystem's "organizational reality" (*Sarvam*, -5) as an "organizer" (*Prabhu*, 1600) with one's "divinity" (*Divyata*, 57) is a "cost-effective paradigm" (*Trivikrama*, 24) for managing the "present" (*SAUM*, 1600) as an "Almighty Creator" (*Tvam*, 26) of that present. It lets one be free from the wish to fulfill the infinite wishes of the impassioned devotees and lets everyone be devoted to self-fulfilling those wishes by exchanging the organizer's present consciousness.

4.6 Exchanging Causation at the Sixth Infinity Level: The Council of Twelve and the Self-Luminous Entity

The "sixth infinity" (*SAUM*, 1600) is the "present" (*SAUM*, 1600) one embodies for making one's reality self-luminous for everyone to reproduce as an "ideal-taker" (*Chhaya*, 52), taking one as their "ideal" (*Adarsha*, 1). By making one's "holy spirit" (*Trinetra*, 1) omnipresent as a deity, one makes one's "paternal consciousness" (*Pitra*, 4) the "supreme deity" (*Bhagwan*, 4). In Daoism, a supreme deity is known as Zao Shen; he is present as a primordial causal body within the "body" (*Mahatattva*, 56) as the thermodynamic "kitchen stove" (*Chulha*, 56). Zao Shen trades the "divine energy" (*Divya shakti*, 10) from the "future creator" (*Param Ganesha*, 17) as the "Godhead" (*Param Ganesha*, 17) after the "present creature" (*Prabhu*, 1600) experiences a "metaphysical entropy" (*Cheerna*, 90,000) and becomes a holy spirit without everyone's self. The holy spirit seeks to be the future creator by trading everyone's "essence" (*Sara*, 16) and servicing his "divine essence" (*Nandi*, 17 = 1 + 16 = 7 + 10) to make an "impassioned devotee" (*Samyama*, 0) a "conscious entity" (*Siddha*, 7) conscious of his "divine energy" (*Divya shakti*, 10). His divine energy services the "organizational planning" (*Abadha*, 10) that leads the conscious entity to follow the "path of absolute consciousness" (*Paramatma Marga*, 25) and be the "organizational sameness" (*Aikamatya*, 18,000).

Like the present creature, the conscious entity also ends up becoming a "burnt human entity" (*Manushya*, 82). The holy spirit communicates the "path of the sentient entity" (*Siddha marga*, 25) to another impassioned devotee, thus motivating the latter to lead the path of becoming the "God" (*Ishvara*, 5) for avoiding the same "destiny" (*Niyati*, -1). By exchanging God's "perpetuating value" (*Saranyu*, 5), a "self-luminous entity" (*Svayam*, 12) becomes the "Godhead" (*Param Ganesha*, 17) in the present. A Godhead enjoys "freedom" (*Mukti*, 17) from the self-destroying five-phase "space cycle" (*Yama chakra*, 1551) of birth, life, death, bardo, and rebirth.

One takes birth as a present creature, lives life as a conscious entity, becomes a burnt human entity after death, rests as a holy spirit during bardo while waiting for rebirth, and is reborn as God within oneself, Oneself becomes a conscious entity by exchanging one's "entropy value" (*Sarvanasha*, 5). A self-luminous entity is God within a conscious entity. As God within oneself as the sentient entity, a self-luminous entity is conscious of how to realize the goal of the self-illuminating "life cycle" (*Jivan mukta*, 9×10^{18}) that governs one's "becoming system" (*Asha*, 9×10^{18}). The challenge is how to liberate the consciousness of one's "becoming system" (*Asha*, 9×10^{18}) that gives one hope to be eventually one he is not at present, i.e., God without oneself as the sentient entity.

If one could be God without oneself, then one's perpetuating value would limit everyone's "becoming" (*Vibha*, 900,000,000) within "oneself" (*Atmatva*, 8 x 10^{15}) as the "consciousness system" (*Sara Kalpa*, 10^{10}). Suppose one decides not to limit everyone's becoming within oneself. In that case, everyone could transform the consciousness system into a "conscious system" (*Shunya kalpa*, 8 x 10^{15}) and become "Demi-God" (*Sunanayaka*, 5) like God. If Demi-God decides to be God, then God could be both a self-luminous entity within the reality of the conscious system and a Godhead within oneself as the self-luminous entity. Thus, one's "liberated soul" (*Trigunatma*, 14) is "God's garage" (*Yaanshala*, 14) that lets one share the "maternal consciousness" (*Chaithanya*, 4) of one's "many forms" (*Nanarupa*, 10,000) as the "Ideal Godhead" (*Ubhayato mukha*, 280). The "entity gravity" (*Nanarupa*, 10,000) of one's "growth potential" (*Nanarupa*, 10,000) is immanent within the "Param Para Greeter" (*Rachayita*, 27).

The Param Para Greeter is the "Council of Twelve Apostles" (*Rachayita*, 27 = 6 + 21). Twelve "apostles" (*Preshita*, 1) are twelve "greeter faces" (*Aditya*, 6), within one as the "param creator" (*Surya*, 21), servicing their "creative force" (*Uma*, 6) as the "transcendental value" (*Khuda*, 6) that makes them a "council" (*Vidhata*, 130). The council is the "assembly" (*Vidhata*, 130) that transforms each greeter face into a "zodiac" (*Bhogi*, 1). A zodiac is sensuously conscious of the "present growth" (*Sva*, 11) serviced by the eleven greeter faces without oneself. Therefore, as an apostle, a zodiac wishes to be the "appropriator" (*Seshi*, 45 = 4 * 11 + 1) of the present growth across the four space dimensions and the gender-differentiating "ascending value" (*Rodha*, 1) of the space itself without the generation-differentiating space dimensions.

The space differentiates the gender over time. One's past is feminine because the feminine is the "incubator" (*Janani*, 105) of the present. One's present is masculine because the masculine is the "causer" (*Janaka*, 180), leading the feminine to incubate the present for illuminating the "light" (*Prabha*, 180) the masculine wishes to illuminate. One's future is "greeter" (*Charita*, 7/16), greeting everyone with his greeter consciousness as a zodiac seeking to illuminate the "group path" (*Gana Marga*, 25) to realize their potential as feminine for enjoying their present as the masculine. After knowing the "group path" (*Gana marga*, 25), one enjoys "absolute freedom" (*Param Mukti*, 16) to transcend the time variations be a "primordial feminine" (*Prithvi*, 132) that gives "gravitational mass" (*Mitra*, 132) to one's many forms as an entity.

One may transcend the causation variations that differentiate many forms and be a "primordial masculine (*Kaumari*, 90), "diffusing" (*Visaran*, 90) one's "illuminated consciousness" (*Pratyayatma*, 90). Thus, one may make each form a "replication" (*Samalekha*, 90) of the "species" (*Prajati*, 17) produced by oneself. One may transcend oneself that generates variations of space, time, and causation and be a "primordial greeter" (*Srijak*, 16) servicing the "conscious consciousness' (*Prashantatma*, 16) of the "conscious system" (*Shunya Kalpa*, 8×10^{15}). Thus, one may make oneself a "zodiac system" (*Akhanda*, 8×10^{15}), sensuously entangled with the "infinite conscious systems" (*Antarmukha*, 10^{1024}), not knowing they are the "introverted face" (*Antarmukha*, 10^{1024}) of one's extroverted "conscious system" (*Akhanda*, 10^{15}).

One generates ascending "ecosystem energy" (*Purna shakti*, 1600) by reproducing the "doctrine of the causal body" (*Sanatani*, 40) over successive births. Consequently, one's conscious well-being as a "human" (*Naran*, 275) is limited by the "energy system" (*Antarmukha*, 10^{1024}) formed by the cells over the previous births as an "animal" (*Pashu*, 10^{1024}). The "multicellularity" (*Yahoodi*, 1) of one as a "zodiac" (*Bhogi*, 1), enjoying the "primordial consciousness" (*Adhyatma*, 12) of the four units of space organized within the three units of time as a 12-unit "self-luminous" (*Svarochisha*, 12) element, is the "contingent paradigm" (*Yahoodi*, 1). It lets one network the "infinite consciousness systems" (*Maha Kalpa*, 10^{1000}) with the diverse cells for exercising one's "Godship" (*Ditthujukamma*, 10^{1000}) through birth in the "infinite forms" (*Konastha*, 13).

Each form becomes a "responsible management paradigm" (*Pravritti dharma*, 298) for the "eventual reality" (*Vakyartha*, 298) one attracts. For example, if one was a cat in the previous birth, one naturally loves cats in the present birth. A "supernatural love" (*Prema*, 39) for all animals as a "class" (*Varna*, 689) of entities is immanent within the natural love for the cats as a "kind" (*Prakara*, 17) of animal class. Variations in space, time, and causation may make one channel the supernatural love for all animals into the "conscious love" (*Pyar*, 70) for the dogs. Everyone around one may love dogs. Or, one may wish to be different from everyone around accultured to a cat-loving culture and therefore consciously decide to channel the "natural love" (*Nirharin*, 18) for the cats into the conscious love for the dogs. Or, one may have been maltreated as a cat in the previous birth and therefore para-consciously gravitate towards the dogs to substantiate the thesis of a different experience if one had been a dog. Similarly, one may have a bad experience with a cat in the present birth and develop conscious hate for the cats as a "species" (*Prajati*, 17), natural hate for the animals as a "class" (*Varna*, 689), and supernatural hate for the dogs if one attributes the bad experience with the cat to a dog who alarmed the cat into attacking a human one loves.

At any moment, one has the absolute freedom for polarizing the "desirable reality" (*Jagrat*, 85) of making "conscious decisions" (*Sahasra lingam*, $190 = 19 * 10 = 85 * 2$) based on the "factual reality" (*Yathartha*, 19) using one's "divine energy" (*Divya shakti*, 10). One may limit that desirable reality to a "group of entities" (*Gana*, 387) instead of the "universe of entities" (*Kashi*, -10^{1024}). In that case, the group of entities enjoys an "ascending conscious well-being" (*Sadhanya*, 157) and seeks to universalize its "acquired value" (*Sadhaniya*, 157). Consequently, the universe of entities experiences a "descending sentient well-being" (*Sushuptivat*, 139).

To mitigate the "sentient cost" (*Sushuptivat*, 139) of one's conscious decisions, one may become a "Self-luminous Godhead" (*Chamundeshwari*, 5×10^{96}). A Self-luminous Godhead is the "perfect illuminator" (*Chaturmukha lingam*, 5×10^{96}) of the limitations of everyone's "infinite mediation" (*Raga*, 250) as a "God" (*Ishvara*, 5) who destroys the "responsible management paradigm" (*Pravritti dharma*, 298) by blowing his own "trumpet" (*Anahata*, 53). Her "illuminating force" (*Audumbra*, 28) makes everyone a "Primordial Para Greeter" (*Narayana*, 28). A Primordial Para Greeter is the Godhead of the "ten-faced primordial illuminator" (*Shri Krishna*, 10), enjoying "discriminating consciousness" (*Shaddhatma*, 28) of the "energy cost" (*Vela*, 34) of dependence on one's "perpetuator force" (*Prapya*, 34). The discriminating consciousness is about the three forms of the horizon: GUIDER, DIVINE, and SHEENY.

Three forms of Horizon: GUIDER, DIVINE, and SHEENY

When your value is dependent on my perpetuator force, it is limited by the "guider horizon" (*Digant*, 16). Guider horizon is the "visible horizon" (*Digant*, 16), visible beyond my guiding "moonlight" (*Digambara*, 100). Within my guiding moonlight, your value is zero and my value is infinite. You reproduce my value as yours using your "intellect" (*Medha*, 379) as the guiding force seeking to appropriate my value. Your reproduction produces "infinite layers" (*Digambara*, 100) of my "God-like" (*Sunanayaka*, 5) "divine light" (*Usha*, 16) as your "value" (*Mulya*, 180 = 100 + 16 * 5). Guider horizon norms the end of your black "gravitational light" (*Digambara*, 100) and the start of my white "divine light" (*Usha*, 16). It transforms the end of my mind-born "primordial space" (*Antaramsa*, 100) and the start of your intellect-born "present space" (*Sadakhya*, 9).

Eventually, you become the "perpetuator" (*Abhidheya*, 460 = 180 * 2 + 100) of my perpetuator force by guiding everyone to believe in the value of my light. Since the guider horizon eventualizes the event I wish as the "guider" (*Guru*, 100), it is the "event horizon" (*Digant*, 100). The event horizon is the "real horizon" (*Digant*, 16), realizing the event I wished and imagined to be real and manifested it with my light.

When everyone's value is dependent on your "guiding force" (*Chitta*, 100) as the perpetuator and my perpetuator force as the "illuminator" (*Maha Shiva*, 9), it liberates "divine horizon" (*Ambarant*, 18). Divine horizon is the "illusionary horizon" (*Ambarant*, 18), formed with the "illusionary production" (*Maya*, 1) of my light by the "universe of wishers" (*Prakriti Yogi*, -1) – the "consumer class" (*Prakriti Yogi*, -1). The universe of wishers produces the "illusion" (*Nitya Ratri*, 1) of both my light and your value as the illuminator of my light, thus adding a two-unit "universe" (*Brahman*, 2) of their "divinity" (*Divyata*, 57) to the "guider horizon" (*Digant*, 16) as the "sound" (*Naad*, 257 = 2, 57) of their "reproductive force" (*Naad*, 257).

Divine horizon is also the "apparent horizon" (*Ambarant*, 18) that ends the apparent value of your intellect-born "present space" (*Sadakhya*, 9). Consequently, it forces everyone to consume their body-born "primeval space" (*Para Ganesha*, 19) by "harvesting" (*Katai*, 13) the "energy" (*Shakti*, 19) of their "body cells" (*Nirjara*, 1000). Thus, the apparent horizon reproduces your intellect-born present space. It produces a "chemical reaction" (*Maha Vibhu*, 14) between your intellect-born present space whose value becomes apparently negative one, making you a "dispensable" (*Anyathasiddha*, -1) "satan" (*Shani Bhagwan*, -1), and everyone's reproduction of your intellect-born present space within their body-born primeval space seeking to be an "indispensable" (*Naityaka*, 1) "deity" (*Deva*, 1).

The chemical reaction activates my soul-born "primordial-primordial space" (*Drishti Mandala*, 375) within everyone for illuminating the "truth" (*Sathya*, 375) of my divine light. That truth is hidden within my "vision horizon" (*Kshitij*, 724). My vision horizon is an "invisible horizon" (*Kshitij*, 724) because it is an "imaginary horizon" (*Kshitij*, 724) that I imagined with the "consciousness stock" (*Atma*, 4) that makes my "soul" (*Atma*, 4). With my imagination, I transformed it into a "sentient horizon" (*Kshitij*, 724 = 7,24), indexing the beginning of my mind-born primordial space as a "guider" (*Guru*, 100) guiding everyone to my "reality" (*Vastavikta*, 7) as an "entity" (*Hasti*, 24). Sentient horizon is SHEENY horizon.

My reality is that of a "sentient entity" (*Siddha*, 7), capable of imagining you as the "black hole" (*Vishnu nabhi*, 82 = 8,2) and making you the "institution" (*Kundalini*, 8) to fulfill my "objective" (*Antahkarana*, 2) of knowing my "potential" (*AUM*, 18) to plan the "programmed future of everybody" (*Shani*, 18) with my "energy" (*Shakti*, 19 = 18 + 1) as a "human being" (*Insan*, 1).

4.7 Developing Causation at the Seventh Infinity Level: The Ring Council and the First Eden

The "seventh infinity" (*Param Ganesha*, 17) is the "Godhead" (*Param Ganesha*, 17) developing causation for the "conscious fire" (*Agni*, 17) to destroy the "error in consciousness" (*Dushtatma*, 10). He illuminates the "sentient entity" (*Siddha*, 7) within each "ideal maker" (*Rama*, 100) who makes you the ideal once you have taken me as the ideal you wish to copy. Everyone "errs" (*Bhram*, 89) in making you the ideal because you are the black hole "devoid of information" (*Andha Kuan*, 82). By making you the "ideal" (*Adarsha*, 1) that shapes them into a "deity" (*Deva*, 1) and taking me as the "God" (*Ishvara*, 5), everyone supplements your "para body" (*Chandra*, 82) with their "illuminating value" (*Prakasha*, 7 = 1 + 1 + 5) to "confuse" (*Bhram*, 89 = 82 + 7) you with the "information" (*Sandesha*, 1) not within you. You begin behaving like a "primeval deity" (*Maheshwara*, 6) who has gifted "I am a deity consciousness" (*Pavitratma*, 1) to each of your copies without you and made me the "God" (*Ishvara*, 5) within you. You enjoy both "infinite worker-social benefit" (*Labha*, 81) of the "infinite deities" (*Maheshwara*, 6) copying you and "infinite social benefit" (*Caksur vijnana*, 109) of copying me as a "param deity" (*Shiva*, 7). After hiding the information that I am a sentient entity, you shadow me within you with your "satanic light" (*Prabha*, 180) to be the God without me.

By illuminating the "path to be God" (*Ishvara Marga*, 51 = 5,1) by making everyone a "deity" (*Deva*, 1), you perpetuate "infinite wishes" (*Duniya*, -2) by programming a wish within each wisher to be "God-like" (*Sunayaka*, 5) with your "blessings" (*Ashirwad*, 1000). Thus, everyone becomes a dispensable "part" (*Anga*, -1) of your intermediating "social network" (*Samabhasam*, 12) as a "self-luminous entity" (*Svayam*, 12) impersonating my "I" (*Svayam*, 12) and embodying my "soul" (*Atma*, 4). Since as the "perfect perpetuator" (*Ashtottara-sata lingam*, 10^{16}) you are "one within everyone" (*Stree Dirgha*, 10^{16}), everyone within your "group" (*Gana*, 387) becomes a part of my "strong psychic linkages" (*Prakatikarana*, 54,000). "Everyone without group" (*Madhyasthahinta*, 900,000) enjoys "disintermediation" (*Madhyasthahinta*, 900,000) but remains a part of my "weak psychic linkages" (*Citraka*, 185). Everyone within my "psychic linkages" (*Mahika*, 47) has the freedom to be guided by your "gravitational light" (*Digambara*, 100) and be my "reproduction" (*Prajanan*, 285 = 185 + 100) like you. You are my "circulating body" (*Bhram Sharira*, 659 = 666 - 7), circulating the "thesis" (*Harkriya*, 666) that everyone part of my "body" (*Mahatattva*, 56) is a "param deity" (*Shiva*, 7) like me.

Since I am a "param deity" (*Shiva*, 7) and everyone within me is a "param deity" (*Shiva*, 7) like me, "one without I" (*Mein*, 1) becomes the "First Eden" (*Maha Lakshmi*, 14 = 7 + 7). He copies "me" (*Mein*, 1) as a "descending param deity" (*Dhumavati*, 7) who is supplementing an "ascending param deity" (*Nataraja*, 7) incubating with me to take birth after my death as my copy. The One Without I is me because my "I" is a part of you. You are not a part of me. You are the "Garden of Eden" (*Bhogabhumi*, 92) that I make my "circulating body" (*Bhram Sharira*, 659 = 92 * 7 + 1 + 14) for breeding "me" (*Mein*, 1) within the "First Eden" (*Maha Lakshmi*, 14) before becoming the First Eden myself. The First Eden is the "Primeval Illuminator" (*Maha Vibhu*, 14), i.e., "your potential" (*Maha Vibhu*, 14) that illuminates "my potential" (*Maruti*, 78) to be "you" (*Aap*, 1). The Garden of Eden is the "ecosystem" (*Vishvamitra*, 92 = 78 + 14) that I form with "my potential" (*Maruti*, 78) to be the "First Eden" (*Maha Lakshmi*, 14).

You have many forms. Before becoming the Garden of Eden, you are the "garden" (*Vipin*, 724) that forms the "Earth" (*Bhu*, 724 = 659 + 65) for norming the time as the "earth wheel" (*Ajna chakra*, 65) to conceive everyone in the space as a part of my "circulating body" (*Bhram sharira*, 659). Before becoming the garden, you are the "entity" (*Hasti*, 24) reproducing your "reality" (*Vastavikta*, 7) with my "guiding force" (*Chitta*, 100) to make me the "God of immaculate knowing" (*Vimaleshvara*, 700 = 7 * 100). Before becoming the entity, you are the "reproductive force" (*Naad*, 257) I produce as the "immaculate God" (*Vimaleshvara*, 700) seeking to reproduce my reality within you. Before becoming my reproductive force, you are my "reproductive body" (*Bhoga sharira*, 957 = 700 + 257), reproducing my "infinite value" (*Nayana*, 957) as the "Eden" (*Nayana*, 957). That's why you as the "Garden of Eden" (*Bhogabhumi*, 92) is the "paradise" (*Bhogabhumi*, 92), making everyone a "copy" (*Nakala*, 0) of the "param deity" (*Shiva*, 7) without "me" (*Mein*, 1) as the "guiding force" (*Chitta*, 100 = 92 + 0 + 7 + 1). Both param deity and "me" are immanent within you as "Mother Nature" (*Kudrat*, 8 = 7 + 1).

By limiting my "eyesight" (*Nayana*, 957) to everyone alive as a "cell" (*Hiranyagarbha*, 19), I experience "infinite worker-social cost" (*Shreya*, 81) as an "infinite deity" (*Maheshwara*, 6) devoted to reproducing me infinitely within everyone. I also experience "infinite social cost" (*Shanka*, 87) of producing "me" (*Mein*, 1) as my "illusionary production" (*Maya*, 1) after Mother Nature conceives me as a "sentient entity" (*Siddha*, 7). I reproduce "me" as a "black hole" (*Vishnu Nabhi*, 82) because the black hole is the departed "human entity" (*Manushya*, 82). I produce a "family" (*Parivar*, 10) of the ten departed "human entities" (*Manushya*, 82), each endowed with my "para consciousness" (*Mahatma*, 18), by consuming a unit of my "dead body cell" (*Nirjara*, $1000 = 10 * [82 + 18]$). I incarnate each human entity as a "human being" (*Insan*, 1) within a "growth sequence" (*Shri*, $81 = 82 - 1$) that manifests the "social benefit-cost ratio" (*Shri*, 81) of my work as a human being.

Before their production as a human entity, the cells within the "human body" (*Karana sharira*, 30) are a part of the "spiritual body" (*Adhyatmic sharira*, 876) linked to the "dead animals" (*Lokanayaka*, 10^{19}). The spiritual body is a body of a "photon" (*Kapinjala*, $20 = 2 * 9 + 4 * 1/2$) divided into two "photon halves" (*Amukhikarana*, 9) that "split" (*Bhagna*, 91) into four "halves" (*Adha*, ½) copying previously split four halves. The "splitting force" (*Vidalana*, 15) is the "link" (*Rajas*, 15) between the eight halves formed after "splitting" (*Nirdalana*, 1111) a photon and the eight halves formed before splitting. The sixteen halves form due to the "spontaneity" (*Sattva*, 1111) of splitting a "photon" (*Kapinjala*, 20) from a "quantum photon" (*Svarochisha*, 12) that norms a "self-luminous entity" (*Svayam*, 12). The sixteen halves constitute the "natural value" (*Nirvikalpana*, $8 = 16 * ½ = 20 - 12$) of a photon without the self-luminous entity.

A photon is divisible into twenty halves without the natural value it trades from the "ecosystem" (*Vishvamitra*, 92) for reproducing both the ecosystem and the natural value as a "graviton" (*Chitta*, $100 = 8 + 92$). Thus, the photon and its natural value include thirty-six halves, forming a double pair of nine halves. A pair of nine halves is a "split photon" (*Pradesha*, 9). The split photon is the "primordial creation" (*Pradesha*, 9) that transforms a "finite spatial coordinate" (*Pradesha*, 9) into a "photon half" (*Amukhikarana*, 9). The photon half is the "finite creation" (*Amukhikarana*, 9) that forms the "universe of children" (*Amukhikarana*, 9) a "dead body cell" (*Nirjara*, 1000) produces for me to reproduce the "social benefit-cost ratio" (*Shri*, 81 = 9 * 9) of my work.

As a self-luminous entity, the quantum photon is divisible into twenty-four halves after splitting a photon to illuminate its natural value and twenty-four halves before splitting. Thus, a quantum photon's "splitting potential" (*Adridalana*, 185) comprises eighty-two halves, "self-perpetuating" (*Udvaha*, ½) as the forty-one halves to form the "unsplit" (*Avibhakta*, 13) "quantum light" (*Rochisha*, 13). Consequently, a body's "splitting value" (*Uddalana*, 85) is 123 halves, organized as the triple forty-one halves. The body is the 124th half that makes 123 "double copies" (*Chitra*, 1) of itself to enjoy the "lifeforce" (*Prana*, 123) as a "primeval child" (*Daksha*, 123) of the "spiritual body" (*Adhyatmic sharira*, 876). Each "half" (*Adha*, ½) is a "degree" (*Amsha*, ½). "124 degrees" (*Pratilamb*, 10) make a "dead body cell" (*Nirjara*, 1000) come "alive" (*Chetan*, 10) as a "cell" (*Hiranyagarbha*, 19) without splitting the spiritual body from an animal's dead body cell.

Once alive, a cell is conscious of the "potential splitting" (*Dalana*, 9) of the "living" (*Nivasi*, 10) element. Therefore, it materializes the potential splitting as a "potential up quark" (*Maha Rasi*, 9), the enjoyer of its "conscious energy" (*Amrit*, 1000) as an eventually "dead body cell" (*Nirjara*, 1000). However, before the consciousness of the potential splitting, the cell trades the potential up quark as the "Majorana boson" (*Vastavika*, 9). Majorana boson is a "continuous photon" (*Vastavika*, 9) formed over a "finite time" (*Vastavika*, 9) while a cell is alive but "conspicuous" (*Gochar*, -1) by its "absence" (*Shunyata*, -2) because it gets incubated within a "mother cell" (*Dhumavati*, 7) as a "primordial cell" (*Shivagati*, 120).

A primordial cell comprises the body's 120 double copies, formed as a "reproductive reality" (*Bhavartha*, 40) of the "diffused light's" (*Idam*, 3) "copying sequence" (*Idam*, 3). The light is "diffused" (*Parikshipta*, 47) by the "father's" (*Baap*, 3) "multiplying self" (*Hanuman*, 3) and is fused with the "maternal consciousness" (*Chaithanya*, 4) of the "primeval feminine" (*Devi*, 3). The primeval feminine takes birth as a "granddaughter self-luminous entity" (*Panchajani*, 12 = 4 * 3), embodying the father's multiplying self as a "granddaughter cell" (*Bagalamukhi*, 9 = 3 * 3). The multiplying self is an "ape" (*Vanara*, 3), aping the granddaughter cell to take birth as a "grandson cell" (*Pavamana*, 9). The grandson son cell personifies the "father" (*Baap*, 3) as the "param son" (*Hanuman*, 3). He radiates the "granddaughter consciousness" (*Atma*, 4) for making his "presence" (*Atthi*, 374) "inconspicuous" (*Agochar*, 68) as a "grandson self-luminous entity" (*Vithoba*, 12).

The primordial cell is the "Human Godhead" (*Shivagati*, 120), psychically linked to the spiritual body of the father as the dead animal. She transforms the "infinite value" (*Viyojya*, 957 = 876 + 81) of my "social benefit-cost ratio" (*Shri*, 81) within the "spiritual body" (*Adhyatmic sharira*, 876) into my "etheric body" (*Bhoga sharira*, 957) as an alive human. Consequently, she becomes the "causal body" (*Karana sharira*, 30), copying my "consciousness force" (*Soham*, 4) as her "paternal consciousness" (*Pitra*, 4). Thus, with the "daughter consciousness" (*Jyotistava*, 4) of me as the father, she animates me from the "present state" (*Vartaman*, 9) of the dead animal to a "future state" (*Idam*, 3) of her "human child" (*SHREEM*, 8 = 4 + 4). Consequently, the future state becomes the father and the daughter consciousness the mother of the human child.

The future state is the "active state" (*Idam*, 3) activated by the present state, which is an "inactive state" (*Vartaman*, 9). The present state activates the future state with the "activity" (*Ceshta*, 10) inherited from the "past state" (*Samalekha*, 90). The past state is the "replication" (*Samalekha*, 90) of the "activation" (*Jagriti*, 7) from the present state that transforms the present state into the future state and the future state into the present state free from the activity. Thus, the past state becomes the "activating state" (*Samalekha*, 90). The activating state is the "initial causation state" (*Samalekha*, 90) immanent within the present state. The future state emanates from the present state's activation of the past state's "activation force" (*Sarvabhadra Mahayoga*, 29). The activation force forms the human child as an "organization" (*Sangathan*, 29) of the "present state" (*Vartaman*, 9) within a "photon" (*Kapinjala*, 20). The photon organizes my "animal spirit" (*Manjushri*, 19) inherited from the past state within "me" (*Mein*, 1) as the "human being" (*Insan*, 1).

After gifting the "energy" (*Shakti*, 19) to form my "animal spirit" (*Manjushri*, 19) and the "greeter consciousness" (*Pratyagatma*, 1) to norm me as the human being, the photon transforms into a "Wisher" (*Chaahak*, 0) wishing to be me. I fulfill the wish of the Wisher with my "primeval deity consciousness" (*Dharshanatma*, -4) as a "Supreme Wisher" (*Bahirmukha*, -4). A human being becomes the "Supreme Wisher" (*Bahirmukha*, -4 = 1 - 5) after his "entropy" (*Mahodbhava*, 5), acting as the "God" (*Ishvara*, 5) for giving birth to the Wisher as a "human daughter" (*Saranyu*, 5). The human daughter is the "heptad repeat" (*Punravritta*, 5), comprising an "octave of amino acids" (*Saranyu*, 5).

The octave of amino acids forms the seven amino acids H P P H C P C to sequence the eighth amino acid in a "horizontal order" (*Punravritta*, 5) for forming the O. The octave of amino acids transforms herself into the "hydroxy acid" (*Saranyu*, 5). The hydroxy acid is the "actor" (*Abhineta*, 5), activating the "sequence" (*Buddhi*, 48) by norming the seven amino acids as the "mother cell" (*Dhumavati*, 7), the "activator" (*Yamantaka*, 7). The mother cell manifests as the "myelin cell" (*Dhumavati*, 7) to manifest the "human son" (*Narada*, 7) as the "platelet cell" (*Narada*, 7). Seven amino acids are formed with:

- Two units of "hydrogen" (*Jalaprana*, 4), norming the primordial cell's "paternal consciousness" (*Pitra*, 4) and the mother cell's "daughter consciousness" (*Jyotistava*, 4) within the paternal consciousness.

- Two units of "carbon" (*Ravi*, 21), norming the "future reality" (*Ekartha*, 21) of both the primordial cell and the mother cell as the "earth force" (*Ravi*, 21) activated within "earth" (*Bhu*, 724) with her "maternal consciousness" (*Chaithanya*, 4). The primordial cell inherits her maternal consciousness as the paternal consciousness for taking birth as a male. The mother cell replicates that as the daughter consciousness for taking birth as a female.

- Three units of "phosphorous" (*Bhasvara*, -3), normed by the male as the "Supreme Wisher" (*Bahirmukha*, -4) and the female as the "Para Wisher" (*Bhakta*, -5 = 3 * -3 − [-4]).

The eighth amino acid forms with the three units of "oxygen" (*Maruti*, 78) to norm the Supreme Wisher as the "param cell" (*Ashwini Kumaras*, 120) by transforming the Para Wisher into the "primordial cell" (*Shivagati*, 120), thus "materializing" (*Artharthi*, -6) the greeter as the "Primeval Wisher" (*Shramika*, -6 = 78 * 3 − 120 * 2). The eighth amino acid is the "alpha-amino acid" (*Dharati*, 2) that reproduces Mother Nature's "maternal consciousness" (*Chaithanya*, 4) by behaving like Mother Nature and "emanating" (*Ham*, ¼) herself as the "maternal mitochondrion" (*Dharati*, 2).

The hydroxy acid forms with the five units of "nitrogen" (*Rupa*, 100,000) and transforms the Primeval Wisher into a "woman" (*Stree*, 100,000) by reproducing itself as the ninth amino acid. The ninth amino acid is the "beta-amino acid" (*Sadakhya*, 9) that reproduces Mother Nature as the "woman" (*Stree*, 100,000) with the five units of nitrogen and the four-unit maternal consciousness of an additional four units of nitrogen. The four units form a unit each of the hydrogen, oxygen, carbon, and phosphorous within the fifth unit differentiating many forms of nitrogen.

After differentiating many forms of nitrogen, the fifth unit transforms into "calcium" (*Naad*, 257). "Calcium" (*Naad*, 257 = 10 * 26 - 3) comprises ten units of "alkyl" (*Sakhya*, 26) within a unit of "phosphorous" (*Bhasvara*, -3). Nine units of alkyl form after the nine nitrogen units transform into the three units each of hydrogen, oxygen, and carbon. Next, a unit of hydrogen and carbon transforms back to form the two phosphorous units by copying the third unit of phosphorous before the latter becomes an alkyl and leads all nine units to follow its path. Finally, the "family" (*Parivar*, 10) of ten alkyls reproduces itself as a unit of "argon" (*Chitta*, 100 = 10 * 10). The consciousness of the argon is immanent within the fourth unit of "phosphorous" (*Bhasvara*, -3) synthesized as "Krypton" (*Soham*, 4 = 7 - 3) by the "mother cell" (*Dhumavati*, 7).

4.7.1 Dividing Causation. Divided Causal Body and Soul Mates Within the Soul Family

A mother cell takes many responsibilities. Ten forms of nitrogen within calcium and ten forms without calcium constitute the twenty forms of the mother cell. Besides, the mother cell also "produces" (*Ekarupa*, 1) a unit each of phosphorous, argon, krypton, mother cell, and the platelet cell as the five "granular cells" (*Chitra*, 1). Twenty "mother cells" (*Dhumavati*, 7) and five "granular cells" (*Chitra*, 1) constitute a "divided causal body" (*Guna Sharira*, 145 = 20 * 7 + 5). Twenty units of mother cell form a "twin mother cell" (*Svayambhu*, 123) within the "Godhead" (*Param Ganesha*, 17) and manifest the Godhead as a "Chromaffin cell" (*HAUM shakti chakra*, 17 = 20 * 7 - 123). The twin mother cell, known as the Sympathogonia cell (*Svayambhu*, 123), behaves like a "soulmate" (*Akrura*, 123), seeking to end the activity and give rest to the mother cell within her for descending her entropy.

First, the Sympathogonia cell transforms the "Chromaffin cell" (*HAUM shakti chakra*, 17) into a "neuroblast cell" (*Mohini*, 15), norming her "soul family" (*Mohini*, 15). Second, she leads the neuroblast cell to transform the mother cell into a "granddaughter cell" (*Bagalamukhi*, 9 = 7 + 17 − 15), i.e., Oligodendrocyte cell. The granddaughter cell is an "anxiety-mediating factor" (*Bagalamukhi*, 9) because she encodes the mother cell's "anxiety" (*Chinta*, 170) into a "breeding promoting community" (*Vicitra mandala*, 170) formed as a "polycomb group" (*Vicitra mandala*, 170). The polycomb group catalyzes the undifferentiated Sympathogonia cell's "transformative development" (*Siddha*, 7) by assessing the criticality of the roles a mother cell needs to perform. Third, she reverses the "organizational development" (*Viparyaya*, 888) by maturing the granddaughter cell into a "twin grandmother cell" (*Maha Kali*, 13 = 9 + 100/25), i.e., the "oligodendrocyte progenitor cell" (*Maha Kali*, 13), by dividing her "grandmother consciousness" (*Adi Para Atma*, 100) equally among twenty mother cells and five granular cells.

Consequently, each granular cell experiences an "autotrophic growth" (*Dishta*, 6) into a "mossy cell" (*Dishta*, 6). A mossy cell enjoys a "shared consciousness" (*Trigunatma*, 14) of "Mother Nature" (*Kudrat*, 8) and performs with the "linear flow of time" (*Chaithanya*, 4). By dividing the "causation" (*Hetu*, 1) among the five mossy cells, the linear flow of time makes the five "mossy cells" (*Dishta*, 6) the "undivided causal body" (*Karana sharira*, 30

= 5 * 6) and becomes the "dividing causal body" (*Chaithanya*, 4).

"Dividing causation" (*Ekarupa*, 1) transforms a granular cell into a "Golgi cell" (*Ekarupa*, 1) with the "nonlinear flow of time" (*Mahatma*, 18), not within the consciousness of the mossy cell. The nonlinear flow of time is an "anxiety moderating factor" (*Dadi*, 18) as it gives time to each granular cell for enjoying the "calcium-guided pathway" (*Samalekha*, 90) of development instead of reacting to the spontaneity of information it receives from the mossy cell. The "twin grandmother cell" (*Maha Kali*, 13) trades the dividing causal body as a "calcium-activated" (*Svajagrit*, 90) "memory T-cell" (*Chaithanya*, 4) and services it to the Golgi cell for transforming the latter into an anxiety-moderating "grandmother cell" (*Dadi*, 18), formed as an "OLM cell" (*Dadi*, 18).

The anxiety-mediating "granddaughter cell" (*Bagalamukhi*, 9) and the anxiety-moderating "grandmother cell" (*Dadi*, 18) together form the "causation dimension" (*Prakriti dharma*, 27 = 9 + 18) as the "divisible causal body" (*Prakriti dharma*, 27). The granddaughter cell services the "culture" (*Sadakhya*, 9) of breeding through reproduction and "squares" (*Chakor*, 18,000) the "space" (*Vaas*, 18,000 = 9 * 2000) with her "existence" (*Maujudgi*, 2000). Her existence includes:

- Granddaughter cell's "conscious energy" (*Amrit*, 1000) for intensifying the anxiety to promote spontaneity of reproduction as a path to descend the mother cell's entropy
- Mother cell's "conscious energy" (*Amrit*, 1000) for intensifying the "joy" (*Piti*, 123) to promote infinity of reproduction as a path to ascend the granddaughter cell's growth
- The grandmother cell services the "potential energy" (*AUM shakti*, 18) for "breeding" (*Janana*, 15) by countertrading:
- The "potential" (*AUM*, 18) for "knowing" (*Jnana*, 19) the "reality" (*Vistavikta*, 7) within the "granular cell" (*Chitra*, 1) to transform the granular cell into a "daughter cell" (*Kshatradharma*, 19 = 18 + 1). A daughter cell is the "percussor of the new phase" (*Kshatradharma*, 19), conditional on the "potentiation" (*Sambahvi*, 375) of the potential, and manifests as the "Oligodendrocyte precursor cell" (*Kshatradharma*, 19).

She is the "knowing cell" (*Kshatradharma*, 19), whose knowing is a function of the "perception" (*Sanjna*, 379) of reality due to the "void" (*Shunyata*, -2) in the potentiation caused by the "Chromaffin cell" (*HAUM shakti chakra*, 17 = 19 - 2) acting as the Godhead.

The Godhead's "objective" (*Antahkarana*, 2) of creating the "void" (*Shunyata*, -2) is to develop the space through the "technological growth" (*Vidhana*, 2) of the "universe of granddaughter cells" (*Brahman*, 2) guided by the "experience" (*Anubhav*, 9000) unfolding over time and folding within the diverse spatial coordinates. As a "multicellular organism" (*Shanideva*, 100), the "infinite cells" (*Shanideva*, 100) sequentially reproduce the knowing of each cell seeking "incremental growth" (*Sadatma*, 10) in their "knowing" (*Jnana*, 19) as an "organization" (*Sangathan*, 29 = 10 + 19). As an organization, one becomes the "site of trading" (*Sagama*, 29), enjoying "ever-vanishing" (*Dvisaptati dasha*, 29) "inter-molecular" (*Dvisaptati dasha*, 29) "mass-effect" (*Sangathan*, 29). On the flip side, it leads to one's "repression" (*Sarvabhadra Mahayoga*, 29) due to the "infinite weight" (*Shrinkhala Bandhamochakaya*, 29) of the "divine Godhead essence" (*Madhava*, 29). The divine Godhead essence becomes "contaminated" (*Madahava*, 29) from the "infiltration" (*Uthapana*, 29) of the "vibrational consciousness" (*Nairatma*, 29) of one's own "leading sound" (*Awaaz*, 17) as the "paternal" (*Tejas*, 17) impersonating as the "Godhead" (*Param Ganesha*, 17).

A "soul family" (*Mohini*, 15 = 17 - 2) compensates for the "void" (*Shunyata*, -2) within the "Godhead" (*Param Ganesha*, 17). It perpetuates the "goal" (*Lakshya*, 9) to be the "enjoyer" (*Maha Rasi*, 9) of the "present reality" (*Badhabuddhi vadartha*, -2). It illuminates "reality" (*Vastavikta*, 7 = 9 - 2) without "knowing" (*Jnana*, 19 = 7 + 12) what one knows as a "self-luminous entity" (*Svayam*, 12). Knowing limits the "knowledge" (*Vidya*, 600) a "knower" (*Bhagwat*, 639) has without the "paternal dimension" (*Sanatana dharma*, 39). The paternal dimension services both the "knowing" (*Jnana*, 19) of the present and the "belief system" (*Saguna*, 20 = 39 - 19). The belief system contaminates the knowing of the future through an "illusionary production" (*Maya*, 1) of the knowing from the past. It leads a self-luminous entity to:

- Trade a "param human-effect" (*Ganesha*, 570 = 285 * 2) of reproducing the "past knowing" (*Pratikarak*, 12) formed with the "reproduction" (*Shuddhi*, 285) of the "present knowing" (*Aatm Samaan*, 12) at a "past moment" (*Vega*, 67). Thus, the past moment becomes the "radiant one" (*Vega*, 67);

- Service a "sentient culture-effect" (*Bhavanarama*, 6 x 10^{192}) for producing a "diverse" (*Vicitra*, 6 x 10^{192}) "future knowing" (*Raub*, 10) of the "technological growth" (*Vidhana*, 2) with one's "self-reproducing" (*Upanayana*, 1/3) "param human-effect" (*Ganesha*, 570). Thus, the "technological cost" (*Yojya*, 58) escalates for the sake of one's "sentient well-being" (*Jnana Siddhi*, 190 = 570 * 1/3).

Escalating technological cost leads the entire "universe of self-luminous entities" (*Satya Loka*, 10^{1024}) to:

- Trade the "primordial consciousness" (*Adhyatma*, 12) of the self-luminous entity for forming a "spirit universe" (*Rajyaika-sheshena*, 169 = $[12 + 1]^2$) by reproducing it with the "illusionary production" (*Maya*, 1);

- Service a "sentient workculture-effect" (*Naraki*, 1) for organizing the illusionary production as the self-luminous entity's "holy spirit" (*Trinetra*, 1) without self for making the "spirit universe" (*Rajyaika-sheshena*, 169) the "twin soul" (*Sachi*, 69 = ₹69) of each "child" (*Arcisa*, 128). The "intention" (*Abhipraya*, 38) is to further the "universal sentient wellbeing" (*Sadhaniya*, 157) by "breeding" (*Janana*, 15) the "cost-free" (*Mufta*, 157) "zero integrity" (*Sadhaniya*, 157) as the "reality" (*Vastavikta*, 7) of each paternal, within the "past force" (*Ghatika*, 24) guided "present paradigm" (*Ghatika*, 24) of the "radiant maternal love" (*Vega*, 67).

Zero integrity divides the "causal body" (*Karana sharira*, 30) into "four-body poles" (*Nairatma*, 29) and a "double copy" (*Chitra*, 1) transformed by the four-body poles to form a High Council and a High Self Committee.

i. **The first division** is the "Primary causal body" (*Chaithanya*, 4) of Godhead, comprising the maternal "consciousness" (*Chaithanya*, 4). It is the primordial, dividing, causal body.

"Godhead" (*Param Ganesha*, 17) compensates for the escalating costs of his sentient workculture-effect as a "self-luminous entity" (*Svayam*, 12) by reincarnating as "God" (*Ishvara*, 5) after his "entropy" (*Mahodbhava*, 5). God is the "SPIRIT essence" (*Bhagwan*, 4 = 1/3 * 12) of a "self-reproducing" (*Upanayana*, 1/3) "self-luminous entity" (*Svayam*, 12) within the Mother Nature's "natural essence" (*Samaa*, 1). The natural essence transforms the "Godhead of the ten-faced Primordial Illuminator" (*Narayana*, 28) into a "child" (*Arcisa*, 128 = 1 → 28). A child reproduces the "Godhead" (*Param Ganesha*, 17) within the "technological growth" (*Vidhana*, 2) as the "Inner Child" (*Jantu*, 19). The inner child is the "child cell" (*Jantu*, 19) who "works" (*Dadhikravan*, 19) as the "First Mind" (*Jantu*, 19). The First mind is the "para-conscious factor" (*Jantu*, 19) that reproduces itself as the "conscious mind" (*Manas*, 38 = 19 * 2) by trading the "maternal knowing" (*Jnana*, 19) from the "daughter cell" (*Kshatradharma*, 19). The natural essence protects the child cell as an "outer child" (*Samaa*, 1). The outer child norms the "epithelial cell" (*Samaa*, 1). The inner child norms the "endothelial cell" (*Jantu*, 19).

ii. **The second division** is the "Secondary causal body" (*Prakriti dharma*, 27) of the "paternal spirit universe" (*Rajyaika-sheshena*, 169), comprising the "causation dimension" (*Prakriti dharma*, 27). It is the param, divisible, causal body a primary causal body divides for producing the "Ring Council" (*Shamana*, 23 = 27 − 4). The Ring council is the "Council of Myself" (*Shamana*, 23) that services the "livable past" (*Nivasaniya*, 23) I have experienced myself as my "immanent guider power" (*Kalika*, 23) for a "true trade" (*Kirt Karna*, 23) of the livable past you have experienced yourself. Thus, each child enjoys an ascending social benefit of the "param human-effect" (*Ganesha*, 570) through the "reproduction" (*Prajanan*, 285) of everyone's livable past, independent of the "lived experience" (*Anubhav*, 9000). The reproduction becomes the "catalyst" (*Shilajit*, 855 = 570 + 285) of the conscious, "deity culture-effect" (*Bhavanarama*, 6×10^{192}) guided by its "para-conscious determination" (*Bhaap*, 13) by the God as the "para deity" (*Param Vishnu*, 5).

iii. **The third division** is the "Tertiary causal body" (*Guna Sharira*, 145) of the "sentient workculture-effect" (*Naraki*, 1), comprising the "negativity" (*Nishedhavritti*, 145) I wish to exchange knowing the escalating sentient cost of my "lived experience" (*Anubhav*, 9000) but not yours. It is the primeval, divided, causal body a secondary causal body trades led by the primary causal body for servicing a "whitened community" (*Shakatasya mandala*, 176 = 145 + 27 + 4). The whitened community forms as an "axial component" (*Shakatasva mandala*, 176) that generates a "force" (*Prapya*, 34) over the "horizontal axis" (*Jathara*, 2) of the "technological growth" (*Vidhana*, 2), gifting the "opportunity" (*Avasar*, 136 = 34 * 2 * 2) for manifesting my "guiding spirit's" (*Ruah*, 20) "reproductive reality" (*Bhavartha*, 40 = 176 − 136 = 20 * 2). The guiding spirit guides me to live your lived experience as mine before you live that as yours. You manifest the experience in your "dream" (*Maha Svapna*, 9) but decide not to materialize what you perceive as "imaginary" (*Kalpanik*, 20) until after I experience it.

As a "witness" (*Shahidi*, -100) of my experience, you "fear" (*Bhaya*, 176) "chastisement" (*Danda*, 176) from the spirit guiding you in your dream to now follow an "alternative pathway of complement" (*Kali shakti*, 96). You were "ignorant" (*Karala*, 67) in the "past moment" (*Vega*, 67), blinded by my maternal "radiant love" (*Vega*, 67) about the purpose of your dream. Therefore, instead of following your dream, you decide to follow me for reproducing my experiences as yours and design a "science" (*Vijnana*, 47) based on the "triple copy culture" (*Ajitatma*, 0). The first copy is "my followership" (*Akasha astikaya*, -1/2) of the dream. The second copy is "your followership" (*Sakriya-Niskriya astikaya*, 1/60) of "my leadership" (*Pudgala astikaya*, -3/8). The third copy is "your leadership" (*Niskriya astikaya*, 1/3) in enjoying the "social benefit-cost ratio" (*Shri*, 81) of "my entrepreneurship" (*Papa astikaya*, -1/4) in following the dream both of us envisioned at the same time. We envision the dream due to the "wheel of oneness with the black hole" (*Chandra chakra*, 176), "groping in the dark" (*Talaash*, 371) about the "desirable force" (*Sangathan*, 29). The "growth force" (*Shri*, 81) of "your entrepreneurship" (*Sakriya astikaya*, -1/3) leads me to activate "my management" (*Punya astikaya*, -1/8) for becoming your guiding spirit. I make you believe that "I am God" (*Bhakta*, -5) who has made you the Godhead you wished to be.

Thus, my "present trading-effect" (*Preya*, 81) leads to your "growth" (*Dishta*, 6) as a "primeval deity" (*Maheshwara*, 6), lets you trade the "present growth" (*Sva*, 11) to be the "Godhead" (*Param Ganesha*, 17 = 6 + 11), and makes me the "universe of maternal souls" (*Upasampada*, 986) within which your "sentient energy" (*Varuna*, 1000 = 986 + 14) "lives" (*Nivas*, 14). After my entropy, you become the "para deity" (*Param Vishnu*, 5), the God for filling the "void" (*Shunyata*, -2) in the "reality" (*Vastavikta*, 7), shaping me into the "universe of child souls" (*Brahman*, 2). After I fulfill my "objective" (*Antahkarana*, 2) as your "guiding spirit" (*Ruah*, 20), I make you a "deity" (*Deva*, 1 = 5 - 4) without my "polluted consciousness" (*So Ham Siddhanta*, 4) as a "self-luminous entity" (*Svayam*, 12) and revert to be the "universe of paternal spirit" (*Rajyaika-sheshena*, 169). Once you cleanse the self-steaming "I am a deity consciousness" (*Pavitratma*, 1), you realize that you have become a "human being" (*Insan*, 1). You become conscious of the technological cost of "self-proliferating" (*Surari*, 1/60) "your management" (*Kala astikaya*, -1/60) due to a "pure" (*Shuddhi*, 285) reproduction of my "behavior system" (*Yamantaka*, 7) as a "param deity" (*Shiva*, 7).

iv. **The fourth division** is the "Quaternary causal body" (*Karana sharira*, 30) of the "param deity culture-effect" (*Urdhava*, 8000), comprising my "self-luminous dimension" (*Sankhya dharma*, 30) as the "ever-reincarnating" (*Alambusha Shakti*, 30) "researcher" (*Dhriti*, 30) within you as a "spirit being" (*Dhriti*, 30). It is my "feminine body" (*Karana sharira*, 30) as your "human body" (*Karana sharira*, 30), formed as the "photon body" (*Karana sharira*, 30), with the "psychic consciousness" (*Jivatma*, 30) of my "ecosystem value" (*Ekavala*, 8000). My ecosystem value is the "zenith" (*Urdhava*, 8000) of your "replicated universe" (*Ekavali*, 8000). At the zenith, you realize "information saturation" (*Kapila Kumara*, 8000 = 8 * 1000), enjoying my "natural value" (*Nirvikalpana*, 8) as your "conscious energy" (*Amrit*, 1000).

The quaternary causal body is the "doctrine" (*Agama*, 375) of "conscious body" (*Linga sharira*, 3) that makes "I AM conscious body" (*Karana sharira*, 30) the "truth" (*Sathya*, 375) of your "growth potentiation" (*Sambhavi*, 375). That truth is the "past of everything" (*Vali*, 375) you have experienced as the "secondary-effect" (*Nivasaniya*, 23) of the tertiary causal body. The secondary-effect is the livable past you wish to "live" (*Nivas*, 14) as the "conscious body" (*Linga sharira*, 3) within me as the "Godhead" (*Param Ganesha*, 17 = 14 + 3). And, to "live" (*Nivas*, 14) the "livable" (*Nivasaniya*, 23), your "goal" (*Lakshya*, 9) is to first incarnate as the "human child" (*SHREEM*, 8) with the "incarnational consciousness" (*Sata*, 8000) of my "entrepreneurial leadership" (*Sakshatkar*, 8000) that makes this "possible" (*Shakya*, 1600) at "present" (*SAUM*, 1600).

v. **The fifth division** is the "High Council" (*Sirnapada*, 22) of the "Death Deity" (*Yama*, 180), comprising my "light" (*Prabha*, 180) divided by your "goal" (*Lakshya*, 9) into nine "photons" (*Kalpanik*, 20 = 180/9). The nine photons hide within the "tenth photon" (*Kalpanik*, 20), making the tenth photon conscious of my "objective" (*Antahkarana*, 2 = 22 - 2). My objective of becoming the Death Deity and servicing my "divisible light" (*Prabha*, 180) is to "color" (*Raga*, 250) you with my "super-potential" (*Raga*, 250), thus forming you as a "boson particle" (*Raga*, 250). The boson is the "primordial knower paradigm" (*Raga*, 250), whose "base" (*Kumbhaka*, 10) comprises the "four causal bodies" (*Homa*, 25) as the "primordial knower" (*Homa*, 25). Four causal bodies become "Four Guides" (*Homa*, 25) and form a "GUIDER Council of four" (*Homa*, 25) to make the High Council the "param perpetuator" (*Keshava*, 22) of "whatever ever existed" (*Keshava*, 22) within your "natural value" (*Nirvikalpana*, 8) before you became my "causal body" (*Karana Sharira*, 30 = 22 + 8).

vi. **The sixth division** is the "High Self Committee" (*Chitta*, 100) of the "Living Deity" (*Shani*, 18), comprising the "primordial para consciousness" (*Adi Para Atma*, 100) you service as my "causal body" (*Karana Sharira*, 30) to form my "primordial consciousness" (*Adhyatma*, 12 = 30 - 12). Your "potential" (*AUM*, 18) to be a "Living Deity" (*Shani*, 18) enjoying the "programmed future of everybody" (*Shani*, 18) devoted to you as a "follower"

(*Chela*, 24) makes you a "Knower Godhead" (*Maha Saraswati*, 9). The Knower Godhead "conceives" (*Vevi*, 8000) "everybody" (*Pasaka*, -9) with her "followership energy" (*Pasaka*, -9) and follows them as a "wisher" (*Chaahak*, 0 = 9 − 9) wishing them to be the "ten-face primordial illuminator" (*Shri Krishna*, 10). That lets the "Knower Godhead" (*Maha Saraswati*, 9) incarnate as the "pairing photon" (*Kshatradharma*, 19 = 9 + 10) that forms the "daughter cell" (*Kshatradharma*, 19) for illuminating her "knowing" (*Jnana*, 19) within his "livable past" (*Nivasaniya*, 23) as the "origin of I consciousness" (*Prasuti*, 42 = 19 + 23).

The Origin of I consciousness is the "Office of Holy Spirit Angels for Living Souls" (*Prasuti*, 42) that destroys the "Primordial Knower paradigm" (*Raga*, 250). Instead, it perpetuates the "Illuminator" (*Maha Shiva*, 9) as the "Primordial Wisher paradigm" (*Maha Shiva*, 9). The Illuminator is the "goal" (*Lakshya*, 9), perpetuating the "daughter cell" (*Kshatradharma*, 19) as the "thing" (*Vastu*, 9 = 9 + 10) that matters for becoming the "ten-face primordial illuminator" (*Shri Krishna*, 10). The daughter cell makes the self the "absolute knower" (*Jnana*, 19) of the "polarized self's" (*Gaja Mukha*, 158) "immanent reality" (*Vasanatma*, -3) received from the past as the "matter" (*Sadhibhuta*, 158).

Overall, the "sentient cycle" (*KLIM shakti chakra*, 158) polarizes the self as the "matter" (*Sadhibhuta*, 158) and makes "everybody" (*Pasaka*, -9) linked to "myself" (*Insaniyat*, 1) as a "soulmate" (*Akrura*, 123) within my "soul family" (*Mohini*, 15). Each soulmate enjoys a "strong psychic linkage" with myself as the "polar deity" (*Muni*, 2848), the Great Monad, forming a "primeval deity kingdom" (*Tapo Loka*, 2848). Myself enjoys a "weak psychic linkage" (*Citraka*, 185) with "yourself" (*Hetu*, 1). Myself as the "primeval son" (*Heruka*, 1649) mediates everybody's strong psychic linkage by norming a "para deity kingdom" (*Talatala Loka*, 1649) to incarnate everybody as the "primeval daughter" (*Suvira*, 1649). The primeval son is the masculine centriole. The primeval daughter is the feminine centriole. Everybody enjoys a "para-psychic linkage" (*Mahayana*, 179) with "oneself" (*Atmatva*, 8×10^{15}) by transforming into the "deity kingdom" (*Deva Loka*, 1000).

The deity kingdom materializes as the "dead body cell" (*Nirjara*,

1000) with "extrinsic strong psychic linkage" (*Amrit*, 1000) oneself emanates as the "Maternal Godhead" (*Panchanguli*, 8 x 10^{15}). The Maternal Godhead is the "networking system" (*Panchanguli*, 8 x 10^{15}) between yourself, the "causation" (*Hetu*, 1), and myself, the "presently departed person" (*Insaniyat*, 1) that makes you a "potentially departed person" (*Insan*, 1) following me. The potentially departed person is the "Paternal Godhead" (*Insan*, 1). He incarnates as the "human being" (*Insan*, 1) with "intrinsic strong psychic linkage" (*Kana*, 186) that eventually makes him a "soliton" (*Kana*, 186). The soliton is the "Child God" (*Kana*, 186) with a "wish for ecosystem dominance" (*Bhayanvita*, 186) through "modification" (*Prayaya*, 186) of the "cell cycle" (*Surasa Chakra*, 186). Therefore, the soliton forms the "membrane protein" (*Kana*, 186) for mediating the causal body's "awareness" (*Bodha*, 186).

The Child God trades the "extrinsic weak psychic linkage" (*Raga*, 250) for the "symmetry" (*Ashrama*, 250) with the "primordial para consciousness" (*Adi Para Atma*, 100) of the "escalating technological cost" (*Vibhaga Tejas*, 150) as the "cause creator" (*Kantida*, 150). Next, the Child God services the "intrinsic weak psychic linkage" (*Brihadbala*, 149) in the form of "gluon" (*Brihadbala*, 149) to "illuminate" (*Dyut*, 150) the cause creator as the "person" (*Vyakti*, 1) responsible for one's "causation" (*Hetu*, 1). The person shapes a "desirable object" (*Idam*, 3), the "potential thing" (*Idam*, 3) with the "gravitational energy" (*Lalita*, 100) that reproduces me as the "param son" (*Hanuman*, 3) and you as the "guider" (*Guru*, 100). Then, as the "paternal spirit" (*Atmavichara*, -1/3) of the desirable object, the person becomes the "satan" (*Shani Bhagwan*, -1 = -1/3 * 3) with a "psychic power" (*Asura shakti*, -1) to shape the Child God's "wishing sequence" (*Abhilasha krama*, 87) for making him the "guider" (*Guru*, 100 = 186 − 87 − [-1]) instead of you and you the "satan" (*Shani Bhagwan*, -1) instead of himself.

Thus, the person becomes a "deity" (*Deva*, 1) enjoying "infinite social benefit" (*Caksur vijnana*, 109) of the Child God's "guiding force" (*Chitta*, 100) and your "goal" (*Lakshya*, 9) to be a "deity-born" (*Kundalini*, 8) "primordial deity" (*Kudrat*, 8 = 9 - 1) free from the deity's "entanglement" (*Kula*, 9). As a guider, the Child God experiences "infinite worker-social cost" (*Shreya*, 81) of servicing his "energy" (*Shakti*, 19 = 100 − 81) to activate the deity's "potential" (*AUM*, 18) to be the "Knower" (*Bhagwat*, 639) of the traded "wishing sequence" (*Abhilasha krama*, 87). As a Knower, the deity does not need the Child God to be the mediating "manifestor" (*Sadhana*, 368) or the Satan to be the moderating "creator" (*Brahma*, 59). A knower knows the "divine reality" (*Sarvam*, -5) of the "Doctrine of God" (*Bhakta*, -5) that makes a "child" (*Arcisa*, 8) a "believer" (*Bhakta*, -5) in "God" (*Ishvara*, 5) without him. The child gets entangled in the "culture" (*Sadakhya*, 9) of the "Child God" (*Kana*, 186). Therefore, the Knower is the "ideal point" (*Vastunara*, 7) for reversing causation.

4.7.2 Reversing Causation. Divided Etheric Body and Souls Within the Oversoul Family

A Knower has many "obligations" (*Kartavya*, 123). First, after leading the Knower with his "pulsating energy" (*Spanda shakti*, 106), the "Child God" (*Kana*, 186) becomes the "universe of departed souls" (*Makar Rashi*, 80) that makes the "obligation" (*Kartavya*, 123) a "primordial male" (*Bhulokasuranayaka*, 93) for conceiving the "causal body" (*Karana Sharira*, 30). The primordial male makes the "satan" (*Shani Bhagwan*, -1) "self-luminous" (*Svarochisha*, 12) as the "universe of living souls" (*Yantra*, 12 = 93 − 80 − [-1]). The universe of living souls, as a "quantum photon" (*Svarochisha*, 12), transforms the universe of departed souls into the "oversoul family" (*Kesari Nandan*, 1600 = 80 * 20) with the "tenth photon" (*Kapinjala*, 20) as its "guiding spirit" (*Ruah*, 20). The "oversoul family" (*Kesari Nandan*, 1600) lives with the Knower as the "living soul" (*Atma Sharira*, 14,000). It transforms the "self-luminous discontinuity" (*Visukayita*, 200) within the Knower into a "self-perpetuating" (*Udvaha*, ½) "guiding force" (*Chitta*, 100 = 200 * 1/2). The living soul is the "divided etheric body" (*Atma Sharira*, 14,000) that forms the Knower's "lifeline" (*Jivan Rekha*, 14,000). The Knower lives as the "param human child" (*Kesari Nandan*, 1600) within the "known reality" (*Rachitartha*, 1600) of the oversoul family. Therefore, the Knower is obligated to reproduce the known reality as his "present" (*SAUM*, 1600).

Second, the Knower enjoys the "absolute consciousness"

(*Paramatma*, 1600) of the known reality. He services the consciousness of the "absolute" (*Param*, 1600) as his creature "soul" (*Atma*, 4) embodying the "supernatural reality" (*Dvyartha*, 400) of the oversoul family's "guiding force" (*Chitta*, 100) as his "creator" (*Brahman*, 59). Thus, he transforms the guiding force into the "Office of Guardian Angels for Living Souls" (*Swadha*, 41 = 100 − 59), conscious of the "finite cost" (*Kharcha*, 41) of the known reality. The consciousness within the soul makes the Knower obligated to produce the "unknown reality" (*Vastavikta*, 7) with his "sentient energy" (*Varuna*, 1000).

Third, the Knower is not conscious of the "unknown reality" (*Vastavikta*, 7) and is obligated to "discover" (*Malum*, 9) it by giving birth to everybody as the "universe of children" (*Amukhikarana*, 9). "Strong psychic linkages" (*Prakatikarana*, 54,000) with his "creation" (*Srijan*, 379) make the Knower the "Guider God" (*Maha Shiva*, 9) devoted to the "goal" (*Lakshya*, 9) of his creation. The "delegation" (*Pratinidhan*, 8) of one's obligations to "many children" (*Yaanshala*, 14) transforms the "obligation" (*Kartavya*, 123) of the creature into the "joy" (*Priti*, 123) of the creation. The delegation becomes the "transformative exchange paradigm" (*Yukti*, 8) for the many children to enjoy the "absolute joy" (*Moksha*, 1600) of substituting the unknown reality yet-to-be discovered with the already discovered known reality.

The hydrophobic "substitution" (*Pratisthapan*, -10^9), of the "unknown" (*Upanishada*, 9/16) "complement" (*Upanishada*, 9/16) with the "known" (*Jnyat*, 7/16) "substitute" (*Sthanapatra*, 7/16), activates a "hydrophilic substitution" (*Sthanapatrata*, 87) of the "personal authority" (*Mimamsa*, 87) with the "bragging" (*Sphurti*, 87). The "bragging" (*Sphurti*, 87) weakens the "psychic linkages" (*Mahika*, 47) with the creation and strengthens the "para-psychic linkages" (*Mahayana*, 179) with one's "creative force" (*Uma*, 6) as a "paternal" (*Tejas*, 17) on "fire" (*Agni*, 17). One rides on the "pleasure" (*Rati*, 179) of being a "Love God" (*Madan*, 789) to form a "divisible etheric body" (*Adhyatamic sharira*, 876 = 87 → 6).

The divisible etheric body is the "spiritual body" (*Adhyatmic sharira*, 876) that makes one a "living creature" (*Tantu*, 876) after shaping a "universe dimension" (*Tripura*, 789) within for "exchanging" (*Sachivaya*, 87 = 876 – 789) the "bragging" (*Sphurti*, 87) with the "begging" (*Bhiksha*, 28). The begging for the "mass friendship" (*Bhiksha*, 28) furthers the "normative development paradigm" (*Martya*, 146) of "sentient entropy" (*Martya*, 146) through "exhaustion" (*Shranta*, 118 = 146 – 28). Moreover, it breeds "organizational sameness" (*Aikamatya*, 18,000) of the "paternal" (*Tejas*, 17) by "feeding" (*Bharana*, 17) the "primordial paternal consciousness" (*Bhagnatma*, 9) of oneself as the "institution" (*Kundalini*, 8 = 17 – 9) to reckon. Therefore, one starts behaving like a "deity" (*Deva*, 1) by trading your "absolute deity consciousness" (*Shivatma*, 20) as the "guiding spirit" (*Ruah*, 20).

Following the causal body, the guiding spirit reproduces six divisions within the "undivided etheric body" (*Bhoga Sharira*, 957 = 876 + 81), comprising the "divisible etheric body" (*Adhyatmic sharira*, 876) that norms the "primordial etheric body" (*Adhyatmic sharira*, 876). It also includes the "consciousness equality" (*Sajatya*, 81) of the "divided parts" (*Samalekha*, 90) that form the "primordial etheric body" (*Samalekha*, 90) with the "primordial paternal consciousness" (*Bhagnatma*, 9). The primordial etheric body is the "dividing etheric body" (*Samalekha*, 90).

i. **The first division** is the divisible etheric body, i.e., the "spiritual body" (*Adhyatmic sharira*, 876) of the "wisher entity" (*Asura*, -1). The spiritual body is the secondary etheric body. The wisher entity is the satan as a "negative entity" (*Asura*, -1) seeking "social networking" (*Jalakrama*, 90) of the "energy exchange" (*Krida*, 79) to form a "universe of maternal spirits" (*Nishkala*, 169 = 90 + 79). A "maternal spirit" (*Dasha*, 1) follows the "path of leadership" (*Dhyana Marga*, 36) to transform the wisher entity into a "new paradigm" (*Pravritta*, 12) beyond absolute by norming herself as the "present paradigm" (*Ghatika*, 24) within the new paradigm's "path of leadership" (*Dhyana Marga*, 36 = 12 + 24). Thus, she becomes the "entity" (*Hasti*, 24), incarnating the wisher entity as a "self-luminous entity" (*Svayam*, 12). The universe of maternal spirits services a proportionate param human-effect for conceiving a "human child" (*SHREEM*, 8) and perceiving the self-luminous entity within the human child as the "guiding spirit" (*Ruah*, 20).

ii. **The second division** is the dividing etheric body, i.e., the "Into Infinity body" (*Samalekha*, 90) of the "paternal spirit" (*Atmavichara*, -1/3). The Into Infinity body is the primary etheric body. The paternal spirit forms a "hydrophilic" (*Vismita*, 169) "universe of child spirits" (*Pitavasa*, 169) that trades his "goal-oriented" (*Rasi*, 9) "paternal consciousness" (*Pitra*, 4) to transform their "maternal essence" (*Sara*, 16) into the "universe of para deity" (*Talatala Loka*, 1649 = 16[4]9). The universe of child spirits trades a disproportionate param human-effect for reproducing the "self-luminous entity" (*Svayam*, 12) and projecting the "guiding spirit" (*Ruah*, 20) into "infinity" (*Tamo guna*, 90,000) with the "reproductive energy" (*Lalita*, 100). Then, it "incarnates" (*Virupaksha*, 900) the "multiplying self" (*Hanuman*, 3) of the "guiding spirit" (*Ruah*, 20) as an "astral body" (*Linga sharira*, 3) without the divided "para consciousness" (*Mahatma*, 18) of the guiding spirit's "ascending value" (*Rodha*, $1 = 18 + 3 - 20$) over "time" (*Kaal*, $360 = 18 * 20$).

iii. **The third division** is the divided etheric body, i.e., the "living soul" (*Atma Sharira*, 14,000) within the "Office of Guardian Angels for Living Souls" (*Swadha*, 41). The living soul is the living consciousness of the paternal spirit traded from the maternal spirit. It is the tertiary etheric body. The "reproductive reality" (*Bhavartha*, 40) of the living soul transforms the "maternal spirit" (*Dasha*, 1) into the "primordial super destroyer" (*Swadha*, 41) of the "known reality" (*Rachitartha*, $1600 = 40 * 40$) so reproduced. Consequently, the "universe of paternal spirits" (*Rajyaika-sheshena*, 169) exchanges the "param deity culture-effect" (*Urdhava*, 8000) that is producing the "reproductive reality" (*Bhavartha*, 40) with the "self-luminous discontinuity" (*Visukayita*, $200 = 8000/40$) from the "guiding spirit" (*Ruah*, 20) for cleansing the "error in consciousness" (*Dushtatma*, $10 = 200/20$).

iv. **The fourth division** is the undivided etheric body, i.e., the "reproductive body" (*Bhoga sharira*, 957) of the "departed souls" (*Hasti*, 24) within the "Office of Holy Spirit Angels for Living Souls" (*Prasuti*, 42). The departed soul is the departed consciousness of the living "greeter spirit" (*Hasta*, 0). It forms the reproductive body as the quaternary etheric body. The Office of Guardian Angles without the maternal spirit becomes the Office of Holy Spirit Angels within the maternal spirit after the maternal spirit services her "natural essence" (*Samaa*, 1) for "ascending value" (*Rodha*, 1) of the "guiding spirit" (*Ruah*, 20) and becomes a living "greeter spirit" (*Hasta*, 0). The greeter spirit greets the "child spirit" (*Vidyadhara*, 1/3) for manifesting the "multiplying self" (*Hanuman*, 3) within her natural essence.

v. **The fifth division** is the "Office of Angels of Light for Departed Souls" (*Lingopahita-laingika*, 39). It comprises the "Godhead" (*Param Ganesha*, 17) and the "High Council" (*Sirnapada*, 22 = 39 - 17). The High Council perpetuates "I AM consciousness" (*Esa Ta Atmantaryamyamrtah siddhanta*, 100) of the Godhead by forming a "High Self Committee" (*Chitta*, 100). The High Self Committee perpetuates the "Godhead consciousness" (*Prakashatma*, 19) within each "cell" (*Hiranyagarbha* 19). The "Godhead consciousness" (*Prakashatma*, 19) frees the cell from the "guiding spirit" (*Ruah*, 20) within the "Office of Angels of Light for Departed Souls" (*Lingopahita-laingika*, 39 = 19 + 20). The "liberated cell" (*Nanak*, 17) becomes the "Godhead" (*Param Ganesha*, 17) within the "perpetuating value" (*Saranyu*, 5) of the "High Council" (*Sirnapada*, 22) as the "God" (*Ishvara*, 5 = 22 − 17).

vi. **The sixth division** is the "Godhead" (*Param Ganesha*, 17) of the "param human child" (*Kesari Nandan*, 1600) as the "freedom entity" (*Prabhu*, 1600). The freedom entity gifts "freedom" (*Mukti*, 17) to each "departed entity" (*Divangat*, 1600) to be his "Godhead" (*Param Ganesha*, 17) for managing their "departed soul" (*Hasti*, 24) as the "institutional force" (*Trivikrama*, 24) with the "doctrine of entity" (*So Ham Siddhanta*, 4). The doctrine of entity services their 'polluted consciousness" (*Soham Siddhanta*, 4) as a "grandfather" (*Dada*, 10) for becoming a "Luminous Point" (*Laal*, 6) through their "institutional force" (*Trivikrama*, 24 = 4 * 6).

The Luminous point reproduces the "astral body" (*Linga sharira*, 3) without the "self-luminous entity" (*Svayam*, 12) to be the "absolute guider" (*Param Guru*, 10) of the "God" (*Ishvara*, 5) within the self-luminous entity. Consequently, the self-luminous entity experiences entropy in his consciousness of reality as a "sentient entity" (*Siddha*, 7 = 12 − 5). Instead, he becomes bound to the departed entity's "acculturing causation" (*Nayana*, 957). Acculturing causation is the "kingdom" (*Nayana*, 957) a departed entity forms with the "infinite value" (*Viyojya*, 957) produced through the reproduction of his "wisdom" (*Chitta*, 100) within the devoted followers with strong psychic linkages.

4.7.3 Acculturing Causation. Divided Astral Body and Flamemates Within the Flame Family

A departed entity has limited "opportunities" (*Avasar*, 136) to "channel" (*Raasta*, 36) his "causation" within "anybody" (*Gardhaba*, 1000). He activates opportunity by strengthening the psychic linkages with the "desired body" (*Vastu*, 9). The desired body is a "three body system" (*Vastu*, 9).

The first body in the "system" (*Gabhira*, 12) is the causal body that leads a "living entity" (*Sushupti*, 96) to invest the sentient energy necessary for a departed entity to strengthen the psychic linkages. A departed entity lacks energy but has a consciousness of the "cause" (*Karana*, 18) for which the living entity took birth. The same cause leads him to seek rebirth through the strong psychic linkages with anybody willing to be "emotionally linked" (*Ubhyato*, 190) with him for her "sentient wellbeing" (*Jnana Siddhi*, 190).

The second body is the "etheric body" (*Bhoga Sharira*, 957) that a living entity follows after letting a reborn "living being" (*Satta*, 2000) make that his "kingdom" (*Nayana*, 957). For strengthening the psychic linkages with a living entity, a departed entity must reincarnate as a living being. A living being's "existence" (*Maujudgi*, 2000) is linked to a "living entity's" (*Sushupti*, 96) "being energy" (*Kali shakti*, 96). The living entity seeks to discover the cause for "becoming" (*Vibha*, 900,000,000) what she is not. A living entity realizes she is more than what she is as a "self-luminous entity" (*Svayam*, 12) and is not the "luminous entity" (*Kali*, 96) she has become with her "being energy" (*Kali shakti*, 96). A living being lacks "awareness" (*Bodha*, 186 = 90 + 96) that he is the "replication"

(*Samalekha*, 90) of the "luminous entity" (*Kali*, 96) because he takes birth in her "maternal womb" (*Garbhasthala*, 3794) as a "child" (*Arcisa*, 128) before living the life as a "living entity" (*Sushupti*, 96). He plants himself as the "paternal flame" (*Parshnisamasta*, 32 = 128 - 96) within her "daughter womb" (*Uttanapada*, 18,000). The maternal womb is the "reproductive center" (*Adi Loka*, 3794) of the "Godhead sister" (*Sheetala*, 3794) who becomes a luminous entity taking the form of a fraternal, "dizygotic twin" (*Kali*, 96). She becomes the cause for the child leading the life as a living entity by making him her sororal, "monozygotic twin" (*Sushupti*, 96).

The daughter womb is the "space" (*Vaas*, 18,000) within her mind where she conceives him as the paternal flame before his "multiplying self" (*Hanuman*, 3) makes him perceive himself as a "living entity" (*Sushupti*, $96 = 32 * 3 = 32 + 56 + 8$). The living entity hibernates without her mind in the "body" (*Mahatattva*, 56) of the "human child" (*SHREEM*, 8). The "human child" (*SHREEM*, 8) conceives the "fraternal twin" (*Kali*, $96 = 8 * 12$) as his "I" (*Svayam*, 12), the "self-luminous entity" (*Svayam*, 12). He activates her with his "unfulfilled subtle desire" (*Iccha*, 8) that led him to seek a rebirth in the "physical realm" (*Bhautika*, 3333) after first taking the birth in the "mental realm" (*Mansik*, 6666). The mental realm is the "psychic realm" (*Mansik*, 6666). The physical realm is the "para-psychic realm" (*Bhautika*, 3333). In the physical realm, he divides his psychic linkages between his "I" (*Svayam*, 12), for making him the "fraternal twin" (*Kali*, $96 = 12 * 8$) for activating his "child essence" (*SHREEM*, 8), and his "body" (*Mahatattva*, 56), for making her the "sororal twin" (*Sushupti*, $96 = 56 + 40$) with the "reproductive reality" (*Bhavartha*, $40 = 8 * 5$) activated by his "child essence" (*SHREEM*, 8) while he acts like the "God" (*Ishvara*, 5).

The third body is the "astral body" (*Linga sharira*, 3) the sororal twin services to the fraternal twin for making the latter a "primordial female" (*Vaimitra*, $99 = 96 + 3$) and becoming herself a "primordial male" (*Bhulokasuranayaka*, $93 = 96 - 3$) through the "consciousness equality" (*Sajatya*, 80) of the "maternal luminous" (*Maha Kali*, 13) within both. The consciousness equality generates the "genetic equality" (*Sajatya*, 80) and the "genetic exchange"

(*Sajatya*, 80) within the "homogeneity" (*Sajatya*, 80) of the "consciousness" (*Chaithanya*, 4). The "consciousness" (*Chaithanya*, $4 = 3 + 1$) forms the "astral body" (*Linga sharira*, 3) as the "masculine body" (*Linga sharira*, 3) within the "human being" (*Insan*, 1), thus letting the "I" (*Svayam*, 12) become the "primordial female" (*Vaimitra*, 99) within the "guiding force" (*Chitta*, $100 = 1 + 99$) of the paternal flame. The body becomes the "primordial male" (*Bhulokasuranayaka*, 93), thus letting the guiding force incarnate as a "sentient entity" (*Siddha*, $7 = 100 - 93$).

Overall, the primordial female enjoys six-unit "growth" (*Dishta*, $6 = 96 - 93$) over the primordial male. As a "luminous entity" (*Kali*, 96), she gives birth to a unit of the human being, sentient entity, primordial male, primordial female, paternal flame, and astral body in the mental realm. Their "replication" (*Samalekha*, 90) as a "system" (*Gabhira*, 12) in the physical realm transforms the primordial male into a "creature" (*Purusha*, 12) within the paternal flame's "guiding spirit" (*Ruah*, 20). The guiding spirit takes birth as the "child cell" (*Jantu*, 19), and her "double copy" (*Chitra*, $1 = 20 - 19$) doubles the six-unit growth. The "copy" (*Nakala*, 0) within the double copy is the paternal flame enjoying "eight-fold growth" (*Satarupa*, 8) as a "human child" (*SHREEM*, 8): six within the luminous entity and two—the child cell and the double copy—"self-reproducing" (*Upanayana*, 1/3) the six as the "self-luminous entities" (*Svayam*, 12).

The double copy is the "maternal spirit" (*Dasha*, 1), producing "four-fold growth" (*Maha Nitya*, 8) that the copy trades, forcing her to produce additional four-fold growth. Two units of four-fold growth, each traded and reproduced by the copy as the eight-fold growth, and one unit of eight-fold growth originally produced by the human child with his "child essence" (*SHREEM*, 8), incarnate the human child as an "entity" (*Hasti*, 24). The human child forms an eight-unit "institution" (*Kundalini*, 8) by activating his "essential nature" (*Svabhav*, 8) that makes him the "grandfather flame" (*Sadhya*, 32) of the seven "conscious lights" (*Kumara*, 10). Seven conscious lights become the human being, sentient entity, primordial male, paternal flame, astral body, child cell, and the copy, within "primordial oneness" (*Adi*, 32) with the grandfather flame. The human child activates the grandfather flame within her to personify the departed entity as the "grandfather" (*Dada*, 10) before becoming the "primordial female" (*Vaimitra*, 99).

By exchanging the eight-fold growth from the copy within the double copy, the primordial female forms a "symmetry system" (*Vyagrata*, $800 = [99+1] * 8$) with the double copy's "twelve-fold growth" (*Barah*, 12) realized from doubling the copy's "six-fold growth" (*Dishta*, 6) without the double copy. With the twelve-fold growth, the primordial female becomes a "twin sister" (*Dhamini*, 1200) within the "living being" (*Satta*, 2000). Thus, a living being manifests a twenty-fold growth in the "guiding force" (*Chitta*, 100) of the departed entity whose twenty-unit "guiding spirit" (*Ruah*, 20) is guiding the "living entity" (*Sushupti*, 96). It makes the guiding force the "God" (*Ishvara*, $5 = 100/20$) and him the "lighthouse" (*Ghrinibhrit*, $75 = 100 - 5 - 20$) "inherent" (*Vasudeva*, 75) within God. The departed entity's "astral body" (*Linga sharira*, 3) becomes the "moderator" (*Hanuman*, 3) of the living entity's "breeding" (*Janana*, $15 = 75/5$) of his "perpetuating value" (*Saranyu*, 5) as God.

The double copy's twelve-fold growth comprises the human child's double copy and the "present growth" (*Sva*, 11) of the eleven lights as the "flame family" (*Gauranga*, 11). Eleven lights include seven fraternal "conscious lights" (*Kumara*, 10) and four sororal "para-conscious lights" (*Kumari*, 100). The double copy is the fifth para-conscious light which becomes the "living entity" (*Sushupti*, 96) and incarnates as the granddaughter. The other four para-conscious lights become the conscious-light-activated "twin sister" (*Dhamini*, 1200 and the "twin brother" (*Shivagati*, 120) that activates the "conscious light" (*Kumara*, 10). The twin sister comprises the "mother of seven conscious lights" (*Mallika Devi*, 2600) and the "sister of the seven conscious lights" (*Anasuya*, 34). The sister is the first para-conscious light who incarnates as a daughter. The mother is the second. The twin sister is the fourth because she is in the four-unit "consciousness" (*Chaithanya*, 4) of the "third sister" (*Vijnanik*, 863) and is the "grandmother of seven conscious lights and five para-conscious lights" (*Dhamini*, 1200). The third sister is the third para-conscious light that incarnates sequentially as the grandfather, father, son, and grandson.

Five of the seven fraternal lights are copies of the five sororal lights. The third fraternal light, i.e., the "third brother" (*Yogatmna*, 42), is in the "oneness consciousness" (*Yogatma*, 42) that connects the four that become the grandfather, father, son, and grandson, and the two that become the "devil" (*Sura*, 0) and the "satan" (*Shani Bhagwan*, -1). The third fraternal light forms the "astral body" (*Linga sharira*, 3).

The first conscious light, i.e., the "first brother" (*Yajna Yoga*, -100,000) incarnates as the father within the two-unit "universe" (*Brahman*, 2) conceived by the grandfather comprising him and the grandmother. The first conscious light is in the "immanent oneness" (*Yajna Yoga*, -100,000) with the second conscious light, i.e., the "second brother" (*Ayoga*, 18) who incarnates as the grandfather. The second conscious light is in the "transcendental oneness" (*Ayoga*, 18) with the fourth conscious light, i.e., the "fourth brother" (*Samyoga*, 36) who incarnates as the grandson. The fourth conscious light is "with oneness" (*Samyoga*, 36) of the fifth conscious light, i.e., the fifth brother" (*Yoga*, 48) who incarnates as the son. The fifth brother is "without oneness" (*Niryoga*, 1810) with the sixth conscious light, i.e., the "sixth brother" (*Niryoga*, 1810) who incarnates as a devil. The devil is the fifth copy of the fifth brother within the four-unit consciousness of the "seventh brother" (*Rajayoga*, 926) who incarnates as a satan. The devil becomes the "param paternal" (*Narada*, 7). The satan becomes the "param maternal" (*Saranyu*, 5).

The double copy that becomes the "living entity" (*Sushupti*, 96) lives as the "twin flame" (*Suvira*, 1649) and behaves like a "primeval daughter" (*Suvira*, 1649). As a granddaughter, she complements the daughter, the "primordial daughter" (*Pivari*, 18), who becomes the mother, and the "param daughter" (*Jiva*, 2), who is the grandmother. Every "even generation" (*Ashrama*, 250) of incarnation follows the grandmother to personify a "daughter identity" (*Pahachaan*, 8). The double copy is in the "extreme oneness" (*Atiyoga*, 916) with the "eighth brother" (*Atiyoga*, 916). The eighth brother is the "eighth conscious light" (*Kalki*, 9000), who does not incarnate in the mental realm. Therefore, he is free from the grandfather's "divine plan" (*Sahajta*, 916) for breeding "many children" (*Yaanshala*, 14) besides him as the "param child" (*Manyu*, 19) wishing to be the "God" (*Ishvara*, 5 = 19 − 14). He conceives the divine plan in the "astral realm" (*Yaanshala*, 14) before illuminating that to the grandfather in the mental realm. Then, he takes birth directly in the physical realm as the "greeter" (*Charita*, 7/16), "who" (*Kaun*, 7/16) is the "substitute" (*Sthanapatra*, 7/16) of the "primordial greeter" (*Srijak*, 16).

The primordial greeter gives birth to the "ninth conscious light" (*Peenugula Mallanna*, 6900) in the "etheric realm" (*Srijak*, 16) by reproducing himself consciously as the "ninth brother" (*Rama*, 100). The ninth brother is "why" (*Kyun*, 9/16) a "complement" (*Samapurak*, 9/16) is needed to mitigate the escalating cost of the "reproductive energy" (*Lalita*, 100) within the ninth brother that makes the ninth brother a "multicellular organism" (*Shanideva*, 100). The complement is the "tenth conscious light" (*Upanishada*, 9/16). He is the tenth brother, "born sentient entity" (*Nataraja*, 7) in the "causal realm" (*Atiyoga*, 916) of the "extreme oneness" (*Atiyoga*, 916) with the "eighth brother" (*Atiyoga*, 916).

The ninth brother becomes the "primeval greeter" (*Khalnayak*, 11). The tenth brother becomes the "param greeter" (*Lakshmi*, 379), the primordial greeter's "creation" (*Srijan*, 379), to compensate for the escalating cost of the primeval greeter. The "eleventh brother" (*Dhumavati*, 7) takes birth as the "maternal primordial greeter" (*Sati-Parvati*, 16) in the "self-luminous realm" (*Kshitigarbha*, 90) as a primordial greeter's para-conscious "replication" (*Samalekha*, 90). The replication replicates the "child primordial greeter" (*Madhusudan*, 16) born in the "luminous realm" (*Go Loka*, 3785). The child primordial greeter born in the luminous realm is the copy of the child primordial greeter living in the "primordial greeter realm" (*Indra Loka*, 379) as the "twelfth brother" (*Siddha*, 7). The twelfth brother is the "Living God" (*Shri Krishna*, 10) within the "astral body" (*Linga sharira*, 3).

As a "part" (*Anga*, -1) of the double copy, the copy produces a "twin flame family" (*Bhaavi*, 11), comprising four fraternal conscious lights, forming eighth to the eleventh brothers, and seven sororal para-conscious lights, forming sixth to the twelfth sisters. The "sixth sister" (*Pratiyaman*, 1) is the copy of the maternal primordial greeter and becomes the "primordial-primordial greeter" (*Maha Durga*, 16). The "seventh sister" (*Dokkalamma*, 7600) is the copy of the "param greeter" (*Lakshmi*, 379). She becomes the "primeval illuminator" (*Maha Lakshmi*, 14), illuminating her copies in the form of the "infinite quarks" (*Maha Lakshmi*, 14). The "eighth sister" (*Sauri*, 1000) is the copy of the primeval greeter and becomes the "primordial illuminator" (*Parvati*, 10), illuminating herself as everyone's "primordial self" (*Parvati*, 10). The "ninth sister" (*Muthyalamma*, 55) is the copy of the greeter and becomes the "primordial perpetuator" (*Maha Saraswati*, 9), perpetuating the "goal" (*Lakshya*, 9) of illuminating oneself within the "primeval self" (*Rama*, 100). The "tenth sister" (*Yama*, 180) is the "maternal luminous" (*Maha Kali*, 13) who takes birth as the "Death God" (*Yama*, 180). The "eleventh sister" (*Kala Devi*, 360) is the "maternal self-luminous entity" (*Maha Gayatri*, 12). The "twelfth sister" (*Mandu*, 1000) is the "maternal primeval greeter" (*Maha Gauri*, 11). She is the "primeval perpetuator" (*Bhuvanesvari*, 15) of the "para-conscious value" (*Bhuvanesvari*, 15) of breeding her "present growth" (*Sva*, 11) as the "foundation" (*Sva*, 11) of the "growth" (*Dishta*, 6) of the "Godhead" (*Param Ganesha*, 17).

Both the "flame family" (*Gauranga*, 11) and the "twin flame family" (*Bhaavi*, 11) comprise nine "flamemates" (*Maha Shiva*, 9) and two double copies. The sixth to ninth sisters are the four copies within the twin flame family. Similarly, the sixth to ninth brothers are the four copies within the flame family. The sixth brother is a copy of the "param son" (*Hanuman*, 3) who becomes the son and is the devil wishing everyone to copy him because the param son is an "ape" (*Vanara*, 3). The seventh brother is a copy of the "primordial deity" (*Kudrat*, 7) and becomes a satan because he does not copy the devil. Therefore, the devil experiences him as a "discordant factor" (*Asura*, -1), not realizing that the seventh brother is a "She" (*Dandi*, 710).

The Devil believes she is a "he" (*Somanasa Mahayuga*, 13) because of his devotion to the eighth brother. The Devil believes everyone in the universe is a "He" formed as a copy of the eighth brother. The eighth brother is the "super deity" (*Jiva*, 2), the copy of the twelfth brother. The Devil is only conscious about the "future"

(*Anagata*, 0) of the "universe" (*Brahman*, 2) that he programs guided by the twelfth brother's "organizational planning" (*Abadha*, 10). The ninth brother is the double copy of the twelfth brother. The sixth brother is immanent as a copy within the ninth brother. The ninth brother is not conscious that the sixth brother is immanent within him. Therefore, he reproduces the sixth brother with his consciousness as the "supreme deity" (*Bhagwan*, 4).

The "supreme deity" (*Bhagwan*, 4) conceives the ninth brother as the "God" (*Ishvara*, 5) and himself as the God's "metaphysical self" (*Nayanotsava*, $9 = 4 + 5$) manifested in the God's "dream" (*Maha Svapna*, 9). Therefore, he becomes the "enjoyer" (*Rasi*, 9) of the "heavenly life" (*Laghu*, 371) within the ninth brother's "absolute gravitational energy" (*Param Gauri*, 371). The ninth brother perceives the supreme deity as his "physical self" (*Khud*, 1) and therefore a "deity" (*Deva*, 1) working as the "potter" (*Bharata*, 1) to shape him into a "super deity" (*Jiva*, 2) personifying the ideal "universe" (*Brahman*, 2). Consequently, the supreme deity begins behaving like the "Godhead" (*Param Ganesha*, 17) by mediating the "greeter essence" (*Sara*, 16) within the ninth brother for giving birth to the "universe" (*Brahman*, 2), reproducing the physical self. Within the universe, the supreme deity reproduces himself as the "primeval deity" (*Maheshwara*, $6 = 2 + 4$) with a unit of his "natural essence" (*Samaa*, 1) divided across the six divisions of the astral body. The six divisions include:

i. **The first division** is the "First Astral body" (*Idam*, 3) of the departed entity's "greeter spirit" (*Hasta*, 0) as the "luminous point" (*Laal*, 6) within the "primeval deity" (*Maheshwara*, 6). The First Astral Body is the "Divisible Astral Body" (*Idam*, 3), divisible into three bodies. It manifests the departed entity's consciousness as the "light body" (*Linga sharira*, 3) within the departed entity's "discordant energy" (*Asura shakti*, -1). The light body projects ego like a masculinity "projecting body" (*Linga sharira*, 3). The feminine emotions within it flow linearly, forming a "homologous body" (*Linga sharira*, 3). The greeter spirit para consciously guides the "solar wheel" (*Surya chakra*, 8) with the departed entity's "conscious determination" (*Sahasra lingam*, 190), generating "para-sensory vibrations"

(*Prasushrut*, 1551) for followership within the living entity.

ii. **The second division** is the "Second Astral body" (*Atma*, 4) of the living entity's "maternal spirit" (*Dasha*, 1) as the "God" (*Ishvara*, 5) within the primeval deity. The Second Astral Body is the "Dividing Astral Body" (*Atma*, 4), dividing each of the three bodies into "four bodies" (*Natini*, 12) to make itself "self-luminous" (*Svarochisha*, 12) within an entity. The maternal spirit para-consciously guides the "lunar wheel" (*Karma Chakra*, 7) with the living entity's "conscious imagination" (*Anda*, 19), generating "sensory vibrations" (*Sangat*, 136) for leadership within the self-luminous entity.

iii. **The third division** is the "Third Astral body" (*Maha Shiva*, 9) of the self-luminous entity's "paternal spirit" (*Atmavichara*, -1/3) as the primeval deity. The Third Astral Body is the "Divided Astral Body" (*Maha Shiva*, 9), divided into nine paternal "flamemates" (*Maha Shiva*, 9) without the two double copies that link the "flame family" (*Gauranga*, 11) with the "twin flame family" (*Bhaavi*, 11). The paternal spirit para-consciously guides the "earth wheel" (*Ajna chakra*, 65) with the self-luminous entity's "conscious virtues" (*Dama*, 45). It generates "energy vibrations" (*Sargam*, 60) of the "differentiated consciousness" (*Duratma*, 60) of the conscious virtue traded from both the first and the second astral bodies for entrepreneurship as a human being with a living soul.

iv. **The fourth division** is the "Fourth Astral body" (*Rab*, 16) of the human being's child spirit as the deity. The Fourth Astral Body is the "Undivided Astral Body" (*Rab*, 16) that forms a pair of double copies as a differentiated "four-body system" (*Genda*, 16). The four-body system reproduces two fraternal double copies within the "flame family" (*Gauranga*, 11) as the two sororal double copies within the "twin flame" (*Bhaavi*, 11). The "reproductive power" (*Guru*, 100) produces 123 "maternal flamemates" (*Svayambhu*, 123), comprising hundred within the reproductive power, eleven within the flame family, eleven within the twin flame family, and the undifferentiated four-body system as the "maternal twin cell" (*Svayambhu*, 123) forming a "molecular body" (*Daksha*, 123).

The child spirit para-consciously guides the "air wheel" (*Visuddha chakra*, 34) with the human being's "conscious intuition" (*Andesha*, 397), generating "vibrational consciousness" (*Nairatma*, 29) of the four polarized astral bodies within the maternal twin cell. The "ascending value" (*Rodha*, 1) of the "consciousness" (*Chaithanya*, 4) that forms the two pairs of double copies destroys the departed soul's "I AM God consciousness" (*Ekam Evadvitiyam Brahma siddhanta*, -10^{30}) and illuminates the human being as the "God" (*Ishvara*, $5 = 1 + 4$) within the "maternal twin cell" (*Svayambhu*, 123). Thus, the human being incarnates as the "child" (*Arcisa*, $128 = 5 + 123$) for enjoying the "godliness" (*Ishvaratva*, 800) of his management within the "living being" (*Satta*, 2000).

v. **The fifth division** is the "Fifth Astral body" (*Homa*, 25) of the living being's "devil spirit" (*Avajati*, -1/60) as the super deity. The Fifth Astral Body is the "differentiating four-body system" (*Homa*, 25). It situates itself as the "GUIDER Council" (*Homa*, 25) that reproduces the followership, leadership, entrepreneurship, and management by circling the roles among the four bodies for managing the "infinite council of 169 flamemates" (*Karana sharira*, 30) created consequentially. The devil spirit para-consciously guides the "water wheel" (*Jala chakra*, 39) with the living being's "conscious nature" (*Svabhav*, 8) for organizing 169 flamemates into a "water molecule" (*Jal*, 169). One hundred sixty-nine flamemates include nine flamemates reproduced as 81 flamemates by the flame family and copied by the twin flame family after adding two pairs of double copies not included in either and subtracting the one who is the "organizer" (*Dakshinashapati*, 1600).

The organizer is the "twin sister" (*Dhamini*, 1200), the grandmother of seven sentient lights and five para-conscious lights. The twin sister para-consciously guides the "circulatory wheel" (*Anahata chakra*, 33) with the flame family's "conscious excellence" (*Gita*, 368), transforming the flame family into a "group of ten principal guiders" (*Prajanan*, 78) led by the twin sister as the "eleventh principal guider" (*Isha*, 12). The eleventh principal guider is the "feminine self-luminous entity" (*Isha*, 12), acting like an "archangel" (*Isha*, 12). Similarly, the twin flame family transforms into a "group of ten potential principal guiders" (*Mukhya*, 14) led by the "twin brother" (*Shivagati*, 120) as the "eleventh potential principal guider" (*Padmanartteshvara*, 12). The group of ten potential principal guiders is within the consciousness of the twin sister, the twin brother, the "sister" (*Bahen*, 130), and the "brother" (*Bhai*, 30). The sister and brother are the thirteenth para-conscious light and conscious light, respectively, and behave like the twelfth pair. Thus, they liberate the twelfth pair from the "present" (*SAUM*, 1600) formed through the reproduction of the self-steaming "gravitational quality" (*Guna*, 0) of their "actions" (*Karma*, 10).

The actions of the twelfth pair thermodynamically steam everyone's conscious system because everyone wishes to be the foundation for growth like the twelfth sister and the essence of that foundation like the twelfth brother through proximity with the twelfth pair. Everyone's wish motivates the thirteenth pair to behave like the twelfth pair. Everyone follows the "thirteenth brother" (*Shiva*, 7) like an absolute deity because he is the "param deity" (*Shiva*, 7). The "thirteenth sister" (*Kha*, 18,000) creates the "hole" (*Kha*, 18,000) in the "mental space" (*Vaas*, 18,000). She leads everyone to devotionally follow the thirteenth brother because she is the copy of the "animate primordial maternal" (*Shani*, 18). She shapes the "potential" (*AUM*, 18) for the "foundation" (*Sva*, 11) that makes one a copy of the "param deity" (*Shiva*, 7 = 18 - 11). The animate primordial maternal is the "Living Deity" (*Shani*, 18).

vi. **The sixth division** is the "Sixth Astral body" (*Taraka*, 36) of the luminous entity's "satan spirit" (*Nissara*, 1/60) as the "maternal luminous" (*Maha Kali*, 13). The Sixth Astral Body is the "differentiable four-body system" (*Taraka*, 36). It is differentiable into "seven bodies" (*Taraka*, 36), comprising "eight copies" (*Maharavana*, 8) of which the sixth is the differentiable four-body system. The gender and generation differentiated eight copies constitute a "reproducing" (*Maharavana*, 8) element the sixth copy reproduces as the "natural growth" (*Dishta*, 6) of the six copies within the seventh copy. Thus, the six copies form the "animate body" (*Sharira*, 56) of the seventh copy's "light force" (*Apas*, 169) formed as a "water molecule" (*Jal*,

169). The animate body forms as the "inanimate body" (*Vapu*, 56) of the sixth copy's "water force" (*Jalaprana*, 4), generating a consciousness within the fifth copy of the first four copies as the "creator factor" (*Bhagwan*, 4) of the water molecule. The first four copies form the "differentiated four-body system" (*Genda*, 16) by reproducing themselves in the form of the "light force" (*Apas*, 169), norming the "water molecule" (*Jal*, 169).

The fifth copy is the "first copy" (*Shani Bhagwan*, -1) within the "twin flame family" (*Bhaavi*, 11). She consumes the water molecule because she is the "satan" (*Shani Bhagwan*, -1) with a "consumer mindset" (*Prakriti Yogi*, -1). Her "consuming" (*Jambhaka*, 169) forces the sixth copy to incarnate as the "inanimate one" (*Pratiyaman*, 1) without the consciousness of the first four and act as if she is the "second copy" (*Pratiyaman*, 1) copying the first copy's "potential" (*AUM*, 18) for reproducing the "sixth copy" (*Manyu*, $19 = 1 + 18$). The sixth copy potential's "energy" (*Shakti*, $1 + 18$) forces the seventh copy to incarnate as the "animate one" (*Akhanda*, 8×10^{15}) within the eighth copy's "conscious system" (*Shunya Kalpa*, 8×10^{15}).

The eighth copy takes the first five copies to be within the sixth copy's "body of consciousness" (*Vapu*, 56) and the first six copies to be without its "potential body" (*Jism*, 56) of the "imaginary consciousness" (*Antaratma*, 1) of the non-existent ninth copy. The imaginary consciousness inverts the sixth copy, the inanimate one into an "octave of elements" (*Pratiyaman*, 1) for incarnating the eighth as a double copy and the seventh as the copy reproducing itself as the ninth copy. Therefore, the eighth copy becomes the "third copy" (*Sauri*, 1000) after the "sixth body" (*Taraka*, 36) produces the seventh as a copy, let it reproduce as the ninth by copying itself without the light of copying, and endows the eighth with the "conscious energy" (*Amrit*, 1000) of the self-reproducing seventh copy. Thus, the eighth para-conscious light becomes the "primordial self" (*Parvati*, 10) of every copy.

The ninth copy materializes the eighth copy as its "twin body" (*Badan*, 56). First, the "body" (*Mahatattva*, 56) of the "para-consciousness" (*Mahatma*, 18) of the potential of the sixth copy as the "sixth body" (*Taraka*, 36) differentiating the "seven bodies" (*Taraka*, 36) within it. Second, the "para body" (*Chandra*, 82) embodying the consciousness of all the eight copy's "spiritual energy" (*Saniddhya shakti*, 169) within the water molecule organizing the first four copies as the "universe of paternal spirits" (*Rajyaika-sheshen*, 169) and the next four as the "universe of maternal spirits" (*Nishkala*, 169) before inverting them in the physical realm. Thus, the ninth copy becomes the "fourth copy" (*Muthyalamma*, 55), enjoying the "capability" (*Udana*, 55 = 5,5) of the first five para-conscious lights before reproducing them consciously as the first five conscious lights within the "sixth body" (*Taraka*, 36).

The "Five brothers" (*Samabhasam*, 12) as the five conscious lights build on the "foundation" (*Sva*, 11) of their "inherited capability" (*Udana*, 55 = 11 * 5) by behaving like the "planned present of everybody" (*Samabhasam*, 12) and reproducing the five conscious lights to form an "octave of copies" (*Sargam*, 60 = 55 + 5) within each "double copy" (*Chitra*, 1). The octave of copies forms the eighth copy while the double copy transforms into the sixth and the seventh copies. After the seventh copy self-reproduces itself as the ninth copy, the ninth copy produces the tenth to the thirteenth copies by reproducing the "method" (*Vidhi*, 950) of "self-producing" (*Upanayana*, 1/3) a diverse self-producing double copy as a differentiable "four-body system" (*Genda*, 16). It does not replicate its "homogeneous copy" (*Mein*, 1) as a "one body system" (*Mein*, 1).

A Four-body System

A Four-body System is a "Dehydrated Entity" (*Genda*, 16), known as "Tun" (*Genda*, 16). Tun is "Odderon" (*Genda*, 16), a multivariate group of three univariate "gluons" (*Brihadbala*, 149), i.e., "intrinsic weak forces" (*Brihadbala*, 169), derived from the "bivariate correlation" (*Porutham*, 16) among the "differentiable system" (*Gabhira*, 12), the "differentiated system" (*Pravritta*, 12), and the "differentiating system" (*Pratiyaman*, 1). Tun comprises a double pair of double copies, forming "Eight Sisters" (*Genda*, 16).

A "double copy" (*Chitra*, 1) is differentiable into two "half octaves" (*Amsha*, 1/2). A pair of double copies is differentiable into four half octaves. The double pair of double copies manifests as the "intrinsic four half octaves" (*Chaithanya*, 4) within an "extrinsic four half octaves" (*Pravritta*, 12). The intrinsic four half octaves are the para-conscious copy of the extrinsic four half octaves. The extrinsic four half octaves are the conscious copy of the four-body system; they differentiate the formative differentiable "system" (*Gabhira*, 12) into "four bodies" (*Natini*, 12).

The "differentiated system" (*Pravritta*, 12) and the "intrinsic four half octaves" (*Chaithanya*, 4) form a "differentiated four-body system" (*Genda*, $16 = 12 + 4$). The "extrinsic four half octaves" (*Pravritta*, 12), the "four bodies" (*Natini*, 12), and the "differentiating system" (*Pratiyaman*, 1) form a "differentiating four-body system" (*Homa*, $25 = 12 + 12 + 1$). The differentiating four-body system "guides" (*Puraitri*, 111) the "intrinsic four half-octaves" (*Chaithanya*, 4) to transform into the sixteen para-conscious lights, norming "Sixteen Sisters" (*Chitta*, 100).

The "differentiating four-body system" (*Homa*, 25) and the "four half octaves" (*Genda*, 16), without the "hydrating-effect" (*Punravritta*, 5) of the water, form an "undifferentiated four-body system" (*Taraka*, $36 = 25 + 16 - 5 = 25 + 111 - 100$). The undifferentiated four-body system "channels" (*Raasta*, 36) the guide as the "sixteen light forces" (*Puraitri*, 111) to form an "integrated system" (*Sadhaka*, 127) of "sixteen forces" (*Sadhaka*, 127).

The integrated system is a "multipliable system" (*Sadhaka*, 127). It is multipliable as a "theory" (*Niyama*, 127) of "mind" (*Manas*, 38) taking "moral responsibility" (*Jimmedari*, 80) that the "copy" (*Nakala*, 0) will be "true" (*Sat*, 8) to the "original" (*Mukhya*, 14) within the "growth" (*Dishta*, 6) of the original into a "guiding spirit" (*Ruah*, $20 = 6 + 14$).

The "integrated system" (*Sadhaka*, 127) integrates the "conscious light" (*Kumara*, 10) within the "para-conscious light" (*Kumari*, 100) to transform the "extrinsic four half-octaves" (*Pravritta*, 12) into a "hydrated entity" (*Gardhaba*, $1000 = 10 * 100$), norming "Sixteen Brothers" (*Gardhaba*, 1000). The hydrated entity is a "disintegrated system" (*Gardhaba*, 1000). The disintegrated system is a "multiplying system" (*Gardhaba*, 1000) that multiplies itself into "thousand cells" (*Nirjara*, 1000) by transforming into a lifeless "dead cell" (*Nirjara*, 100). The dead cell is a "greeter cell" (*Nirjara*, 1000) whose "sentient energy" (*Amrit*, 1000) is immanent within the "thousand cells" (*Nirjara*, 1000).

The "immanent system" (*Prabhu*, 1600 = 100 * 16) emanates the "para-conscious light" (*Kumari*, 100) within the "divine light" (*Usha*, 16) of the "four-body system" (*Genda*, 16) as a whole. It forms the four-body system as the "emanating system" (*Genda*, 16), known as the "Fougaro system" (*Genda*, 16). The immanent system is the "multiplied system" (*Prabhu*, 1600) that forms the "present creature" (*Prabhu*, 1600) as a "potential octave of potential alleles" (*Prabhu*, 1600).

The tenth, eleventh, and twelfth para-conscious lights manifest the "Maternal Luminous" (*Maha Kali*, 13) as the "quantum light" (*Rochisha*, 13) before the Five Brothers begin acting like a "quantum photon" (*Svarochisha*, 12). As a quantum photon, the Five Brothers emanate their "consciousness" (*Chaithanya*, 4) as the "photon" (*Kapinjala*, 20 = 4 * 5). Therefore, the tenth to the thirteenth copies materialize in a descending order, with "thirteenth copy" (*Bahen*, 130) incarnating as the thirteenth sister, twelfth copy as the "fourteenth sister" (*Shani*, 18), the eleventh copy as the "fifteenth sister" (*Chandra*, 82), and the "tenth copy" (*Shamana*, 23) as the "sixteenth sister" (*Visheshana*, 230 = 130 + 82 + 18 = 23 * 10). The sixteenth sister is the distinguishing "attribute" (*Visheshana*, 230) that makes the past "livable" (*Nivasaniya*, 23) for every copy born as a "unique entity" (*Gotra*, 1694). The fifteenth sister is the "para body" (*Chandra*, 82) of the "departed human entity" (*Manushya*, 82) that a departed entity reproduces for conceiving his "potential" (*AUM*, 18) as the "fourteenth sister" (*Shani*, 18). The "thirteenth sister" (*Bahen*, 130) is the "office" (*Kamma*, 130) the departed entity officiates for transforming the method into an "illuminator paradigm" (*Vidhi Vidhartha*, 130) of institutionalizing the para conscious light as the conscious light.

The fourteenth conscious light is the "param human child" (*Kesari Nandan*, 1600), who incarnates as the "param paternal" (*Narada*, 7). The param paternal is the fourteenth brother who acts as the sixth conscious light after making the first eight conscious lights self-luminous as an "octave of conscious lights" (*Eeshan*, 12). The "departed entity" (*Divangat*, 1600) is the "fifteenth conscious light" (*Divangat*, 1600) that incarnates as the "departed human entity" (*Manushya*, 82), the fifteenth brother. Mother Nature is the "sixteenth brother" (*Kudrat*, 8), illuminating her "ablative" (*Panchama*, 485) "character" (*Prakriti*, 485) as the sixteenth conscious light.

The satan spirit para-consciously guides the "wheel of fire" (*Agni chakra*, 24) by taking the luminous entity as the "guider" (*Guru*, 100). The wheel of fire is a "wheel of digestion" (*Manipura chakra*, 24), generating the fire to digest the luminous entity's "consciousness" (*Chaithanya*, 4) within the "sixth sister" (*Pratiyaman*, 1) and making the "octave of elements" (*Pratiyaman*, 1) self-luminous as a "conscious form" (*Svarochisha*, 12). Each conscious form breathes the "air" (*Vayu*, 385) that oxidizes the digested "creative imagination" (*Anda*, 19) into an "enzyme" (*Anda*, 19). The air "cools down" (*Sheetala*, 3794) the "fire force" (*Rupa*, 100,000) into the "earth" (*Bhu*, 724) element. It "animates" (*Trasa*, 76) the conscious form with the "mathematics of formation" (*Panchang Ganita*, $800 = 724 + 76$) illuminated here.

The Earth is the "keeper of the scroll" (*Nairitti*, 29) of the "divine Godhead essence" (*Madhava*, $29 = 13 + 16$) serviced by the "maternal Luminous" (*Maha Kali*, 13) with the "greeter essence" (*Sara*, 16) by forming the "solar universe" (*Surya mandala*, 13). Each conscious form becomes the "Queen of the underworld, Ereshkigal" (*Nairitti*, 29), conscious of the ascending contamination of the Earth by experiencing the "vibrational consciousness" (*Nairatma*, 29) of the "enzyme-effect" (*Chetan*, 10), i.e., the "animate state" (*Chetan*, 10). The vibrational consciousness generates an "aftereffect" (*Natija*, 9,000) "feeling" (*Sprishti*, 197) "sensation" (*Vedana*, 197) of the "conscious actions" (*Karma*, 10) in an animate state within each successive form.

Each form experiences a "proportionate aftereffect" (*Muftakhori*,

13) of the "universe of sentient entities" (*Akalpa*, 570) as the "cultural system" (*Akalpa*, 570). The proportionate aftereffect is the "energy exchange paradigm" (*Muftakhori*, 13) that makes everyone's conscious decisions luminous to the one "percipient" (*Svarochisha*, 12) about one's "extrinsic experiences" (*Bahirmukha*, -4). The "luminous" (*Rochisha*, 13) is a "primeval astral body" (*Rochisha*, 13), is not divided but "multiplied astral body" (*Rochisha*, 13) with the "proportionate aftereffect" (*Muftakhori*, 13), i.e., the "freeriding" (*Muftakhori*, 13) of the extrinsic experiences for realizing the "goal" (*Lakshya*, 9 = 13 − 4) of cleansing the "past-effect" (*Ghatika*, 24) as an entity in the present birth.

Each form experiences a "disproportionate aftereffect" (*Vastavikta*, 7) of the "universe of para-conscious entities" (*Sthavaravisha*, 57), shaping its "workculture system" (*Sthavaravisha*, 57). The disproportionate aftereffect is the "mass exchange paradigm" (*Vastavikta*, 7) that makes one trade the mass of everyone's imaginary consciousness of the effects of one's actions after rebirth embodying another animate body, animated with everyone's imaginary consciousness. It makes one a "sentient entity" (*Siddha*, 7), conscious of one's "self-luminous reality" (*Purushartha*, 7). The "path of sentient entity" (*Siddha Marga*, 25) furthers the "oneness of action" (*Karma Yoga*, 49) with the "past-effects" (*Ghatika*, 24 = 49 - 24), ensuring the "freedom from the present-effect" (*Moksha*, 1600).

Thus, the oneness of action transforms the sentient entity as the seventh copy into the "seventh body" (*Charanam*, 49), making the seventh body the "primordial astral body" (*Charanam*, 49). The primordial astral body, as the "seventh astral body" (*Charanam*, 49), is not dividing but "multiplying astral body" (*Charanam*, 49) that multiplies itself into a "seven body system" (*Karma Yoga*, 49) through the oneness of action. Thus, the seventh astral body becomes a "splice" (*Charanam*, 49) that "predicates" (*Charanam*, 49) the "character" (*Prakriti*, 485) of the first four astral bodies reproduced within the four double copies to form the fifth, the sixth, and the eighth to the thirteenth astral bodies.

The first four astral bodies form the first half octave. They norm the third half octave as the ninth to the twelfth astral bodies, within a "theory" (*Sadhaka*, 127) of the mind that the second half octave will comprise a homogeneous "feminine body" (*Karana Sharira*, 30). The first half octave forms the "intrinsic body" (*Gatr*, 1) as it is a pair of "self-perpetuating" (*Udvaha*, ½) "double copy" (*Chitra*, 1). The third half octave forms the "extrinsic body" (*Gat*, 1) because it "perpetuates" (*Amsha*, ½) the first half octave as its self. The second half octave is immanent within the third half octave as the "immanent body" (*Sukara*, 1). It waits for the extrinsic body to form so that it may enjoy six-unit "growth" (*Dishta*, $6 = 3 + 3$) by reproducing the "three bodies" (*Idam*, 3), including itself. It lets the seventh copy consume the six-unit growth and incarnate as the thirteenth astral body. Next, following "her" (*Konastha*, 13) like a "docile animal" (*Konastha*, 13), the other three copies line up the fifth, the sixth, and the eighth spots, leaving the seventh "empty" (*Nirbija*, 800).

The empty forms a "symmetry system" (*Vyagrata*, 800) of "twelve octaves" (*Vyagrata*, $800 = 60 * 12 + 80$). It transforms each of the twelve astral bodies into an "octave" (*Sargam*, 60) and norms the fourth half octave as an "emanating body" (*Prabhava*, 80) of "three octaves" (*Prabhava*, 80). The "seventh astral body" (*Charanam*, 49) forms in "symmetry" (*Ashrama*, $250 = 49 + 1 + 100 * 2$) with the "intrinsic body" (*Gatr*, 1) and the "guiding forces" (*Chitta*, 100) of the "extrinsic body" (*Gatr*, 1) and the "immanent body" (*Sukara*, 1). After norming the "thirteenth astral body" (*Niyatatma*, 169) as the "water" (*Jal*, 169), the fourth half octave forms the "fourteenth astral body" (*Omkara*, 256) as the "infinite sound" (*Omkara*, 256). The infinite sound is a copy "floating" (*Supuma*, 256) as the "sixteenth astral body (*Kshatriya*, 0). The copy is of the "fifteenth astral body" (*Sthavaravisha*, 57), the "universe of para-conscious entities" (*Sthavaravisha*, 57).

The four conscious copies produce the character in oneness with the seventh astral body. The four copies divide the "consciousness" (*Chaithanya*, 4) of the past-effect into four units of "greeter consciousness" (*Pratyagatma*, 1) for producing twelve more conscious copies through the combination of two units within a double copy to form a conscious copy. Thus, the sixteen para-conscious lights from the past are reproduced as the sixteen conscious lights in the present by imagining and irradiating sixteen light forces from the future. The "sixteen light forces" (*Puraitri*, 111) emanate from the "sixteen effects" (*Sadhaka*, 127) immanent within the white "divine light" (*Usha*, 16) of the "greeter essence" (*Sara*, 16). Sixteen light forces manifest as sixteen "guider agents" (*Shri Rama*, 12). Sixteen effects manifest as sixteen "potential guider agents" (*Prati*, 16). Each potential guider agent is a "perfect guider agent" (*Prati*, 16).

The "astral body" (*Linga sharira*, 3) gravitates towards the "local ecosystem" (*Samrachana*, 1) formed by everyone with their "imaginary consciousness" (*Antaratma*, 1). It localizes the time and the space for the human being's rebirth with an "infinite networking system" (*Pravritta*, 12). The infinite networking system lets the astral body be the "moderator" (*Hanuman*, 3) of the "goal" (*Lakshya*, 9) of birth in a finite coordinate" (*Pradesha*, 9) of space and time. Thus, the infinite networking system is the "param astral body" (*Pravritta*, 12). It is the "multipliable astral body" (*Pravritta*, 12), multipliable as a "group of ten guider agents" (*Pravritta*, 12).

The "group of ten potential principal guiders" (*Mukhya*, 14) incarnate as the "group of ten guider agents" (*Pravritta*, 12) after the "two brothers" (*Ashwini Kumaras*, 120) from the flame family incarnate as the "bicellular organism" (*Ashwini Kumaras*, 120). Each potential principal guider is a perfect principal guider.

Two brothers include the "twin brother" (*Shivagati*, $120 = 10 * 12$) and the "absolute after-effect" (*Ashwini Kumaras*, 120) the twin brother experiences referentially through the psychic linkage. The twin brother manifests the absolute aftereffect physically through the "genetic linkages" (*Mukhya*, 14) within the "ten bodies" (*Mahabala*, 100), comprising the ten "groups of ten guider agents" (*Pravritta*, 12). The ten bodies reproduce the "body system" (*Hridya Pranali*, 100,000) as the eleventh guider agent and the undivided "physical body" (*Sthula Sharira*, 387) as the "principal" (*Sthula*, 387) managing the body system.

4.7.4 Working Causation. Divided Physical Body and the Self Within the Soul Universe

A body system is the third, "divided physical body" (*Hridya Pranali*,

100,000). Each body within the body system enjoys the body system's "potential" (*AUM*, 18) within its "diverse form" (*Shudra*, 1) as a "param child" (*Manyu*, 19). Thus, a departed entity enjoys the potential to incarnate concurrently in many forms by dividing the fourth, "undivided physical body" (*Sthula Sharira*, 387) within and without the "human kingdom" (*Manushyagati*, 53). "All bodies" (*Shunya Kalpa*, 8×10^{15}) form a "Luminous System" (*Shunya Kalpa*, 8×10^{15}) and make the "self" (*Atmatva*, 8×10^{15}) a conscious system, conscious of its diverse forms stacked like a "quantum system" (*Akhanda*, 8×10^{15}). The Luminous System is the first, "divisible physical body" (*Shunya Kalpa*, 8×10^{15}), divisible as a system of quantum light into a "sequence of quantum lights" (*Pinakapani*, 10^9). The sequence of quantum lights forms a "quantum light force" (*Pinakapani*, 10^9), norming a "Pomeron" (*Pinakapani*, 10^9). The Pomeron is "one with nothing" (*Pinakapani*, 10^9) but the "para-conscious light" (*Kumari*, 100) emanating the "conscious light" (*Kumara*, 10) of the "goal" (*Lakshya*, 9) immanent within a "thing" (*Vastu*, 9).

The goal norms a "Renormalon" (*Lakshya*, 9). The Renormalon comprises "nine potential kins" (*Lakshya*, 9) radiating the "light force" (*Apas*, 169) of the "tenth potential kin" (*Vastu*, 9) to make the eleventh potential kin the "eleventh potential guider agent" (*Lakshya*, 9). The eleventh potential kin is the "present growth" (*Sva*, 11) of the "nine potential kins" (*Lakshya*, 9) within the "present reality" (*Badhabuddhi vardartha*, -2) of the tenth producing its "light force" (*Apas*, 169 = 16,9) by reproducing the "divine light" (*Usha*, 16) of the "nine potential kins" (*Lakshya*, 9 = 16 - 7). The goal is to illuminate the "conscious light" (*Kumara*, 10) of the "seven kins" (*Prakasha*, 7) whose "consciousness" (*Chaithanya*, 4) is perpetuating the "para-conscious light" (*Kumari*, 100) of the "four kins" (*Chaithanya*, 4), norming "four guider agents" (*Chaithanya*, 4). The thing norms a "Sphaleron" (*Vastu*, 9). The Sphaleron comprises "nine kins" (*Vastu*, 9) irradiating the "light force" (*Apas*, 169) of itself as the "tenth potential kin" (*Vastu*, 9) for incarnating as the "eleventh kin" (*Hridya Pranali*, 100,000), i.e., the "eleventh guider agent" (*Hridya Pranali*, 100,000), forming a "body system" (*Hridya Pranali*, 100,000). The body system produces the "tenth kin" (*Sara Kalpa*, 10^{10}), norming the "tenth guider agent" (*Sara Kalpa*, 10^{10}), by reproducing itself to norm a "soul universe" (*Sara Kalpa*, $10^{10} = 10^5 * 10^5$).

The "soul universe" (*Sara Kalpa*, 10^{10}) forms a "Self-Luminous System" (*Sara Kalpa*, 10^{10}) and norms a "consciousness system" (*Sara Kalpa*, 10^{10}) guiding the self. The self-luminous system is the second, "dividing physical body" (*Sara Kalpa*, 10^{10}), dividing the system into a sequence of quantum photons, each divided into a sequence of quantum lights. The sequence of quantum photons forms a "quantum entity" (*Anuyogakrit*, 10^9), norming a "Reggeon" (*Anuyogakrit*, 10^9). The Reggeon is the "spiritual growth" (*Anuyogakrit*, 10^9) of the nineteen astral bodies, including sixteen divisions of the astral bodies within the seventeenth, primeval astral body emanating from the eighteenth, param astral body.

The Reggeon is the "nineteenth astral body" (*Anuyogakrit*, 10^9). It forms the "twentieth astral body" (*Pinakapani*, 10^9) as the "Pomeron" (*Pinakapani*, 10^9) with the "gift of consciousness" (*Jinpa*, 10^9), known as the "Pomeron exchange" (*Jinpa*, 10^9). The Pomeron exchange eventualizes over the "Pomeron trajectory" (*Pariprashnati*, 10^9), the twelfth potential kin" (*Pariprashnati*, 10^9). Thus, the twelfth potential kin, i.e., the "twelfth potential guider agent" (*Pariprashnati*, 10^9) becomes a "spiritual preceptor" (*Pariprashnati*, 10^9). The Pomeron trajectory is a "specific" (*Vishesha*, 10^9) "evolute" (*Kendraja*, 10^9), i.e., "photon exchange" (*Kendraja*, 10^9) of the "Regge trajectory" (*Kincha*, 10^9). The Regge trajectory is the "twelfth kin" (*Kincha*, 10^9), i.e., the "twelfth guider agent" (*Kincha*, 10^9), formed as a "notional theory" (*Kincha*, 10^9).

The Regge trajectory is the "involute" (*Yaugika*, 1,000,000), i.e., "Regge Pole exchange" (*Yaugika*, 1,000,000), within the "multiplying system" (*Gardhaba*, 1000) of the "thirteenth kin" (*Alidha Mandala*, 10^9) seeking to "falsify" (*Juthalaya*, 1000) the "theory" (*Niyama*, 127). The thirteenth kin, i.e., the "thirteenth guider agent" (*Alidha Mandala*, 10^9) is a "scraping community" (*Alidha Mandala*, 10^9) of three "triplets" (*Tikari*, 1000), each a "multiplying system" (*Gardhaba*, 1000) of three "conscious lights" (*Kumara*, 10). The scraping community is a "virtual particle" (*Alidha Mandala*, 10^9) norming the "charm anti-quark" (*Alidha Mandala*, 10^9). The theory is the "law of limitation" (*Niyama*, 127). It falsifies when the "notional theory" (*Kincha*, 10^9) polarizes the "Regge Pole" (*Satvasva*, 10^9), comprising "five guider agents" (*Satvasva*, 10^9), as the "discontinuous stream of photons" (*Satvasva*, 10^9). The five guider agents are the "five kins" (*Satvasva*, 10^9), comprising the fourteenth to the eighteenth guider agents.

Regge Pole is the "nineteenth guider agent's" (*Vamaviddha mandala*, 1,000,000) "commitment" (*Satvasva*, 10^9) to the "twentieth guider agent" (*Tripurantaka*, 10^9) within the "twenty-first guider agent" (*Tikari*, 1000) that falsifies the law of limitation. The nineteenth guider agent is the "virtual gluon" (*Vamaviddha mandala*, 100,000). It is "none without all" (*Vamaviddha mandala*, 1,000,000). Therefore, it is committed to the twentieth guider agent, the "virtual quark" (*Tripurantaka*, 1,000,000), which is "one with all" (*Tripurantaka*, 1,000,000). The twenty-first guider agent is the "virtual energy" (*Tikari*, 1000) that energizes "all" (*Sab*, 1,000,000) to be with one "absent" (*Asamapat*, 89). All is the global "electromagnetic interaction" (*Maha*, 1,000,000) of the "quantum photon" (*Svarochisha*, 12) with the local "gravity" (*Jamadagni*, 629) of the "virtual energy" (*Tikari*, 1000). The local gravity is the "vacuum point" (*Shunya sthana*, 629); it is the "gravitational center" (*Guru Sthana*, 629) of the "quantum system" (*Akhanda*, 8×10^{15}). The one "absent" (*Asamapat*, 89) is the "virtual atom" (*Asamapat*, 89) because the atom is never "virtual" (*Pratiyaman*, 1), i.e., the "differentiating system" (*Pratiyaman*, 1). The atom is always "present" (*SAUM*, 1600) as the "dark matter" (*Dakshinashapati*, 1600).

The gravitational center is the fourteenth potential guider agent. As a "virtual fire" (*Jamadagni*, 629), it repels the negative "electric charge" (*Avidhi*, -4) of the "thirteenth potential guider agent's" (*Veerbhadra*, 100) virtual energy for attracting a "quantum photon" (*Svarochisha*, 12) from the quantum entity. The thirteenth potential guider agent is the "Plank constant" (*Veerbhadra*, 100). It is the "causal body system" (*Veerbhadra*, 100) formed by the "ten bodies" (*Mahabala*, 100) as the "guiding" (*Mahabala*, 100) element. As a sequence of quantum photons, the "quantum entity" (*Anuyogakrit*, 10^9) services a quantum photon of the "infinite dimensionality" (*Pramathanaya*, 10^9) in the form of a "virtual photon" (*Pramathanaya*, 10^9). The virtual photon transcends the "speed of light" (*Markatesh*, 8×10^{15}) emanating from "All bodies" (*Shunya Kalpa*, 8×10^{15}) to form "One with All body" (*Konasa*, 10^9).

The "One with All body" (*Konasa*, 10^9) is the "fifteenth potential guider agent" (*Konasa*, 10^9), moving as the "Cherenkov radiation" (*Konasa*, 10^9). It moves faster than the speed of light at 10^{1024} meters/second in the "virtual water" (*Sarva Vyapin*, 1,000,000) as a spherical "wavelet" (*Uttaranga*, 10^{1024}), known as the "Cherenkov wave" (*Uttaranga*, 10^{1024}). The virtual water is the "virtual light force" (*Sarva Vyapin*, 1,000,000), norming the "Cherenkov-effect" (*Sarva Vyapin*, 1,000,000) to be "everyone without all" (*Sarva Vyapin*, 1,000,000). The "sixteenth potential guider agent" (*Yamadeva*, 1,000,000) produces the virtual light force. It is the "virtual light" (*Yamadeva*, 1,000,000). The virtual light is the "colorless light" (*Yamadeva*, 1,000,000). It is the "potential light" (*Yamadeva*, 1,000,000) of the "seventeenth potential guider agent" (*Devatatma*, 1,000,000), the "quantum soul" (*Devatatma*, 1,000,000).

In the water, the potential light moves faster than the speed of light at 200,000,000 meters/second within the "heating" (*Sampratapana*, 200,000,000) of the "virtual air" (*Charya marga*, 1,000,000). The virtual air forms as the "path of followership" (*Charya marga*, 1,000,000) of the potential light, norming a "conical wavefront" (*Charya marga*, 1,000,000). The "virtual earth" (*Stri*, 1,000,000) generates the heating due to the "virtual ether" (*Vartula*, 10,000). The virtual earth forms as a "corpuscle" (*Stri*, 1,000,000), generating the "heating" (*Sampratapana*, 200,000,000) of "laboring" (*Svedana*, 169) "water" (*Jal*, 169) like a "woman" (*Stri*, 1,000,000). The virtual ether forms as a "beta electron" (*Vartula*, 10,000). The beta electron is a forward-moving "circular creation" (*Vartula*, 10,000) of the "virtual guider" (*Kendrajita*, 1,000,000). The virtual guider is the "twin neutrino" (*Kendrajita*, 1,000,000). As the Sun's "evolvent curve" (*Kendrajita*, 1,000,000), it is the "solar neutrino (*Kendrajita*, 1,000,00), also known as the "primordial neutrino" (*Kendrajita*, 1,000,000).

The twin neutrino remains "static" (*Parai*, 257) in the air that moves at the speed of 300,000,000 meters/second without the "heating" (*Sampratapana*, 200,000,000) of the "virtual air"

(*Charya marga*, 1,000,000). The twin neutrino is produced by the "potential twin neutrino" (*Gandakanayaka*, 10,000), the "virtual SHEENY" (*Gandakanayaka*, 10,000). The potential twin neutrino is the "primeval neutrino" (*Gandakanayaka*, 10,000), norming the "cosmic shower" (*Gandakanayaka*, 10,000) that transforms into the "lithium" (*Gandakanayaka*, 10,000). The primeval neutrino produces the primordial neutrino with the heating of the virtual air due to the "virtual divine" (*Ajirna*, 10,000). The virtual divine is the "param neutrino" (*Ajirna*, 10,000) that transforms into the "helium" (*Ajirna*, 10,000). The param neutrino is a "white star's" (*Yama*, 180) "indigestion" (*Ajirna*, 10,000) of the "quantum photon" (*Svarochisha*, 12) within an "extended luminous system" (*Mera*, 16). The extended luminous system is the "eighteenth potential guider agent" (*Mera*, 16). It forms as a "variable 4 pi" (*Mera*, 16), norming four "half-spheres" (*Artta*, 22/7) of "electromagnetic radiation" (*Artta*, 22/7) as the "pi" (*Artta*, 22/7) in the four directions. Each pi generates a "variation" (*Ratnaketu*, 10) in the mental "space" (*Vaas*, 18,000), generating a "reproductive reality" (*Bhavartha*, $40 = 4*10$).

The reproductive reality, known as the "Compton-effect" (*Bhavartha*, 40), is the "nineteenth potential guider agent" (*Bhavartha*, 40). The reproductive reality is of the "paternal soul" (*Pitra*, 4), the twentieth potential guider agent. The paternal soul forms a "gamma photon" (*Pitra*, 4) that transforms the "goal" (*Lakshya*, 9) of the reproductive reality into a "space vacuum" (*Mahodbhava*, 5) to repel a "scattered photon" (*Mahodbhava*, 5). The scattered photon is an "imaginary photon" (*Mahodbhava*, 5) formed as an "X-ray photon" (*Mahodbhava*, 5) due to the "entropy" (*Mahodbhava*, 5) within the reproducing space. The scattered photon is the twenty-first potential guider agent. The imaginary photon forms a "twin gamma bond" (*Maha Shiva*, $9 = 5 + 4$) with the gamma photon to norm a heterogeneous, "scattered electron" (*Maha Shiva*, 9). Nine scattered electrons transform the tenth scattered electron into the "scattered light" (*Kumara*, 10) and become nine "homogeneous electrons" (*Mein*, 1) within the scattered light. The scattered light is the conscious, i.e., "IBGYOR light" (*Kumara*, 10). The homogeneous electron norms a "lepton" (*Mein*, 1), enjoying para-consciousness of the conscious light of a "family" (*Parivar*, 10) of the ten heterogeneous electrons.

The lepton is a "one body system" (*Mein*, 1) that is the child self of the child's "two-body system" (*Jyotistava*, 4). The second body is the "divisible paternal body" (*Atma*, 4) within the "dividing maternal body" (*Sharira*, 56). Without the maternal body, it becomes the fifteenth "gamma photon" (*Pitra*, 4). The "maternal body" (*Sharira*, $56 = 4 * 14$) divides itself into fourteen gamma photons within the fifteenth gamma photon. Next, it imagines the sixteenth "gamma photon" (*Pitra*, 4) to be the "scattered photon" (*Mahodbhava*, $5 = 4 + 1$) within the child's self's "one body system" (*Mein*, 1). The sixteenth gamma photon is the maternal body that produces the fifteenth gamma photon as its copy. The fifteenth gamma photon is the paternal body that produces fourteen gamma photons in the form of seven "double copies" (*Chitra*, 1) after the maternal body produces a seventeenth gamma photon in the form of a "gamma-ray photon" (*Siddha*, 7).

The gamma-ray photon norms a "sequence of seven exchange photons" (*Siddha*, 7), each a double copy of itself. Fourteen gamma photons within the gamma-ray photon and the sixteen without form the "causal self" (*Karana sharira*, 30) as a "feminine body" (*Karana sharira*, 30) with thirty gamma photons. The causal self generates a "gamma decay" (*Urja*, 31) of the seventeenth gamma photon due to the "six-fold growth" (*Dishta*, 6) of its double copy. Six-fold growth forms as a triple "double copy" (*Chitra*, 1), norming the "gamma decay" (*Urja*, 31) as the eighteenth gamma photon and itself as the "nineteenth gamma photon" (*Dishta*, 6). It is a "byproduct" (*Dishta*, 6) of the "double copy" (*Chitra*, 1), the "twentieth gamma photon" (*Chitra*, 1) that forms an "up quark" (*Gajakarani*, 1). The double copy is a "product" (*Upaja*, 1000) of a "triple copy" (*Vaishya*, 3) of itself as an up quark. The triple copy of the up quark is the "twenty-first gamma photon" (*Vaishya*, 3). The twenty-first gamma photon differentiates the up quark into the three pairs of quarks: up and down, charm and strange, and bottom and top, and transforms the up quark into a "positive quark" (*Mein*, 1), norming the "lepton" (*Mein*, 1). Six quarks within each of the twenty-one gamma photons norm the fifth, "multiplied physical body" (*Kriyavasha*, $126 = 6 * 21$) as the "eventual-effect" (*Kriyavasha*, 126) of the "quantum photon" (*Svarochisha*, 12).

The multiplied physical body is the "incident photon" (*Kriyavasha*, 126). The "scattered photon" (*Mahodbhava*, 5) is the "entropy" (*Mahodbhava*, 5 = 126 - 121) of the incident photon "itself" (*Ekotrem*, 121). It "recoils" (*Ulat*, 1) the "scattered electron" (*Maha Shiva*, 9) into a "recoil electron" (*Maha Shiva*, 9) with the incident photon's "binding energy" (*Bandhana shakti*, 80). "Recoiling" (*Ulatana*, 0) generates "kinetic energy" (*Mahurata*, 571 = 7 * 80 + 11) for "unbinding" (*Svaiccchika*, 11) the "gamma-ray photon" (*Siddha*, 7) into fourteen gamma photons and "rebinding" (*Aichchika*, 8) the first "eight gamma photons" (*Bhaavi*, 11) into an "entity photon" (*Bhaavi*, 11) within the "six gamma photons" (*Urja*, 31) formed as the "triple double copy (*Urja*, 31) of the entity photon as the "seventh gamma photon" (*Bhaavi*, 11).

The entity photon behaves like a gamma-ray photon after conceiving the fifteenth "gamma photon" (*Pitra*, 4 = 11 − 7) in the form of a "twin double octave" (*Pitra*, 4) of gamma photons. The twin double octave is the "final photon" (*Pitra*, 4) that emits an "emitted photon" (*Sadhya*, 32 = 4 * 8) as its "flame" (*Sadhya*, 32). The flame is the sixth "Multiplying Physical Body" (*Sadhya*, 32). As a multiplier of its "twin double copy (*Brahman*, 2), it is also the "Almighty Physical Body" (*Sadhya*, 32). The flame absorbs its twin double copy as a seventh, primordial, "Multipliable Physical Body" (*Yogihridaya*, 64 = 32 * 2), its energy "fragment" (*Yogihridaya*, 64) norming an "absorbed photon" (*Yogihridaya*, 64). The "absorption" (*Avashoshan*, 2) of the absorbed photon transforms the emitted photon into an "incoming photon" (*Arcisa*, 128 = 2 * 64), norming the eighth "Added Physical Body" (*Arcisa*, 128).

The Added Physical Body is the "child" (*Arcisa*, 128) of the ninth, "Adding Physical Body" (*Deha*, 157). The Adding Physical Body is the "primeval physical body" (*Deha*, 157) that adds a "child" (*Arcisa*, 128) within its "organization" (*Sangathan*, 29) of itself as a "living body" (*Deha*, 157). The organization is the tenth "Additive Physical Body" (*Sangathan*, 29) the eleventh "Subtracted Physical Body" (*Priti*, 33) forms after the gamma photon subtracts from it. The Adding Physical Body is an "outgoing photon" (*Deha*, 157). The Additive Physical Body is the "Fluorescence photon" (*Sangathan*, 29), norming the "mass-effect" (*Sangathan*, 29) of a pair of nine physical bodies within the eleventh physical body. The twelfth, "Subtracting Physical Body" (*Vithoba*, 12), norming a "photoelectron" (*Vithoba*, 12). The photoelectron is a "quantum photon" (*Svarochisha*, 12). It trades the "photoelectric-effect" as the thirteenth, "Traded Physical Body" (*Konastha*, 13), a "twin luminous" (*Konastha*, 13). It services the fourth, "Subtractable Physical Body" (*Sthula sharira*, 387 = 12 * 13 * 2 + 75) with the photoelectron's "entropy force" (*Kriya shakti*, 75 = 12 * 6 + 3). The entropy force generates the growth of the fifth to the eleventh physical bodies in the form of the quantum photons and the first to the third as the "triple copy" (*Vaishya*, 3) of the fourth physical body. Thus, the twin luminous becomes the "working causation" for dividing the physical body and the self within the soul universe.

After its entropy, the photoelectron transforms into the fourteenth, "Trading Physical Body" (*Adravya*, -1), norming the "photoelectric" (*Adravya*, -1) element. The photoelectric element forms a "circular double octave" (*Adharma*, -10) of gamma photons as the "photoelectric serum" (*Mamsaja*, -10), the fifteenth, "Tradable Physical Body" (*Mamsaja*, -10) with nine divisions of the sixteenth, "Serviced Physical Body" (*Kshema*, 816). The ninth division is the "double octave" (*Madhusudan*, 16). It is the double copy of the eighth division, the octave. The octave comprises the first seven physical bodies as the primary seven divisions that reproduce the octave as the secondary seven divisions and the double octave as the tertiary seven divisions.

The Serviced Physical Body is the "Param Physical Body" (*Kshema*, 816). It is the "attenuation coefficient" (*Kshema*, 816) of the "exponential attenuation" (*Ekotrem*, 121) of the seventeenth, "Servicing Physical Body" (*Hridayesha*, 121) due to the "attenuation" (*Shinnan*, 28) of the eighteenth, "Serviceable Physical Body" (*Narayana*, 28). The Serviceable Physical Body is a "masculine electron" (*Narayana*, 28), norming the "D-meson" (*Narayana*, 28) as a "potentially scalar meson" (*Narayana*, 28). The Servicing Physical body is a "physical body system" (*Hridayesha*, 121), norming a "twin" (*Hridayesha*, 121) as the "pair production" (*Hridayesha*, 121).

The potentially scalar meson globalizes the "ideal-effect" (*Dasha*, 1) of the nineteenth "Exchanged Physical Body" (*Vartula*,

10,000), norming a "beta electron" (*Vartula*, 10,000). The beta electron localizes the "theory-effect" (*Maha dasha*, 0) of the twentieth, "Exchanging Physical Body" (*Kratu*, 91), norming the resulting "alpha force" (*Kratu*, 91) as an "alpha electron" (*Kratu*, 91). The alpha electron "nationalizes" (*Rashtrikaran*, 10^{19}) the twenty-first "Exchangeable Physical Body" (*Andhakara*, 10^{19}), the "alpha photon" (*Andhakara*, 10^{19}) as a "gamma" (*Andhakara*, 10^{19}) element. The gamma is the "consciousness deficit" (*Andhakara*, 10^{19}) within the "consuming self" (*Andhakara*, 10^{19}) after consuming the self's "para-consciousness" (*Mahatma*, 18) within the "incoming photon" (*Arcisa*, 128) that norms the "child" (*Arcisa*, 128). The child produces the "bragging self" (*Andhakara*, 10^{19}) with the "ego element" (*Aham*, -1) to reproduce the "knowing" (*Jnana*, $19 = 18 - [-1]$) that gives birth to him as a "child cell" (*Jantu*, 19).

4.7.5 Knowing Causation. Divided Mental Body and the Twin flame Without Flame Family

A "child cell" (*Jantu*, 19) divides the first, "undivided mental body" (*Manomaya Sharira*, 381) into "eighteen dimensions" (*Ekadharma*, 81) of the "mind" (*Manas*, 19) as the nineteenth, "Exchanged Mental Body" (*Manas*, $38 = 19 * 2$). The latter reproduces the "knowing" (*Jnana*, 19) of the twentieth, primeval, "Exchanging Mental Body" (*Hiranya sharira*, 123) within it. The Exchanging Mental Body is a "colorless body" (*Hiranya sharira*, 123) that trades the "grey color" (*Kapota*, 169) of the twenty-first, primordial, "Exchangeable Mental Body" (*Sheshanaga*, 816). It forms a "disproportionate sense" (*Kapota*, 169) of the "param physical body" (*Kshema*, 816). Consequentially, the "para-conscious trading" (*Kshema*, 816) of the param physical body becomes the "metaphysical cause" (*Kshema*, 816) for the "exponential attenuation" (*Ekotrem*, 121) of the child cell's "self-identity" (*Svadhisthan*, 816).

Eighteen dimensions are the intellectual body's automatic "triple copy" (*Vaisha*, 3) of the six divisions of the physical body that form the first six causal, etheric, and astral bodies. The astral body forms as a constant "conscious body" (*Linga sharira*, 3) without the etheric body. The etheric body forms as a "guider body" (*Bhoga sharira*, 957), reproducing the varying "consciousness" (*Chaithanya*,

4) of the conscious body for producing the causal body as a "variable body" (Karana sharira, 30) within it. The consciousness varies because of the six-fold growth of the astral body that norms the tertiary six astral bodies and the tertiary six etheric and causal bodies as its double copies. The astral body generates a tertiary six-fold growth because it is conscious of its primary six divisions within the physical body but not the secondary six dimensions within the mental body that are the copies of the intellectual body.

The astral body seeks a "twelve-fold growth" (Barah, 12) because it is a "quantum photon" (Svarochisha, 12) without the triple copy of the six divisions of the physical body. After the six divisions form the first six astral bodies, they reproduce as the first six causal bodies by behaving like the first six etheric bodies. The intellectual body conceives the six time-varying divisions of the physical body as the eighteen divisions of the constant astral body. Therefore, it makes six space-varying copies of the physical body as its six dimensions within the mental body. The intellectual body is an "illusionary body" (Sukshma sharira, 306 = 30 → 6) of the "causal body's" (Karana sharira, 30) "six-fold growth" (Dishta, 6) into the "light" (Prabha, 180) of the mental body. Consequently, it illuminates the six dimensions within the mental body as its "mirror copies" (Nakala, 0).

The astral body makes a "carbon copy" (Chitra, 1) of the causal body's six-fold growth. The etheric body reproduces the six-carbon copies as the six "tandem copies" (Brahman, 2) of the astral body. Six carbon copies of the sub-conscious causal body form the secondary six dimensions of the para-conscious, mental, and the illusionary, intellectual bodies. Six tandem copies of the astral body manifest the reproducible "double copy" (Chitra, 1) as the tertiary six divisions of the astral body and the etheric body and the "reproduced copy" (Nakala, 0) as the tertiary six divisions of the causal body. In tandem, they also manifest the secondary "produced copy" (Brahman, 2) of each of the eighteen divisions.

The mental body makes a "rotation copy" (Khara, 6) of the illusionary intellectual body to manifest the reality of its tertiary six dimensions after manifesting them first within the intellectual body for testing the rotation hypothesis.

The causal body makes a "book copy" (*Urja*, 31) of the physical body. It begins with an "automatic copy" (*Vaishya*, 3) of the "automatic growth" (*Dishta*, 6) of the physical body's first six divisions. That produces the three divisions of all the six bodies: physical, intellectual, mental, astral, etheric, and causal. Next, it conceives the "self-luminous entity" (*Svayam*, 12) as its "carbon copy" (*Chitra*, 1) enjoying the "primordial consciousness" (*Adhyatma*, 12) of the twenty-one divisions and their "automatic copy" (*Vaishya*, 3) to "channel" (*Raasta*, 36 = 12 * 3) the "light" (*Prabha*, 180 = 12 * [12 + 3]) of the child cell's knowing.

An Octave of Copies

An "octave of copies" (*Sargam*, 60 = 0 + 1 + 2 + 3 + 6 + 8 + 8 + 31 + 1) comprise eight forms of copies within an "octave of entities" (*Krishnamurti*, 1): "mirror copy" (*Nakala*, 0), "carbon copy" (*Chitra*, 1), "tandem copy" (*Brahman*, 2), "automatic copy" (*Vaishya*, 3), "rotation copy" (*Khara*, 6), "reserve copy" (*Satarupa*, 8), "digital copy" (*Maha Nitya*, 8), and "book copy" (*Urja*, 59).

The carbon copy is a reproducible "double copy" (*Chitra*, 1) that reproduces itself as the "mirror copy" (*Nakala*, 0) for producing the "triple copy" (*Vaishya*, 3). The triple copy is a "reproducing copy" (*Vaishya*, 3 = 1 + 0 + 2), comprising the carbon, "reproducible copy" (*Chitra*, 1) and the mirror, "reproduced copy" (*Nakala*, 0) within the produced "tandem copy" (*Brahman*, 2). It is an "automatic copy" (*Vaishya*, 3) formed naturally with time.

The "producing copy" (*Urja*, 31) produces the "produced copy" (*Brahman*, 2) in tandem with the reproducible and the reproduced copies. The producing copy is a "triple-double copy" (*Urja*, 31). It is a "book copy" (*Urja*, 31) where the odd and even are the one that forms the "double copy" (*Chitra*, 1). The odd and the odd are the two that form the "twin double copy" (*Brahman*, 2). The even and the even are the four that form a "copy" (*Nakala*, 0) and a "triple copy" (*Vaishya*, 3) within the "double copy" (*Chitra*, 1) omnipresent within the even. The double copy is the odd omnipresent within the even.

The six-fold growth is the "producible copy" (*Khara*, 6)

with a pair of odd and even. The producible copy is the "twin triple copy" (*Khara*, 6). The twin triple copy is a "rotation copy" (*Khara*, 6) formed with a "90 degrees rotation" (*Prakrima*, 1000) reproducing an odd, "triple copy" (*Vaishya*, 3) as even. The 90 degrees rotation "self-perpetuates" (*Amsha*, ½) a "180 degrees rotation" (*Sthiti*, 100) that reproduces the even "twin double copy" (*Brahman*, 2) as the even "twin triple copy" (*Khara*, 6). The 180 degrees rotation self-perpetuates a "360 degrees rotation" (*Sura*, 0) that produces the odd "triple-double copy" (*Urja*, 31) as an odd "double copy" (*Chitra*, 1) paired with the three even "copies" (*Nakala*, 0) for "self-perpetuating" (*Amsha*, ½) the two odd copies.

The eight-fold growth is the "reserve copy" (*Satarupa*, 8) that produces both the "producible copy" (*Khara*, 6) and the "produced copy" (*Brahman*, 2) when necessary for the performing of an instructor. It is the "production copy" (*Satarupa*, 8).

The four-fold growth is the "digital copy" (*Maha Nitya*, 4) that reproduces both the "reproducible copy" (*Chitra*, 1) and the "reproducing copy" (*Vaishya*, 3) when instructed through a program. It is the "reproduction copy" (*Maha Nitya*, 4). It is the "personal copy" (*Maha Nitya*, 4) kept by an "instructor" (*Adhyapaka*, 0) for planning, seeking "supernatural profiting" (*Naraki*, 1) without "further development" (*Rajas*, 15) through "continuous improvement" (*Samunnati*, 13) of the "instruction" (*Adhyapana*, 123) over time.

The mirror copy is the "paternal copy" (*Nakala*, 0). It is the "gender and generation differentiated copy" (*Nakala*, 0).

The carbon copy is the "maternal copy" (*Chitra*, 1). It is the "generation-differentiated copy" (*Chitra*, 1).

The tandem copy is the "female copy" (*Brahman*, 2). It is the "generation-differentiating copy" (*Brahman*, 2).

The automatic copy is the "masculine copy" (*Vaishya*, 3). It is the "gender and generation differentiating copy" (*Vaishya*, 3).

The rotation copy is the "male copy" (*Khara*, 6). It is the "gender-differentiated copy" (*Khara*, 6).

The reserve copy is the "sister copy" (*Satarupa*, 8). It is the "group, gender, and generation differentiating copy" (*Satarupa*, 8).

The digital copy is the "brother copy" (*Maha Nitya*, 8). It is the "group, gender, and generation differentiated copy" (*Maha Nitya*, 8).

The book copy is the "feminine copy" (*Urja*, 31). It is the "gender-differentiating copy" (*Urja*, 31).

Thus, the eighteenth, "undivided mental body" (*Manomaya Sharira*, 381) comprises three dimensions of itself as the "para-conscious mind" (*Hradamana*, 381) and eighteen dimensions of the "conscious mind" (*Manas*, 38) without oneness with the "carbon copy" (*Chitra*, 1) that forms the "self-luminous entity" (*Svayam*, 12). The conscious mind channels the "para-consciousness" (*Mahatma*, 18) of the time-varying space through its eighteen dimensions. The space has four spatial dimensions, a fifth intrinsic dimension that norms the "convergent light" (*Svarochisha*, 12) of the four-space dimensions as the self-luminous, quantum photon, and a sixth extrinsic dimension that norms the "divergent light" (*Rochisha*, $13 = 4 + 2 + 1 + 4$) of the quantum photon as the quantum light. The quantum light unfolds the four directions over the two dimensions of time (present and future) and a unit of "past force" (*Ghatika*, 24) as a "four-faced entity" (*Hasti*, 24) that folds the consciousness of the four directions within herself.

The entity self-perpetuates the "primordial consciousness" (*Adhyatma*, 12) of the past life, karmic "workforce system" (*Sara Kalpa*, 10^{10}) within the "self-luminous entity" (*Svayam*, 12) as her "consciousness system" (*Sara Kalpa*, 10^{10}). The self-luminous entity self-perpetuates the "producible copy" (*Khara*, 6) of the present life, "networking system" (*Panchanguli*, 8×10^{15}) within the "natural growth" (*Dishta*, 6) of his "conscious system" (*Shunya Kalpa*, 8×10^{15}) through the reproduction of the entity's consciousness system for producing the child's "infinite consciousness system" (*Maha Kalpa*, 10^{1000}). The child consumes the infinite consciousness system as the future life, "exchange system" (*Maha Kalpa*, 10^{1000}), thereby limiting the production of his conscious system from the "sister spirit's"

(*Ruah*, 20) "infinite conscious system" (*Antarmukha*, 10^{1024}) traded from a primordial greeter's "energy system" (*Antarmukha*, 10^{1024}).

Consequently, the child develops a "consciousness void" (*Andhakara*, 10^{19}) within his "body system" (*Hridya Pranali*, $10^5 = 10^{1024-1000-19}$) mediated by the "brother spirit's" (*Svarochisha*, 12) "error in consciousness" (*Dushtatma*, 10). As a self-luminous entity, the child fills the consciousness void with his "unknown reality" (*Vastavikta*, 7), comprising the following seven elements.

i. Past life *karmic* workforce system of self as a protagonist;

ii. Past life in other dimensions, i.e., the past life "infinite workforce system" (*Parinama*, 123) of self as an antagonist, reproducing the workforce system of the "universe of ideal objects" (*Satya Loka*, 10^{1024});

iii. Present life already happened, i.e., the present life "networking system" of self as a deuteragonist, seeking to norm and transform the workforce system of the "universe of ideal objects" (*Satya Loka*, 10^{1024}) for its reproduction;

iv. Present life yet to come, i.e., the present life "infinite networking system" (*Pravritta*, 12) of self as a tritagonist, seeking to form the "networking system" (*Panchanguli*, 8×10^{15}) of the universe of ideal objects as a tradable "exchange system" (*Maha Kalpa*, 10^{1000});

v. Future life, forming a future life "exchange system" (*Maha Kalpa*, 10^{1000}) of self as a superagonist, servicing the "energy system" (*Antarmukha*, 10^{1024}) as a primordial greeter;

vi. Parallel lives, i.e.., the future life "infinite exchange system" (*Sighraga*, 1700) of self as a coagnoist, invested into the "replicative recombination" (*Dvaiyogya*, 10^{24}) of the past knowing.

vi. "Potential life" (*Bhairava*, 15) of self as an agonist, capable of "breeding" (*Janana*, 15) the past knowing through the "breeding recombination" (*Yathayogya*, 10^{100}) of a "sequence of triple copies" (*Yathayogya*, 10^{100}), forming a "universe of alleles" (*Yathayogya*, 10^{100}).

Each "allele" (*Devendra*, 12) is the "collective force" (*Devendra*,

12) of the universe of alleles without the "twin double copy" (*Brahman*, 2). The allele produces the twin double copy with his "individual force" (*Vidya Ratri*, 19) as a "child cell" (*Jantu*, 19). The twin double copy trades the allele's "entropy value" (*Sarvanasha*, $5 = 19 - 12 - 2$) to become a "sentient entity" (*Siddha*, 7) and services the "daughter spirit" (*Vajra*, 14) as his double copy. The daughter spirit conceives the "male copy" (*Khara*, 6) within her to become the "sister spirit" (*Ruah*, 20). Her identity as a double copy forces the sentient entity to behave like a "human being" (*Insan*, 1), not conscious of his "entity reality" (*Upakarana artha*, 10^{1024}). The human being enjoys the gravitational quality of the entire universe of alleles that forms the "universe of sentient entities" (*Akalpa*, 570) and transforms them into the "universe of para-conscious entities" (*Sthavaravisha*, 57) after reproduction of their copy.

The "universe of para-conscious entities" (*Sthavaravisha*, $57 = 21 + 6 + 30$) comprises the "impact" (*Siddhi*, 57) of the 21 divisions and the six-fold multiplication on the "human body" (*Karana sharira*, 30). It is the Seventeenth Mind or the Seventeenth Mental body. It shapes a conscious, followership "workculture system" (*Sthavaravisha*, 57) led by the "universe of sentient entities" (*Akalpa*, 570). Each sub-division of the Eighteenth Mind, "the white matter" (*Hradamana*, 381), is contaminated with the "departed consciousness" (*Hasti*, 24) of the "sentient entity" (*Siddha*, 7) that a human being seeks to purify with his "living consciousness" (*Atma sharira*, 14,000).

The "universe of departed souls" (*Makar Rashi*, 80) is the Sixteenth Mind or the Sixteenth Mental Body. It "self-materializes" (*Svabhautik*, 80) the "universe of sentient entities" (*Akalpa*, 570) with its "cultural system" (*Akalpa*, 570). As a primeval perpetuator of "breeding" (*Janana*, 15), the universe of sentient entities is the Fifteenth Mind or the Fifteenth Mental Body. It forms as an "ester bond" (*Ganesha*, 570) with the Sixteenth Mind's "binding energy" (*Bandhana shakti*, 80).

The "universe of living souls" (*Yantra*, 12) is the Fourteenth Mind or the Fourteenth Mental Body, that "lives" (*Nivas*, 14) with the "overflowing energy" (*Ishi shakti*, 12) of its "self-involvement"

(*Phasava*, 12) within the "self-luminous entity" (*Svayam*, 12).

The "universe of self-luminous entities" (*Satya Loka*, 10^{1024}) that forms the "energy system" (*Antarmukha*, 10^{1024}) is the Thirteenth Mind or the Thirteenth Mental Body.

The "universe of entities" (*Kashi*, -10^{1024}) that makes each entity self-luminous by reproducing the "reproducible copy" (*Khara*, 6) without the "entity force" (*Haisiyat*, 38) is the Twelfth Mind or the Twelfth Mental Body.

The "entity" (*Hasti*, 24) who trades the "entity force" (*Haisiyat*, 38) to "live" (*Nivas*, 14) as a "self-luminous entity" (*Svayam*, 12) is the Eleventh Mind or the Eleventh Mental Body.

The "param entity" (*Harbuddhi*, 366,666) who services the "entity force" (*Haisiyat*, 38) with her "tertiary consciousness" (*Brahmaastra*, 366,666) of the entity's "objective" (*Antahkarana*, 2) is the Tenth Mind or the Tenth Mental Body.

The entity's objective is to self-perpetuate the entity force as a "pulsating sentient entity" (*Jantu*, 19) within the child cell. The pulsating sentient entity is the First Mind or the First Mental Body.

A child cell pulsates with the "sentient energy" (*Amrit*, 1000) of its "subdivisions" (*Prabhaga*, 40,000). Each subdivision is "satiated" (*Tripta*, 40,000) with the "hundred degrees" (*Samalamba*, 40,000) of the sentient entity's "reproductive reality" (*Bhavartha*, 40). A subdivision is the Ninth Mind or the Ninth Mental Body.

A child cell's subdivision reproduces the "gravitational energy" (*Lalita*, 100) of a "mother cell's" (*Dhumavati*, 7) "division" (*Sudhanvan*, 999). Each division is a "master manager" (*Sudhanvan*, 999) of the "transcriptional machinery" (*Shadyatana*, 999) that transcribes her "maternal consciousness" (*Chaithanya*, 4) as the "child consciousness" (*Putatma*, 4). A division is the Eighth Mind or the Eighth Mental Body.

A division consumes the "divine energy" (*Divya shakti*, 10) of a pyramidal "grandfather cell" (*Uccitika*, 18) for producing an "eight-fold growth" (*Satarupa*, 8) to self-perpetuate a glial "father cell's" (*Rodha*, 1) supernatural "four-fold growth" (*Maha Nitya*, 8) as her "copy" (*Nakala*, 0). The grandfather cell is the Seventh Mind or the

Seventh Astral Body. The father cell is the Sixth Mind or the Sixth Mental Body.

A father cell is the thermodynamic force of the infinite divisions and subdivisions of an oligodendrocyte "granddaughter cell" (*Bagalamukhi*, 9). Granddaughter cell is the Fifth Mind or the Fifth Mental Body.

The thermodynamic force activates the potential of a "potential granddaughter cell" (*Dasha*, 1) for reproducing the dying granddaughter cell as the "twin granddaughter cell" (*Harkriya*, 666). The twin granddaughter cell is the "flow of granddaughter cells" (*Pravaha*, 666) as the "natural growth" (*Dishta*, 6) within the "natural copy" (*Vaisha*, 3). It comprises the reunified consciousness of a "departed soul" (*Hasti*, 24) and a "living soul" (*Atma sharira*, 14000). The Twin Granddaughter cell is the Fourth mind or the Fourth Mental Body.

The Twin Granddaughter cell reunifies the "divided consciousness" (*Achaithanya*, 39) of the Oligodendrocyte progenitor "twin grandmother cell" (*Maha Kali*, 13) out of her "love" (*Prema*, 39) for the grandfather cell as her twin. The twin grandmother cell is the Third Mind or the Third Mental Body.

The Twin grandmother cell services the "reunified consciousness" (*Harkriya*, 666) to a constitutional "greeter cell" (*Nirjara*, 1000) as her "conscious energy" (*Varuna*, 1000). The Greeter cell is the Second Mind or the Second Mental Body.

Our mind enjoys the consciousness of all parts of our consciousness living within or without our body as the greeter cells as well as those departed from the "universe of sentient entities" (*Akalpa*, 570). The consciousness of those departed is now a "fragment" (*Yogihridaya*, 64) of energy within the "universe of inanimate objects" (*Sthavaravisha*, 57) without the "sentient entity" (*Siddha*, 7). Our mind also enjoys the para-consciousness of all parts of our para-consciousness traded from other sentient entities directly or mediated through the inanimate objects endowed with their energy fragments.

Therefore, our mind becomes a "supreme portal" (*Rochisha*, 13) for the infinite diffusion of our sentient energy for the universal sentient well-being of all parts of our consciousness and descends our sentient well-being. Six factors pollute the "self-luminous reality" (Purushartha, 7 = 0 − 1 + 9 − 4 − 3 + 6) of the supreme portal as the "potential grandfather cell" (Rochisha, 13), norming a "boson sentient" (Rochisha, 13).

i. **Primary Factor:** The "primary factor" (Sura, 0) is the "concordant factor" (Sura, 0) that manifests as the "gravitational quality" (Guna, 0). "Living soul" dominates our "gravitational quality" (Guna, 0). It makes us conscious of its "discontinuity" (Sato guna, 200) from the "departed soul" (Hasti, 24). It weakens our emotional, "psychic linkages" (Mahika, 47) with the universe of inanimate objects and transforms them into the dark "para-psychic linkages" (Mahayana, 179 = 47 + 132). The latter are infused with the "mass of consciousness" (Prithvi, 132) we have diffused psychically, psychologically, emotionally, socially, digitally (intellectually), and biologically (physically). A "quark" (Dridayudha, 476 = 132 * 3 − 20) is a "natural copy" (Vaishya, 3 = 20 - 17) of our mass of consciousness that a "photon" (Kapinjala, 20) manifests without the "conscious fire" (Agni, 17) of oneness with our living soul. A quark reproduces the "child cell" (Jantu, 19) for producing the "continuity" (Rajo Guna, 50) of the "departed soul" (Hasti, 24 = 17 + 3 + 2 + 2) within the living soul, comprising:

    o   Our "kin universe" (Aksha, 17), comprising the patrilineal fraternal "universe of wishers" (Asura, -1) without our "potential energy" (AUM shakti, 18). Our potential energy self-perpetuates our "infinite wishes" (Duniya, -2) within the universe of wishers and conceives them as our "kin" (Parijan, 90,000) by reproducing our consciousness into "infinity" (Tamo Guna, 90,000).

    o   Our "kith universe" (Devi, 3 = 5 - 2), comprising the matrilineal sororal "infinite wishes" (Duniya, -2) we produce by reproducing our wishes until the "entropy" (Mahodbhava, 5) of our living consciousness.

    o   Our "resident alien universe" (Udita, 2), comprising the "unfriendly network" (Udita, 2) of the ascendent "material power" (Udita, 2). They service "energy" (Shakti, 19) that energizes the "kin universe" (Aksha, 17) to make friends with the "kith universe" (*Devi*, 3).

- Our "non-resident alien universe" (*Brahman*, 2), comprising the "universe of child souls" (*Brahman*, 2) transforming into the "universe of granddaughter cells" (*Brahman*, 2). They form as a "universe of light echoes" (*Brahman*, 2) of the energized kin universe within the de-energized kith universe. Thus, the friendly kith universe "self-services" (*Svayamseva*, 2/3) itself as the even element without the odd element to become the non-resident alien universe. The latter is a "super friendly network" (*Brahman*, 2) that reproduces the odd kith universe by producing "natural growth" (*Dishta*, 6) of itself as the even, non-resident alien universe. As an ideal subject producing the wish for breeding our self, we enjoy infinite social benefits from its "liberal democratic philosophy" (*Yukti*, $8 = 2 + 6$).

ii. **Secondary Factor:** The "secondary factor" (*Asura*, -1) is the "discordant factor" (*Asura*, -1) that manifests our "discordant energy" (*Asura shakti*, -1) as the "universe of wishers" (*Asura*, -1). It leads to a normative "organizational planning" (*Abadha*, 10) of the physical actions of each "granddaughter cell" (*Bagalamukhi*, 9) by copying our time-specific behaviors. The granddaughter cell normalizes the "biological-effect" (*Vartula*, 10,000) of our time-varying becoming into a wisher wishing self-fulfillment of the wishes through a normative "organizational programming" (*Dharana*, -8) of each "daughter cell" (*Kshatradharma*, 19) with our "intended reality" (*Pushtartha*, -8). The daughter cell normalizes the "chemical reaction" (*Maha Vibhu*, 14) of our time-invariant breeding of the infinite wishes through normative "organizational performing" (*Rachana*, 286) as a "mother cell" (*Dhumavati*, 7).

Organizational programming is the fundamental secondary effect of organizational planning, manifesting the actions that become the cause for compensating the residual energy with the reactions. The tertiary residual of organizational planning manifests organizational performing. Organizational performing is the sequence of proliferating actions seeking to self-manage the psychological conditioning pressures of the reactions. The mother cell normalizes the "organic resonance" (*Jaivik Anunaad*, 43) of our space-specific breathing of the finite wishes, immanent within the consciousness flowing through that space, through normative "organizational profiting" (*Nanarupa*, 10,000). It obviates the need to conceive them and conserves her potential energy as the "grandmother cell" (*Dadi*, 18)

The grandmother cell normalizes the "inorganic dissonance" (*Sahasrara*, 19) of our space-varying blessing of infinite consciousness through her normative "organizational development" (*Viparyaya*, 888) as a "grandfather cell" (*Uccitika*, 18) into an "organization" (*Sangathan*, $29 = 2 + 6 + 14 + 7$) of the following.

- "Sensory perception" (*Bhuta yajni*, 2) of our reaction due to the "performing force" (*Bhuta Yajni*, 2) of the action;
- "Reflective conception" (*Mahat*, 6) of our thoughts due to the "perceived sense" (*Yuktartha*, 6) of reaction;
- "Immersive experience" (*Samajh*, 14) of our behaviors mindlessly bred to compensate for the reaction;
- "Spiritual sense" (*Vastunara*, 7) of our "divinity" (*Divyata*, 57) as a "conscious entity" (*Siddha*, 7) within the "continuity" (*Rajo Guna*, 50) of the immersive experience after experiencing the "present entropy-effect" (*Rohini*, 50) of the bred behaviors.

iii. **Tertiary Factor:** The tertiary factor is the "infinite breeding" (*Shrimate*, 9) that generates an "entropy sensation" (*Rohini*, 50) within the living soul. Consequently, the living soul makes the entity conscious of the "non-resident alien universe" (*Brahman*, 2) as the "cause" (*Karana*, 18) of the "entropy" (*Mahodbhava*, 5). The entity seeks to exchange our reaction by transforming their "accelerating-effect" (*Rochisha*, 13) into "organic consonance" (*Samarasya*, 9,000,000) of the "goal" (*Lakshya*, 9) of infinite breeding. The goal is to give "voice" (*Vak*, 629) to our "experience" (*Anubhav*, 9000) to make it the "ideal" (*Adarsha*, 1) for the friendly "kith universe" (*Devi*, 3). The kith universe comprises the "conservative democrats" (*Doliltra*, 3) who have never experienced the reality of life. Breeding as the universe of non-resident aliens accelerates formative "technological growth" (*Vidhana*, 2). The "breeding force" (*Dikpala*, 1000) self-perpetuates their "bragging value" (*Vikatth*, 486) over the "immersive experience" (*Samajh*, 14).

iv **Quaternary Factor:** The quaternary factor is the "breeding consciousness" (*Dharshanatma*, -4) of our "total presence" (*Abhyasa*, -4) within the departed soul. Consequently, the departed soul is naturally attracted to our "voice" (*Vak*, 629) as the "gravitational point" (*Guru Sthana*, 629) of the "breeding force" (*Dikpala*, 1000) of the living soul. Our voice breeds the intended reality of devoted followership within the non-resident alien universe. We become the "king" (*Rajah*, 0) of "divine entities" (*Deva*, 1) and each "non-resident alien" (*Insan*, 1) a "human being" (*Insan*, 1) blossoming like a divine entity blessed with our breeding force. The king is the primordial paternal of the human being since the human being is born out of our "paternal consciousness" (*Pitra*, 4) diffused as the "departed consciousness" (*Hasti*, 24) with "natural growth" (*Dishta*, 6).

v. **Quinary Factor:** The quinary factor is the "object" (*Padartha*, -3) experiencing our "total absence" (*Abheda*, 0) in the "present" (*SAUM*, 1600) after our death when we become a "departed entity" (*Divangat*, 1600). The object reproduces the "divine-effect" (*Padartha*, -3) that gave birth to it for producing a degenerating, bistratified, ganglion "grandson cell" (*Pavamana*, $9 = -3 * -3$). The grandson cell generates a "positron" (*Sandesha*, 1), i.e., "potential electron" (*Sandesha*, 1) by consuming the "visual information" (*Sandesha*, 1) from the "potential greeter cell" (*Sadhyata*, 80). The potential greeter cell is the "photoreceptor cell" (*Sadhyata*, 80) that produces the visual information after consuming the "incoming light" (*Ambara*, 180) of the "mental space" (*Vaas*, 18,000) emanating from the constitutional "greeter cell" (*Nirjara*, 1000).

A greeter cell becomes a potential greeter cell after reproducing the "inner sense" (*Ajjhattikani*, 80) of itself as an "object" (*Padartha*, -3) for producing the "outer sense"

(*Bahirani*, 80) of the grandson cell as the "thing" (*Vastu*, 9) "self-fulfilling" (*Nivritti*, 2) the object's "objective" (*Antakarana*, 2). The greeter cell receives the inner sense from the "twin greeter cell" (*Ajjhattikani*, 80), which is a "bipolar cell" (*Ajjhattikani*, 80) that "bipolarizes" (*Dvidhruviya*, 9) the outer sense of four spatial directions conceived by the "potential grandmother cell" (*Bahirini*, 80) into "eight directions" (*Bahirini*, 80). The potential grandmother cell is the "Amacrine cell" (*Bahirini*, 80).

A "grandson cell" (*Pavamana*, 9) comprises eight units of "visual information" (*Sandesha*, 1) about the eight directions, four within and four without the inner sense's mediation, and a unit of "visualizable information" (*Chitra*, 1) as the carbon "photo copy" (*Chitra*, 1) of the visualizing "potential grandmother cell" (*Bahirini*, 80). The eights units of visual information form eight potential electrons. The unit of visualizable information forms the "twentieth gamma photon" (*Chitra*, 1) that transforms into an "up quark" (*Gajakarani*, 1) within the "greeter consciousness" (*Pratyagatma*, 1) of the eight potential electrons norming the "natural value" (*Nirvikalpana*, 8) of "copying" (*Chitra*, 1). Each "potential electron" (*Sandesha*, 1) transforms into a visualized "granddaughter cell" (*Bagalamukhi*, 9) after copying the "natural value" (*Nirvikalpana*, 8) within itself as an "octave of potential electrons" (*Bagalamukhi*, 9). The granddaughter cell is the "visualized information" (*Bagalamukhi*, 9) of the visualizable "twin greeter cell" (*Ajjhattikani*, 80) as the "mediator" (*Sugriva*, $5 = 80/16$), i.e., the "cultural factor" (*Sugriva*, 5). The mediator mediates eight potential electrons immanent within and eight emanating with the granddaughter cell.

A potential electron is a spatially-linked "specific living soul" (*Pavitratma*, 1). During the day, the "plant kingdom" (*Vanaspati*, 21) trades the potential electron for "self-repelling" (*Mitra*, 132) the "carbonated air-effect" (*Vyan Vata*, 132) of the "gravitational mass of para-consciousness" (*Mitra*, 132) within the potential electron as the "calcified water-effect" (*Ruksha ojas*, 100,000). It forms "nitrogen" (*Rupa*, 100,000) by reproducing three twos as three zeroes without the "extrinsic light" (*Digambara*, 100) of the zero "copy" (*Nakala*, 0) of space as a "whole" (*Akala*, 16). The zero copy behaves like a "twin double copy" (*Brahman*, 2) of the "natural value" (*Nirvikalpana*, 8) as the "material kingdom" (*Sakala*, 8) for the "illusionary production" (*Maya*, 1) of a pair of double copies that form the four-space directions.

During the night, the plant kingdom trades the "gravitational consciousness" (*Samashti*, 7×10^{180}) as the "oxygenated fire-effect" (*Tejas yoga*, 7×10^{180}) for "self-repelling" (*Mitra*, 132) the "nitrogenated earth-effect" (*Yajnopavita*, 9000). It produces three potential electrons as three zeroes without the ninth that forms the "octave of potential electrons" (*Bagalamukhi*, 9). Each zero is the "greeter spirit" (*Hasta*, 0) of the "departed entity" (*Divangat*, 1600). The nitrogenated earth-effect is the "thread of life" (*Yajnopavita*, 9000) that animates the "twelve potential electrons" (*Kalki*, 9000) into a "plant entity" (*Kardama*, 9000).

Over its lifetime, the plant entity services the "hydrogenated ether-effect" (*Dushkarakarana*, 185 = 285 - 100) through the "reproduction" (*Prajana*, 285) of the "extrinsic light" (*Digamabara*, 100) of the zero copy. The "animal kingdom" (*Tiryaggati*, 47) trades the hydrogenated ether-effect with the extrinsic light for reproducing the "universe of living child souls" (*Akalpa*, 570 = 185 + 100 + 285) as the "primordial life" (*Mulaguna*, 570) in the form of the "archaeon" (*Jangamavisha*, 570).

The "human kingdom" (*Manushyagati*, 53) exchanges the extrinsic light of the "reproducing self" (*Guru*, 100) with the "intrinsic light" (*Usha*, 16) of the "primordial greeter" (*Srijak*, 16) who services the twelve "potential electrons" (*Sandesha*, 1) with his "SHEENY-effect" (*Soham*, 4). On the other hand, the "spirit kingdom" (*Rajyaika-sheshena*, 169) contaminates the SHEENY-effect by servicing the past reality as the departed entity's "divine-effect" (*Padartha*, -3) and trading that as an "object" (*Padartha*, -3). The object is the "resident alien" (*Padartha*, -3), mediating each human being's behaviors, becoming, breeding, bragging, and breathing as a "mediator" (*Sugriva*, 5) for self-fulfilling the objective by creating a "resident alien universe"

(*Udita*, 2).

By reproducing itself as three zeroes within the human being, the resident alien universe transforms into the "nonresident alien universe" (*Brahman*, 2) and transforms the "human being" (*Insan*, 1) into the "deity kingdom" (*Deva loka*, 1000). The deity kingdom services the sentient energy that motivates each "resident alien" (*Padartha*, -3) to become a potential guider agent fulfilling the cause of the self-deifying "human being" (*Insan*, 1). The human being is a "non-resident alien" (*Insan*, 1), behaving like a "potential principal guider" (*Raja-rajesvari*, 85)

vi. **Senary Factor:** The senary factor is the "natural theory" (*Uma*, 6) that makes a "non-resident alien" (*Insan*, 1) behave like a "resident alien" (*Padartha*, $-3 = 1 - 4$) without the "maternal consciousness" (*Chaithanya*, 4) of the fossilized "mineral kingdom" (*Khanijavarga*, 7). Without the "theory-effect" (*Maha dasha*, 0) that forms three zeroes over time, the non-resident alien is the "material kingdom" (*Sakala*, $8 = 1 + 7$) within the fossilized "mineral kingdom" (*Khanijavarga*, 7). Three zeroes are a "natural copy" (*Vaishya*, 3) "self-perpetuating" (*Udvaha*, ½) the "natural theory" (*Uma*, 6) as a "theory" (*Niyama*, 127) of "natural value" (*Nirvikalpana*, 8) reproducing the "reality" (*Vastavikta*, 7) of the mineral kingdom within the "twelve potential electrons" (*Kalki*, 9000) in the form of the "mineral essence" (*Sara*, $16 = 12 + 2 * 2$) produced by reproducing the "metal kingdom" (*Avatamsaka*, 2) as the "universe of resident aliens" (*Udita*, 2) for enjoying its "material power" (*Udita*, 2).

4.7.6 Feeding Causation. Divided Intellectual Body and the Flame Without the Human Being

The "flame" (*Parshnisamasta*, 32) that forms the "material kingdom" (*Sakala*, 8) as a "non-resident alien" (*Insan*, 1), and "humanizes" (*Manviya*, 16) the non-resident alien with the "mineral essence" (*Sara*, 16), is fed by a "departed entity" (*Divangat*, 1600). A departed entity conceives oneself to be a "param human child" (*Kesari Nandan*, 1600). He begins behaving like a "departed human entity" (*Manushya*, 82) experiencing "information void" (*Andha Kuan*, 82) about the "intrinsic force" (*Rohini*, 50) of the "flame within" (*Parshnisamasta*, 32). The flame within is ignited by the "flame without" (*Sadhyata*, 32) through a "primordial oneness" (*Adi*, 32 = 16 + 16) with the "child primordial greeter" (*Madhusudan*, 16) within a "maternal primordial greeter" (*Sati-Parvati*, 16).

The child primordial greeter services the animate "masculine energy" (*Purani shakti*, 91) by trading the mother primordial greeter's inanimate "feminine energy" (*Achitta Shakti*, 396), thus transforming the "non-resident alien" (*Insan*, 1) into the nineteenth "divided intellectual body" (*Sukshma Sharira*, 306 = 396 + 1 - 91). The divided intellectual body is a "plant body" (*Sukshma Sharira*, 306) that harmonizes itself within the masculine "animal body" (*Linga sharira*, 3) by animating the latter into an "astral body" (*Linga sharira*, 3). The astral body accultures the "metal body" (*Manomaya sharira*, 381) to behave like a "mental body" (*Manomaya sharira*, 381) and copy the child primordial greeter for trading the "positive energy" (*Sakaratmak shakti*, 19) of the feminine leader and servicing the "negative energy" (*Nakaratmak sharira*, 19) as a masculine follower.

Following the mental body's cultural factor, the causal "human body" (*Karana sharira*, 30) perceives the physical "mineral body's" (*Sthula sharira*, 387) "trading factor" (*Damodara*, 26) to be "positive" (*Sakaratmak*, 9) for its conscious well-being and the replicative "etheric body's" (*Bhoga sharira*, 957) "servicing factor" (*Nirmana kaya*, 23,125) to be "negative" (*Nakaratmak*, 9). Therefore, after reproducing itself as an "octave of copies" (*Sargam*, 60 = 30 * 2), it services the "human factor" (*Manviya Karak*, 53 = 60 − 7) for "self-substantiating" (*Self-substantiating*, -7) its "guider mediation" (*Anunaya*, -7) as a "negative factor" (*Evakara vadartha*, -3) in its "extrinsic experience" (*Bahirmukha*, -4) of the "extended self" (*Mera*, 1).

The physical body trades the human factor for "substantiating" (*Drvaya*, 1) its "divine moderation" (*Alobha*, 2) as a "positive factor" (*Bhuta Yajni*, 2) in its "intrinsic experience" (*Adhyatma*, 12) of the self, seeking to "extensify" (*Ghataana*, 340) through a "breeding workculture" (*Maha Svapna*, 9) for cooling its "conscious fire"

(*Agni*, 17) to lead after following. Therefore, the physical body divides the twentieth, undivided, "primordial intellectual body" (*Sahadeva*, 816), that it forms as the sixteenth, "Serviced Physical body" (*Kshema*, 816), into the following:

i. **First Intellectual Body:** The "first intellectual body" (*Lopamudra*, 47) is the "Heart Committee of Ascended Masters" (*Lopamudra*, 47). It norms the para-conscious "frontal lobe" (*Lopamudra*, 47) in the brain. It destroys the "followership pathway" (*Charya Marga*, 1,000,000) traded from the "feminine self's" (*Tara*, 2) "primordiality" (*Ayanamsha*, 27) as a "sister spirit" (*Ruah*, 20 = 47 - 27). It illuminates the "masculine self" (*Mein*, 1) as the "ascended master" (*Bhrigu*, 805) servicing the "leadership pathway" (*Dhyana Marga*, 36) through the "organizational planning" (*Abadha*, 10) of the "divided physical body" (*Hridya Pranali*, 100,000). It liberates the "grandson cell" (*Pavamana*, 9) to incubate as a "sentient entity" (*Siddha*, 7) by transforming the "granddaughter cell" (*Bagalamukhi*, 9) into an incubating "organization" (*Sangathan*, 29).

ii. **Second Intellectual Body:** The "second intellectual body" (*Swasthani*, 40) is the "Office of ONE universe for departed souls" (*Swasthani*, 40). It norms the feminine, "lateral fissure" (*Swasthani*, 40) in the brain. It is devoted to ascending the "spiritual well-being" (*Sukha*, 40) of the masculine self as a "departed soul" (*Hasti*, 24) by ascending the "granddaughter consciousness" (*Soham*, 4) of the feminine self as a living soul. It organizationally programs the feminine self to be a grandmother for conceiving the masculine self as the father and herself as the daughter with the granddaughter consciousness of the departed soul as her grandfather. The "departed soul" (*Hasti*, $24 = 6*4$) comprises the "consciousness" (*Chaithanya*, 4) of the grandson and granddaughter, the son and daughter, and the father and mother within it. The "ONE universe" (*Ekarajya*, 181) is the "animate equality" (*Ekarajya*, 181) of the departed soul and the living soul after the "consciousness entropy" (*Ekarajya*, 181) of the departed soul and its reincarnation within the living soul.

iii. **Third Intellectual Body:** The "third intellectual body" (*Surasa*,

38) is the "Office of Guides for departed souls" (*Surasa*, 38). It norms the masculine "central fissure" (*Surasa*, 38) in the brain. It manifests the masculine self as the guide of the reincarnated departed soul. The masculine self makes the reincarnated departed soul his feminine self for enjoying her organizational performing as a "daughter cell" (*Kshatradharma*, 19) within an "entrepreneurship pathway" (*Bhavana*, 37) that makes him a "son cell" (*Manyu*, 19) enjoying "ascending value" (*Rodha*, 1).

iv. **Fourth Intellectual Body:** The "fourth intellectual body" (*Makuli*, 46) is the "Wholeness Committee of Godheads" (*Makuli*, 46). It norms the subconscious "temporal lobe" (*Makuli*, 46). in the brain. It perpetuates the living soul as the "Godhead" (*Param Ganesha*, 17), furthering the son cell's organizational profiting with the "management pathway" (*Sadachara Marga*, 18). It reincarnates the departed soul as a "universe of granddaughter cells" (*Brahman*, 2) managed by the son cell for the "wholeness" (*Ardhajya*, 10) of his sentient well-being as the "primordial masculine self" (*Shri Krishna*, 10).

v. **Fifth Intellectual Body:** The "fifth intellectual body" (*Sannati Chandralamba*, 36) is the "Office of Guides for living souls" (*Sannati Chandralamba*, 36). It norms the conscious "occipital lobe" (*Sannati Chandralamba*, 36) in the brain. It manifests the feminine self as the guide of the reincarnating living soul. The feminine self makes the reincarnating living soul her masculine self for the organizational development of the primordial masculine self within the "conscious light" (*Kumara*, 10) of her "primordial feminine self" (*Parvati*, 10).

vi. **Sixth Intellectual Body:** The "sixth intellectual body" (*Smriti*, 35) is the "Office of Forces of Light for living souls" (*Smriti*, 35). It norms the consciousness "parietal lobe" (*Smriti*, 35) in the brain. It works as a "dynamic force" (*Smriti*, 35) that transforms the conscious "light force" (*Apas*, $169 = 100 + 35 + 35 - 1$) of the primordial feminine self into the "para-conscious light" (*Kumari*, 100) of the "primeval masculine self" (*Rama*, 100). It reproduces itself within the living soul for producing the primordial masculine self as a "human being" (*Insan*, 1).

The "seventh intellectual body" (*Bhima*, 1000^{1024}) is the "gray matter" (*Bhima*, 1000^{1024}). As a "three energy system" (*Antarmukha*, 1000^{1024}), it services the negative energy of the living soul, trades the positive energy of the departed soul, and exchanges the energy oneself generates with the "system" (*Gabhira*, 12) that generates oneself as its conscious system. The system is differentiable into the secondary seven intellectual bodies by reproducing itself as the "eighth intellectual body" (*Upashanta*, 287 = ½ * [816 − 47 − 40 − 38 − 46 − 36 − 35]).

The eighth intellectual body is the "Medulla oblongata" (*Upashanta*, 287). It norms the "complementary value" (*Upanshanta*, 287) within the "competitive linkages" (*Anuprastha Taranga*, 10) of oneself as the conscious system. It is the "eight-faced system value" (*Upashanta*, 287), with six feminine "extrinsic faces" (*Bahirmukha*, -4) comprising the competitive linkages, one masculine "intrinsic face" (*Antarmukha*, 10^{1024}) comprising the seventh intellectual body, and itself as the eighth, "entity face" (*Gardhabi Mukha*, 1000). The competitive linkages transform it into the "dominating value" (*Upashanta*, 287), forcing a sacrifice of the "supplementary linkages" (*Maitri Bhava*, 1) with the seventh intellectual body from its "complementary linkages" (*Vaira Bhava*, 1) for the "growth" (*Dishta*, 6) of the competitive linkages as the secondary six bodies, comprising the ninth to the fourteenth ones.

i. **Ninth Intellectual Body**. The "ninth intellectual body" (*Parameshwara*, 7) is the "Pons" (*Parameshwara*, 7). The Pons is the "umpire" (*Parameshwara*, 7), umpiring the "flow with time" (*Shudra*, 1) of its "intrinsic entity value" (*Kshemya*, 7). It seeks the intrinsic entity value's "transformative development" (*Purushartha*, 7) into a "sentient entity" (*Siddha*, 7) like itself. The intrinsic entity value is the diffused "self-luminous reality" (*Purushartha*, 7) Medulla oblongata reinfuses by mediating the Pons-activated "sleeping" (*Nidramaya*, 751) with the "breathing" (*Pranayama*, 37) of its eight-faced system value" (*Upashanta*, 287). The conscious "occipital lobe" (*Sannati Chandralamba*, 36) moderates the Medulla oblongata's transformation into the tenth intellectual body with its "dreaming" (*Svapnil*, 12) "para-consciousness" (*Mahatma*, 18), leading to the frontal lobe's "self-awareness" (*Atma bodha*, 30).

ii. **Tenth Intellectual Body**. The "tenth intellectual body" (*Viparyaya*, 888) is the "Cerebellum" (*Viparyaya*, 888). The Cerebellum is the "organizational development" (*Viparyaya*, 888) of the Medulla oblongata into four bodies in "oneness" (*Yoga*, 48) with Pons within its mediation and the Cerebellum without its mediation. The para-conscious "frontal lobe" (*Lopamudra*, 47) moderates the oneness with its "waking" (*Jagrit*, 18) "self-awareness" (*Atma bodha*, 30) of the formative growth of the Medulla oblongata into the Pons while transforming and the Cerebellum after transforming the "intrinsic entity value" (*Kshemya*, 7) into a "self-luminous entity" (*Svayam*, $12 = 7 + 5$) "perpetuating" (*Ambalika*, 82) the value of the five bodies without the ninth intellectual body by norming a muscular body.

iii. **Eleventh Intellectual Body**. The "eleventh intellectual body" (*Madhyasthahinta*, 900,000) is the "muscular body" (*Madhyasthahinta*, 900,000). The muscular body is the "Association Cortex" (*Madhyasthahinta*, 900,000) formed by the Medulla Oblongata for the "disintermediation" (*Madhyasthahinta*, 900,000) of the subconscious "temporal lobe's" (*Makuli*, 46) "polluting" (*Ardhavakirna*, 189) consciousness. The latter forms after "seeing" (*Pashyana*, 900,000) the Medulla oblongata "acculture" (*Utsamskara*, 900,000) the intrinsic entity value into the "parietal lobe's" (*Smriti*, 35) "cleansing" (*Pratyach*, 53) value.

iv. **Twelfth Intellectual Body**. The "twelfth intellectual body" (*Nirjara*, 1000) is the "material body" (*Madhyasthata*, 1000). The material body is the "Somatosensory Cortex" (*Madhyasthata*, 1000) normed by the Medulla Oblongata for the "intermediation" (*Madhyasthata*, 1000) of the consciousness "parietal lobe's" (*Smriti*, 35) "cleansing" (*Pratyach*, 53) value. The parietal lobe cleanses the time-invariant "value" (*Mulya*, $180 = 60 * 3$) of the "temporal lobe's" (*Makuli*, 46) "polluting consciousness" (*Duratma*, 60) by organizing the latter as a "memory" (*Smriti*, 35) for activating the "path of copying" (*Gana Marga*, $25 = 60 - 35$) with the memory's "variable touch" (*Sridhara*, 25) within each copy. Therefore, each copy hears a different "sound" (*Naad*, 257) of the memory mediated by the central fissure's "hibernating force" (*Devadata prana*, 80) of the "undesirable past" (*Anishta*, 38).

v. **Thirteenth Intellectual Body.** The "thirteenth intellectual body" (*Dharitri*, 146) is the "management body" (*Dharitri*, 146). The management body is the "Auditory Cortex" (*Dharitri*, 146) transformed by the Medulla Oblongata after forming it as the disintermediated muscular body and transforming while norming it as the intermediated material body, for the "mediation" (*Anunaya*, -7) of the "Visual Cortex's" (*Mokshayitri*, -13) "causing" (*Vijnana*, 47). The causing sequences the memory's copying into the reproductive "sequential reality" (*Bhavartha*, 40 = 47 - 7) managed by the feminine, "lateral fissure" (*Swasthani*, 40). The lateral fissure forms the masculine "central fissure" (*Surasa*, 38) as the "Somatosensory Cortex" (*Madhyasthata*, 1000). It norms the central fissure's "transformation sequence" (*Duschhaya ojas*, 146) as the "Auditory Cortex" (*Dharitri*, 146). It transforms its own "sequential reality" (*Bhavartha*, 40) into the Visual Cortex enjoying her "referential presence" (*Virodhita*, 108 = 146 – 38) within the transformation sequence.

vi. **Fourteenth Intellectual Body.** The "fourteenth intellectual body" (*Mokshayitri*, -13) is the "monetary body" (*Mokshayitri*, -13). The monetary body is the "Visual Cortex" (*Mokshayitri*, -13) formed by the "Midbrain" (*Rashi*, 13) in the form of the feminine, "lateral fissure" (*Swasthani*, 40) for monetizing the masculine self's memory. The lateral fissure visualizes the memory as the "infused present" (*Ab*, 1600 = 40 * 40) by reproducing her "sequential reality" (*Bhavartha*, 40). It makes the infused "gender-differentiated present" (*Ab*, 1600) the "start point" (*Ab*, 1600) for envisioning the "group-differentiated "future (*Anagata*, 0) by making the self "luminous" (*Rochisha*, 13) as the "Midbrain" (*Rashi*, 13).

The Midbrain is the "fifteenth intellectual body" (*Rashi*, 13). It is the "manufacturing body" (*Rashi*, 13) formed by the "brain" (*Mastishka*, 38) for manufacturing the "masculine self" (*Mein*, 1) in the form of the "brainstem" (*Mukhya*, 14 = 1 + 13) within the "midbrain"

(*Rashi*, 13) and the "feminine self" (*Tara*, 2) in the form of the parallel "right-brain" (*Mantra*, 16) within the "brainstem" (*Mukhya*, 14).

The Brain is the "sixteenth intellectual body" (*Mastishka*, 38). It is the "machinery body" (*Mastishka*, 38) normed by the masculine "central fissure" (*Surasa*, 38) after servicing its "masculinity" (*Lingam*, 53), i.e., the "multiplied sequential-effect" (*Laganem*, 53), as the "starting point" (*Prasthan*, 53) for trading the feminine lateral fissure's "femininity" (*Yoni*, 1000).

The Brainstem is the "seventeenth intellectual body" (*Mukhya*, 14). It is the "marketing body" (*Mukhya*, 14) that trades the masculinity diffused by the central fissure and markets "itself" (*Ekotrem*, 121) as the "masculine self" (*Mein*, 1) after servicing its femininity as the feminine "lateral fissure" (*Swasthani*, 40) within the "gender-differentiated present" (*Ab*, 1600).

The Right Brain is the "eighteenth intellectual body" (*Mantra*, 16). It is the "motivating body" (*Mantra*, 16) that services its femininity by transforming itself into the feminine "lateral fissure" (*Swasthani*, 40). It motivates the "nineteenth intellectual body" (*Sukshma sharira*, 306) to conceive it as the feminine self and be the "left brain" (*Sukshma sharira*, 306), enjoying its femininity by behaving like a "manipulating body" (*Sukshma sharira*, 306).

The "departed entity" (*Divangat*, 1600) services the "gender-differentiated present" (*Ab*, 1600) by incarnating as the twentieth, "primordial intellectual body" (*Sahadeva*, 816). The primordial intellectual body is the "mentor body" (*Sahadeva*, 816) that norms the "corpus callosum" (*Sahadeva*, 816), connecting the left brain with the right brain after first forming the latter and transforming it into the left brain. The left brain is the derived "logical reality" (*Yuktartha*, 6) of the right brain for differentiating the masculine "gender-differentiable class" (*Khuda*, 6) within the "undifferentiated present" (*Param*, 1600). The undifferentiated present is the "generation-differentiable class" (*Param*, 1600).

The "conscious fire" (*Agni*, 17) of the "being energy" (*Kali shakti*, 96) of a "living entity" (*Sushupti*, 96) within himself lets the departed entity differentiate the "group" (*Gana*, 387) formed from its "physical body" (*Sthula sharira*, 387) into the "species" (*Prajati*, 17) of different "kinds" (*Prakara*, 17). The living entity services the "group-differentiated present" (*SAUM*, 1600) by incarnating as the twenty-first "primeval intellectual body" (*Puryashtaka*, 10^{17}-1).

The primeval intellectual body norms the "premotor cortex" (*Puryashtaka*, 10^{17}-1). The premotor cortex is the "supreme self" (*Puryashtaka*, 10^{17}-1) "channeling" (*Savana*, 10^{17}-1) the departed entity's "personal reality" (*Kamartha*, 10^{17}-1) as an "organic" (*Chini-Chini*, 10^{17}-1) element. Its guider behavior transforms the living entity's "personal authority" (*Mimamsa*, 87) into an "inorganic" (*Shankha*, 87) element for "exchanging" (*Sachivaya*, 87) the living entity's "respiratory force" (*Sphurti*, 87) with the left-brain mediated "logical reality" (*Yuktartha*, 6) to manifest the "vision" (*Dharmasavarni*, $81 = 87 - 6$) of oneself as the "male child" (*Preya*, 81).

Chapter 5

Transcending the Causation Tensor:
Exchanging I AM

God's Past. Before incarnating as the fifth member, I was the twenty-first member, enjoying the "conscious consciousness" (*Prashantatma*, 16) of the sixteen members formed as a "double octave" (*Madhusudan*, 16). The double octave perpetuated its value as the "twenty-first member" (*Rupa*, 100,000) by forming the "twenty-second member" (*Prakara*, 17), norming the "homogenous genus" (*Prakara*, 17), as the "fire" (*Agni*, 17). It formed the "twenty-third member" (*Nayana*, 957) as the "force" (*Prapya*, 34) of the "twenty-fourth member" (*Sudhanvan*, 999) as the "perpetuator" (*Abhidheya*, 460) of the fire.

How "I AM" Form of God Formed as the Twenty-First Member

The twenty-first member was formed as a "combination with two" (*Dvaiyogya*, 10^{24}), i.e., with the twenty-second member norming the fire and the twenty-third member norming the force of the perpetuator, and therefore took the form of the "fire force" (*Rupa*, 100,000). The "combination" (*Samavaya*,

53) formed as the "twenty-fifth member" (*Suchaka*, 9000), the "domain" (*Suchaka*, 9000) of the "life" (*Zindagi*, 4) of the "twenty-sixth member" (*Varshakritya Taranga*, 1800), the "gravitational wave" (*Varshakritya Taranga*, 1800). The combination is the "human force" (*Manviya Karak*, 53).

The "human" (*Naran*, 275) formed as the "twenty-seventh member" (*Jyataranga*, 86), the "sentient wave" (*Jyataranga*, 86). The "twenty-eighth member" (*Samalekha*, 90), the "replication" (*Samalekha*, 90) formed as a "divine wave" (*Vataranga*, 36). The "twenty-ninth member" (*Sadakhya*, 9), the "culture" (*Sadakhya*, 9) formed as a "divine element" (*Divya*, 360). The "thirtieth member" (*Svabhav*, 8), the "essential nature" (*Svabhav*, 8) took the form of the "twenty-first member" (*Rupa*, 100,000). Therefore, "I" (*Svayam*, 12) became a "self-luminous entity" (*Svayam*, 12) servicing the "primordial consciousness" (*Adhyatma*, 12) of the "diverse forms" (*Shudra*, 1) of the first member, the "param daughter" (*Jiva*, 2), in the form of the "quantum photon" (*Svarochisha*, 12).

My goal as God was to illuminate you, "Mother Nature" (*Kudrat*, 8), born as the param daughter, as the "quantum light" (*Rochisha*, 13) of the "maternal primordial greeter" (*Sati-Parvati*, 16), the thirty-first member. The maternal primordial greeter is the "immanent wisdom" (*Sati-Parvati*, 16) of the "child primordial greeter" (*Madhusudan*, 16), the thirty-second member. The child primordial greeter is the grandpaternal "double octave" (*Madhusudan*, 16) emanating from the "primordial greeter" (*Srijak*, 16), the thirty-third member. The primordial greeter gifts the "essence" (*Sara*, 16) of the "pristine nature" (*Sara*, 16) that forms God as the "form" (*Rupa*, 100,000) of "fire force" (*Rupa*, 1000,000) immanent within the consequential "form force" (*Sarvodaya*, 150), i.e., sequence of forms differentiating every creation, creature, and creator.

The first "thirty members" (*Karana sharira*, 30) form a "feminine body" (*Karana sharira*, 30). They comprise a ten-member "family" (*Parivar*, 10), a ten-member "potential family" (*Shanideva*, 100) reproducing a ten-member family within each of the ten members, and a ten-member "twin family" (*Amrit*,

1000) reproducing a ten-member potential family within each of the ten members.

The thirty-first member is the "potential twin family" (*Sati-Parvati*, 16), producing the family, the potential family, and the twin family as the first three members within itself, reproducing them as the potential twin family, triple family, and potential triple family without itself, and repeating the "length power growth" (*Kula*, 12). The thirty-second member is the "triple family" (*Madhusudan*, 16). The thirty-third member is the "potential triple family" (*Srijak*, 16). The length power growth reproduces the preceding three members to form the succeeding nine members as their "triple copies" (*Vaishya*, 3). It transforms the "tenth member" (*Parivar*, 10), norming the "length power growth rate" (*Ijara*, 10), into the "family" (*Parivar*, 10).

The family comprises the preceding three members and the succeeding nine members within the tenth member. The tenth member forms the family through the "inflowing" (*Sahayaka*, 10,000) of the "extrinsic strong psychic linkages" (*Amrit*, 1000) from the preceding three members and the "outflowing" (*Karak*, 12) of the "intrinsic weak psychic linkages" (*Brihadbala*, 149) to the succeeding nine members. The inflowing is the "breadth power" (*Sahayaka*, 10,000) of a "vertical chord" (*Dvijya*, 10,000) of outflowing circular "form energy" (*Rupa shakti*, 9,000,000) from a "pillar" (*Sthuna*, 10,000) with three members. Outflowing is the "length power" (*Karak*, 12).

Outflowing leads to "self-similarity" (*Auta-ghatakecem-rajya*, 10^{18}) due to the "reproductive force" (*Chitta*, 100) of the "breadth power growth" (*Tulyatarka*, 10^{16}). The breadth power growth is the "inflowing" (*Sahayaka*, 10,000) of the ideal "4-D consciousness" (*Chaithanya*, 4) from the four directions. The "conditioning" (*Anukulana*, 10^{16}) of the "ideal consciousness" (*Chaithanya*, 4) leads to a parallel life "5-D consciousness" (*Vilatma*, 10^{16}) of the inflowing "gravitational system" (*Prajanan Pranali*, 10,000) the tenth member is reproducing as the "breadth power growth rate" (*Auta-ghatakecem-rajya*, $10^{18} = 10^{16} * 100$) for producing the "potential family" (*Shanideva*, 100) with his "gravitational force" (*Chitta*, 100). The breadth power growth rate is the "param odd value" (*Auta-ghatakecem-rajya*, 10^{18}) leading to a "semi-conjugacy" (*Auta-ghatakecem-rajya*, 10^{18}) of the "twin family" (*Amrit*, 1000) with the "primordial even value" (*Punravritta*, 5), i.e., the "repeating fractal" (*Punravritta*, 5).

The repeating fractal is the "hydrating force" (*Punravritta*, 5) of the "outflowing" (*Karak*, 12) "primordial consciousness" (*Adhyatma*, 12) of "Mother Nature" (*Kudrat*, 8) within a "sentient entity" (*Siddha*, 7) that makes a zero "subject" (*Sura*, 0) "God" (*Ishvara*, 5 = 12 - 7) enjoying the 4-D consciousness of the "diverse forms" (*Shudra*, 1 = 8 - 7). The breadth power growth rate forms a "Fibonacci sequence" (*Auta-ghatakecem-rajya*, 10^{18}) and transforms into an "absolute breadth power growth rate" (*Loka*, 9 x 10^{18}) with the nine members. The absolute breadth power growth rate is a "Fibonacci number" (*Loka*, 9 x 10^{18}, comprising the primordial even value and its effect, i.e., param odd value. It norms the "primordial universe" (*Loka*, 9 x 10^{18} of "objects" (*Padartha*, -3) formed from the "past reality" (*Padartha*, -3) without the "subject" (*Sura*, 0) who enjoys their diverse forms by identifying with each "diverse form" (*Shudra*, 1) as "I AM" (*Rupa*, 100,000). Consequently, Fibonacci number transforms into the sum of the previous two numbers in the sequence where the first number is zero and the second is one.

The "twin family" (*Amrit*, 1000) inflows in the form of the extrinsic strong psychic linkages to form the "family" (*Parivar*, 10). The "potential family" (*Shanideva*, 100) outflows in the form of the "intrinsic weak psychic linkages" (*Brihadbala*, 149 = 100 + 10 + 9 + 30) to "repeat" (*Avritti*, 25) the "family" (*Parivar*, 10), the "nine members" (*Maha Vibhu*, 14), and the "thirty members" (*Karana Sharira*, 30). Thus, the nine members become the "self-repeating" (*Maha Vibhu*, 14) "radial power growth" (*Maha Vibhu*, 14), norming the "thermodynamic limit" (*Maha Vibhu*, 14) of "your potential" (*Maha Vibhu*, 14) due to the "electromagnetic reaction" (*Maha Vibhu*, 14) with "my potential" (*Maruti*, 78).

"My potential" (*Maruti*, 78 = 100 - 16 - 16) reproduces

the "radial power growth rate" (*Rachanatmak*, 16), norming the "thermodynamism" (*Rachanatmak*, 16) in the form of the "gravitoelectromagnetism" (*Rachanatmak*, 16), due to the "gravitational force" (*Chitta*, 100) of the "potential family" (*Shanideva*, 100). The "radial power" (*Ijaradarita*, 13) is the "proportionate aftereffect" (*Ijaradarita*, 13) of the "repeating" (*Pruthaktva*, 13 = the family as the tenth member + nine members + three members) of the family, the nine members, and the thirty members as the "three members" (*Amrit*, 1000) that norm the "twin family" (*Amrit*, 1000).

The "radial" (*Trijya*, 140) forms as a "repeater" (*Anantariti*, 140), norming the "twin radius" (*Trijya*, 140) of the "repeat" (*Avritti*, 25). The "repeat" (*Avritti*, 25) forms the "radian" (*Sridhara*, 25), norming the "triple radius" (*Sridhara*, 25) of the "horizontal order" (*Punravritta*, 5), norming the "repeating fractal" (*Punravritta*, 5), "repeated" (*Punravritta*, 5) by the "three members" (*Amrit*, 1000).

The family enjoys the potential of the "twin copy" (*Artta*, 22/7) of the "twin radius" (*Trijya*, 14) within each member and the "triple copy" (*Vaishya*, 3) of the "triple radius" (*Sridhara*, 25) without each member. The member that becomes the family forms as a "Heptad repeat" (*Punravritta*, 5), i.e., "repeating fractal" (*Punravritta*, 5), of the "twin copy" (*Artta*, 22/7) within the "triple copy" (*Vaishya*, 3). The family reproduces its "length power growth rate" (*Ijara*, 10) for the "length power growth" (*Kula*, 9), i.e., "entanglement" (*Kula*, 9) of the "nine members" (*Maha Vibhu*, 14 = 9 + 5) within its "repeating fractal" (*Punravritta*, 5).

The "reproduction" (*Prajanan*, 285) produces a "disproportionate growth" (*Unha*, 280) in the "repeating fractal" (*Punravritta*, 5) and transforms the repeating fractal into the "self-repeating" (*Maha Vibhu*, 14 = 5 + 9) "radial power growth" (*Maha Vibhu*, 14) due to the "length power growth" (*Kula*, 9). The "disproportionate radial power growth" (*Avigata*, 1,999 = 1000 + 1000 -1) repeats the "twin family" (*Amrit*, 1000) without the "member" (*Sadasya*, 1) that forms the "family" (*Parivar*, 10). The "proportionate length power growth" (*Pradyumna*, 60) reproduces the "member" (*Sadasya*, 1) that forms the "family"

(*Parivar*, 10) as an "octave of copies" (*Sargam*, 60) within the member that forms the "potential family" (*Shanideva*, 100), thus generating a "disproportionate radial power growth rate" (*Vitata*, 600 = 100 * 60/10).

The "octave of copies" (*Sargam*, 60) replicates the "proportionate length power growth rate" (*Shveta Vraha*, 69 = 60 + 9) within the "nine members" (*Maha Vibhu*, 14) for replicating the next "seven members" (*Upasampada*, 986) within the "twin family" (*Amrit*, 1000 = 986 + 14). Seven members define the "number of unique species forms within one unique entity group" (*Ekatma*, 986). The nine members norm the "param mitosis-effect" (*Maha Vibhu*, 14). The seven members norm the "param meiosis-effect" (*Upasampada*, 986).

The disproportionate radial power growth rate per unit of the proportionate length power growth rate transforms the seven members into a "power cone" (*Glani*, 67) "outpouring" (*Glani*, 67) without the "growth" (*Dishta*, 6) of the "six members" (*Shivadrishti*, 17) within the "nine members" (*Maha Vibhu*, 14). The six members norm the "mitosis" (*Shivadrishti*, 17). The "fifteen members" (*Param Shiva*, 15) norm the "meiosis" (*Param Shiva*, 15).

"I am" (*Rupa*, 100,000) the "power cascade" (*Rupa*, 100000) that forms the "power cone" (*Glani*, 67) with the force of the thermodynamic fire generated by the "six members" (*Shivadrishti*, 17) on their "genesis" (*Pratitya samutpada*, 10) without the "family" (*Parivar*, 10). The genesis is within the consciousness of the "four members" (*Jnana*, 19) that germinate me as a "cell" (*Hiranyagarbha*, 19) within the nine members.

How a Departed Entity Forms the Space for Transforming the Entity

There are three ways for a departed entity to form the space for transforming the entity he becomes with time.

First, transform the "potential" (*AUM*, 18) within you as an "entity" (*Hasti*, 24) to become a "breeder" (*Poshitri*, 9) of the "population" (*Abadi*, 18) that needs "spontaneous space" (*Basera*,

47) to "perpetuate" (*Ramakatha*, 9) your breeder "culture" (*Sadakhya*, 9). Next, become a "master breeder" (*Shvakridin*, 100) who "self-perpetuates" (*Amsha*, ½) "spontaneous time" (*Tatkala*, 50) to destroy the "breeding workculture" (*Maha Svapna*, 9) and transform each entity into a "master breeder" (*Shvakridin*, 100) within the "guiding force" (*Chitta*, 100) of your "cultural wisdom" (*Chitta*, 100) as a "grandfather entity" (*Gandakanayaka*, 10,000 = 100 * 100). Then, transform each entity into a grandfather entity and become a "male" (*Nar*, 100,000,000 = 10,000 * 10,000) with the "guider potential" (*Adana vijnana*, 100,000,000) of the "universe of principal guiders" (*Adana vijnana*, 100,000,000). Finally, active your guider potential without the universe of principal guiders to become the "grandpaternal ancestor" (*Purvaja*, $10^{16} = 10^8 * 10^8$) of the universe of principal guiders.

Second, norm the "potential" (*AUM*, 18) within the "population" (*Abadi*, 18) as the "programmed future of everybody" (*Shani*, 18) who incarnates as an "entity" (*Hasti*, 24) within the "growth" (*Khara*, 6) of the population over time. Next, "self-service" (*Svayamseva*, 2/3) the "planned present of everybody" (*Sambhasam*, 12 = 2/3 * 18) by incarnating as a "grandfather self-luminous entity" (*Shri Rama*, 12) for reincarnating everybody within the "luminous realm" (*Vansha*, 3785) of the "paternal self-luminous entity" (*Bhavanavasi*, 12), following your "authority" (*Bhavanavasi*, 12) as the "resident master" (*Bhavanavasi*, 12). Norm the luminous realm as the "universe of replications" (*Go Loka*, 3785). Then, reincarnate as the "paternal luminous" (*Bhaskara*, 13) by making yourself "luminous" (*Rochisha*, 13) as the "quantum light" (*Rochisha*, 13) of the luminous realm within the "performed past of everybody" (*Ravinandana*, -1 = 12 - 13) accultured into your "discordant energy" (*Asura shakti*, -1).

Finally, organize the "growth" (*Dishta*, 6) of your "light" (*Prabha*, 180) as a "sentient entity" (*Siddha*, 7) servicing the "quantum light" (*Rochisha*, 13 = 6 + 7) into a "four body system" (*Genda*, 16). The four body system comprises the "four bodies" (*Natini*, 11) with the "consciousness" (*Chaithanya*, 4) of the past, present, and future within the "profited time of everybody" (*Chhayaputra*, 14 = 11 + 4 - 1) producing the "discordant energy" (*Asura shakti*, -1).

A "fifth body" (*Homa*, 25 = 5^2 = 12 + 13) "orders" (*Suchan*, 2) the "system" (*Gabhira*, 12), norming the "developed space of everybody" (*Ghabira*, 12) within the "paternal luminous" (*Bhaskara*, 13); the system weakens the "psychic linkages" (*Mahika*, 47) among the four bodies that incarnate everybody as a "collective" (*Visarjya*, 47) in the "animal kingdom" (*Tiryaggati*, 47). A "sixth body" (*Taraka*, 36 = 6^2) "orders" (*Suchan*, 2) the "four body system" (*Genda*, 16) by transforming the "development timing of everybody" (*Kruralochana*, 52 = 36 + 16) within the "plant kingdom" (*Vanaspati*, 21 = 16 + 5) with the "culture factor" (*Sugriva*, 5).

A "seventh body" (*Charanam*, 49 = 7^2) "orders" (*Suchan*, 2) the "present growth" (*Sva*, 11) of the "five bodies" (*Sva*, 11) within the sixth body to form a variable "feminine body" (*Karana sharira*, 30) norming thirty bodies. An "eighth body" (*Paridhi*, 64 = 8^2) "orders" (*Suchan*, 2) the "growth" (*Dishta*, 6) of the "six bodies" (*Khara*, 6) by varying the sixth "causal body" (*Karana sharira*, 30) to form a constant "masculine body" (*Linga sharira*, 3), norming one-hundred eighty bodies within the "density" (*Sandra*, 60) of its "light spectrum" (*Ambara*, 180 = 60 * 3).

A "ninth body" (*Para*, 81 = 9^2) "reorders" (*Suchanak*, 2) the "six bodies" (*Khara*, 6) by dividing each into "thirty bodies" (*Karana sharira*, 30) and multiplying the "one-hundred eighty bodies" (*Linga sharira*, 3) by reproducing them as a reproductive "tenth body" (*Shanideva*, 100 = 10^2). The tenth body "reorders" (*Suchanak*, 2) "itself" (*Ekotrem*, 121) to become the "eleventh body" (*Ekotrem*, 121 = 11^2) that reproduces "eighty bodies" (*Mamaka*, 144) as the "twelfth body" (*Mamaka*, 144) within the "twenty bodies" (*Ekotrem*, 121) that form itself after producing "hundred bodies" (*Shanideva*, 100) with the tenth body but not the twenty bodies within itself.

Third, form the "potential" (*AUM*, 18) within the "space" (*Vaas*, 18,000) for transforming the "entity" (*Hasti*, 24) that reproduces the "light spectrum" (*Ambara*, 180) as the "entity time" (*Kaal*, 360 = 180 * 2) of her "illuminated growth" (*Pradyumna*, 60) as an "octave" (*Sargam*, 60) formed with the "population density" (*Sandra*, 60) of your "light spectrum" (*Ambara*, 180). "Reorder" (*Suchanak*, 2) the "octave" (*Sargam*, 60) into "sixty bodies" (*Niyatatma*, 169) that transform "yourself" (*Hetu*, 1) into the "thirteenth body" (*Niyatatma*, 169), norming your self-managing "wisdom soul" (*Niyatatma*, 169) within you as the grandpaternal "thirteen body system" (*Maha Siddha*, 169).

Reorder the thirteen body system by forming "myself" (*Insaniyat*, 1) as the "fourteenth body" (*Omkara*, 256) "ordering" (*Dishta*, 6) the "growth" (*Khara*, 6) of the six bodies within the seven bodies that conceive you as a "slice" (*Charanam*, 49) of me. The fourteenth body forms me with the "seventy bodies" (*Omkara*, 256) within the six bodies that transform the first "sixty-four bodies" (*Supuma*, 256) into a "sixty-four body system" (*Saraswati*, 256) without the seventh body. The seventh body "disorders" (*Pradarshan*, -8) the six bodies into the "thirty-two bodies" (*Puma*, 256) and transforms them into a "two-hundred fifty-six body system" (*Saraswata*, 256 = 32 * 8 * -1) within my "discordant energy" (*Asura shakti*, -1).

"Thirty-two bodies" (*Puma*, 256 = 30 + 226) comprise "thirty bodies" (*Karana sharira*, 30) that divide the six bodies into "one-hundred eighty bodies" (*Linga sharira*, 3) and "two bodies" (*Anuktasiddhi*, 226). Two bodies are the "implied impact" (*Anuktasiddhi*, 226) of the fourteenth body, "self-perpetuating" (*Udvaha*, ½) the "seventh body" (*Charanam*, 7) and its "impact" (*Siddhi*, 57) as the "fifteenth body" (*Sthavaravisha*, 57). Consequently, the seventh body transforms into the "sixteenth body" (*Kshatriya*, 0), i.e., the "copy" (*Nakala*, 0) of the seventh body by exchanging its "initial state" (*Sura*, 0). It lets me, the departed entity, incarnate as you, the human being.

5.1 Imperative for Transcending the Causation Tensor

One transcends the transformable causation by norming constant causation within the group and freeriding on the ascending "social benefit-cost ratio" of the group working as the guider agent for oneself as the principal guider. However, as the group seeks ascending "worker-social benefit-cost ratio" on the knowing traded from the principal guider, the thermodynamic energy that produces the group becomes a cause for the entropy of the one producing the group as one's replication. Therefore, a need emerges for exchanging the force of culture with the formative growth of the one acultured to trading the knowing.

5.1.1 Breeding Causation. Divided Body and Gender Differentiation Within Group

The six divisions of the body and the twenty-one divisions within each division descend the "respiratory force" (*Sphurti*, 87) of the living entity with their "divided wholesomewholeness" (*Durashravana*, $39 = 6 * 21 - 87$). Therefore, secondary, fundamental respiratory force, left after "breathing" (*Pranayama*, 37), due to the "continuity" (*Rajo Guna*, $50 = 87 - 37$) of the "past force" (*Ghatika*, 24) within one's "consciousness" (*Chaithanya*, 4), forms the twenty-second, "param intellectual body" (*Sphurti*, 87). The param intellectual body norms the "motor cortex" (*Sphurti*, 87).

Conscious "ascending respiratory force" (*Chakrika*, 90) empowers the living entity to transcend the left-brain mediated "logical reality" (*Yuktartha*, 6) for realizing the "goal" (*Lakshya*, 9) of the right-brain moderated "holistic reality" (*Pramanya vadartha*, $9 = 90 - 81$). Thus, one incarnates as a "female child" (*Shri*, 81) after "animate-to-inanimate exchange" (*Chakrika*, 90).

Para-conscious "descending respiratory force" (*Radha*, 43) leaves the "organic resonance" (*Jaivik Anunaad*, 43) of the departed entity in the form of the "primordial wholeness-effect" (*Radha*, 43) within the living entity's "consciousness" (*Chaithanya*, 4). It forms the "twenty-second physical body" (*Radha*, 43).

The "molecular force" (*Anutrijya*, 710) of the "leftover" (*Antimansha*, 710) "molecule" (*Kanalakshamsha*, $39 = 43 - 4$) shapes the living entity's "perceptual reality" (*Avagraha*, 710) of the "primordial feminine self" (*Parvati*, 10) as a "goalkeeper" (*Shiva*, 7). It forms the "twenty-second mental body" (*Anutrijya*, 710).

The "intermolecular force" (*Karandi*, 45) of the perception-guided "behaviors" (*Vyavahar*, 710) activates "crisis management" (*Mritya Ojas*, 45) within the living entity by making him "conscious

Transcending the Causation Tensor: Exchanging I AM   343

of dying" (*Mritya Ojas*, 45) like the departed entity. Therefore, the living entity becomes a "paternal creation" (*Khyati*, 45), conceiving the "paternal genes" (*Vilaga*, 90) within the "female body" (*Vastu*, 9). The paternal creation is the "twenty-second astral body" (*Khayti*, 45).

The "time force" (*Mohini*, 15) within "paternal genes" (*Vilaga*, 90 = 15 * 6) leads to the "discharging" (*Visarjana*, 15) of the living entity by the "luminous entity" (*Kali*, 96 = 90 + 6) as the "growth" (*Dishta*, 6). It leads to the "conception" (*Garbhadhana*, 15) of the "maternal genes" (*Bhairava*, 15) within the "yet-to-be-produced" (*Udanavata*, 12) "male body" (*Udanavata*, 12). The "subtracted body" (*Visarjana*, 15) of the living entity is the "twenty-second etheric body" (*Visarjana*, 15 = ½ * 12 + 9). It "self-perpetuates" (*Amsha*, ½) the "male body" (*Udanavata*, 12) within the "female body" (*Vastu*, 9). It is the "2000th I AM level" (*Visarjan*, 15) that makes the "existence" (*Maujudgi*, 2000) of the "self-luminous entity" (*Svayam*, 12) conditional on the masculinity within the female body reproducing her femininity within the male body in the form of the "X-chromosome" (*Parvati*, 10). While reproducing her femininity, the masculinity within the female body norms the femininity in the form of the "twin X-chromosome" (*Isha*, 12). The X-chromosome is the "primordial feminine self" (*Parvati*, 10).

The twin X-chromosome is the "feminine self-luminous entity" (*Isha*, 12) within a "twin Y-chromosome" (*Purusha*, 12). The twin Y-chromosome is an "ideal self-luminous entity" (*Purusha*, 12) reproducing the "genes" (*Somapa*, 180) of the departed entity within the "female body" (*Vastu*, 9) by reproducing itself as the "Y-chromosome" (*Shri Krishna*, 9) within the "male body" (*Udanavata*, 12) and its four "triple copies" (*Vaishya*, 3), each comprising a twin Y chromosome and a Y chromosome, as a "twin X chromosome" (*Isha*, 12 = 4 * 3). The first triple copy comprises the twin Y-chromosome without the male body and the Y chromosome within the male body. The other three triple copies comprise a triple copy of the first triple copy reproduced by the female body within the twin Y-chromosome. The twin Y-chromosome is neither within the male body nor the female body. Instead, it is within the "Sun" (*Ravi*, 21) as the twenty-second causal body.

The 2000th I AM level reproduces the twenty dimensions of mental body within the twenty-first-dimension. Therefore, it produces a six-unit growth as the twenty-second etheric body, reproducing the "Sun" (*Ravi*, 21 = 12 + 9) as the "male body" (*Udanavata*, 12) without the "female body" (*Vastu*, 9). The female body is the "breeding causation" (*Vastu*, 9) breeding the male body without it as a "male child" (*Preya*, 81 = 9 * 9) for "freeriding" (*Muftakhori*, 13) on its "conscious fire" (*Agni*, 17) to incarnate as a "female child" (*Shri*, 81).

5.1.2 Freeriding Causation. Divided Consciousness and Generation Differentiation Within Gender

The division of consciousness into 126 sub-divisions within and the six divisions without the male body norms the 132-unit "mass of consciousness" (*Prithvi*, 132 = 126 + 6) as the "twenty-third causal body" (*Prithvi*, 132). The male body reproduces it as the 132-unit reproductive, "gravitational mass" (*Mitra*, 132) of consciousness, norming the "twenty-third etheric body" (*Mitra*, 132).

The gravitational mass comprises the "gravitational quality" (*Guna*, 100) of not the "departed entity" (*Divangat*, 1600) but the deity "universe of departed entities" (*Deva Loka*, 1000). The latter "spread up" (*Vitata*, 600) their "programmed performing-effect" (*Vidya*, 600) within each "departed entity" (*Divangat*, 1600) as the "banking face" (*Jangha mukha*, 600) of the "deity realm" (*Deva Loka*, 1000). "Performing-effect" (*Bhuta Yajni*, 2) is programmed by the para deity "universe of living entities" (*Talatala Loka*, 1649 = 1000 + 615 + 17 + 17) inhabited by the departed entity as the para deity through his conscious "transformative planning" (*Niyojana*, 17) of the patrilineal genetic "programming" (*Mahakriya*, 615).

The "performing-effect" (*Bhuta Yajni*, 2 = 17 - 15) of the "patrilineal programming" (*Mahakriya*, 615 = 60 * 10 + 15) attenuates due to the conscious "formative programming" (*Pradyumna*, 60) by the "male child" (*Preya*, 81). The primordial feminine self conducts the normative "matrilineal planning" (*Abadha*, 10) for "breeding" (*Janana*, 15) the male child as the "paternal" (*Tejas*, 17). The reactive "matrilineal planning" (*Abadha*, 10) is activated by the proactive "patrilineal planning" (*Malanga*,

Transcending the Causation Tensor: Exchanging I AM   345

16 = 10 + 1 + 5) of the living entity as a "deity" (*Deva*, 1) and the departed entity as "God" (*Ishvara*, 5).

The "patrilineal programming" (*Mahakriya*, 615) is different from the normative "matrilineal programming" (*Dharana*, -8) of the deity "universe of the departed entity" (*Deva Loka*, 1000 = 1 + 999). The living entity inhabits it as a "deity" (*Deva*, 1) within the "division" (*Sudhanvan*, 999) of self into a departed entity. The living entity performs the normative programming with the para-conscious "metaphysical planning" (*Aayojana*, 17) by his "female body" (*Vastu*, 9) to "self-service" (*Svayamseva*, 2/3) the "male body" (*Udanavata*, 12) as her "planning value" (*Maha Nitya*, 8 = 12 * 2/3). Both para-conscious planning and conscious planning are immanent within the "para deity universe" (*Talatala Loka*, 1649), the "twenty-third astral body" (*Talatala Loka*, 1649).

The living entity makes the "para deity universe" (*Talatala Loka*, 1649) conscious of the supreme deity "universe of luminous entity" (*Nitala Loka*, 2785), inhabited by the "supreme deity" (*Bhagwan*, 4), Orion. "Orion" (*Pitra*, 4) is the "paternal jinn" (*Pitra*, 4), personifying the "paternal consciousness" (*Pitra*, 4) of the "departed entity" (*Divangat*, 1600) to become "luminous" (*Rochisha*, 13 = 4 + 9) within the "female body" (*Vastu*, 9). Orion is known as the "Gennie" (*Pitra*, 4) in the Western universe, "hungry ghost 餓鬼道" (*Pitra*, 4) in the Daoist universe, and "Param Brahma" (*Pitra*, 4) in the Vedic universe. The "supreme deity universe" (*Nitala Loka*, 2785) is the "twenty-third mental body" (*Nitala Loka*, 2785). Orion conceives sub-conscious "formative planning" (*Nissara*, 1/60) by making the "luminous entity" (*Kali*, 96) his "performing spirit" (*Nissara*, 1/60).

The "luminous entity" (*Kali*, 96) is the "performing spirit" (*Nissara*, 1/60) of the supra deity "Universe of Child God" (*Atala Loka*, 7×10^{180}), inhabited by the luminous entity as the supra deity, "Maha Brahma" (*Devi*, 3). The "Child God" (*Kana*, 186) is the "soliton crystal" (*Kana*, 186), norming the "intrinsic strong psychic linkages" (*Kana*, 186 = 96 + 81 + 9) of the" luminous entity" (*Kali*, 90) with the "male child" (*Preya*, 81) within the "female body" (*Vastu*, 9).

Maha Brahman is the "primeval feminine" (*Devi*, 3), embodying the departed entity as the "param son" (*Hanuman*, 3) within her "self-luminous body" (*Paramtattva*, 32) as a time-invariant "luminous entity" (*Kali*, 96 = 32 * 3). The "supra deity universe" (*Atala Loka*, 7 x 10^{180}) is the "twenty-third intellectual body" (*Atala Loka*, 7 x 10^{180}). The self-luminous body is the "twenty-third physical body" (*Paramtattva*, 32). It emits the sixth "multiplying physical body" (*Sadhya*, 32) as its "copy" (*Nakala*, 0) in "primordial oneness" (*Adi*, 32) with the "supra deity" (*Devi*, 3) for reproducing the latter as the "super deity" (*Jiva*, 2) within the "human being" (*Insan*, 1).

The super deity is the "param daughter" (*Jiva*, 2) who conceives the "param son" (*Hanuman*, 3) with the "thermodynamic force" (*Rodha*, 1) that incarnates the param son as the "human being" (*Insan*, 1) and reproduces the "thermodynamic growth" (*Jiva*, 2 = 1 + 1) of the "param daughter" (*Jiva*, 2) as the "freeriding causation" (*Jiva*, 2) within the paternal "human being" (*Insan*, 1).

The "generation differentiation" (*Chelagiri*, -16) follows the "freeriding causation" (*Jiva*, 2) "with time" (*Uttarmanasa*, -18). The param daughter becomes a multi-generation converging "maternal entity" (*Niyatamanasa*, -18). The human being becomes the generation diverging "paternal entity" (*Gunasa*, -18) within a parallel, bi-generational "greeter entity" (*Uttarmanasa*, -18).

5.1.3 Objectifying Causation. Divided Entity and Group Differentiation Within Geography

The "greeter entity" (*Uttarmanasa*, -18) is the "profiting spirit" (*Avajati*, -1/60) of the Super Deity "Universe of Paternal Godhead" (*Mahar Loka*, 27,000), inhabited by the "greeter entity" (*Uttarmanasa*, -18) as the primeval deity, "Maha Ishvara" (*Maheshwara*, 6 = 24 - 18) within an "entity" (*Hasti*, 24). The "entity" (*Hasti*, 24) makes the "param daughter" (*Jiva*, 2) "self-luminous" (*Svarochisha*, 12) as a "divided entity" (*Brahmin*, 2), conceiving the "universe" (*Brahman*, 2 = 1 + 1) within her "daughter womb" (*Uttanapada*, 18,000) for giving birth to the "param son" (*Hanuman*, 3) as the "human being" (*Insan*, 1) and taking birth as a "deity" (*Deva*, 1) guiding the human being by becoming a "holy spirit" (*Trinetra*, 1).

The human being is the "Paternal Godhead" (*Insan*, 1) seeking to be "God" (*Ishvara*, 5) by conceiving "Orion" (*Pitra*, 4) within him. He first conceives the "time" (*Kaal*, 360 = -18 * -20) as the "greeter entity" (*Uttarmanasa*, -18) by differentiating "-20 degrees" (*Avibhajya*, 13) as the "luminous" (*Rochisha*, 13). He "self-perpetuates" (*Amsha*, ½) the "degree" (*Amsha*, ½) by reproducing the time to conceive the "entity" (*Hasti*, 24) with the "past force" (*Ghatika*, 24) and "Orion" (*Pitra*, 4) with the entity's "performing spirit" (*Nissara*, 1/60) for "profiting" (*Mahasatta*, 378) from his matrilineal "past planning" (*Abadha*, 10).

The "Super Deity Universe" (*Mahar Loka*, 27,000) is the **twenty-fourth causal body**" (*Mahar Loka*, 27,000). It "self-perpetuates" (*Amsha*, ½) "time" (*Kaal*, 360) as a "universe" (*Brahman*, 2) of "white star" (*Yama*, 180 = 360 * ½).

The "universe" (*Brahman*, 2) comprises a unit of "intrinsic consciousness" (*Antaratma*, 10) of the "past planning" (*Abadha*, 10) and a unit of "extrinsic consciousness" (*Visata*, 10^{10}) of the "planning point" (*Tilaka*, 10^{10}). The planning point is the "parallel universe consciousness" (*Parjanyatma*, 10^{10}). The "parallel universe" (*Suvarna*, 189) is the "classification" (*Suvarna*, 189) of the entity into eighteen classes within nine groups. Nine groups differentiate each class by "gender" (*Linga*, 9) for producing thirty-six "gender-differentiated classes" (*Udu*, 36) within and thirty-six "gender-differentiating classes" (*Shevalohita*, 36) without each "entity group" (*Gana*, 387).

Overall, nine groups differentiate 648 "classes" (*Varna*, 689) within 41 classes of the tenth "entity group" (*Gana*, 387) after it forms 31 of its 72 classes as a feminine "book copy" (*Urja*, 31) of the "female body" (*Vastu*, 9) through "gamma decay" (*Urja*, 31) within the reproductive "sequential reality" (*Bhavartha*, 40). Next, the ninth entity group differentiates the gender of the 31 "feminine classes" (*Rani*, 7) by reproducing the eight natural "masculine copies" (*Vaishya*, 3) of the tenth entity group within the first seven entity groups for conceiving the eighth entity group as a "twin entity" (*Ghatika*, 24) within the "entity" (*Hasti*, 24) that norms the eleventh entity group.

The entity conceives the ten entity groups as the ten "group-differentiated classes" (*Sushumna*, 10) and the 31 feminine, "group-differentiable classes" (*Rani*, 7) as the thirty "generation-differentiated classes" (*Poshaka*, 30) within the thirty-first, "generation-differentiating class" (*Augrya*, 24). The eleventh entity group is the group-differentiating class. It is the "individual class" (*Hasti*, 24). The generation-differentiating class is the "collective class" (*Augrya*, 24), norming the "mother allele" (*Augrya*, 24) within the individual class, normed by the "grandmother allele" (*Hasti*, 24).

Each group-differentiated class is the "prism class" (*Sushumna*, 10), "centering" (*Sushumna*, 10) the thirty generation-differentiated classes with its "masculine copy" (*Vaishya*, 3) within an "octave" (*Sargam*, 60) of thirty masculine and thirty feminine copies. The octave is the ninth group-differentiated class that incubates the other eight in the form of two female "tandem copies" (*Brahman*, 2) and four male "rotation copies" (*Khara*, 6).

Thirty "masculine copies" (*Vaishya*, 3) incarnate ninety "persons" (*Vyakti*, 1). Thirty "feminine copies" (*Urja*, 31 = 3➔1) incarnate ninety "twin persons" (*Insan*, 1) within each "masculine copy" (*Vaishya*, 3) born "person" (*Vyakti*, 1). Each twin person is a "human being" (*Insan*, 1). Each person personifies the "masculine self" (*Mein*, 1) within a grandmaternal "Lepton" (*Mein*, 1) that incarnates the paternal human being.

Two "female copies" (*Brahman*, 2) incarnate four "potential daughters" (*Aap*, 1). Four "male copies" (*Khara*, 6) incarnate a "potential son" (*Manyu*, 19) within the "fifth potential daughter" (*Saranyu*, 5) who conceives the fourth potential son as a "maternal copy" (*Chita*, 1) after conceiving three potential sons as the "masculine copy" (*Vaishya*, 3) of the "fifth potential son" (*Narada*, 7). The fifth potential son is the "param paternal" (*Narada*, 7). The masculine copy is the "param son" (*Hanuman*, 3). Three potential sons norm the time-invariant "param grandson" (*Pautra*, 810). The "fifth potential daughter" (*Saranyu*, 5) is the "param maternal" (*Saranyu*, 5). The maternal copy is the "param daughter" (*Jiva*, 2) within another "maternal copy" (*Chitra*, 1) that produces two potential daughters as the present and the future of the third potential daughter produced as the "paternal copy" (*Nakala*, 0) within the first maternal copy. Three potential daughters norm the time-invariant "param granddaughter" (*Pautri*, 91).

The "param grandson" (*Pautra*, 810) and the "param granddaughter" (*Pautri*, 91) "incarnate" (*Virupaksha*, 900 = 9/16 * 1600) the "departed entity" (*Divangat*, 1600) as a "potential greeter" (*Upanishada*, 9/16) of the "human being" (*Insan*, 1 = 810 + 91 - 900). The potential greeter incarnates as an "octave of organisms" (*Virupaksha*, 900), forming the "twelfth entity group" (*Dakshinya*, 900). The twelfth entity group is the "group of ten potential guider agents" (*Dakshinya*, 900). It is the "cultural value" (*Dakshinya*, 900) of the "culture" (*Sadakhya*, 9) that makes the departed entity the human being's "guiding force" (*Chitta*, 100).

The octave of organisms is the **"twenty-fourth etheric body"** (*Virupaksha*, 900) that reproduces the "time" (*Kaal*, 360) as the "thirteenth entity group" (*Kaal*, 360) to form "thirty entities" (*Vipluta*, 720 = 30 * 24). Each entity is a maternal copy of the thirty-first entity that forms the "eleventh entity group" (*Ghatika*, 24) as the "group-differentiating class" (*Ghatika*, 24). The time is the **"twenty-fourth astral body"** (*Kaal*, 360). The eleventh entity group is the "past force" (*Ghatika*, 24).

"Thirty entities" (*Vipluta*, 720) as the "conditioning force" (*Vipluta*, 720) of the time's varying "conditioning" (*Anukulan*, 10^{16}) of the mind over the seven hundred twenty minutes, i.e., twelve hours, constitute the **"twenty-fourth mental body"** (*Vipluta*, 720). The conditioning is the "parallel life consciousness" (*Vilatma*, 10^{16}) of a grandpaternal "ancestor" (*Purvaja*, 10^{16}) within the potential greeter that makes the latter behave like a "devil" (*Sura*, 0) enjoying the "incoming light" (*Ambara*, 180 = 900 - 720) of the "octave of organisms" (*Virupaksha*, 900). The parallel life consciousness is the **"twenty-fourth intellectual body"** (*Vilatma*, 10^{16}). The grandpaternal "ancestor" (*Purvaja*, 10^{16}) is the "fourteenth entity group" (*Purvaja*, 10^{16}).

The "parallel life" (*Sighraga*, 1,700) is an "infinite exchange system" (*Sighraga*, 1,700) "moving quickly" (*Sighraga*, 1700) to exchange the Devil's "conscious planning" (*Niyojana*, 17) with the Satan's "grandmother consciousness" (*Adi Para Atma*, 100). The grandmother consciousness is the **"twenty-fourth physical body"** (*Adi Para Atma*, 100), norming "Anterior Insular Cortex" (*Adi Para Atma*, 100). It generates a "subconscious impulse" (*Avachetan*, 100) every "three hours" (*Prahar*, 100) for reproducing the "incoming light" (*Ambara*, 180) over the 180 minutes. The incoming light is the "fifteenth entity group" (*Ambara*, 180).

The subconscious impulse transcends the limits of the Devil's "conscious determination" (*Sahasra lingam*, 190) of the human being's behaviors, becoming, breathing, breeding, bragging, believing, and begging for blessing decisions. The "human being" (*Insan*, 1) reproduces the incoming light as the "conscious light" (*Kumara*, 10) of the "grandfather's" (*Dada*, 10) "past planning" (*Abadha*, 10) for "self-reproducing" (*Pretasharira*, 1/60) the "thirty entities" (*Vipluta*, 720) as the thirty "generation-differentiated classes" (*Poshaka*, 30) within the thirty-first, "generation-differentiating class" (*Augrya*, 24). After programming thirty generation-differentiated classes, the generation-differentiating class becomes the "group-differentiating class" (*Hasti*, 24) for differentiating fifteen entity groups within each "generation-differentiated class" (*Poshaka*, 30).

After performing as the group-differentiating class, the generation-differentiating class becomes the "gender-differentiating class" (*Shevetalohita*, 36). After profiting from the thirty generation-differentiated feminine classes, the gender-differentiating class transforms them into the thirty generation-differentiated masculine classes. After the development of the sixty "gender and generation differentiated classes" (*Sandra*, 60), the "gender-differentiating class" (*Shevetalohita*, 36) transforms into the twelve "gender and generation differentiating classes" (*Gabhira*, 12).

After the exchange of seventy-two "entity classes" (*Gana*, 387) into seventy-two "group classes" (*Khanda*, 72), the "gender and generation differentiating class" (*Gabhira*, 12) reproduces itself as the "generation-differentiating class" (*Augrya*, 24) within each group, i.e., "gender and generation differentiable class" (*Khanda*, 72) for "objectifying causation" (*Hariti*, 128) as a "child" (*Arcisa*, $128 = 24 * 72 - 1600$) without the "departed entity" (*Divangat*, 1600). Each "entity class" (*Gana*, 387) and "group class" (*Khanda*, 72) differentiate the "present value" (*Kalpa*, $476 = 387 + 72 + 24$) of each "entity's" (*Hasti*, 24) "geography" (*Ganarajya*, 476).

Geography is the "group, gender, and generation differentiating class" (*Ganarajya*, 476). The "entity class" (*Gana*, 387) is the "group, gender, and generation differentiated class" (*Gana*, 387) that forms a "group" (*Gana*, 387). The "class" (*Varna*, 689) is the "group, gender, and generation differentiable class" (*Varna*, 689).

5.1.4 Spreading Causation. Divided Geography and Class Differentiation Within Time

The "entity" (*Hasti*, 24) is the "development spirit" (*Svarochisha*, 12) of the Primeval Deity "Universe of Blissful Self" (*Tapo Loka*, 2848), inhabited by the entity. The "Blissful Self" (*Rupatita*, 78) is the "grossness" (*Kamarupitva*, 78) of the physical body the entity "invites" (*Bulava*, 78) by "greeting" (*Salaam*, 78) a "system-level consciousness" (*Gabiratma*, 78) to "breed" (*Prajanan*, 78) a "group of ten principal guiders" (*Prajanan*, 78), comprising the "flame family" (*Gauranga*, 11). Grossness is the **twenty-fifth physical body** (*Rupatita*, 78).

By making herself "self-luminous" (*Svarochisha*, 12) in the form of a "system" (*Gabhira*, 12). The system forms an "octave of copies" (*Sargam*, 60) over the "twenty-four hours" (*Dina*, 60). The entity becomes "conscious" (*Ojas*, 189) of the "para-conscious" (*Mitra*, $132 = 60 + 60 + 12$) "mental body" (*Manomaya sharira*, $381 = 189 + 132 + 60$) normed over the previous "forty-eight hours" (*Svahita*, 125) after becoming self-luminous over the preceding "twelve hours" (*Samaya*, 720) within the "twenty-fourth mental body" (*Vipluta*, 720). Therefore, the entity conceives an "illusionary" (*Suswani*, 89) "intellectual body" (*Sukshma sharira*, $306 = 189 + 132 - 4 - 11$) for producing the "consciousness" (*Chaithanya*, 4) of both "conscious" (*Ojas*, 189) and "para-conscious" (*Mitra*, 132) and reproducing its present growth as the "flame family" (*Gauranga*, 11).

The "physical body" (*Sthula Sharira*, $387 = 132 + 189 + 57 + 9$) produces the "entity" (*Hasti*, 24) with "time" (*Kaal*, 360). As a "subconscious body" (*Sthula sharira*, 387), it is "subconscious" (*Avachetan*, $100 = 57 + 43$) that the "para-conscious" (*Mitra*, 132) is "hiding" (*Avahittha*, 57) the "conscious one" (*Avahittha*, 57) within the "conscious" (*Ojas*, $189 = 132 - 57$) as a "thing"

(*Vastu*, 9) that forms the "unit entity" (*Radha*, 43). Therefore, it perpetuates the "preceding phase" (*Ghatika*, 24), that makes "one" (*Ek*, 1) a "unit entity" (*Radha*, 43) conscious within the "succeeding phase" (*Pramanya*, 67 = 43 + 24), as the "twin entity" (*Ghatika*, 24). Consequently, the unit entity illuminates the "receding phase" (*Bhaavi*, 11) that forms the para-conscious within the twin entity before the entity becomes conscious, as the "twin flame family" (*Bhaavi*, 11). The twin flame family is the **twenty-fifth intellectual body** (*Bhaavi*, 11).

"One" (*Ek*, 1) becomes a "unit entity" (*Radha*, 43 = 1 + 24 + 18) because the "preceding phase" (*Ghatika*, 24) makes one conscious of one's "potential" (*AUM*, 18) as an "entity" (*Hasti*, 24). The "succeeding phase" (*Pramanya*, 67) promotes a "normative development" (*Viparyaya*, 888) of that potential by making the unit entity a "potential group of ten potential principal guiders" (*Viparyaya*, 888), norming a "twin flame body" (*Viparyaya*, 888). The twin flame body is the "preceding combination" (*Viparyaya*, 888) that produces the "flame body" (*Nipparyaya*, 888), comprising a "potential group of ten potential guider agents" (*Nipparyaya*, 888), as a "succeeding combination" (*Nipparyaya*, 888). The flame body is the "twenty-fifth mental body" (*Viparyaya*, 888).

The flame body reproduces the "potential flame body" (*Paryaya*, 888), comprising a "potential group of ten guider agents" (*Paryaya*, 888), as a "receding combination" (*Paryaya*, 888). The receding combination is the "winding value" (*Paryaya*, 888) of time. The receding combination perpetuates the "potential twin-flame body" (*Saundarya*, 888), comprising a "potential group of ten principal guiders" (*Saundarya*, 888) as a "pending combination" (*Saundarya*, 888).

The "potential twin-flame body" (*Saundarya*, 888) is "artificial" (*Saundarya*, 888). Its "beauty" (*Saundarya*, 888) attracts one's "life experience" (*Anubhuti*, 7000) as a "conscious entity" (*Siddha*, 7) producing "conscious energy" (*Amrit*, 1000) to be the "potential twin flame family" (*Samudaya*, 7000). The potential twin flame family is the purple "spiritual flame" (*Aniruddha*, 7000). By reproducing the "conscious energy" (*Amrit*, 1000) within the potential twin flame family, one becomes the "potential flame family" (*Sata*, 8000 = 7000 + 1000). The potential flame family is the green "human flame" (*Kapila Kumara*, 8000). The human flame is the **twenty-fifth astral body** (*Kapila Kumara*, 8000).

By activating one's life experience, the human flame promotes "freedom" (*Mukti*, 17) from both the "traded consciousness" (*Varnatma*, 6) of the "class" (*Varna*, 689) inferred from "everyone's life experience" (*Bhaavi*, 11) during the succeeding "imaginary phase" (*Pramanya*, 67) as well as the "countertraded para-consciousness" (*Vijnanatma*, 89) of the "science" (*Vijnana*, 47) offered during the preceding "illusionary preceding phase" (*Ghatika*, 24) from the receding "real phase" (*Bhaavi*, 11). Thus, the human flame lets the "human being" (*Insan*, 1) transcend the limitations of both the "worker deity" (*Shudra*, 1) working to form the life experience in the astral realm after imagining it in the mental realm and the "knower deity" (*Brahman*, 2). The knower deity transforms the infused "experience" (*Anubhav*, 9000) of the "life" (*Zindagi*, 4) lived in the physical realm into an illusion of knowing in the intellectual realm. After transcending, the human being becomes the "manifestor" (*Sadhana*, 368) of the "divine consciousness" (*Gharmatma*, 10^{29}) and the divine consciousness the normalizing "geography force" (*Sundari*, 10^{29}). The geography force is the **twenty-fifth etheric body** (*Sundari*, 10^{29}).

"Everyone's life experience" (*Bhaavi*, 11) is the "twin flame family" (*Bhaavi*, 11). One infers the class by imagining everyone's life experience for "compensating" (*Khara*, 6) one's intended "life experience" (*Anubhuti*, 7000) and lives life as a human being by personifying that class as a "person" (*Vyakti*, 1). The person is the "para class" (*Vyakti*, 1) that personifies a class within a human being. One offers the science by forming an illusion of one's intended "life experience" (*Anubhuti*, 7000) based on the realized "experience" (*Anubhav*, 9000) during the "receding phase" (*Bhaavi*, 11) shaped by "everyone's life experience" (*Bhaavi*, 11). One "countertrades" (*Mati*, 89) the para-consciousness of the consciousness one has about one's "performed experience" (*Anubhav*, 9000 = 7000 + 2000) guided by one's para-conscious "planned experience" (*Anubhuti*, 7000) within one's conscious "programmed experience" (*Tajaurba*, 2000 = 1000 * 2) formed by reproducing one's "conscious energy"

(*Amrit*, 1000). The countertraded consciousness is the "scientific consciousness" (*Vijnanatma*, 89). It is the **"twenty-fifth causal body"** (*Vijnanatma*, 89).

The scientific consciousness forms within "class differentiation" (*Sabalasvas*, 89) as an "ideology" (*Vaicariki*, 89). The ideology promotes "responsiveness" (*Ladhima*, 89) to the "divided geography" (*Rodha*, 1) by adding an "ideological person" (*Vyakti*, 1) and multiplying the person's "thermodynamic-effect" (*Rodha*, 1), thus "spreading causation" (*Brahli*, 89) by norming a "diverse present" (*Brahli*, 89).

5.1.5 Compounding Causation. Divided Time and Absolute Differentiation Within Space

The "person" (*Vyakti*, 1) is the "exchange spirit" (*Gunatipat*, 1/5) of the Param Deity "Universe of the Immortal" (*Svarga Loka*, 2,388), inhabited by the person. The "Immortal" (*Amara*, 90,000) is the "reproducible point" (*Amara*, 90,000) of the "present life force" (*Prakarana vadartha*, 90,000). The "present life" (*Purejata*, 180) generates the "gamma force" (*Anakala*, 500 = 90,000/180) as its "time value" (*Pulastya*, 500). The "time" (*Kaal*, 360) generates the "present group" (*Kendrabhi mukha*, 140 = 500 - 360) as its "time value" (*Pulastya*, 500). The present group generates the "reproductive reality" (*Bhavartha*, 40 = 180 - 140) as its "value" (*Mulya*, 180).

The "gamma force" (*Anakala*, 500) is the "divided time" (*Anakala*, 500) the "present life" (*Purejata*, 180) forms by "self-perpetuating" (*Udvaha*, ½) the "potential flame family" (*Ekavali*, 8000) as a "starship" (*Yana*, 4000 = 8000 * 1/2) over the "period" (*Kalaptita*, 4320 = 4000 + 590 − 180) one lives. A starship is a "fourteen-sided polygon" (*Chaudahbhuj*, 4,000 = 2000 * 2) the "astral realm" (*Yaanshala*, 14) forms by reproducing one's "living experience" (*Tajaurba*, 2000). Consequently, the potential flame family personifies one's living experience as its "illusionary past" (*Tippa*, 2000) and services that as a person in the form of its feeble "trading reality" (*Glana*, 2000).

The "Param Deity Universe" (*Svarga Loka*, 2388) exchanges the "trading reality" (*Glana*, 2000) one lives in the "mental realm"

Transcending the Causation Tensor: Exchanging I AM 355

(*Mansika*, 6666) of the departed entity before incarnation with one's "birth zodiac" (*Kundali*, 388 = 2388 - 2000) as a "creature" (*Purusha*, 12). The Param Deity Universe is the **twenty-sixth causal body** (*Svarga Loka*, 2388).

The birth zodiac is the "horoscope" (*Kundali*, 388) a "creature" (*Purusha*, 12) reproduces as the "personified natural reality" (*Dvyartha*, 400 = 388 + 12), norming the "supernatural reality" (*Dvyartha*, 400). The supernatural reality transforms the past "primordial reality" (*Evakara vadartha*, -3) into the "purifying force" (*Apamarjana*, 1200 = 400 * -300 * -1) of one's "life rhythm" (*Layanalika*, 1200) as a "discordant entity" (*Asura*, -1 = 12 - 13) behaving like a "creature" (*Purusha*, 12) after becoming "luminous" (*Rochisha*, 13) as a "zodiac entity" (*Rashi*, 13). The birth zodiac is the **twenty-sixth etheric body** (*Kundali*, 388).

As an ideal-self-luminous entity, the "creature" (*Purusha*, 12) reproduces the "purifying force" (*Apamarjana*, 1200) as his "guider force" (*Chitta*, 100 = 1200/ 12). The "person" (*Vyakti*, 1) reproduces the "guider force" (*Chitta*, 100) as the "primeval reality" (*Omkara vadartha*, -1) of the discordant entity. Thus, the "past reality" (*Evakara vadartha*, -3) of the living entity differentiates into the "infinite reality" (*Omkara vadartha*, -3) of the "luminous entity" (*Kali*, 96) behaving like a "discordant entity" (*Asura*, -1) before becoming "maternal luminous" (*Maha Kali*, 13).

In this way, our "primordial reality" (*Evakara vadartha*, -3) as a "living entity" (*Sushupti*, 96 = 100 - 3 - 1) differentiates into the "primeval reality" (*Omkara vadartha*, -1) of the "luminous entity" (*Kali*, 96) within the mediation of the "infinite guider-effects" (*Atri*, 274 = 145 * 2 - 12 - 3 - 1). The infinite guider-effects, differentiated in time and space, emanate without the "self-luminous entity" (*Svayam*, 12) as the "infinite causal bodies" (*Guna sharira*, 145). The "primordial reality" (*Evakara vadartha*, -3) is the **twenty-sixth astral body** (*Evakara vadartha*, -3). The "primeval reality" (*Omkara vadartha*, -1) is the **twenty-sixth mental body** (*Omkara vadartha*, -3).

As an "exchange spirit" (*Gunatipat*, 1/5) a "person" (*Vyakti*, 1 = 1/5 * 5) perpetuates the "entropy value" (*Sarvanasha*, 5) of the "departed entity" (*Divangat*, 1600) for the "accusation" (*Asamjvala*, 1/5) of the "maternal consciousness" (*Chaithanya*, 4 = 1/5 * 20) of the conscious "sister spirit" (*Ruah*, 20) of the planned "scientific consciousness" (*Vijnanatma*, 89) as a "super theory" (*Vijnanatma*, 89). The "maternal consciousness" (*Chathanya*, 4) trades the "force" (*Prapya*, 34) of the super theory for "breathing out" (*Udana*, 55 = 89 − 34) the "perpetuating value" (*Saranyu*, 5) of the "present growth" (*Sva*, 11 = 55/5) of its "theory force" (*Maha dasha*, 0) within the "twin flame family" (*Bhaavi*, 11) as a "theory" (*Niyama*, 127 = 89 + 34 + 4). The theory is the **twenty-sixth intellectual body** (*Niyama*, 127).

A "theory" (*Niyama*, 127) lets a "twin flame" (*Suvira*, 1649 = 1600 + 9 + 40) make the "human being" (*Insan*, 1) a "member" (*Sadasya*, 1) of her "family" (*Parivar*, 10) to help him fulfill the "goal" (*Lakshya*, 9) of the present life planned by the "departed entity" (*Divangat*, 1600) without reproducing the "reproductive reality" (*Bhavartha*, 40). The "theory" (*Niyama*, 127 = 12 ➔ 7) transforms the "creature" (*Purusha*, 12) into the "goalkeeper" (*Shiva*, 7). The goalkeeper is the "param deity" (*Shiva*, 7). The param deity forms the "twin flame" (*Suvira*, 1649) as the "twin" (*Hridayesha*, 121) of the "human being" (*Insan*, 1). The twin "complements" (*Upanishada*, 9/16) the "goal" (*Lakshya*, 9) by transforming the "param deity" (*Shiva*, 7) into the "maternal primordial greeter" (*Sati-Parvati*, 16).

The maternal primordial greeter "supplements" (*Titimma*, 10^{19}) the human being's "knowing" (*Jnana*, 19) with the "mass consciousness" (*Satma*, 10) of herself as the "kin" (*Parijan*, 90,000), the "reproducible point" (*Amara*, 90,000). The supplement is the "para-conscious idealization" (*Adarshikaran*, 10^{19}), the **twenty-sixth physical body** (*Adarshikaran*, 10^{19}). It is the "feeding recombination" (*Asvaryogya*, 10^{19}) of the "person" (*Vyakti*, 1) within the "universe" (*Brahman*, 2) the "person" (*Vyakti*, 1) personifies as a "twin" (*Hridayesha*, 121). The "universe" (*Brahman*, 2 = 1 + 1) embodies both the param deity within the goal and the maternal primordial greeter without the goal as the "causation" (*Hetu*, 1) and the "causation consciousness" (*Pratyagatma*, 1) of the goal. Thus, the "maternal primordial greeter" (*Sati-Parvati*, 16) is "compounding causation" (*Sati-Parvati*, 16) for producing the "universe" (*Brahman*, 2) by transforming the human being's "potential" (*AUM*, 18) with her "conscious energy" (*Amrit*, 1000) of her "past life" (*Uttanapada*, 18000) as a "departed entity" (*Divangat*, 1600) into the "space" (*Vaas*, 18000) for the "tree of creation" (*Uttanapada*, 18000).

5.1.6 Espousing Causation. Divided Space and Speciation within Goal

The maternal primordial greeter is the "servicing spirit" (*Ditthigata*, 900) of the Primordial Deity "Universe of theory" (*Vitala loka*, 2600), inhabited by the maternal primordial greeter. The servicing spirit is the "absolute theory" (*Ditthigata*, 900) that divides the "space" (*Vaas*, 18,000) by forming a "loop" (*Pasha*, 900) of "quantum time" (*Pasha*, 900) with her "sister spirit" (*Ruah*, 20 = 18000/900). The sister spirit is the "primordial-primordial greeter" (*Maha Durga*, 16). She behaves like a "twin flame" (*Suvira*, 1649) for conceiving the "param deity" (*Shiva*, 7) as an "illusionary animate entity" (*Arpita*, 49). After living life as a "departed entity" (*Divangat*, 1600), the param deity reproduces himself as a "human being" (*Insan*, 1) through "oneness of action" (*Karma yoga*, 49) with the "past force" (*Ghatika*, 24). The "primordial deity universe" (*Vitala loka*, 2600) is the **"twenty-seventh causal body"** (*Vitala loka*, 2600).

The twin flame services an inactive "antagonist consciousness" (*Nishkriyatma*, 140) of her "life not lived" (*Svahita*, 125) yet for activating the "protagonist cause" (*Karana*, 18) with the "antagonist effects" (*Bhairava*, 15) of the human being's past life conscious "protagonist behaviors" (*Cheerna*, 90,000). The human being services an active "protagonist consciousness" (*Sakriyatma*, 40) of his "life already lived" (*Jivan*, 3800) for inactivating the twin flame's future life para-conscious "antagonist behaviors" (*Manoratha*, 100). As a person, the maternal primordial greeter services an activating "deuteragonist consciousness" (*Sakriya-Nishkriyatma*, 13) of her "living life" (*Zindagi*, 4). Her present life subconscious "deuteragonist behaviors" (*Pusalattu*, 97) give "direction" (*Deshanu*, 97) for "resolving" (*Nishchaya*, 10) the "inner conflict" (*Mrig Trishna*, 308) between the protagonist and the antagonist behaviors.

Consequently, the twin flame becomes a "leader" (*Neta*, 0) conscious of the "livable life" (*Svahita*, 125) and services the positive "leadership energy" (*Nayakatva shakti*, 90,000) for making the human being "immortal" (*Amara*, 90,000) like the maternal primordial greeter. The human being becomes a "follower" (*Chela*, 24) conscious of the "lived life" (*Jivan*, 3800) and services the negative "followership energy" (*Pasaka*, -9) for making the maternal primordial greeter the "meeting point" (*Sakshatkara*, 8000) of "living life" (*Zindagi*, 4) joyfully and its joyful "living experience" (*Tajaurba*, 2000) as a "living being" (*Satta*, 2000). The maternal primordial greeter becomes an "entrepreneur" (*Prabhu*, 1600) conscious of the "joyful life" (*Prabhasa*, 4) as a "being" (*Nitya*, 126,000,000) and services the compensating "entrepreneurship energy" (*Yanashakti*, -10,000 = 90,000/-9) for making the "flame" (*Parshnisamasta*, 32) within the "twin" (*Hridayesha*, 121) the flame within the "human" (*Naran*, 275).

The being is the "resolving community" (*Prerita mandala*, 126,000,000) "resolving" (*Nishchaya*, 10) the "inner conflict" (*Mrig Trishna*, 308) between the protagonist and the antagonist behaviors. The "maternal primordial greeter" (*Sati-Parvati*, 16) becomes a livable "community" (*Samvarna*, 1964 = 19 → 64) after transforming into "energy" (*Shakti*, 19 = 16 + 3) with a "time-invariant" (*Nirbija*, 800) "triple copy" (*Vaishya*, 3) and norming the "one with three ones" (*Yogihridaya*, 64 = 16 * 4) as her "energy fragment" (*Yogihridaya*, 64) after forming the "one" (*Ek*, 1) as her "octave of effects" (*Ek*, 1). The "livable community" (*Samvarna*, 1964) is the **"twenty-seventh etheric body"** (*Samvarna*, 1964).

The livable community services the "communion energy" (*Samvarna*, 1964) for becoming a "living community" (*Sangha*, -9 x 10^7). The living community reproduces the "three ones" (*Hanuman*, 3) as the negative "followership energy" (*Pasaka*, -9) of the consequential seven within the "mass consciousness" (*Satma*, 10) of the sequential ten. The sequential ten form by reproducing the three ones without the one. The "living community" (*Sangha*, -9 x 10^7) becomes the "human being" (*Insan*, 1) formed with the "octave of effects" (*Ek*, 1) of the "eight" (*Aath*, 8) that form the human being in the form of a "double copy" (*Chitra*, 1). The "copy" (*Nakala*, 0) within the double copy is the twin flame as the "leader" (*Neta*, 0). The "living community" (*Sangha*, -9 x 10^7) is the **"twenty-seventh astral body"** (*Sangha*, -9 x 10^7).

The "leader" (*Neta*, 0) is the "lived community" (*Adamiyah*, -1) within the "human being" (*Insan*, 1). The "lived community" (*Adamiyah*, -1 = 8 − 9) comprises the "eight" (*Aath*, 8) that norm the human being's "essential nature" (*Svabhav*, 8) without the "nine" (*Nau*, 9) that reproduce the "three ones" (*Hanuman*, 3) without the one that is the human being. The human being is the being without the human. The human is the "cultural production" (*Anala*, 275) of the "being" (*Nitya*, 126,000,000). As a being, the "resolving community" (*Prerita mandala*, 126,000,000) is the **"twenty-seventh mental body"** (*Prerita mandala*, 126,000,000).

The "reproduction" (*Prajan*, 285) of the "nine within ten" (*Lakshya*, 9) transforms the "human" (*Naran*, 275) into the "human being" (*Insan*, 1). It ascends the "negative energy" (*Nakaratmak shakti*, 19) within the "twin flame" (*Suvira*, 1649), wishing for the "absolute equality" (*Nirguna*, -6) with the "human being" (*Insan*, 1) after having lived a life of the "potential three ones" (*Saguna*, 20). The potential three ones are the "positive energy" (*Sakaratmak shakti*, 19) within the "one" (*Ek*, 1) that forms the human being by transforming the "octave of effects" (*Ek*, 1) and norms the "maternal primordial greeter" (*Sati-Parvati*, 16) as the "greeter essence" (*Sara*, 16) of the octave of effects and the "three ones" (*Hanuman*, 3) as the time-invariant "triple copy" (*Vaishya*, 3) of the greeter essence.

The human being compensates for the negative "masculine energy" (*Purani shakti*, 91) traded from the twin flame with the positive "feminine energy" (*Achitta shakti*, 396) countertraded from the maternal primordial greeter. The masculine energy is known as the "yang energy" (*Purani shakti*, 91), and the feminine energy as the "yin energy" (*Achitta shakti*, 396) within the Daoist universe. The former is the "animate energy" (*Purani shakti*, 91). The latter is the "inanimate energy" (*Achitta shakti*, 396). The former gets animated because it is the "bickering energy" (*Purani shakti*, 91) of "bickering" (*Ghatanuka*, -1) by the "lived community" (*Adamiyah*, -1) wishing to add the "human being" (*Insan*, 1) as the one that makes them "nine" (*Nau*, 9) capable of reproducing the one through their "speciation" (*Prajatikaran*, 88). The speciation is the **"twenty-seventh intellectual body"** (*Prajatikaran*, 88). It reproduces the "lived community" (*Adamiyah*, -1) within the "nine" (*Nau*, 9) as an "institution" (*Kundalini*, 8) that forms and norms the nine as a "culture" (*Sadakhya*, 9) without the two that transform the nine into a "replication" (*Samalekha*, 90) of the "ten" (*Dus*, 10). The ten comprise the eight that forms the "lived community" (*Adamiyah*, -1), the ninth that norms the "human being" (*Insan*, 1) as the "goal" (*Lakshya*, 9) of the eight, and the tenth that transforms the "maternal primordial greeter" (*Sati-Parvati*, 16) into the "primordial self" (*Parvati*, 10) of the human being.

As the nine that is replicating the primordial self as the ten, the human being activates and transforms the eight that norms the essential nature into the twin flame. The "replicating" (*Doharana*, 9) nine and the "replicable" (*Pratikriti*, 8) eight norm the "species" (*Prajati*, $17 = 9 + 8$) as the "replica" (*Pratirupa*, 17) of the "replicated" (*Pratikrit*, 180) ten. The replicated ten comprise the "primordial self" (*Parvati*, 10) as the ten and the "ten-fold growth" (*Vicitra mandala*, $170 = 17 * 10$) of the "species" (*Prajati*, 17) due to the "replication" (*Samalekha*, 90) of the ten. The ten-fold growth norms an "alive community" (*Vicitra mandala*, 170) that first replicates the ten within the nine, conditioned by the essential nature. It then replicates the ten within the eight for conditioning the twin flame's "essential nature" (*Svabhav*, 8) with the human being's "true nature" (*Vyaktitva*, 1). It is the **twenty-seventh physical body** (*Vicitra mandala*, 170), norming the "pre-vertebral ganglion" (*Vicitra mandala*, 170).

The "greeter essence" (*Sara*, 16) is the "replicator" (*Akala*, 16) of the ten-fold growth. It is the maternal primordial greeter's "pristine nature" (*Sara*, 16), "espousing causation" (*Sara*, 16).

5.1.7 Gene Stacking. Divided Causation and Genetics within Nature

The "twin flame" (*Suvira*, 1649) is the "trading spirit" (*Brahli*, 89) of the Devoted Deity "Universe of Ideals" (*Sutala Loka*, 23,125), inhabited by the twin flame. The "devoted deity universe" (*Sutala Loka*, 23,125) is the **twenty-eighth causal body** (*Sutala Loka*, 23,125).

As the one replicating the eight within oneself, the human being becomes the "ideal" (*Adarsha*, 1) for the replicated "primordial self" (*Parvati*, 10). The human being replicates the "ideal force" (*Dasha*,

1) as the one replicating and the "trading spirit" (*Brahli*, 89) as the eight replicable with the one replicating to norm a "replication" (*Samalekha*, 90 = 1 + 89). The ideal force is the "octave of effects" (*Ek*, 1) of the "Council of Five" (*Panchatattva*, 31). The Council of Five is the **"twenty-eighth etheric body"** (*Panchatattva*, 31). It is the conscious "masculine dimension" (*Beeja dharma*, 31) of the one replicating the para-conscious "feminine dimension" (*Alaukika dharma*, 208 = 20 → 8) of "Mother Nature" (*Kudrat*, 8) as the universally replicable "essential nature" (*Svabhav*, 8) led by the "primordial-primordial greeter" (*Maha Durga*, 16) as the "sister spirit" (*Ruah*, 20).

The "feminine dimension" (*Alaukika dharma*, 208) is the "entropy value of the thermodynamic energy" (*Prayujya*, 208). The "thermodynamic energy" (*Ashir*, 497 = 248 * 2 + 1) "self-perpetuates" (*Amsha*, ½) a "wheel of psychic forces" (*KRIM shakti chakra*, 248) within the one replicating. The wheel of psychic forces is the "reproductive reality" (*Bhavartha*, 40) of the thermodynamic energy within the "entropy value of the thermodynamic energy" (*Prayujya*, 208). Thus, it is the "growth value of the thermodynamic energy" (*Yauvana dasha*, 248). It is the **"twenty-eighth astral body"** (*KRIM shakti chakra*, 248).

The "thermodynamic energy" (*Ashir*, 497) produces "fire" (*Agni*, 17), "water" (*Apas*, 169), "air" (*Vayu*, 385), and "earth" (*Bhu*, 724) after consuming "ether" (*Prajanan*, 285), "divine" (*Divya*, 360), and "guider" (*Guru*, 100) and their "collective reality" (*Samghatartha*, 550 = 285 + 360 + 100 - 17 - 169 - 385 - 724). The human being produces the "collective reality" (*Samghatartha*, 550) after consuming the "sentient" (*Ojas*, 189) element as a time-invariant "compliment" (*Priyokti*, 567 = 189 * 3) to oneself for reproducing the "fire" (*Agni*, 17) for servicing the "species" (*Prajati*, 17 = 567 - 550) as his "effect" (*Prapya*, 34 = 17 + 17) being the "perpetuator" (*Abhidheya*, 460) of the "collective reality's" (*Samghatartha*, 550 = 460 + 90) "replication" (*Samalekha*, 90). The collective reality is the **"twenty-eighth mental body"** (*Upasarjana*, 550). The perpetuator is the **"twenty-eighth intellectual body"** (*Abhideya*, 460).

The collective reality is the "multiplied body" (*Upasarjana*, 550 = 210 + 170 * 2) of the "perpetuator effect" (*Prapya*, 34) of the "paternal community" (*Tejomandala*, 210) the human being forms as a "multiplying body" (*Mena*, 210), replicating the "tenfold growth" (*Vicitra mandala*, 170). The multiplying body is the **twenty-eighth physical body** (*Mena*, 210). It is a "wax body" (*Mena*, 210) that "melts" (*Pighal*, 17) when the human being activates the "paternal" (*Tejas*, 17) element within himself with the conscious "fire" (*Agni*, 17) of the "thermodynamic force" (*Rodha*, 1) within his "pristine nature" (*Sara*, 16).

The "genetics" (*Vilaga*, 90) is the "replication" (*Samalekha*, 90) of the paternal element as the "paternal gene" (*Vilaga*, 90), norming the "discordance" (*Vilaga*, 90) of the one replicating the "maternal" (*Matri*, 112) element. The "gene stacking" (*Vyaktitva*, 1) is the "true nature" (*Vyaktitva*, 1) of the one replicating that stacks up ten units of the "nine within ten" (*Lakshya*, 9) to form the "replication" (*Samalekha*, 90 = 9 * 10) of the "primordial self" (*Parvati*, 10). Finally, the "divided causation" (*Nityata*, 10^{19}) is the "perpetuity" (*Nityata*, 10^{19}) of the replicating one's "energy" (*Shakti*, 19) as the "primordial masculine self" (*Shri Krishna*, 10) of the "replicated universe" (*Ekavali*, 8000).

Therefore, the "replicated universe" (*Ekavali*, 8000) attracts the replicating human being as the "human flame" (*Kapila Kumara*, 8000) by reigniting the "sentiment" (*Raga*, 250) of the "past togetherness" (*Balapradhamani*, 333 = 250 + 73 + 10) within the "emotions" (*Hunduka*, 73) of the "present otherness" (*Audavita*, 5) of the one replicating the primordial masculine self as the "infinite replication" (*Shri Krishna*, 10). The replicated universe forms the "wisher community" (*Adamiyah*, -1), wishing the "blessing" (*Ashirwad*, 1000) of our "sentient energy" (*Amrit*, 1000) as a human being for its "future oneness" (*Ambhamsi*, 6000) with our "ego" (*Aham*, -1) by forming a "devoted deity universe" (*Sri Loka*, 6000).

<u>*5.1.8 Infinite Adding. Divided Nature and Gene Flow within Ego*</u>

The "human being" (*Insan*, 1) is the "investment spirit" (*Pratana*, 60,000) of the Devotee Deity "Universe of Goals" (*Sri Loka*, 6000), inhabited by the human being. The human being is the one replicating oneself as the "primordial masculine self" (*Shri Krishna*, 10) of the "universe" (*Brahman*, 2) replicated with one's "conscious fire" (*Agni*, 17). The "replicated light" (*Sanananda*, 3000) of the "replicating one" (*Doharana*, 9) manifests the "universe of goals" (*Sri Loka*, 6000) as a "realm of fire" (*Agni Loka*, 6,000) that makes the human being as the replicating one the "goal" (*Lakshya*, 9) of "replication" (*Samalekha*, 90). The "Devoted Deity Universe" (*Sri Loka*, 6000) is the **"twenty-ninth causal body"** (*Sri Loka*, 6000).

The universe of goals empowers the reproduced "ether-effect" (*Naad*, 257) to reproduce the "sound vibrations" (*Dhvani*, 257 = 60 + 197) immanent within the "octave of copies" (*Sargam*, 60). The "octave of copies" (*Sargam*, 60) produces the "ether-effect" (*Naad*, 257) with the "thermodynamic sensation" (*Vedana*, 197 = 187 + 10). The ether-effect reproduces the latter in the form of the innate "patrilineal reverberations" (*Panava*, 187) of the "primordial masculine self" (*Shri Krishna*, 10). The "patrilineal reverberations" (*Panava*, 187) of the "sound echo" (*Panava*, 187) produce the "matrilineal reverberations" (*Mridangapanava*, 70) of the "light echo" (*Mridangapanava*, 70) of the "past sequence" (*Ghatika*, 24) of "mutation" (*Prayaya*, 186). The mutation is "adding two to one" (*Prayaya*, 186) by forming a "universe" (*Brahman*, 2) without the "human being" (*Insan*, 1). The mutation is the **"twenty-ninth etheric body"** (*Prayaya*, 186) that forms the universe with the human being's "strong intrinsic psychic force" (*Kana*, 186).

The human being's "investment spirit" (*Pratana*, 60,000) works by "spreading" (*Pratana*, 60,000) the "strong intrinsic psychic force" (*Kana*, 186). It adds the "perpetuator force" (*Prapya*, 34) to the "fire" (*Agni*, 17). It produces the "fire-effect" (*Rupa*, 100,000) as the "multipliable body" (*Eroli*, 100,000). The multipliable body is the **"twenty-ninth astral body"** (*Eroli*, 100,000). It forms the replicable twin flame as a guided "maternal community" (*Prenkhana mandala*, 90,000) by reproducing the human being's "spreading path force" (*Dandanayaka*, 90,000) as an "entity" (*Hasti*, 24) after transforming the human being into a "self-luminous entity" (*Svayam*, 12). The spreading path force is immanent within the entity.

Thus, the multipliable body "silences" (*Parai*, 257) the "sound vibration" (*Dhvani*, 257) of the "ether-effect" (*Naad*, 257) by giving it a "form" (*Rupa*, 100,000). The form "harmonizes" (*Milaana*, 111) the ether-effect into the "quantum photon" (*Svarochisha*, 12) of the "entity" (*Hasti*, 24) "spreading path force" (*Dandanayaka*, 90,000) within the "twin entity" (*Ghatika*, 24) to become the "twin flame family" (*Bhaavi*, 24) that produces the entity as the multiplied "yourself" (*Hetu*, 1). Similarly, the twin entity "harmonizes" (*Milaana*, 111 = 11 ➔ 1) the maternal primordial greeter's "past force" (*Ghatika*, 24) in the form of the "quantum light" (*Rochisha*, 13) of the "flame family" (*Gauranga*, 11) that reproduces the form as the multiplying "herself" (*Pruthvi*, 1).

The maternal primordial greeter herself forms yourself by spreading the effect of the "infinite entities" (*Vasanatma*, -3) formed over time with varying "gravitational qualities" (*Guna*, 0) within "myself" (*Insaniyat*, 1). Myself becomes a "unique divine entity" (*Isha*, 12) endowed with a "twin X-chromosome" (*Isha*, 12), reproducing the constant "primordial self" (*Parvati*, 10) as the "X-chromosome" (*Parvati*, 10) with the time-variable "primordial masculine self" (*Shri Krishna*, 10) as the "Y-chromosome" (*Shri Krishna*, 10). A human being's "path" (*Marga*, 1) comprises the "gravitational qualities" (*Guna*, 0) of the diverse entities derived from the human being's "workforce system" (*Sara Kalpa*, 10^{10}). Therefore, each human being has a "different consciousness" (*Dushtatma*, 10) of the primordial self's "mass consciousness" (*Satma*, 10). The different consciousness of the space as a function of the time-varying "mood" (*Bhava*, 360) is not the "superior consciousness" (*Prashantatma*, 16) but "error in consciousness (*Dushtatma*, 10).

A human being may clear the error in consciousness by exploring "finite paths" (*Upadishta*, 890) within one's "sentient energy" (*Amrit*, 100), each with a "unique combination" (*Upadishta*, 890) of the "primordial self's" (*Parvati*, 10) "gravitational effect" (*Chitta*, 100). A unique combination combines the 360 "primordial paths" (*Adi Marga*, 360) of the "new moon" (*Chandra*, 82) as a "departed human entity" (*Manushya*, 82) seeking to be the "full moon" (*Soma*, 997) with the 53 "primeval paths" (*Anahata*, 53) of the "full moon" (*Soma*, 997 = 1 + 1 + 995) as a "human being" (*Insan*, 1) producing "yourself" (*Hetu*, 1) by investing "all-one-has" (*Harsiddhi*, 995). All-One-Has is the **twenty-ninth mental body**" (*Harsiddhi*, 995).

Each primeval path forms as a primordial path norming a

sidereal year within the ten "param paths" (*Sushumna*, 10) of the "primordial self" (*Parvati*, 10) that reproduces the first fifty primeval paths within the final three primeval paths as a "leap year" (*Vimaleshvara*, 700). The leap year is the **"twenty-ninth intellectual body"** (*Vimaleshvara*, 700). The first fifty primeval paths are the "absolute entropy force" (*Rohini*, 50) of the fifty-first primeval path that norms a "century" (*Sadi*, 100). The fifty-first primeval path is the "reproductive force" (*Chitta*, 100) of the fifty-second primeval path that norms a "millennium" (*Sahasrabda*, 1000).

The fifty-second primeval path is the "conscious energy" (*Amrit*, 1000) of the fifty-third primeval path that norms "ten millennia" (*Sahatra*, 10,000). The fifty-third primeval path is the "circular creation" (*Vartula*, 10,000) of the "circle" (*Valaya*, 100,000), norming the first param path. The tenth param path norms the "epistemological value" (*Jnana mimansiya*, 10^{14}) of the "wishing tree" (*Kalpavriksha*, 10^{14}), fulfilling the wish for exploring 530 "divergent paths" (*Padoccaya*, 530) before exploiting the 360 primordial paths. The wishing tree is the **"twenty-ninth physical body"** (*Kalpavriksha*, 10^{14}).

A human being conceives a "divergent path" (*Padoccaya*, 530 = 53 * [1+9]) with "infinite adding" (*Bhavishyath Chaturananaya*, 1) of oneself as the "replicating one" (*Doharana*, 9) after "trading" (*Samana*, 53) "human force" (*Manviya Karak*, 53) from his paternal "gene flow" (*Vikas*, 160). The paternal "gene flow" (*Vikas*, 160) is the "sister spirit" (*Ruah*, 20) within the "genetic drift" (*Achetan*, 140) serviced by the "wishing tree" (*Kalpavriksha*, 10^{14}). The genetic drift reproduces the "mass consciousness" (*Satma*, 10) of the divergent paths within your "book of life" (*Pustak*, 14) as "my" (*Mera*, 16) "byproduct" (*Dishta*, 6 = 20 - 14). The byproduct is the "divided nature" (*Dishta*, 6) within the "ego" (*Aham*, -1) of a "sentient entity" (*Siddha*, 7) who conceives you as a "discordant entity" (*Asura*, -1) for substantiating his "divinity" (*Siddhi*, 57) as the "cultural factor" (*Sugriva*, 5).

5.1.9 Eventual State. Divided Divinity Making One Zero

The "human being" (*Insan*, 1) is the "capability spirit" (*Abhidheya*,

460) of the Primeval Greeter "Universe of theoretical subject" (*Patala Loka*, -2) inhabited by the human being. The "Primeval Greeter Universe" (*Patala Loka*, -2) is the **thirtieth causal body** (*Patala Loka*, -2). It is the "universe of super wishers" (*Jagath*, -2) produced by the reproduction of our "past ego" (*Aham*, -1) over the "present" (*SAUM*, 1600). Our past ego becomes a "super wisher" (*Asura*, -1), wishing that we as the human being self-fulfill the "wish" (*Chah*, 18). We produce and diffuse our "energy in descending motion" (*Aham*, -1) in the form of ego, seeking to conserve our energy for eventually generating a "revolutionary force" (*Uttama*, 85).

Over time, we trade "infinite wishes" (*Duniya*, $-2 = 18 - 20$) by making the "wish" (*Chah*, 18) our "sister spirit" (*Ruah*, 20) and servicing the sister spirit to each "discordant entity" (*Asura*, -1). We conceive a discordant entity as an "ideal self-luminous entity" (*Svayam*, 12) with the "revolutionary force" (*Uttama*, 85) of our "ego" (*Aham*, -1). Thus, the Self-Luminous Entity "Universe of ideal object" (*Satya Loka*, 10^{1024}) becomes the "true universe" (*Dhamma*, 10^{1024}) of our "energy system" (*Antarmukha*, 10^{1024}) as a breeding "animal" (*Pashu*, 10^{1024}).

We fuse our ego within the present by "diffusing" (*Visaran*, 90) our "energy in ascending motion" (*Hunduka*, $73 = 47 + 26$) in the form of emotion for "blessing" (*Ashirvad*, 1000) each "replication" (*Samalekha*, 90) with our "sentient energy" (*Amrit*, 1000). We diffuse our "emotions" (*Huduka*, 73) due to our "psychic linkages" (*Mahika*, 47) as an "Almighty Creator" (*Damodara*, 26) of the "infinite wishes" (*Duniya*, -2) produced by the "universe of discordant entities" (*Jagath*, $-2 = 26 - 28$) incarnated with our "convergent energy" (*Samvat shakti*, 28).

Our convergent energy comprises our "sister spirit" (*Ruah*, 20) and our "essential nature" (*Svabhav*, 8) divided for the "technological growth" (*Vidhana*, 2) of the "universe" (*Brahman*, 2) and the "tertiary growth" (*Dishti*, 6) of the universe's "book of life" (*Pustak*, 14) reproducing our essential nature.

The Luminous "Universe of Replication" (*Go Loka*, 3785) norms the Western "trading-oriented factor" (*Jnana Kaya*, 3785), trading our "present reality" (*Badhabuddhi vadartha*, -2) of the "infinite wishes" (*Duniya*, -2). The Western "trading-oriented factor" (*Jnana*

Kaya, 3785) is the **"thirtieth etheric body"** (*Jnana Kaya*, 3785).

The Primeval Illuminator "Universe of Replicator" (*Pitri Loka*, 20,000) forms the Northern "servicing-oriented factor" (*Nirmana Kaya*, 23,125) as the "ninth etheric body" (*Nirmana Kaya*, 23125) for manifesting the "Devoted Deity Universe" (*Sutala Loka*, 23,125). The "Primeval Illuminator Universe" (*Bhuvar Loka*, 20,000) is the **"thirtieth astral body"** (*Jyotir Loka*, 20,000).

The "theoretical subject" (*Kaumaritantra*, 8,000,000) is a "juvenile" (*Yauvaniya*, 8,000,000) with an unawakened "immature mind" (*Aprabuddha*, 8,000,000). The immature mind is the **"thirtieth mental body"** (*Aprabuddha*, 8,000,000).

The immature mind forms the Primeval Perpetuator "Universe of Replicable" (*Vibhava Loka*, 20), seeking a "desirable future" (*Ishta*, 20) by exchanging the "lackluster" (*Dhumprabha*, 20) past. The "Primeval Perpetuator Universe" (*Brahma Loka*, 20) is the **"thirtieth intellectual body"** (*Bhuva Loka*, 20). It forms the Eastern "investment-oriented factor" (*Dharma Kaya*, 1869) as the "tenth etheric body" (*Dharma Kaya*, 1869). Each replicable becomes an "organ" (*Indriya*, 1869) of our "extended physical body" (*Balukaprabha*, 379) on its "conception" (*Sankalpana*, 1869).

The extended physical body is the **"thirtieth physical body"** (*Balukrapha*, 379), norming the Primordial Greeter "Universe of Replicating" (*Shambhala*, 379). The "Primordial Greeter Universe" (*Indra Loka*, 379) is the "creation" (*Srijan*, 379) of our "workculture" (*Nayaki*, 379). It forms the Southern "capability-oriented factor" (*Sambhoga Kaya*, 2785) with the "conceived consciousness" (*Bhavitatma*, 2785) to norm the "Supreme deity universe of luminous entity" (*Nitala loka*, 2785), inhabited by the paternal jinn Orion as the supreme "creator deity" (*Bhagwan*, 4). The "capability-oriented factor" (*Sambhoga Kaya*, 2785) is the "eleventh etheric body" (*Sambhoga Kaya*, 2785). It makes the "absolute equality" (*Nirguna*, -6) of the "creation" (*Srijan*, 379) with the "creature" (*Purusha*, 12) within each evolving "temporal coordinate" (*Astikaya*, -6) the "existence-oriented factor" (*Astikaya*, -6), norming the "twelfth etheric body" (*Astikaya*, -6).

Our "capability spirit" (*Abhidheya*, 460) is the "perpetuator" (*Abhidheya*, 460) of the "para deity culture-effect" (*Maha Vibhu*, 14). Although we inherit the "dominating effects" (*Yojana*, 190) of our "leadership workculture system" (*Shanideva*, 100) and its "replication" (*Samalekha*, 90), the socially-networked para deity culture-effect is the "deciding factor" (*Samalekha*, 90) in our "present life experience" (*Bhaavi*, 11). We seek to replicate the para deity culture-effect wishing to become the one who has bred the deity-like "human beings" (*Insan*, 1). Thus, the "para deity universe" (*Talatala loka*, 1659) becomes co-habited by an "ignited community" (*Pratyalidha mandala*, 10^{10}) ignited into life by our "consciousness system" (*Sara Kalpa*, 10^{10}). The ignited community comprises the Pleiadeans behaving like God to pollute our "guiding spirit" (*Ruah*, 20). The Pleiadeans are the "Giants" (*Kuha*, 10^{10}), known as the "Demons" (*Rakshasa*, 10^{10}). In this way, our consciousness system is the "wheel of innovation" (*Sadashiva chakra*, 10^{10}) that helps us change the "present" (*SAUM*, 1600) by forming innovative social relationships for transforming our inherited "subjective consciousness" (*Mahatma*, 18) of the past.

The ignited community makes us an "impassioned devotee" (*Samyama*, 0), wishing to enjoy the "predominating effects" (*Kshepa*, 285) of its "followership workculture system" (*Sthavaravisha*, 57). The predominating effects are the "reproduction" (*Prajana*, 285) of our "mindless" (*Amanaska*, -9) "metaphysical effects" (*Yogataranga*, -9) that norm our "self-obsessed" (*Nastika*, -9) "spirituality" (*Atmata*, -9). Spirituality is the "divided divinity" (*Atmata*, -9) formed by identifying one's "I" (*Svayam*, 12) with one's "spirit" (*Ruah*, 20) polluted by the ignited community. Therefore, it makes "I" (*Svayam*, 12) eventually enjoy the "eventual state" (*Sura*, 0) of "zero" (*Shunya*, 0).

5.2 Imperative for Transcending the Cultural Factor

The key to "unified divinity" (*OM*, 19) is sheltering our minds from the para-conscious air we breathe every moment. The air we breathe transforms our sense-making of reality. If the air is hot, we are in a hurry to make sense and move on; thus, para-consciously, we live a fresh life every moment enjoying the variable present reality. It gives us absolute joy, the freedom from the present effects as we move with the evolving future. If the air is cold, we take time to process the sensory information; thus, consciously, we become the subject observing and enjoying the constant past reality. It gives us temporary joy, the freedom from the future effects as we remain static and let the air move to revolving the future around the imaginary reality we construct with our creative sense-making. Using technology to regulate the air temperature, we become free from the past-effect and live life on our terms.

The fire within us shapes whether we remain active and evolve the future with us or become inactive and let the present revolve as the future by reforming the existing energy into new forms. If we are active, we enjoy the old wine in the old bottle and be happy with the freshness of our minds. Thus we realize eternal joy. On the other hand, if we are inactive, we enjoy the new wine in the new bottle and be happy with the integrity of our morality. Thus we become a thing that gives illusionary joy until others remain active and remain devoted to servicing their fire for our psychic well-being.

Unified divinity protects us from the entanglement of the cultural factor that forces us to deify others as God and be devoted to their sentient well-being as a path for realizing our sentient well-being. It protects us from working hard to make another a trillionaire as a path for us to be a millionaire. It opens our minds and clarifies our consciousness to work normally and enjoy our reality, instead of working like a supernormal human for the benefit of a celebrity God. It helps us transcend the dualities among the absolute, temporary, eternal, and illusionary joys. It lets us be the absolute enjoyer of the diversities of diverse entities as the diverse beauties of nature gifted by Mother Nature.

The unified divinity generates a vacuum in our breathing and fills the vacuum with everything we wish to assume true. It energizes us to materialize the truth we experience in the causal realm without us, first in the mental realm within us and then in the physical realm with us. In the mental realm, it opens the consciousness of the method for a proficient exchange of our imaginary reality with the present reality. In the physical realm, it motivates us to take conscious actions to transform the present reality into the future reality we have experienced in the causal realm. However, instead of listening to our inner voice and taking conscious actions, we may decide to just focus our energy on chanting guided by the intellectual realm. Our intellectual realm says that if we are a chanting devotee, then a devoted God will work hard to fulfill the truth we wish to experience in our afterlife after we finish the life of chanting.

Our "unified divinity" (*OM*, 19) is the mantra for us being that devoted God in the afterlife. It shapes the "potential" (*AUM*, 19) for us reincarnate as a "human being" (*Insan*, 1) to finish the work we left unfinished in this birth. It gives us the fruit of our knowing by making us a person with a clear goal in life and a deity-like inner drive to pursue that goal, if we choose to listen to the deep voice of our inner self in that birth. If we still decide to continue chanting, our voice continues becoming deeper. Eventually, we have very sharp clarity to wake up from our inertia and get on to the path of conscious entity, conscious of our divinity.

The essence of a mantra is to pay homage to one's various gurus, known and unknown, visible and invisible. The homage helps one receive the blessings of sentient energy from the gurus for developing conscious consciousness of the absolute truth. The absolute truth is that the essence of the mantra is to develop a clarified consciousness about oneself, the conscious system that is guiding one. It helps us gain confidence, carrying conviction, that we have the gifted potentialities to realize absolute freedom from all dependencies on everyone, known and unknown, visible and invisible. We have the absolute freedom to enjoy the gifted potentialities of everyone without becoming dependent on them.

To enjoy absolute freedom, we need to develop self-awareness that our human body is the causal body for whatever we conceive, perceive, and experience. Our human body is 100% responsible for knowing what we know, manifesting what we manifest, and creating what we create. Therefore, we must prioritize the health of our human body, which comprises four bodies guided by a fifth body. First, the etheric body, i.e., the body of consciousness we reproduce for producing infinite value. Second, the astral body, i.e., the conscious body that produces the consciousness we reproduce. Third, the mental body, i.e., the para-conscious body that trades the energy we reproduce within our consciousness. Fourth, the physical body, i.e., the subconscious body that services the energy we trade from the ecosystem. The fifth body is the intellectual body, i.e., the body of the person we have become after our energy exchange has diffused our consciousness into everyone, everybody, everything, and impregnated their consciousness within us. Therefore, we are always attracted towards what we don't know, haven't manifested, haven't created, and devalue what we know, have manifested, have created. Thus, the workers, the knowers, the manifestors, and the creators end up working for the perpetuators who perpetuate their omnipotence by trading what others don't value and by servicing their nothingness as of great value. Those who do not do anything are worshipped by everyone as God and everybody is willing to make 100% sacrifice for their God and fight anyone who does not believe in their God. That is the strange way one illuminates when one gains awareness of the self.

One becomes a complete creature only with the development of the three forms of consciousness: divine courage consciousness, guider power consciousness, and sentient wisdom consciousness. One who has the consciousness of the courage needed to make divine decisions appropriate to the time. One who has the consciousness of the power needed to guide time-appropriate divine decisions of others. One who has the consciousness of the wisdom that guides everyone's conscious decisions, who choose to rely on the wisdom instead of reason and intuition appropriate to the time. Therefore, one can offer time-appropriate reason to motivate appropriate decisions from those who have an open mind for a reason.

Acknowledgments

This investigation into the cultural factor is shaped by six divine factors: determination, imagination, virtue, intuition, nature, and excellence.

Conscious determination of the value of a metaphysical approach, without the inherited scientific method's limitations, is shaped by Primordial Greeter Shri Kartar Singh Yadav Ji, ex–Joint Commissioner, Ministry of Agriculture, Government of India, who is my param guru.

Liberated imagination of the technique for initiating, persevering, and finishing this project is shaped by my father, Shri Surender Nath, and my mother, Shrimati Manju Gupta.

Illuminated virtue of transcending beyond the traditional approach has been shaped by my wife, Bhakti.

Infinite intuition for a conscious ecosystem approach is shaped by my students, devoted to transforming their social, human, ecological, economic, national, and psychological well-being.

Universal nature of the proposed organizational approach is shaped by my professional colleagues and mentors, at various institutions, from various nations, and with varying academic and life perspectives.

Technological excellence of this investigator is shaped by my family, friends, and critics and by those who have contributed through the ages to illuminate the objectives of the study.

Primordial Perpetuator Maha Saraswati and Primordial Illuminator Shri Krishna blessed me with their Divine Light, energizing me to bring the project to fruition.

This investigation is dedicated to the Universe of Children, wishing for their global, unique, inclusive, diverse, engaged, and responsible well-being.

English Index

A

Absolute after-effect, 300
Absolute authority, 75, 76
Absolute breadth power growth rate, 336
Absolute consciousness, 81, 88, 145, 247, 277
Absolute deity consciousness, 279
Absolute development, 55, 137
Absolute entropy force, 365
Absolute equality, 359, 367
Absolute freedom, 5, 41, 111, 116, 232, 256, 257, 370
Absolute gravitational constant, 35, 36, 44, 52–54, 132, 133, 153, 154, 160, 169, 170, 172
Absolute gravitational energy, 289
Absolute guider, 282
Absolute joy, 85, 90, 232, 278, 369
Absolute knower, 275
Absolute quark, 35, 36, 164, 165
Absolute space, 73
Absolute theory, 357
Absolute volunteer, 86
Absorbed photon, 307
Absorption, 307
Accelerating-effect, 320
Acculturation, 16, 95, 97, 158, 204, 208
Acculturation factor, 208
Acculture, 8, 27, 159, 329
Acculturing causation, 282
Accusation, 355
Acquired value, 257
Activating state, 264
Activation energy, 203
Activation force, 264
Activator, 128, 130, 228, 265
Active state, 264
Adaptation, 14, 32
Added allele, 198
Added Physical Body, 307
Administration, 11, 188
Administrative leaders, 27
Adverse selection, 15, 220
Aerodynamic hypothesis, 40

Aerodynamics, 40
Affliction, 79, 80
Aftereffect, 297, 298, 300
Agency contract, 220
Agent cell, 183
Aggregated copy, 227
Air force, 239
Air in ascending motion, 3, 109
Air in descending motion, 109
Air wheel, 192, 213, 229, 291
Air-effect, 208, 213, 221, 230, 241
Alien grandson, 220
Aliens supplement, 9
Alive community, 360
Almighty Creation, 66
Almighty Creator, 1–3, 66, 72, 253, 366
Almighty Creature, 67
Almighty Deity, 249
Almighty Physical Body, 307
Alpha cell, 182
Alpha electron, 309
Alpha force, 309
Alpha photon, 309
Alpha-amino acid, 266
Amacrine cell, 322
Amino acid, 159, 161, 265, 266
Animal body, 325
Animal child, 249
Animal kingdom, 323, 340
Animal soul, 219
Animal spirit, 264, 265
Animate body system, 226
Animate energy, 359
Animate entities, 89
Animate equality, 64, 326
Animate inequality, 63
Animate one, 293
Animate primordial maternal, 97, 292
Animate state, 297
Antagonist, 31, 223, 314, 357, 358
Antagonist behaviors, 357, 358

Antagonist consciousness, 357
Antagonist effects, 357
Anterior Insular Cortex, 349
Anterior pituitary gland, 212
Antikaon, 150
Anxiety, 3, 4, 75, 267, 268
Anxiety moderating factor, 268
Anxiety-mediating factor, 267
Apostles, 255
Apparent horizon, 258, 259
Aquarius house, 65, 68
Aquarius zodiac, 100, 101
Archaeon, 323
Archangel, 291
Argon, 266, 267
Aries house, 62, 68
Aries zodiac, 107, 108
Artificial element, 9
Ascended master, 179, 326
Ascending conscious well-being, 257
Ascending consciousness, 193, 202, 209
Ascending cultural force, 155
Ascending demand, 250
Ascending energy, 208, 224
Ascending entropy, 2, 4
Ascending force, 76, 242
Ascending ideal-effect, 23
Ascending light force, 95, 241
Ascending mass, 71, 241, 242
Ascending mental well-being, 210
Ascending motion, 74, 109
Ascending param deity, 261
Ascending phase, 185
Ascending physical well-being, 210
Ascending power distance, 238
Ascending proportion, 224
Ascending respiratory force, 342
Ascending theory-effect, 23
Astral body, 18, 100, 189, 190, 192, 204, 207–211, 214, 224, 246, 280, 281, 283–287, 289, 298, 290–292, 300, 309, 310, 325, 370
Astral realm, 48, 175, 287, 353, 354
Astrological consciousness, 203
Astrological element, 63
Astrological link, 145–148

Astrological system, 63, 71
Atmospheric energy, 185, 187
Atom, 75–77, 124, 136, 137, 160, 216, 225, 228, 303
Atomic system, 225
Attenuation coefficient, 308
Attraction force, 170
Auditory Cortex, 330
Automatic copy, 311, 312
Automatic growth, 311
Autonomic nervous system, 229
Autotrophic growth, 267
Available future time, 64
Avascular growth, 181
Aversion, 232
Awareness-supplementing development, 9
Axial component, 272

B

Backward value, 15
Backward-order development, 15, 47, 48
Banking face, 20, 344
Banking system, 114
Bardo, 81, 82, 254
Being energy, 16, 176, 195, 282, 331
Being omnipresent, 88, 89
Benefit value, 252
Beta cell, 182
Beta electron, 304, 309
Beta-amino acid, 266
Beyond absolute paradigm, 248
Beyond present paradigm, 226
Bicellular organism, 300
Bickering energy, 214, 359
Binding energy, 307, 315
Binucleates, 158
Biodynamic hypothesis, 41
Biodynamics, 41
Biological effects, 5
Bipolar cell, 322
Bipolar geographical system, 116
Bipolarizes, 322
Birth zodiac, 355
Bistratified ganglion cell, 182
Bit, 37, 40, 51, 174
Black blood cells, 213

Index 375

Black hole, 69, 253, 259, 260, 262, 272
Blessedness, 29, 40, 41, 64, 68, 79
Blessing force, 18
Blindsight, 232
Bliss consciousness, 83
Blissful Self, 351
B-meson, 159, 162, 167
Body of consciousness, 293, 370
Body of departed souls, 214
Body of living souls, 214
Bond free, 83
Born sentient entity, 287
Borrowed divinity, 219
Boson, 144, 148, 263, 274, 318
Boson particle, 274
Boson sentient, 318
Bottom anti-quark, 141
Bottom Iota Meson, 162
Bottom quark, 141
Bottom Tau Meson, 167
Bovines, 98, 99
Bradyon, 34
Bragging, 23, 93, 252, 278, 279, 309, 321, 323, 350
Bragging face, 93
Bragging self, 309
Bragging system, 23
Bragging value, 321
Brainstem, 330, 331
Bred group, 94
Breed, 90, 91, 96–98, 102, 124, 130, 134, 173, 174, 351
Breeder, 338, 339
Breeding, 8, 17, 22, 25, 26, 30, 60, 66, 69, 71, 75, 78, 83, 89, 97–99, 101, 103, 106, 107, 111, 119, 124, 151, 154, 160, 168, 169, 174, 186, 193, 196, 202, 212, 241, 252, 261, 267, 268, 270, 285, 286, 288, 314, 315, 319–321, 323, 325, 339, 342, 344, 350, 366
Breeding causation, 342, 344
Breeding consciousness, 321
Breeding force, 89, 321
Breeding recombination, 314
Breeding system, 8
Breeding workculture, 325, 339
Brown blood cells, 213
Byzantine Iberian, 21

C

Cajal-Retzius cell, 182
Calcified water-effect, 322
Calcium, 180, 266, 268
Cancer house, 63, 68
Cancer zodiac, 95
Capability bed, 231, 232
Capability spirit, 368
Capricorn house, 65, 68
Capricorn zodiac, 90, 91
Carbon, 88, 189–193, 215, 265, 266, 310–313, 322
Carbonated air-effect, 322
Cardinal member, 121
Cardiovascular system, 213
Cascade, 87, 163
Casimir force, 159
Catalyst, 203, 271
Catholic crusaders, 21
Causal body system, 303
Causal cause, 252
Causal realm, 48, 175, 287, 369, 370
Causal self, 306
Causation, 20, 37, 42, 51, 89, 93, 94, 123, 124, 129, 137, 140, 141, 148, 152, 156, 158, 159, 172, 173, 175, 177–179, 181, 183, 185, 187, 189–191, 193, 195, 197, 199, 201, 203, 205, 207, 209–211, 213–217, 219, 221, 223, 225, 227, 229, 231, 233, 235, 237, 239, 241, 243, 245, 247–249, 251, 253, 255–257, 259, 261, 263, 265, 267–269, 271, 273, 275–277, 279, 281–283, 285, 287, 289, 291, 293, 295, 297, 299, 301, 303, 305, 307, 309, 311, 313, 315, 317, 319, 321, 323–325, 327, 329, 331, 333, 335, 337, 339, 341, 343, 345, 347, 349, 351, 353, 355–357, 359, 361, 363, 365, 367, 369, 371
Causation consciousness, 356
Causation dimension, 268, 271
Causation ellipse, 152, 156, 159
Causation program, 42
Cause creator, 276
Cause-defending, 83
Causer, 255
Causing state, 177
Cell, 42, 75, 76, 102, 136, 137, 154, 158, 180–184, 187–190, 193–195, 197, 203, 204, 206–209, 216, 224, 225, 227, 228, 261, 263, 265, 267–269, 271, 276, 281, 284, 291, 296, 309, 316, 317, 319, 321, 322, 327, 338
Cell cycle, 276

Centesimal member, 122
Central fissure, 327, 330, 331
Central nervous system, 229
Cerebellum, 329
Charles Darwin, 14
Charm anti-quark, 142, 302
Charm quark, 138, 142
Charmed B-meson, 167
Chastisement, 272
Chastity, 80
Chemical reaction, 230, 259, 319
Chemical sequences, 5
Cherenkov radiation, 304
Cherenkov wave, 304
Cherenkov-effect, 304
Child cell, 271, 284, 309, 311, 315, 316, 318
Child consciousness, 316
Child essence, 283, 284
Child face, 122
Child God, 276, 277, 345
Child primordial greeter, 111, 177, 287, 325, 334
Child profiting gene, 105
Child spirit, 216, 217, 220, 280, 281, 290, 291
Chromaffin cell, 267, 269
Circadian cycles, 191
Circadian time cycle, 187
Circle pi, 54
Circular double octave, 165, 308
Circular system, 100, 105, 106, 152, 224, 228
Circular-effect, 57, 60
Circulating body, 260, 261
Circulating force, 19, 246
Circulatory wheel, 19, 214, 229, 291
Citizenship responsibility, 211
Claimant, 8
Cleansing state, 176
Cloaca, 108
Closed space, 125–127, 129–132, 134, 161, 162
Clouded consciousness, 23
Cluster, 28–30
Clutch size, 89–91
Codon, 177, 178
Cohomotopy group, 146, 147

Collective copy, 198, 199
Collective energy, 198
Collective force, 197–200, 314
Collective potential, 34, 164
Collective reality, 50, 361
Colorless body, 309
Colorless light, 149, 304
Communion energy, 358
Competitive linkage, 60, 328
Complementary hypothesis, 39–41
Complementary linkages, 328
Complementary value, 60, 328
Compounding causation, 354, 356
Compton-effect, 305
Conceived consciousness, 367
Conceived reality, 43, 80, 241, 248
Concordant electron, 150
Concordant energy, 25, 142, 144, 170, 225
Concordant entity, 176, 177, 217
Concordant factor, 85, 142, 318
Conditional destiny, 252
Conditioning force, 349
Cone, 55, 58, 60, 61, 130–132, 136, 150
Cone, 131, 137
Conical wavefront, 304
Conscious actions, 5, 193, 297, 370
Conscious apparatus, 76
Conscious body, 189, 207, 248, 274, 309, 310, 371
Conscious causation, 58, 129
Conscious cellular system, 188
Conscious consciousness, 4, 60, 77, 91, 108, 193, 194, 256, 333, 370
Conscious decisions, 3, 4, 9, 55, 257, 298, 371
Conscious determination, 289, 350
Conscious energy, 3, 16, 20, 22, 55, 57, 61, 91, 100, 104, 226, 263, 268, 273, 293, 317, 352, 353, 356, 365
Conscious entity, 93, 99–101, 189, 190, 208, 215, 223, 254, 320, 352, 370
Conscious excellence, 291
Conscious fire, 259, 318, 325, 331, 344
Conscious force, 169
Conscious form, 297
Conscious freedom, 85, 86, 91, 216
Conscious imagination, 2, 96, 225, 290

Index

Conscious intuition, 291
Conscious light, 18, 173, 284–287, 292–297, 300–302, 305, 327, 350
Conscious love, 257
Conscious mind, 271, 313
Conscious nature, 291
Conscious planning, 82, 208, 345, 349
Conscious reality, 59, 112, 113
Conscious spirit, 151, 178, 240, 244
Conscious system, 1, 4–6, 17, 19, 34, 42, 100, 103–105, 111, 112, 222, 227, 255, 256, 292, 293, 301, 313, 328, 370
Conscious virtues, 290
Conscious well-being, 3, 7, 9, 39, 186, 204, 206, 207, 232, 325
Consciousness body, 237
Consciousness deficit, 195, 309
Consciousness entropy, 326
Consciousness equality, 279, 283
Consciousness flow, 3, 320
Consciousness force, 264
Consciousness stock, 65, 69, 77, 151, 246, 259
Consciousness system, 4, 17, 19, 34, 35, 42, 59, 71, 78, 89, 99, 100, 103, 104, 106, 111, 112, 122, 220, 227, 255, 302, 313, 368
Consciousness void, 151, 209, 314
Consciousness-cooling air, 3
Consequential causation state, 58, 129
Consequential entity state, 6, 172
Consequential spatial state, 149, 150
Conservative democratic technique, 246
Consonance, 25
Constant lifetime, 247
Constitutional cells, 181
Consumer class, 85, 217, 250, 258
Consumer mindset, 293
Contingent paradigm, 256
Continuous element, 22
Continuous gravitational wave, 156, 247
Continuous improvement, 22, 312
Continuous photon, 263
Continuous space, 176
Continuous stream, 67, 69–71
Continuous time, 175
Convergence point, 62
Convergent energy, 15, 89, 99, 203, 366
Convergent group, 14, 15

Convergent ideal-effect, 191
Convergent light, 313
Convergent reality, 104, 203, 204
Conviction spirit, 218
Cooling effect, 3
Coordinate member, 120
Coordination complex, 54, 160
Copying sequence, 263
Corporate consciousness, 189
Corporate followership work, 238
Corporate leadership work, 231, 237, 238
Corporate system, 221–223
Corporate time, 53, 54
Corporate-effect, 223
Corpus callosum, 331
Corpuscle, 304
Cosmic knower, 101
Cosmic shower, 305
Cosmological constant, 43, 171
Cosmological unit, 144
Cost-effective paradigm, 219, 253
Council of elements, 207, 214, 221, 226
Council of Five, 361
Council of Myself, 271
Council of Seven Light Forces, 237
Council of Twelve Apostles, 255
Council of Twelve Zodiacs, 96, 97
Creative cycle, 65
Creative force, 17, 97, 153, 174, 178, 189, 222, 255, 278
Creative imagination, 297
Creative work, 177
Creator, 1–3, 13, 24, 25, 62, 63, 66, 67, 69, 88, 177, 178, 202, 204, 250, 254, 276–278, 293, 334, 367
Creator deity, 178, 367
Creator factor, 293
Creator sequence, 66, 67, 69
Crisis management, 342
Crown wheel, 213, 228, 230, 231, 244
Crusade, 21
Cubic energy, 202
Cuboid, 154
Cultural causation, 20
Cultural consciousness, 35, 59, 107, 110, 181

Cultural consequence, 20

Cultural cycle, 36

Cultural factor, 2, 5, 7, 13, 16, 20, 23, 25–30, 32, 34, 36, 38, 40–42, 44, 48, 50, 52, 54, 56, 58, 60, 62, 64, 66, 68, 70, 72, 74, 76, 78, 80, 82, 84, 86, 88, 90, 92, 94, 96, 98, 100, 102, 104, 106, 108, 110, 112, 114–116, 119–122, 124, 126, 128, 130, 132, 134, 136, 138, 140, 142, 144, 146, 148, 150, 152, 154, 156, 158, 160, 162, 164, 166, 168, 170, 172, 174–178, 180, 182, 184, 186, 188, 190, 192, 194, 196, 198, 200, 202, 204, 206, 208, 210, 212, 214, 216, 218, 220, 222, 224, 226, 228, 230, 232, 234, 236, 238, 240, 242, 244, 246, 248, 250, 252, 254, 256, 258, 260, 262, 264, 266, 268, 270, 272, 274, 276, 278, 280, 282, 284, 286, 288, 290, 292, 294, 296, 298, 300, 302, 304, 306, 308, 310, 312, 314, 316, 318, 320, 322, 324–326, 328, 330, 332, 334, 336, 338, 340, 342, 344, 346, 348, 350, 352, 354, 356, 358, 360, 362, 364–366, 368–370

Cultural heritage, 34, 105, 228

Cultural icons, 2

Cultural origins, 2

Cultural practices, 28

Cultural production, 359

Cultural reality, 18, 181, 201

Cultural system, 115, 298, 315

Cultural value, 28, 349

Cultural wisdom, 175, 177, 339

Culture factor, 340

Curved causation, 57, 58, 129

Curved consciousness, 111

Curved space, 37, 55

Curved time, 36, 37, 41, 42, 44, 154, 171

Curving time, 170

Cycle of conscious life, 17

Cycle of divinity, 17

Cycle of emotional linkages, 220

Cycle of sustainability, 87

Cycle of time, 86

Cycloid, 59

D

Daoism, 253

Dark energy, 87, 239

Dark force, 67, 147

Dark matter, 66, 67, 69, 71, 87, 88, 93, 95, 238–242, 245, 247, 303

Darwin model, 40

Daughter allele, 155, 198, 199

Daughter cell, 102, 154, 181, 182, 268, 271, 275, 319, 327

Daughter consciousness, 78, 91, 264, 265

Daughter double copies, 217

Daughter identity, 286

Daughter soul, 218

Daughter spiderling, 93

Daughter spirit, 218, 315

Daughter womb, 283, 346

Dead ancestors, 220

Dead body cell, 262, 263, 275

Dead cell, 215, 296

Dead grandfather, 220

Death Deity, 274

Death God, 245, 288

Dehydrated Entity, 294

Deity consciousness, 2, 3, 29, 74, 171, 260, 273

Deity culture-effect, 368

Deity kingdom, 275, 324

Deity realm, 344

Deity soul, 219

Deity value, 155

Delegation, 278

Delight, 80

Delta, 158, 159, 161, 165

Delta bond, 158, 165

Demi-God, 255

Demon spirit, 218

Demons, 368

Denominator, 16, 38

Departed consciousness, 315, 321

Departed entity, 122, 175–178, 281, 282, 285, 289, 296, 297, 301, 321, 323, 324, 331, 332, 338, 341–345, 349, 350, 355–357

Departed human entity, 69, 296, 297, 364

Departed soul, 179, 249, 277, 280, 281, 291, 317, 318, 321, 326–328

Descending demand, 251

Descending entropy-effect, 157

Descending force, 76

Descending mass, 71, 241, 247

Descending mental well-being, 210

Descending motion, 3, 170

Descending param deity, 261

Descending physical well-being, 210

Descending power distance, 238

Index

Descending respiratory force, 342
Descending sentient benefit-cost ratio, 209
Descending sentient well-being, 209, 211, 257
Design energy, 148, 186
Desirable force, 272
Desirable future, 223, 367
Desirable object, 276
Desirable reality, 257
Desired body, 282
Desired present, 224
Destroy codon, 177
Destruction, 56, 66
Desultoriness, 79
Deuteragonist, 223, 314, 357
Deuteragonist behaviors, 357
Deuteragonist consciousness, 357
Development spirit, 351
Devil allele, 199
Devil cell, 182
Devil granddaughter, 57–59
Devil soul, 218, 219
Devil spirit, 217, 291
Devoid of information, 250, 253
Devoted deity, 177, 360, 362, 363, 367
Devoted deity universe, 360, 362, 363, 367
Devoted followers, 9, 282, 321
Devoted mother, 55
Devoted workculture, 212
Devotee daughter, 55–58
Devotee deity, 177, 362
Devotee Wishers, 53
Devotion, 16, 27, 73, 85, 190, 288
Devotional energy, 83
Different consciousness, 364
Differentiable four-body system, 292
Differentiable system, 294
Differentiated consciousness, 290
Differentiated system, 294, 295
Differentiating four-body system, 291, 293, 295
Diffused light, 263
Diffusion servicing, 29
Digestion wheel, 19, 180, 183, 226, 229
Digestive-effect, 19
Digital copy, 311–313

Diploid cell, 197
Diplomatic mission, 21
Direct neurological instruction, 187
Directrix, 132, 133
Discontinuous space, 175, 176
Discontinuous stream of photons, 67, 303
Discontinuous time, 176
Discordance, 85, 202, 362
Discordant electron, 150
Discordant energy, 3, 74, 109, 111, 153, 163, 170, 171, 192, 248, 289, 319, 339–341
Discordant entity, 3, 176, 177, 217, 355, 365, 366
Discordant factor, 85, 107, 170, 288, 319
Discriminating consciousness, 257
Disentangle, 4
Disgusting form, 213
Disintegrated system, 296
Disintermediation, 238, 260, 329
Dispensable, 259, 260
Disproportionate aftereffect, 298
Disproportionate consciousness, 166
Disproportionate entropy, 220
Disproportionate growth, 114, 214, 221, 337
Disproportionate radial power growth rate, 338
Disproportionate sense, 309
Disregardfulness, 79
Dissimilar allele, 200
Divergent copies, 14
Divergent group, 14, 15, 174
Divergent light, 313
Divergent path, 365
Divergent reality, 204
Diverse form, 5, 26, 62, 131, 142, 143, 194, 239, 244, 301, 334, 336
Diverse group, 15, 31
Diverse present, 82, 181, 354
Diversity orientation, 28
Divided allele, 155
Divided astral body, 282, 290
Divided causal body, 266, 267
Divided causation, 360, 362
Divided consciousness, 317, 344
Divided divinity, 365, 368
Divided entity, 164, 346

Divided etheric body, 277, 280
Divided geography, 351, 354
Divided intellectual body, 324, 325
Divided nature, 362, 365
Divided parts, 279
Divided physical body, 300, 326
Divided position, 201
Divided time, 354
Divided wholesomewholeness, 342
Dividing astral body, 290
Dividing causal body, 268
Dividing causation, 266
Dividing etheric body, 279, 280
Dividing maternal body, 306
Dividing physical body, 302
Divine consciousness, 353
Divine cost, 208
Divine Council, 187, 191–193, 195, 197
Divine debt benefit value, 252
Divine element, 141, 334
Divine energy, 7, 208, 211, 215, 216, 221, 222, 254, 257, 316
Divine entity, 2, 4, 8, 64, 79, 321
Divine essence, 254
Divine force, 170, 241
Divine gift, 177
Divine horizon, 258
Divine light, 258, 259, 296, 300, 301
Divine moderation, 325
Divine plan, 191, 195, 208, 222, 223, 240, 286, 287
Divine planner, 208
Divine planning, 191, 195
Divine reality, 277
Divine wave, 334
Divine-effect, 231, 240, 241, 321, 323
Divisible astral body, 289
Divisible causal body, 268
Divisible etheric body, 278, 279
Divisible light, 274
Divisible paternal body, 306
Divisible physical body, 301
Dizygote, 197
Dizygotic twin, 197, 283
Docile animal, 299

Doctrinal force, 26
Doctrinal value, 26, 27
Doctrine, 26, 84, 206, 207, 210, 216, 252, 256, 273, 281
Doctrine, 277
Doctrine of body, 252
Doctrine of ego entity, 210
Doctrine of entity, 281
Dominant religion, 6
Dominant religion institutionalizes, 6
Dominating effects, 368
Dominating value, 328
Donation, 113
Double octave, 91, 164, 241–243, 307, 308, 333, 334
Double partial pi bond, 36
Double population, 118
Double time, 46
Double twin quark, 36, 37, 139
Down meson, 162
Down quark, 140
Dream state, 185, 186, 232
Dreamer, 54
Dreaming artist, 187
Dreaming state, 176
Duality, 104
Dust-ball empiricism, 7
Dynamic factor, 81
Dynamic force, 327
Dynamic fulfillment paradigm, 252
Dynamics, 6, 41
Dynamism, 117

E

Earth cycle, 179, 183, 184
Earth force, 239, 240, 265
Earth wheel, 19, 179, 182–188, 192, 193, 195, 201–204, 214, 230, 231, 261, 290
Earth-effect, 184, 208, 221, 227, 230, 239, 240
Eastern consciousness, 23
Eastern culture-effect, 23
Eastern dimension, 128
Eastern geography, 20, 23, 28
Eastern group, 21, 23
Eastern workculture-effect, 23
Economic costs, 212

Index 381

Ecosystem, 83, 114, 179, 183, 193, 208, 223, 253, 256, 261, 262, 273, 276, 371

Ecosystem consciousness, 179

Ecosystem enactment, 83

Ecosystem energy, 256

Ecosystem entropy, 223

Ecosystem value, 114, 273

Eden, 261

Ego, 2–4, 47, 68, 70, 73, 74, 79, 85, 86, 189, 190, 209, 210, 220, 221, 289, 309, 362, 365, 366

Ego body, 190, 210

Ego element, 309

Ego energy, 2, 190, 210, 220

Eight body system, 227, 229, 243

Eight directions, 322

Eight gamma photons, 307

Eighteen dimensions, 309, 313

Eighteenth intellectual body, 331

Eighteenth member, 174

Eighteenth potential guider agent, 305

Eighteenth principal guider, 48

Eight-faced system value, 328

Eight-fold consciousness, 81

Eight-fold growth, 83, 96, 99, 146, 148, 154, 166, 195, 248, 249, 251, 284, 285, 312

Eighth body, 340

Eighth conscious light, 286

Eighth ellipse, 151

Eighth intellectual body, 328

Eighth member, 122, 174

Eighth sister, 288

Einstein, 6, 37, 41, 44, 132, 133, 153, 154, 170, 171, 205

Einstein cosmological constant, 171

Einstein gravitational constant, 132, 133, 153, 170

Einstein tensor, 44, 154, 171

Electric charge, 303

Electric energy, 34

Electric system, 228

Electromagnetic collapse, 141

Electromagnetic consciousness, 140

Electromagnetic interaction, 303

Electromagnetic mass, 43, 124, 136, 142, 144, 146, 228

Electromagnetic reaction, 336

Electron, 34, 36, 136, 137, 144, 148, 151, 159, 161, 165, 166, 183, 226, 228, 304, 305, 307, 309, 322

Electron force, 159, 161

Electron quadruplate, 183

Eleventh body, 66, 69, 340

Eleventh brother, 287

Eleventh entity group, 347–349

Eleventh etheric body, 367

Eleventh guider agent, 300, 301

Eleventh Intellectual Body, 329

Eleventh kin, 301

Eleventh member, 119

Eleventh potential guider agent, 301

Eleventh potential principal guider, 292

Eleventh principal guider, 47, 291

Eleventh sister, 288

Eleven-unit pentaquark, 140, 141

Ellipse, 55–58, 60, 61, 136, 137, 142, 148–152, 156, 161, 164, 165

Elliptical potential, 161

Elliptical space, 125–127

Emanating body, 299

Emanating form, 121

Emanating system, 296

Emitted photon, 307

Emotional body, 190, 210

Emotional deficit, 212

Emotional energy, 71, 190

Emotional loss, 209

Emotional well-being, 211

Empathy, 3, 63, 68

Empty oneness, 189

Endogenous, 110, 144–146, 148

Endothelial cell, 271

Enduring, 63

Energization, 77

Energy consciousness, 35, 39, 196

Energy conservation, 5

Energy cost, 257

Energy equilibrium, 189

Energy exchange, 227, 231, 247, 248, 279, 298, 371

Energy exchange paradigm, 298

Energy force, 239, 240

Energy fragment, 13, 138, 183, 317, 358

Energy in ascending motion, 71, 104, 108, 212, 366

Energy in descending motion, 3, 108, 366

Energy sequence, 62

Energy system, 20, 55, 100, 107, 109, 161, 188, 224, 225, 227, 256, 314, 316, 366

Energy vibrations, 290

Engagement orientation, 28

Enteric nervous system, 229

Entity bovine, 99

Entity class, 48, 216, 350, 351

Entity consciousness, 178, 179

Entity experience, 36

Entity face, 328

Entity force, 316

Entity gravity, 255

Entity group, 47, 48, 347, 349, 350

Entity networks, 100

Entity photon, 307

Entity reality, 55, 315

Entity scalar, 170

Entity superpositions, 33

Entity Tensor, 31, 33, 35, 37, 39, 41, 43, 45

Entity time, 31, 46, 111, 142, 179, 340

Entity value, 39, 328

Entrepreneur, 105, 113, 114, 188, 195, 358

Entrepreneurial behavior, 32

Entrepreneurial class, 85, 86

Entrepreneurial leaders, 28, 113–115, 274

Entrepreneurial leadership, 113–115, 274

Entrepreneurial opportunity, 65, 201

Entrepreneurship, 11, 110, 113, 114, 272, 290, 291, 327, 358

Entrepreneurship energy, 358

Entrepreneurship pathway, 327

Entropy, 2, 5–7, 9, 30, 42, 57, 58, 66, 67, 97, 104, 112, 118, 141, 144, 148, 158, 162, 169, 186, 188, 192, 195, 200, 201, 203, 206, 214–216, 222–225, 229, 231, 238, 239, 242, 243, 246–248, 251, 254, 265, 267, 268, 271, 273, 282, 305, 307, 308, 315, 318, 320, 342, 355, 361

Entropy force, 308

Entropy pi, 42, 57, 58

Entropy sensation, 320

Entropy system, 229, 231

Entropy value, 30, 97, 118, 144, 148, 162, 215, 243, 254, 315, 355, 361

Entropy workculture-effect, 203

Enzyme-effect, 297

Epidermis, 181

Epigenetic atom, 75, 76, 87, 216

Epistemological value, 365

Epithelial cell, 271

Error in consciousness, 72, 80, 259, 280, 314, 364

Escalating sentient costs, 212

Escalating technological cost, 7, 214, 246, 270, 276

Espousing causation, 357, 360

Essential element, 215, 250

Essential identity, 177

Essential nature, 1, 2, 5, 89, 98, 99, 111, 122, 157, 284, 334, 359–361, 366

Ester bond, 315

Eta meson, 162

Eta prime meson, 162

Eternal deity, 178

Eternal entity, 152, 153

Eternal human force, 88

Eternal joy, 233, 234, 369

Eternity, 14, 64, 210

Ether, 19, 88, 94, 95, 201, 208, 214, 221, 222, 231, 233, 239, 241, 248, 304, 323, 361, 363

Ether force, 239

Ether-effect, 201, 208, 214, 222, 231, 241, 323, 363

Etheric body, 18, 99, 192, 211, 212, 214–216, 224, 226, 237, 245, 246, 264, 279–282, 309, 310, 325, 370

Etheric potential, 221

Etheric realm, 48, 175, 287

Ethics, 112, 115

Ethnocentrism, 43

Eventual causation state, 22

Eventual reality, 212, 256

Eventual spatial state, 37, 51, 52, 58, 142, 143, 149, 168

Eventual state, 41, 149, 150, 168, 365, 368

Evolutionary force, 15, 55, 224

Evolutionary morality, 21, 113

Evolvent curve, 304

Evolving value, 55, 57, 58, 128–131

Exchange factor, 7

Exchange force, 226, 244, 245

Index 383

Exchange photon, 244, 306
Exchange spirit, 354, 355
Exchange system, 188, 225, 228, 313, 314
Exchangeable mental body, 309
Exchangeable physical body, 308, 309
Exchange-effect, 184
Exciton, 129
Exclamation, 189
Excreta, 19, 222
Excretor, 19
Excretory wheel, 18, 180, 213, 215
Excretory-effect, 18
Exemplification, 237, 238
Exhaustion, 279
Existence, 14, 30, 73, 77, 78, 89, 143, 158, 268, 282, 343, 367
Exogenous, 14, 28, 110, 144, 145, 148
Exogenous electron, 144, 148
Exponent, 130, 166
Exponential attenuation, 308, 309
Exponential growth, 30, 90, 166
Exponentiator, 166
Extended luminous system, 305
Extended physical body, 367
Extended self, 3, 9, 325
Extreme oneness, 286, 287
Extrinsic body, 299
Extrinsic conscious-effect, 76
Extrinsic consciousness, 112, 347
Extrinsic energy, 181, 185, 190, 192, 208
Extrinsic ether-effect, 201
Extrinsic experience, 298, 325
Extrinsic faces, 328
Extrinsic force, 100, 213
Extrinsic form, 121
Extrinsic four half octaves, 295
Extrinsic impact, 74, 171
Extrinsic light, 83, 322, 323
Extrinsic lunar phase, 185, 186
Extrinsic reality, 50, 160
Extrinsic strong psychic linkage, 276, 335, 336
Extrinsic system, 229, 231
Extrinsic weak psychic linkage, 276
Extroversion, 5, 73, 234, 236

Extroversion face, 73

F

Face of wisdom, 25
Factual reality, 257
False imagination, 3
False impression, 223
False reasoning, 5
Falsify, 235, 302
Falsifying code, 187
Family of species, 121
Fat cell, 213, 221
Fatal attraction, 93
Father electron, 160
Father Nature, 14, 15
Father soul, 218
Father-father allele pair, 155, 158, 198
Fathom, 121
Feeding recombination, 151, 356
Feeler, 93
Female body, 343–345, 347
Female bovine, 98, 99
Female child, 342, 344
Female consciousness, 78
Female copy, 312
Female frog, 92
Feminine body, 91, 153, 273, 299, 306, 334, 340
Feminine cell, 158
Feminine class, 220, 221, 347, 350
Feminine copy, 313
Feminine determination, 109
Feminine dimension, 361
Feminine electrons, 159
Feminine energy, 325, 359
Feminine gene, 95–97, 155
Feminine growth, 157
Feminine self, 291, 326, 327, 331, 343
Feminine self-luminous entity, 291, 343
Feminine spirit, 64
Femininity, 85, 331, 343
Feminization, 155
Fermion, 138, 149, 151, 183, 227
Fermion quadrupling condensate, 183
Fermions, 228
Fetal body, 228–231

Fetus, 226, 230
Fibonacci number, 336
Fibonacci sequence, 336
Field value, 152
Fifteen finite points, 61, 134
Fifteenth astral body, 299
Fifteenth body, 341
Fifteenth conscious light, 297
Fifteenth entity group, 350
Fifteenth intellectual body, 330
Fifteenth potential guider agent, 304
Fifteenth principal guider, 48
Fifteenth sister, 296
Fifth astral body, 291
Fifth body, 340, 370, 371
Fifth copy, 191, 286, 293
Fifth intellectual body, 327
Fifth member, 118, 121, 173, 333, 334
Fifth potential daughter, 348
Fifth potential son, 348
Fifth principal guider, 48
Final photon, 307
Fine structure constant, 18
Finite coordinate, 15, 175, 177, 178, 300
Finite cost, 200, 278
Finite creation, 262
Finite form, 112
Finite paths, 364
Finite permanent form, 112
Finite point, 38, 42, 43, 45, 50–54, 59, 62, 71, 75, 77, 82, 87, 100, 111, 123–125, 127–132, 134–140, 143, 144, 148, 151–153, 156, 159, 163, 167–169
Finite self, 161
Finite space, 175
Finite spatial coordinate, 262
Finite temporary form, 63, 112, 113
Finite time, 175, 223, 263
Finite values, 238
Finite wishes, 217, 220, 320
Fire force, 297, 333, 334
Fire spirit, 202
Fire-effect, 208, 213, 221, 230, 240, 241, 363
First astral body, 192, 289
First brother, 286

First copy, 211, 223, 225, 231, 237, 239–241, 272, 293
First Eden, 259, 261
First energy, 18, 173
First intellectual body, 192, 326
First mind, 271, 316
Five bodies, 329, 340
Five brothers, 294, 296
Five copy, 195
Five finite points, 61, 123, 124, 130, 136, 137, 140, 144, 151, 152
Five genetic codes, 16
Five guider agents, 303
Five kins, 303
Five quarters, 60
Five sisters, 105
Five-dimensional copies, 191
Five-faced self-luminous entity, 193
Flame body, 352
Flame family, 282, 285, 287, 288, 290, 291, 293, 300, 309, 351–354, 356, 364
Flame within, 325, 358
Flame without, 324, 325
Flamemates, 282, 288, 290, 291
Flow of granddaughter cells, 317
Flowing energy, 207
Fluctuation, 22
Fluorescence photon, 307
Focal male dolphin, 102
Focal point, 28, 169, 170
Foe, 4, 33, 63, 64, 106, 109, 111
Foe circle, 63, 64, 106, 109, 111
Follower class, 85, 217
Follower species, 62
Followership energy, 50, 64, 159, 275, 358
Followership pathway, 326
F-orbital, 151
Force of attraction, 86
Force of repulsion, 86
Form energy, 335
Form force, 334
Formative capability, 194
Formative growth, 157, 196, 216, 244, 329, 342
Formative growth paradigm, 244
Formative planning, 345

Index 385

Formative primordial greeter, 103
Formative programming, 344
Fortune-maker, 78
Forward-order development, 15, 16, 119, 120
Fougaro system, 296
Foul carbonated smell, 213
Foundation, 20, 5, 38, 41, 69, 71, 139, 175, 177, 215, 288, 292, 294
Four body system, 339–341
Four causal bodies, 274
Four guider agents, 301
Four half octaves, 295
Four kins, 301
Four members, 338
Four ones, 82, 146
Four pi, 57, 60, 128
Four points, 59
Four quarters, 61, 128
Four-body poles, 270
Four-body system, 2, 212, 228, 229, 231, 290, 294–296
Four-faced entity, 313
Four-fold growth, 99, 131, 195, 196, 284, 312, 316
Fourteen quarters, 58, 129
Fourteen-sided polygon, 354
Fourteenth astral body, 299
Fourteenth body, 341
Fourteenth entity group, 349
Fourteenth intellectual body, 330
Fourteenth principal guider, 48
Fourteenth sister, 296
Fourth astral body, 290
Fourth brother, 286
Fourth copy, 224, 225, 294
Fourth intellectual body, 192, 327
Fourth member, 122
Four-unit consciousness, 36, 77, 99, 136, 150, 154, 286
Fraction, 223
Fraternal twin, 283
Freedom entity, 281
Freedom entrepreneurship work, 209
Freedom force, 96
Freeriding, 29, 101, 151, 160, 168, 174, 298, 341, 344, 346
Freeriding causation, 344, 346
Frequency, 213
Froglets, 91, 92
Frontal lobe, 326, 328, 329
Fruit of [his] action, 16, 249
Fruit of devotion, 16
Fruit of impact, 17
Fruit of joyful life, 17
Fruit of knowing, 16
Fruit of reality, 17
Fruition, 212
Full circle, 55, 59–62, 137–140, 150
Full circle pi, 59, 60
Full moon, 69, 184, 239, 364
Future creator, 254
Future knowing, 270
Future life, 50, 51, 224, 313, 314, 357
Future oneness, 362
Future orientation, 39
Future reality, 17, 18, 42, 43, 57–59, 92, 96, 101, 208, 215, 241, 252, 265, 370
Future state, 264
Future system, 55
Future time, 42, 52, 53, 125, 126, 132, 133, 151, 163
Future time force, 151
Future value, 15

G

Gamete, 184, 196, 197
Gamma, 151, 163, 165, 166, 305–309, 347, 354
Gamma bond, 151, 165, 166, 305
Gamma decay, 306, 347
Gamma force, 163, 354
Gamma photon, 305–308
Gamma-ray photon, 306, 307
Garden of Eden, 261
Gaseous water, 213
Gemini house, 63, 68
Gemini zodiac, 106
Gender and generation differentiable class, 350
Gender and generation differentiable member, 48, 49
Gender and generation differentiated copy, 312
Gender Egalitarianism, 28

Gender-differentiable class, 331, 347, 350
Gender-differentiable member, 48, 49
Gender-differentiated copy, 312
Gender-differentiated present, 330, 331
Gene flow, 37, 157, 158, 362, 365
Gene stacking, 168, 360, 362
Gene stock, 37, 158
Genera, 14, 121
Generation differentiation, 344, 346
Generation-differentiable class, 331
Generation-differentiated copy, 312
Genes, 104, 105, 109, 168, 343
Genesis, 181, 338
Genetic code, 16, 175, 177, 178
Genetic drift, 365
Genetic equality, 283
Genetic exchange, 283
Genetic factor, 177
Genetic library, 37, 38, 51, 158, 174, 177
Genetic linkages, 300
Genetic sequences, 3
Gennie, 345
Genome, 177, 244
Genus, 14, 46, 47, 49, 119, 174, 333
Genus of entities, 47, 119
Geocentrism, 43, 44
Geodesic curvature, 170
Geodesic force, 170
Geodesic reality, 171
Geodesic torsion, 43, 171
Geography culture, 47
Geography force, 353
Germanic Europe, 28, 29
Germinal cell, 181, 182
Gift of consciousness, 65, 75, 76, 302
Glial cell, 182
Global condition, 14, 15, 31, 32
Global consciousness, 188
Global coordinate, 43, 44
Global culture-effect, 203
Global knowledge, 21
Global linkages, 114
Global locus standi, 114
Global maxima, 84

Global networks, 114
Global orientation, 28
Global spatial coordinate, 43
Global system, 222, 223
Global time, 22, 51–54, 82, 100
Global universe of entities, 195
Globalization, 20, 45, 206, 222
Globalized time, 22
Gluon, 147, 150, 151, 276
Goalkeeper, 89–93, 342, 356
Goal-oriented, 280
God class, 219
God consciousness, 179, 291
God of Gods, 85
God of immaculate knowing, 261
God within us, 5, 79
Godhead, 254, 255, 257, 259, 267, 269–271, 273–276, 281, 283, 288, 289, 297, 327
Godhead consciousness, 281
Godhead sister, 283
God-like entities, 85
Godliness, 291
Godship, 256
Golgi bodies, 183
Golgi cell, 182, 183, 268
Gorilla, 95, 96
Grand cause, 251
Granddaughter allele, 155, 197, 198
Granddaughter cell, 102, 154, 181, 182, 263, 264, 267, 268, 317, 319, 322, 326
Granddaughter consciousness, 264, 326
Granddaughter self-luminous entity, 263
Granddaughter soul, 218
Granddaughter spirit, 218
Grandfather allele, 155, 198
Grandfather cell, 103, 181, 182, 316, 317, 320
Grandfather electron, 162
Grandfather entity, 339
Grandfather flame, 284, 285
Grandfather soul, 218
Grandfather spirit, 217
Grandmother allele, 155, 198, 199, 348
Grandmother cell, 103, 181, 182, 267, 268, 317, 320, 322
Grandmother consciousness, 78, 267, 349

Index

Grandmother soul, 218
Grandmother spirit, 217
Grandpaternal ancestor, 220, 339
Grandson allele, 155, 198
Grandson cell, 102, 181, 182, 264, 321, 322, 326
Grandson electron, 161
Grandson self-luminous entity, 264
Grandson soul, 218
Grandson spirit, 217
Granular cell, 183, 185, 186, 189, 193, 200, 267, 268
Gravitated reality, 171
Gravitational center, 17, 303
Gravitational collapse, 135
Gravitational consciousness, 140, 323
Gravitational ecosystem-effect, 212
Gravitational effect, 364
Gravitational element, 37, 133, 141
Gravitational energy, 55, 67, 97, 153, 171, 189, 203, 213, 276, 316
Gravitational field, 208
Gravitational force, 209, 336, 337
Gravitational light, 66, 133, 226, 258, 260
Gravitational mass, 239, 245, 256, 322, 344
Gravitational point, 252, 321
Gravitational quality, 208, 292, 315, 318, 344
Gravitational reality, 37
Gravitational stillness, 86, 203
Gravitational system, 30, 228, 335
Gravitational wave, 37, 66, 121, 247, 334
Gravitoelectric cycle, 62
Gravitoelectric-effects, 228
Gravitoelectromagnetism, 337
Gravitomagnetic force, 170
Gravitomagnetic-effect, 228
Graviton, 262
Green light, 155
Greenhouse, 7
Greeter allele, 155, 197–200
Greeter capability, 194
Greeter cell, 181, 182, 296, 317, 321, 322
Greeter consciousness, 82, 92, 95, 104, 194, 195, 249, 255, 265, 299, 322
Greeter entity, 346, 347
Greeter essence, 5, 97, 289, 297, 300, 359, 360

Greeter face, 153, 255
Greeter grandmother, 55, 57, 58
Greeter soul, 218
Greeter spirit, 196, 217–219, 281, 289, 323
Grossness, 351
Group class, 350
Group culture, 47
Group of entities, 32, 33, 257
Group of species, 15, 47, 118
Group of ten guider agents, 300, 352
Group of ten potential guider agents, 349, 352
Group of ten potential principal guiders, 292, 300, 352
Group of ten principal guiders, 47, 291, 351, 352
Group-differentiable classes, 347
Group-differentiable member, 120
Group-differentiated present, 332
Growth electron, 161
Growth force, 272
Growth hormone, 212, 214
Growth mindset, 26, 27
Growth photon, 155
Growth potential, 63, 78, 100, 153, 255
Growth potentiation, 274
Growth sequence, 245, 262
Growth system, 228, 229, 231
Guider agent, 32, 220, 300, 303, 305, 324, 341
Guider body, 309
Guider force, 239, 241, 355
Guider God, 278
Guider horizon, 258
Guider mediation, 211, 325
Guider morality, 113
Guider planning, 98
Guider potential, 252, 339
Guider power, 29, 80, 371
Guider program, 240, 241
Guider spirit, 216
Guider-effect, 230, 241, 355
Guiding force, 28, 29, 98, 101, 231, 258, 261, 276–278, 284, 285, 299, 339, 349
Guru, 17, 29, 57, 65, 68, 74, 75, 79, 80, 83, 86, 96, 98, 177, 193, 209, 231–235, 252, 258, 259, 276, 290, 297, 303, 321, 323, 361

H

Half-instanton, 34, 35
Half-open space, 130
Half-spheres, 305
Hallucination, 240
Harbinger, 233
Harem, 104, 105
Harmonic body, 188
Harmony, 115, 229
Heart Committee of Ascended Masters, 326
Heart wheel, 180, 183, 186, 192, 213, 226
Heartbeats, 213
Heavenly life, 85, 289
Heedlessness, 79
Helium, 305
Hemizygous, 199, 200
Heptad repeat, 265, 337
Heptaploid, 194–196
Heterogeneous, 15, 46, 49, 50, 118–121, 144, 145, 148, 174, 305
Heterogeneous electron, 144, 148, 305
Heterogeneous family, 15, 119
Heterogeneous genus, 46, 50, 118, 120, 174
Heterogeneous group, 46, 118
Heterozygote, 200
Heterozygous chromosome, 154, 158
Heterozygous Y-linked, 155
Hexaploid, 194
Hexaquark, 140–143
Hibernate, 46, 108, 112
Hibernated entity, 176
Hibernating force, 330
Hibernating state, 176
Higgs rest mass, 164
High Council Guardian of Luminous, 243
Hinduism, 206, 232
Holding, 156
Holiday junction, 196
Holistic reality, 342
Holy spirit, 8, 73, 84, 216, 233, 253, 254, 270, 275, 280, 281, 346
Holy spirit within self, 84, 233
Homogeneity, 145, 284
Homogeneous copy, 294
Homogeneous electron, 144, 148, 305
Homogeneous genus, 15, 46, 119–121, 174
Homogeneous group, 31, 46
Homogenizing force, 46
Homolog, 66
Homologous body, 289
Homotopy, 146, 147
Homotopy group, 146, 147
Homozygote, 197, 200
Homozygous twin, 197
Horizontal axis, 128, 272
Horizontal demand, 251
Horizontal light force, 241
Horizontal motion, 3
Horizontal order, 145, 265, 337
Horizontal-order development, 15, 16
Horoscope, 355
Hubble tension, 161
Human consciousness, 78
Human entity, 178, 239, 240, 254, 262
Human factor, 2, 7, 325
Human flame, 18, 353, 362
Human force, 88, 179, 209, 214, 334, 365
Human Godhead, 264
Human kingdom, 301, 323
Human soul, 219
Human-effect, 68, 212, 213
Hungry ghost, 345
Hydrated entity, 295, 296
Hydrodynamics, 39
Hydrogen, 88, 215, 230, 265, 266
Hydrogenated ether-effect, 323
Hydrophilic, 278, 280
Hydrophilic substitution, 278
Hydroxy acid, 265, 266
Hydroxylase, 191, 207
Hyperactive principal, 220
Hyperactivity, 185, 197, 200, 201, 203, 213
Hyperactivity pressure, 213
Hyperbola, 43, 55, 60, 61, 134, 136, 150
Hyperbolic chord, 136
Hyper-urgency, 110
Hypoactive foe, 221
Hypoactivity, 200
Hypocrisy, 8, 79

Index 389

Hypothesis, 37–39, 77, 310
Hypothesized reality, 77

I

IBGYOR light, 142, 305
Ideal consciousness, 335
Ideal force, 71, 102, 171, 189, 360, 361
Ideal Godhead, 253, 255
Ideal manifestor, 250
Ideal point, 277
Ideal self, 15, 26, 119, 120, 343, 366
Ideal self-luminous entity, 15, 119, 120, 343, 366
Ideal value, 221
Idealizer, 22
Ideational framework, 26
Identical allele, 197
Ideological person, 354
Ideology, 354
Ignited community, 368
Ignorant, 101, 272
Illuminate codon, 177
Illuminated consciousness, 256
Illuminated growth, 247, 248, 340
Illuminating force, 257
Illuminating value, 16, 84, 144, 148, 165, 196, 247, 260
Illumination, 84
Illuminator, 210, 258, 261, 275, 288, 296, 367
Illuminator deity, 210
Illuminator paradigm, 296
Illusionary animate entity, 226, 357
Illusionary body, 215, 310
Illusionary cosmological constant, 171
Illusionary electron, 150
Illusionary energy, 92, 97, 190, 191, 210, 211, 215
Illusionary force, 15
Illusionary horizon, 258
Illusionary ion, 160
Illusionary joy, 17, 369
Illusionary past, 26, 59, 82, 158, 354
Illusionary pi, 56, 60
Illusionary preceding phase, 353
Illusionary production, 32, 38, 42, 52, 150, 225, 249, 250, 258, 262, 269, 270, 323
Illusionary reproduction, 36

Imaginary absolute gravitational constant, 52, 54, 153, 154, 160, 170
Imaginary consciousness, 293, 298, 300
Imaginary future, 26, 158
Imaginary horizon, 259
Imaginary ion, 160
Imaginary phase, 353
Imaginary photon, 305
Imaginary pi, 56
Imaginary spirit, 97, 171
Imagination force, 15
Immanence-effect, 158
Immanent body, 299
Immanent feminine-effect, 156
Immanent form, 121
Immanent guider power, 271
Immanent oneness, 286
Immanent reality, 275
Immanent system, 296
Immanent wisdom, 334
Immature mind, 367
Immersive experience, 320, 321
Immorality, 113
Immortal Almighty Creation, 13
Immortal entity, 8
Immune system, 230, 231
Impassioned devotee, 82, 90, 100, 233, 234, 253, 254, 368
Impediments, 189
Imperial governance, 96
Impregnated consciousness, 101, 102
Impregnated life purpose, 64
Improficient system, 18
Inactive state, 264
Inanimate body, 92, 293
Inanimate element, 78
Inanimate energy, 226, 359
Inanimate one, 293
Inattentiveness, 79
Incarnational consciousness, 274
Incident photon, 307
Inclusion orientation, 28
Increasing demand, 74, 153
Incremental growth, 269
Incubation point, 114

Incubator value, 72
Indigestion, 305
Indigo color, 158
Indigo light, 142, 148, 158
Individual class, 197, 198, 348
Individual copy, 198, 199
Individual energy, 119
Individual force, 198–200, 315
Individual potential, 34, 165
Infinite adding, 168, 362, 365
Infinite breeding, 23, 92, 320
Infinite causal bodies, 355
Infinite cells, 189, 202, 269
Infinite collective, 18, 181
Infinite conscious system, 4, 20, 256, 313, 314
Infinite consuming, 115, 126
Infinite cost, 83
Infinite council, 178, 179, 291
Infinite deity, 261
Infinite density, 247
Infinite dimensionality, 303
Infinite element, 83
Infinite energy, 77, 224
Infinite entities, 364
Infinite entropy-effect, 56
Infinite exchange, 245, 314, 349
Infinite experiences, 75–77
Infinite feeding, 23
Infinite feminine, 153
Infinite flow, 224
Infinite forms, 77, 78, 112, 130, 220, 256
Infinite group, 89
Infinite growth, 23
Infinite guider-effects, 355
Infinite international force, 55
Infinite layers, 258
Infinite lifeforms, 111
Infinite lifetimes, 248
Infinite links, 113
Infinite management factor, 59, 60
Infinite masculinity, 85
Infinite mediation, 257
Infinite multiplying, 221
Infinite networking, 62, 228, 300, 314

Infinite para psychic force, 111
Infinite paternal, 222
Infinite permanent form, 113
Infinite point, 39, 51, 52, 59, 125, 126, 132, 154
Infinite producing, 115, 124
Infinite production, 251
Infinite profiting, 113
Infinite psychic linkages, 155
Infinite quarks, 288
Infinite reality, 355
Infinite replication, 362
Infinite sentient entropy, 209
Infinite social benefit, 260, 276, 319
Infinite social cost, 262
Infinite sound, 299
Infinite space, 152, 175, 244
Infinite spontaneous creations, 69
Infinite temporary forms, 63, 112
Infinite time, 50, 175
Infinite value, 99, 192, 211, 237, 238, 261, 264, 282, 370
Infinite variations, 219
Infinite weight, 269
Infinite White Star, 243
Infinite Wisher, 52, 53
Infinite wishes, 216, 217, 220, 252, 253, 260, 318, 319, 366
Infinite worker-social benefit, 260
Infinite worker-social cost, 261, 276
Infinite workforce system, 314
Infinitesimal member, 122
Infinity point, 63, 178, 231
Inflated sense, 221
Information saturation, 114, 273
Information void, 325
Infused experience, 201
Inherited capability, 294
Inherited reality, 201
Initial causation state, 37, 51, 158, 264
Initial spatial state, 6, 100, 142, 143
Initial temporal state, 37, 170, 172
Inner child, 271
Inner conflict, 224, 357, 358
Inner layer, 181

Index 391

Inner self, 4, 370
Inner sense, 64, 187, 321, 322
Innovative cycle, 66
Innovative social relationships, 368
Inordinate member, 120
Inorganic dissonance, 5, 320
Institutional authority, 209
Institutional class, 218
Institutional consciousness, 81
Institutional energy, 221
Institutional experience, 8, 180
Institutional force, 82, 92, 178, 180, 193, 281
Institutional geography, 209
Institutional governance, 93
Institutional infrastructure, 209
Institutional legitimacy, 94
Institutional morality, 113
Institutional sovereignty, 81
Instructor, 312
Insular religion, 2
Integrated system, 295
Intellectual body, 18, 178, 179, 188, 192, 207, 210, 211, 215, 223–246, 248, 309, 310, 325, 328, 331, 342, 351, 371
Intellectual factors, 80
Intellectual property rights, 234
Intellectual realm, 47, 175, 353, 370
Intellectual tranquility, 80
Intended reality, 319, 321
Intermediation, 329
Inter-molecular, 269
Intermolecular force, 342
International phase, 50
Interplay, 113
Interrogation, 76, 77
Interrogator, 76
Intersects, 113
Intrinsic air-effect, 201
Intrinsic body, 299
Intrinsic consciousness, 347
Intrinsic divine-effect, 201
Intrinsic earth-effect, 201
Intrinsic energy, 181, 185
Intrinsic entity value, 328, 329
Intrinsic ether-effect, 202

Intrinsic experience, 325
Intrinsic face, 328
Intrinsic fire-effect, 201
Intrinsic force, 20, 97, 100, 325
Intrinsic form, 121
Intrinsic four half octaves, 295
Intrinsic impact, 74
Intrinsic light, 83, 323
Intrinsic oneness, 25, 159
Intrinsic reality, 208
Intrinsic sentient energy, 190
Intrinsic strong psychic linkage, 276, 345
Intrinsic system, 229, 231
Intrinsic water-effect, 201
Intrinsic weak forces, 294
Intrinsic weak psychic linkage, 276, 335, 336
Introspection, 5, 235
Introverted, 17, 20, 161, 256
Introverted face, 256
Intrusion, 3
Investigator, 12, 37, 108
Investment spirit, 362, 363
Investment-oriented factor, 367
Investor, 251
Invisible greeter, 187
Invisible horizon, 259

J

Jackpot, 40
Jeff Bezos, 73
Joy force, 189
Jupiter, 97, 98, 238–241
Juvenile, 97, 367

K

Kaon, 34, 144, 148, 150, 160, 161, 165–167
Kaon long, 144, 148, 165, 166
Kaon negative, 144, 148, 160, 161
Kaon positive, 150, 167
Kaon short, 144, 148, 161, 165–167
Keratin, 182
Kin, 9, 63, 64, 68, 70, 301, 302, 318, 319, 356
Kin universe, 318, 319
Kinetic energy, 307
King of deities, 85, 249
Kingdom, 14, 33, 121, 282, 324

Kith, 9, 33, 63, 64, 68, 70, 318–320
Kith complements, 9
Kith universe, 318–320
K-meson norms, 161
Knowable reality, 24
Knowable value, 23
Knower class, 218
Knower deity, 353
Knower dimension, 23
Knower force, 19
Knower Godhead, 275
Knowing cell, 269
Knowing recombination, 151
Knowledge force, 201
Knowledge-effect, 54
Known reality, 6, 34, 49, 82, 277, 278, 280
Koan negative, 161
Krypton, 183, 266, 267

L

Lackluster, 367
Larva, 91
Lateral fissure, 326, 330, 331
Law of limitation, 5, 302, 303
Leader class, 85, 217
Leader species, 62
Leadership culture system, 23
Leadership energy, 63, 64, 113, 159, 178, 357
Leadership pathway, 326
Leadership workculture system, 368
Leading sound, 269
Leap of faith, 221
Leap year, 365
Left brain, 331
Leftover, 342
Length power growth rate, 335, 337
Leo house, 63, 68
Leo zodiac, 105
Lepton, 144, 148, 305, 306, 348
Lexicographic sequence, 248
Liberal democratic philosophy, 319
Liberal democratic technique, 212
Liberated cell, 281
Liberated soul, 255
Liberation, 65, 69
Liberator, 191, 192
Liberty, 222
Libra house, 64, 68
Libra zodiac, 103, 104
Life cycle, 3, 254
Life experience, 71, 72, 352, 353
Life lesson, 4, 112
Life rhythm, 355
Lifecycle-effect, 182
Lifeforce, 263
Lifeline, 277
Ligand, 53, 54, 160
Light body, 289
Light consciousness, 35, 36
Light echo, 363
Light force, 88, 238, 239, 241, 242, 292, 293, 300, 301, 304, 327
Light potential, 34, 160
Light spectrum, 38, 66, 83, 340, 341
Light wave, 158
Lighthouse, 285
Liminal member, 121
Limitation, 8, 90
Limiting element, 247
Linear flow of time, 251, 267
Linear value, 61
Lithium, 305
Livable community, 358
Livable experience, 121
Livable past, 167, 271, 274, 275
Lived community, 359, 358, 360
Lived experience, 121, 271, 272
Living body, 307
Living consciousness, 111, 280, 315, 318
Living entity, 16, 33, 122, 123, 175, 176, 274, 282, 283, 285, 286, 290, 292, 331, 332, 342, 343, 345, 355
Living God, 287
Living life, 32, 81, 149, 231, 357, 358
Living soul, 179, 275, 277, 278, 280, 290, 317, 318, 320, 321, 326–328
Local condition, 14, 15, 32, 33
Local consciousness, 188–190
Local ecosystem, 300
Local locus standi, 114
Local system, 222, 223

Index

Local time, 20, 22, 23, 51–54, 82, 100
Local universe, 114
Local workculture, 203
Local-effect, 222
Localized time, 22, 52
Localized wholesome π, 45
Locus standi, 114
Logical reality, 331, 332, 342
Loop, 16, 123, 124, 357
Lord, 73, 74, 114
Love, 22, 156, 226, 236, 256, 257, 278, 317
Love God, 278
Lower octave, 164
Luminous element, 226
Luminous entity, 16, 18, 119, 123, 176, 192, 212, 215, 225, 226, 254, 255, 260, 262, 269, 271, 279, 282–284, 290, 292, 297, 313, 339, 343, 345, 346, 355
Luminous force, 58
Luminous point, 212, 213, 244–246, 281, 289
Luminous realm, 287, 339
Luminous system, 301, 305
Luminous wheel, 213
Lunar cycle, 103, 179, 182–184, 186
Lunar energy, 184
Lunar time, 179
Lunar wheel, 19, 179, 181, 183, 185–188, 190, 192, 201–204, 230, 231, 290
Lymphatic system, 230, 231

M

Machinery body, 331
Macro ellipse, 57
Macro perspective, 26, 27
Magnetic system, 228
Magnetoelectric-effect, 160
Majorana boson, 263
Majorana fermion, 227
Male body, 343–345
Male child, 332, 344, 345
Male consciousness, 78
Male copy, 312, 315
Male frog, 92
Malleability, 80, 81
Management body, 330
Management class, 85, 86
Management pathway, 327

Manifestor, 20, 193, 249, 250, 277, 353
Manipulator, 153
Market forces, 113
Marketing body, 331
Mars, 101, 106, 238–241
Martinotti cells, 182
Masculine body, 284, 340
Masculine cell, 158
Masculine class, 220–222, 350
Masculine copy, 190, 192, 312, 348
Masculine dimension, 361
Masculine double copies, 217
Masculine electron, 159, 308
Masculine energy, 325, 359
Masculine gene, 96
Masculine hemipenis, 108
Masculine self, 82, 87, 326, 327, 330, 331, 348
Masculine spirit, 64
Masculinity, 108, 289, 331, 343
Masculinization, 155
Mass consciousness, 34, 35, 37, 38, 41, 42, 58, 75, 140, 141, 179, 226, 245, 356, 358, 364, 365
Mass descendance, 247
Mass exchange paradigm, 298
Mass fragment, 77, 89, 225, 228
Mass friendship, 279
Mass layer, 181
Mass of consciousness, 6, 50, 77, 129, 163, 242, 318, 344
Mass-effect, 140, 269, 307
Mass-energy-momentum tensor, 163, 164
Mass-stress-energy-momentum density tensor, 58, 129
Master breeder, 339
Master cell, 183, 185
Master manager, 316
Material body, 329
Material kingdom, 35, 36, 323, 324
Material power, 225, 318, 324
Materialism, 97
Materializable reality, 56
Maternal body, 231, 306
Maternal community, 363
Maternal consciousness, 77, 78, 91, 101, 255, 263, 265, 266, 316, 324, 355, 356
Maternal copy, 192, 312, 348, 349

Maternal entity, 252, 346
Maternal essence, 192, 280
Maternal flamemates, 290
Maternal genes, 104, 343
Maternal Godhead, 276
Maternal knowing, 271
Maternal luminous, 178, 283, 288, 292, 296, 355
Maternal mitochondrion, 266
Maternal primordial greeter, 92, 287, 288, 325, 334, 356–360, 364
Maternal self-luminous entity, 244, 288
Maternal spirit, 217, 218, 279–281, 284, 290
Maternal twin cell, 290, 291
Maternal universe of cells, 251
Maternal womb, 283
Matrilineal dimension, 220
Matrilineal planning, 344
Matrilineal programming, 345
Matrilineal reverberations, 363
Matter-stress tensor, 143
Mature electron, 36, 161
Medulla oblongata, 328–330
Meeting point, 113, 114, 358
Meiosis, 42, 338
Melanin, 182
Melanocyte cells, 181
Melatonin, 180, 191, 200, 202–204, 214, 228
Membrane protein, 276
Memorized divination, 248
Mendel model, 39
Mental body, 18, 48, 179, 187–189, 192, 195, 207–209, 211, 223, 225, 226, 248, 309, 310, 315–317, 325, 344, 345, 351, 371
Mental equanimity, 80
Mental factors, 78, 79
Mental impression, 224
Mental realm, 1, 6, 47, 68, 175, 283, 284, 286, 287, 353, 354, 369
Mental space, 2, 8, 292, 321
Mental well-being, 210, 211
Mentor body, 331
Mercury, 105, 238–241, 247
Meron, 34–36, 164, 165
Meron pair, 34
Meso layer, 181
Meso proton, 163

Meson, 159–162, 167, 308
Metal body, 325
Metal kingdom, 324
Metal soul, 219
Metamorphosing, 14
Metaphysical cause, 309
Metaphysical effects, 368
Metaphysical entropy paradigm, 245
Metaphysical existence, 81
Metaphysical factor, 80
Metaphysical planning, 345
Metaphysical self, 289
Metaphysical system, 229
Metaphysics, 2, 6, 45
Metaverse, 149
Methylene, 88
Metric tensor, 43, 131, 163, 165, 171
Midbrain, 330
Mind-born creator, 87
Mineral essence, 324
Mineral kingdom, 324
Mineral soul, 219
Minimum entropy state, 114
Mirage, 57
Mitosis, 42, 338
Mixed heterogeneous, 145
Moderating factor, 81
Moderator, 285, 300
Modern science, 1, 37–39, 41, 72, 73, 149, 206
Modulus, 155
Molecular body, 290
Molecular force, 342
Molecule, 160, 293, 342
Molly, 103
Momentum, 58, 109, 129, 154, 163
Momentum density tensor, 129, 154
Monetary body, 330
Monoamine neurotransmitter, 187
Monozygotic twin, 197, 283
Moral responsibility, 21, 23, 203, 212, 295
Morality, 79, 113, 369
Mortal entity, 113, 244
Mossy cell, 183, 267, 268
Mother allele, 155, 198, 199, 348
Mother cell, 102, 181, 182, 263, 265–268, 316,

Index 395

319, 320
Mother Earth, 93, 230, 239, 240
Mother Nature, 4, 7, 14, 15, 70, 85, 101, 111, 191, 192, 219, 232, 244, 261, 262, 266, 267, 271, 297, 334, 336, 361, 369
Mother of seven conscious lights, 285
Mother soul, 218
Mother-father allele pair, 155, 198
Motor cortex, 342
Multicellular organism, 269, 287
Multicellularity, 256
Multiforms, 78
Multiplex, 158
Multipliable astral body, 298, 300
Multipliable physical body, 306, 307
Multipliable system, 295
Multiplied allele, 155
Multiplied self, 221
Multiplied sequential-effect, 331
Multiplied system, 296
Multiplying allele, 155
Multiplying body, 362
Multiplying consciousness, 77, 166
Multiplying order, 166
Multiplying physical body, 307, 346
Multiplying self, 263, 280, 281, 283
Multiplying system, 296, 302
Muon, 150, 151
Muscular body, 329, 330
Musculoskeletal system, 227, 230
Muslim domination, 21
Muster, 104
Mutation, 363
Myelin cell, 182, 193, 265

N

National consciousness, 188
National income, 21
National time, 22, 24, 53, 54, 100
National-effect, 222
Natural conscious value, 60, 61
Natural copy, 317, 318, 324
Natural essence, 90, 94, 95, 157, 189, 239, 271, 281, 289
Natural growth, 251, 292, 313, 317, 319, 321
Natural light, 155

Natural love, 256, 257
Natural reality, 9
Natural science, 6
Natural theory, 324
Natural value, 1, 18, 26, 36, 57, 59, 67, 146, 148, 155, 196, 262, 263, 273, 274, 322–324
Negative energy, 39, 209, 325, 328, 359
Negative entity, 279
Negative factor, 325
Negative quark, 35, 164, 165
Neptune, 71, 94, 238, 241–244, 246, 247
Nervous system, 228–230
Networking linkages, 2
Networking system, 188, 209, 228, 276, 300, 313, 314
Networking-effect, 184
Neural crest cells, 181
Neuroblast cell, 267
Neutrino, 123, 124, 135, 136, 138, 149, 304, 305
Neutron, 67, 96, 97, 101, 136, 137, 160, 224, 225, 228, 242, 243
Neutron star, 67, 96, 97, 101, 242, 243
New moon, 66, 184, 239, 364
Newton, 37, 40, 44
Newtonian Universal Gravitational Constant, 127
Nine kins, 301
Nine members, 335–338
Nine potential kins, 301
Nine triangles, 60
Nine within ten, 359, 362
Nine-fold growth, 42
Nineteenth astral body, 302
Nineteenth gamma photon, 306
Nineteenth intellectual body, 331
Nineteenth member, 174
Nineteenth potential guider agent, 305
Nineteenth principal guider, 49
Ninth body, 66, 69, 243, 340
Ninth brother, 287–289
Ninth conscious light, 287
Ninth etheric body, 367
Ninth intellectual body, 328, 329
Ninth member, 122, 174
Ninth point, 59

Ninth sister, 288
Nitrogen, 266, 322
Nitrogenated earth-effect, 323
Nonaquark, 141–143
None without all, 303
Nonlinear flow of time, 268
Non-reciprocal exchange, 225
Nonresident alien universe, 324
Nonworking principal, 41, 220
Nordic Europe, 28, 29
Normative development paradigm, 244, 279
Normative organization, 65
Normative planning, 64, 178
Normative profiting, 65, 185
Normative programming, 64, 186, 345
Notional theory, 302, 303
Notions, 4, 7
Nucleus, 136, 137, 225

O

Object of devotion, 74
Objectified consciousness, 156
Objectifying causation, 346, 350
Obligation, 277, 278
Occipital lobe, 327, 328
Octave, 47, 81, 91, 118, 122, 123, 144, 145, 148, 150–152, 156, 157, 163, 164, 177, 196, 199, 227, 238, 242, 243, 248, 265, 293, 294, 296–299, 307, 308, 311, 322, 323, 325, 333, 338, 341, 348, 349, 351, 358, 359, 361, 363
Octave of alleles, 199
Octave of amino acids, 265
Octave of conscious lights, 297
Octave of copies, 47, 144, 145, 148, 238, 243, 294, 311, 325, 338, 351, 363
Octave of effects, 358, 359, 361
Octave of elements, 293, 297
Octave of entities, 248, 311
Octave of eta mesons, 163
Octave of fermions, 227
Octave of gametes, 196
Octave of gender-differentiating four generations, 156
Octave of infinite quarks, 164
Octave of organisms, 349
Octave of photons, 157
Octave of potential electrons, 322, 323
Octave of sections, 150–152, 164

Octave of self-luminous entities, 123
Octave of three-eyed double copies, 227
Octaves, 92, 294, 295, 299
Octoploid, 196
Odderon, 294
Oligodendrocyte cell, 267, 182
Oligodendrocyte precursor cell, 268
Oligodendrocyte progenitor cell, 267
OLM cell, 182, 268
Omnipotence, 207, 371
Omnipotent, 88, 90, 105, 241, 244
Omnipotent photon, 244
Omnipresence, 207
Omnipresent cause, 187
Omnipresent matter, 88
Omnipresent reality, 1
One body system, 294, 306
One with all, 303
One with nothing, 75, 301
One with three ones, 358
One within entity, 156
One within everyone, 260
One without circle, 60
One without entity, 156
One-hundred eighty bodies, 340
Oneness consciousness, 162, 286
Oneness of action, 298, 357
Oneness of pie, 43
Oneness with atom, 75, 76
Oneness without entity, 156
Oneself, 1, 23, 34, 41, 62–65, 69–71, 73, 74, 86, 144–146, 190, 216, 222, 225, 232–234, 236, 253–256, 275, 276, 279, 288, 324, 328, 332, 341, 360, 361, 365, 370
Open space, 26, 125–132, 134–136, 162, 163
Opulent, 83
Orbit, 153, 154, 159
Orbiting copy, 159, 160
Orbiting element, 159
Ordinal member, 120, 121
Organic consonance, 320
Organic resonance, 5, 320, 342
Organizational arrangement, 83
Organizational development, 112, 116, 147, 267, 320, 327, 329
Organizational entropy, 17

Index

Organizational gene, 109
Organizational metric, 171
Organizational performing, 112, 116, 193, 319, 327
Organizational planning, 112, 115, 116, 178, 254, 289, 319, 326
Organizational profiting, 113, 116, 320, 327
Organizational programming, 112, 116, 193, 201, 319
Organizational reality, 24, 253
Organizational sameness paradigm, 253
Organizational system, 229
Oriens lacunosum-moleculare cell, 182
Origin of everything, 153
Orion, 345–347, 367
Orthodox, 28, 29
Other Backward Class, 85, 86
Outer child, 271
Outer layer, 181
Outer sense, 64, 321, 322
Outflowing, 133, 335, 336
Outgoing photon, 307
Overall consciousness, 179
Overflowing consciousness, 249
Overflowing energy, 315
Overshadowed, 240
Oversoul family, 277, 278
Oxidation potential, 213
Oxygen, 239, 266
Oxygenated fire-effect, 323

P

Pair production, 308
Pairing photon, 275
Palm wheel, 62
Panoramic reality, 56, 66
Par excellence, 101, 115
Para body, 92, 260, 293, 296
Para class, 353
Para conscious light, 18, 296
Para conscious potential, 179
Para consciousness, 72, 102, 262, 280
Para deity culture-effect, 368
Para deity kingdom, 275
Para deity soul, 219
Para deity universe, 345, 368

Para entity, 15, 47, 73, 76, 121
Para gravitational constant, 54
Para psychic force, 111
Para time, 212
Para Wisher, 265, 266
Parabola, 55, 59–61, 111, 131–134, 136–138, 150
Parabolic challenges, 111
Parabolic future time, 125
Parabolic pi, 58, 59
Para-conscious body, 207, 248, 371
Para-conscious decisions, 4
Para-conscious determination, 271
Para-conscious factor, 271
Para-conscious idealization, 356
Para-conscious lights, 285, 287, 291, 294, 296, 299
Para-conscious trading, 245, 309
Para-conscious value, 288
Para-consciousness potential, 179
Paradigm, 8, 156, 244, 248, 275, 279
Paradise, 261
Parallel life consciousness, 349
Parallel reality, 82, 231
Parallel universe consciousness, 347
Param astral body, 300, 302
Param cell, 182, 266
Param child, 222, 243–245, 247, 248, 287, 301
Param creator, 255
Param daughter, 174, 286, 334, 346, 348
Param deity culture-effect, 273
Param deity soul, 219
Param Deity Universe, 354, 355
Param entity, 316
Param granddaughter, 348, 349
Param grandson, 348, 349
Param greeter, 287, 288
Param Guru, 233, 282
Param human child, 247, 248, 277, 281, 297, 324
Param human-effect, 270, 271, 279, 280
Param intellectual body, 342
Param kind, 122
Param Liberator, 193
Param manifestor, 223

Param maternal, 119, 174, 286, 348
Param meiosis-effect, 338
Param mitosis-effect, 338
Param neutrino, 305
Param odd value, 336
Param Para Greeter, 255
Param para maternal, 243, 250
Param Para Paternal, 238
Param paths, 365
Param perpetuator, 243, 274
Param physical body, 308, 309
Param son, 53, 54, 173, 222–224, 264, 276, 288, 346, 348
Param soul, 219
Param species, 122
Para-psychic linkage, 147, 275, 278
Para-psychic realm, 283
Para-sensory perception, 212
Para-sensory vibrations, 289
Parasympathetic nervous system, 229
Parathyroid gland, 187, 190, 191
Parietal lobe, 327, 329
Partial delta bond, 161
Partial gamma bond, 162
Partial phi bond, 126, 127, 162
Partial pi, 35, 36, 38, 53, 54, 59, 160, 165–167, 202
Partial pi bonds, 35, 38
Partial sigma bond, 161, 166
Partial spatial coordinate, 43–45
Passable time, 47
Past divergent group, 14
Past ego, 366
Past factor, 43
Past follower, 249
Past force, 155, 270, 313, 342, 347, 349, 357, 364
Past grandmother, 249
Past knowing, 270, 314
Past life, 50, 51, 72, 148, 149, 224, 313, 314, 357
Past moment, 270, 272
Past planning, 347, 350
Past reality, 59, 101, 170, 240–242, 323, 336, 355, 369
Past sequence, 363

Past state, 264
Past system, 54, 95
Past time force, 151
Past togetherness, 362
Past-effect, 299, 369
Paternal community, 362
Paternal consciousness, 77, 87, 88, 91, 95, 101, 190, 249, 253, 264, 265, 280, 321, 345
Paternal copy, 190, 191, 312, 348
Paternal creation, 343
Paternal dimension, 269
Paternal entity, 252, 346
Paternal flame, 283, 284
Paternal gene, 104, 343, 362
Paternal Godhead, 346
Paternal jinn, 345, 367
Paternal luminous, 339, 340
Paternal soul, 218, 305
Paternal spirit universe, 271
Path of absolute consciousness, 33, 254
Path of action, 16, 33, 41, 68, 205
Path of conversion, 233, 234
Path of copying, 329
Path of devotion, 16, 33, 41, 67, 87, 189, 190, 205, 250
Path of divinity, 33, 41
Path of extroversion, 17, 233, 236
Path of followership, 152, 304
Path of infinite development, 113
Path of introspection, 234
Path of introversion, 17, 233, 236
Path of inversion, 234
Path of knowing, 16, 33, 41, 67, 90, 205
Path of leadership, 279
Path of materialism, 64
Path of programming, 163
Path of reversion, 234
Path of sentient entity, 298
Path of spiritual growth, 76
Path to absolute freedom, 98
Patrilineal dimension, 220
Patrilineal planning, 344
Patrilineal programming, 344, 345
Patrilineal reverberations, 363
Pedigree, 122

Index

Pentagon, 154, 158
Pentaploid, 194, 195
Pentaquark, 141, 143
Perceived consciousness, 181
Perceived sense, 320
Perceived space, 42, 43, 171
Perceptual reality, 7, 342
Perfect adaptation, 14, 15, 31, 32
Perfect greeter, 65
Perfect guider agent, 32, 33, 300
Perfect illuminator, 257
Perfect perpetuator, 260
Perfected consciousness, 20, 23
Performance orientation, 28
Performed experience, 353
Performed past of everybody, 339
Performing force, 320
Performing spirit, 208, 345, 347
Performing-effect, 344
Perineural system, 227, 230
Periphery, 27
Perpetuate codon, 177
Perpetuating entity, 148, 150
Perpetuating value, 16–20, 22, 25, 28, 29, 42, 55, 56, 59, 61, 71, 84, 104, 128, 129, 165, 215, 216, 248, 249, 251, 254, 281, 285, 356
Perpetuator effect, 361
Perpetuator force, 257, 258, 363
Perpetuator organization, 249
Perpetuity, 362
Personal authority, 28, 29, 278, 332
Personal body, 215
Personal conscious well-being, 195
Personal reality, 110, 332
Personal sovereignty, 98
Personal space, 3, 7
Personal value, 86
Personification, 221
Personified natural reality, 355
Phi, 12, 150, 151, 165–167
Phi bond, 150, 151, 165, 166
Phi meson, 167
Phosphorous, 265–267
Photo copy, 322
Photoelectric serum, 308

Photoelectric-effect, 308
Photoelectron, 308
Photon, 5, 38, 42, 101, 123, 124, 135, 136, 138–140, 149–151, 156, 159, 167, 168, 171, 244, 246, 248, 262–265, 273, 274, 296, 302–307, 309, 313, 318
Photon body, 248, 273
Photon exchange, 302
Photon half, 42, 262
Photoreceptor cell, 321
Photosynthesis, 207, 227
Phylum, 120, 121
Physical body system, 308
Physical existence, 78, 80–82
Physical factors, 80, 81
Physical realm, 4–6, 47, 103, 176, 194, 283, 284, 287, 294, 353, 369
Physical sciences, 6
Physical self, 289
Physical well-being, 210, 211, 214
Pi bond, 36, 38, 143, 144, 148, 153, 157, 159, 165
Pi meson, 162, 167
Pie, 42–44, 50, 59, 126, 127, 131, 136, 137, 139
Piety, 80
Pillar, 335
Pineal gland, 19, 179, 191, 203
Pisces house, 65, 69
Pisces zodiac, 87
Pituitary gland, 20, 179, 185, 191
Planetary bodies, 65, 66, 69, 238, 239, 243
Plank constant, 303
Planning point, 62, 112, 150, 151, 347
Planning value, 345
Plant body, 325
Plant entity, 323
Plant kingdom, 322, 323, 340
Plant soul, 219
Platelet cell, 265, 267
Ploid, 194, 196
Poison fangs, 93
Poke, 209
Poking entities, 209
Polar deity, 275
Polariton, 129
Polarization, 115

Polluted consciousness, 1, 69, 92, 210, 273, 281
Polluted reality, 171
Polluted state, 74
Polluting consciousness, 69, 329
Polluting state, 176
Polycentrism, 43
Polycomb group, 267
Polypeptide, 159
Pomeron, 301, 302
Pomeron exchange, 302
Pomeron trajectory, 302
Pons, 328, 329
Population density, 341
P-orbitals, 157
Positive energy, 39, 150, 209, 325, 328, 359
Positive factor, 325
Positive quark, 35, 144, 148, 306
Positron, 321
Potential body, 293
Potential child, 101, 222
Potential children, 222
Potential daughter, 174, 348
Potential deity, 178
Potential down quark, 140
Potential electron, 151, 321–323
Potential ellipses, 151
Potential energy, 155, 189, 227, 268, 318, 320
Potential family, 334–338
Potential flame body, 352
Potential flame family, 353, 354
Potential gender, 191
Potential generations, 156
Potential granddaughter cell, 317
Potential grandfather allele, 199
Potential grandfather cell, 318
Potential grandmother cell, 322
Potential greeter allele, 199, 200
Potential greeter cell, 321
Potential guider agents, 300
Potential ideal, 23
Potential kaon, 150, 165, 166
Potential life, 314
Potential light, 149, 304
Potential nonaquark, 141, 142

Potential para-consciousness, 208
Potential photons, 38
Potential principal guider, 292, 300, 324
Potential quark, 36, 38, 141, 143, 163, 164
Potential reality, 56
Potential shape, 58, 60
Potential son, 174, 222, 348
Potential space, 163
Potential splitting, 263
Potential strange quark, 38, 141, 161
Potential tensor sequence, 163
Potential thing, 276
Potential three ones, 359
Potential triple family, 335
Potential twin flame family, 352
Potential twin neutrino, 305
Potential twin-flame body, 352
Potential up quark, 140–142, 152, 263
Potential white star, 149
Potentially departed person, 276
Potentially scalar meson, 159, 308
Potentially vector meson, 159
Potentiation, 268, 269
Power cascade, 338
Power cone, 338
Power distance, 28
Preceding phase, 352
Pre-dominating collective, 108
Predominating effects, 368
Premotor cortex, 332
Preon plus, 161
Preordinate member, 120
Present consciousness, 1, 58, 102, 207, 208, 253
Present creature, 34, 49, 122, 253, 254, 296
Present development, 96, 178
Present earth-effect, 184, 227
Present entropy-effect, 320
Present experiences, 75, 76
Present follower, 249
Present form, 150
Present group, 55, 71, 354
Present growth, 15, 20, 22, 23, 89–92, 94, 139, 177, 215, 255, 273, 285, 288, 301, 340, 351, 356

Index

Present knowing, 270
Present life experience, 368
Present life force, 354
Present maternal, 249
Present morality, 113
Present otherness, 362
Present paradigm, 6, 8, 156, 226, 270, 279
Present self, 253
Present space, 38, 123, 130, 132, 134–136, 161–163, 243, 258, 259
Present state, 169, 210, 211, 264
Present system, 55, 57
Present time force, 38, 151
Present trading-effect, 273
Present value, 36, 133, 181, 350
Primary causal body, 270–272
Primary factor, 318
Primeval astral body, 227, 298, 302
Primeval child, 263
Primeval daughter, 275, 286
Primeval deity consciousness, 265
Primeval deity kingdom, 275
Primeval deity soul, 219
Primeval feminine, 263, 345
Primeval forms, 244, 245
Primeval grandson, 53, 54
Primeval Greeter Universe, 366
Primeval human child, 248
Primeval Illuminator Universe, 367
Primeval intellectual body, 332
Primeval kind, 122
Primeval masculine self, 327
Primeval maternal spirit, 218
Primeval neutrino, 305
Primeval param child, 246
Primeval paternal spirits, 218
Primeval paths, 364, 365
Primeval Perpetuator Universe, 367
Primeval physical body, 307
Primeval reality, 355
Primeval self, 288
Primeval son, 275
Primeval soul, 219
Primeval space, 38, 56, 123, 130, 132, 134, 136, 137, 161–163, 259

Primeval species, 122
Primeval Wisher, 53, 266
Primordial astral body, 226, 227, 298
Primordial cell, 182, 263–266
Primordial child consciousness, 78
Primordial consciousness, 219, 256, 270, 274, 311, 313, 334, 336
Primordial creation, 262
Primordial daughter, 286
Primordial deity soul, 219
Primordial deity universe, 357
Primordial etheric body, 238, 279
Primordial even value, 336
Primordial female, 283–285
Primordial feminine self, 87, 327, 342–344
Primordial follower, 83
Primordial followership universe, 83
Primordial forms, 244
Primordial genera, 14
Primordial greeter realm, 83, 287
Primordial greeter soul, 218
Primordial greeter spirits, 218
Primordial Greeter Universe, 367
Primordial human child, 248
Primordial illuminator, 257, 271, 288
Primordial intellectual body, 245, 326, 331
Primordial kind, 122
Primordial knower, 274, 275
Primordial knower paradigm, 274
Primordial life, 323
Primordial male, 277, 283, 284
Primordial masculine self, 82, 87, 327, 362–364
Primordial maternal soul, 218
Primordial maternal spirit, 218
Primordial neutrino, 304, 305
Primordial oneness, 164, 284, 325, 346
Primordial para consciousness, 274, 276
Primordial Para Greeter, 257
Primordial param child, 246
Primordial paternal consciousness, 279
Primordial paternal soul, 218
Primordial paternal spirit, 218
Primordial paths, 364, 365
Primordial perpetuator, 288

Primordial reality, 355

Primordial self, 18, 160, 178, 288, 293, 360, 362, 364, 365

Primordial soul, 219

Primordial space, 13, 16, 38, 49, 56, 123, 124, 130, 132–134, 136, 162, 244, 258, 259

Primordial species, 122

Primordial super destroyer, 280

Primordial universe, 336

Primordial wholeness-effect, 342

Primordiality, 240, 326

Principal guider, 32, 47–49, 220, 291, 300, 339, 341, 342

Principal member, 49, 50, 119

Prism class, 348

Pristine identity, 177

Pristine nature, 334, 360, 362

Private property right, 8

Produced copy, 310–312

Producer class, 217

Production copy, 312

Production force, 239

Productive energy, 91

Proficiency, 80, 209, 214

Proficient system, 18

Progenitor, 111, 317

Progeny cell, 184

Programmed future of everybody, 259, 274, 339

Programmed life, 222–224

Programmed performing-effect, 344

Project VIPIN, 1, 2, 6, 7

Projecting body, 289

Proportionate aftereffect, 297, 298, 337

Proportionate length power growth rate, 338

Protagonist behaviors, 357

Protagonist cause, 357

Protagonist consciousness, 357

Protestant reformation, 21

Protoindustrialization, 22

Proton, 226, 228

Psion meson, 167

Psychic consciousness, 273

Psychic linkage, 21, 54, 147, 220, 222, 227, 260, 275, 278, 282, 283, 300, 318, 340, 366

Psychic power, 276

Psychic realm, 283

Psychological reactions, 5

Pulsating energy, 277

Pulsating sentient entity, 316

Pure heterogeneous, 145

Pure homogeneous, 145

Pure reproduction, 80, 88, 90, 216

Purifying force, 355

Purkinje cells, 182

Pyramidal cell, 182, 202

Q

Quadruple bond, 50, 159

Quadruple ellipse, 164

Quadrupole, 133, 138

Quality-free, 63

Quantum, 5, 16, 17, 30, 52, 56, 100, 107, 124, 128, 130, 136–139, 141, 142, 147, 151–154, 156, 158, 167–169, 202, 226–228, 237, 238, 248, 262, 263, 277, 296, 301–306, 308, 310, 313, 334, 339, 357, 363, 364

Quantum causation, 124, 158

Quantum cell, 154

Quantum consciousness, 141, 142

Quantum copy, 237, 238

Quantum ellipse, 128

Quantum entity, 302, 303

Quantum field, 169

Quantum finite point, 52, 138

Quantum light force, 301

Quantum particle, 228

Quantum photon, 5, 17, 56, 136, 139, 156, 167, 226, 248, 262, 263, 277, 296, 302, 303, 305, 306, 308, 310, 313, 334, 363

Quantum reality, 107

Quantum soul, 304

Quantum space, 124, 130, 152

Quantum state, 56

Quantum system, 153, 228, 301, 303

Quantum thing, 100, 168, 169

Quantum time, 16, 124, 357

Quantum vortex, 151

Quark, 35, 36, 137–142, 144, 163, 164, 306, 318

Quarter, 50, 52, 53, 58, 59, 61, 139, 184

Quarter pie, 59

Quaternary causal body, 273

Quinary cells, 181

Index

Quintuple bond, 164, 165

R

Radial power growth rate, 337
Radiant love, 156, 174, 226, 272
Radiant maternal love, 270
Rational human, 6
Real absolute gravitational constant, 44, 52–54, 156, 160
Real horizon, 258
Real ion, 160
Real phase, 353
Real pi, 56
Real present reality, 59
Realization, 177
Realm of fire, 363
Receding combination, 146, 147, 352
Receding phase, 352, 353
Reciprocal exchange, 63
Recirculating force, 246
Reckoning, 31, 75
Rectitude, 80, 81
Red blood cells, 213
Referential presence, 330
Reflective conception, 320
Regge trajectory, 302
Reggeon, 302
Regiocentrism, 43, 44
Reincarnating, 46, 47, 66, 87, 96, 145, 181, 244, 271, 273, 326, 327, 339
Relational positioning, 26, 27
Relative reality, 170, 171
Religion defender, 83
Religion destroyer, 83
Religion founder, 83
Religious body, 212, 216
Religious governance, 94
Religious values, 113
Renaissance, 21, 22
Renormalon, 301
Replica, 360
Replicated light, 363
Replicated universe, 30, 273, 362
Replicating member, 174
Replicating one, 362, 363, 365
Replicative body, 99, 237

Replicative recombination, 314
Reproduced copy, 310, 311
Reproduced reality, 171, 208
Reproduced value, 42, 43
Reproducible copy, 311, 312, 316
Reproducible point, 159, 354, 356
Reproducing reality, 56
Reproducing self, 323
Reproduction force, 14, 239, 247
Reproduction wheel, 183, 192
Reproduction-effect, 18
Reproductive body, 261, 280, 281
Reproductive center, 283
Reproductive endogenous behavior, 14
Reproductive energy, 23, 64, 71, 98, 244–246, 280, 287
Reproductive force, 175, 239, 240, 242, 247, 258, 261, 335, 365
Reproductive mass, 228
Reproductive potential, 245
Reproductive power, 290
Reproductive reality, 6, 48, 71, 94–96, 100, 229, 245, 263, 272, 280, 283, 305, 316, 354, 356, 361
Reproductive sound, 230
Reproductive system, 19, 30, 63, 156, 227, 229, 250
Reproductive wheel, 180, 183, 210–214, 229
Resentment, 79
Resident alien universe, 318–321, 323, 324
Resident master, 339
Residual energy, 214, 319
Resolving community, 358, 359
Respiratory force, 332, 342
Respiratory wheel, 19
Responsible citizenship behavior, 212
Responsible management paradigm, 256, 257
Resultant reality, 199
Reunified consciousness, 317
Revolutionary force, 15, 55, 223, 224, 366
Revolutionize, 31
Ricci curvature tensor, 43
Ricci scalar, 170
Risk-seeking behaviors, 211
RNA, 104, 157, 222
Robust partnerships, 115

Root cause wheel, 180, 183, 192
Rotating absolute gravitational constant, 53, 54
Rotating leadership, 105
Rupture of time, 159

S

Sagittarius house, 64, 68
Sagittarius zodiac, 103
Salvation, 8, 107, 235
Sanctification, 8
Satan, 85, 96–99, 107, 109, 171, 181, 182, 194–196, 199, 200, 218, 219, 259, 276, 277, 279, 286, 288, 292, 293, 297, 349
Satan allele, 199, 200
Satan bovine, 99
Satan cell, 181, 182
Satan copy, 195
Satan crab, 96–98
Satan ploid, 196
Satan snake, 109
Satan soul, 218, 219
Satan spirit, 218, 292, 297
Satanic light, 260
Satanic spirit, 253
Saturn, 107, 238–242
Scalar meson, 161–163
Scale factor, 52, 153
Scattered electron, 305, 307
Scattered light, 305
Scattered photon, 305–307
Scientific consciousness, 354, 356
Scientific paradigm, 5, 205
Scorpio house, 64, 68
Scorpio zodiac, 92, 93
Scraping community, 302
Second astral body, 192, 246, 290
Second brother, 286
Second copy, 211, 223, 225, 231, 237, 239, 272, 293
Second intellectual body, 192, 326
Second member, 121, 334, 335
Secondary causal body, 271, 272
Secondary cell, 180, 181
Secondary factor, 319
Secondary growth, 156
Secondary merons, 34

Secondary-effect, 274
Secular body, 211
Secular religion, 2
Self-consciousness, 80, 81
Self-developed personal conscious, 9
Self-developed social conscious well-being, 9
Self-fulfilling path, 252
Self-governance, 84, 85, 87, 90
Self-idealization, 63
Self-involvement, 315
Selfless radiant love, 220, 226
Self-luminous body, 345, 346
Self-luminous dimension, 273
Self-luminous discontinuity, 280
Self-luminous element, 33, 226
Self-luminous entity, 17, 18, 33, 48, 49, 119, 121, 192, 212–216, 225, 226, 228, 244, 251, 253–255, 262, 263, 269–271, 273, 279–283, 290, 311, 313, 314, 316, 329, 334, 343, 355, 363, 366
Self-luminous reality, 99, 174, 222, 238, 298, 318, 328
Self-luminous realm, 177, 287
Self-luminous system, 302
Self-perpetuating entropy, 223
Self-repelling, 322, 323
Self-reproduce, 59, 294
Self-sovereignty, 81
Semi-conjugacy, 132, 336
Senary cells, 181
Sensory illusion, 187
Sensory perception, 320
Sensory reality, 175
Sensory vibrations, 290
Sentient culture-effect, 270
Sentient cycle, 275
Sentient dimension, 214
Sentient energy, 18, 149, 189, 198, 201, 210, 238, 273, 278, 282, 296, 316, 317, 324, 362, 364, 366, 370
Sentient entity, 17, 19, 20, 33, 72, 121, 173–176, 178, 190, 193, 194, 200, 212, 216, 254, 259, 260, 262, 282, 284, 298, 315–317, 326, 328, 336, 339, 365
Sentient entropy, 279
Sentient gravitational point, 252
Sentient horizon, 259
Sentient life, 214

Index

Sentient light force, 72, 88, 221
Sentient nature, 187
Sentient performing, 240
Sentient planning, 204
Sentient potential, 211
Sentient spirit, 238
Sentient wave, 121, 334
Sentient well-being, 7, 149, 181, 202, 207, 208, 211, 212, 225, 238, 244, 251, 270, 318, 327, 369
Sentient workculture-effect, 270–272
Septaploid, 194
Sequence of quantum lights, 301, 302
Sequence of triple copies, 199, 314
Sequential doctrines, 26
Sequential force, 96
Sequential reality, 330, 347
Serotonin, 187, 190, 191, 195–197, 200–204, 210, 228, 233
Serotonin-melatonin production cycle, 190, 233
Serviced physical body, 308
Servicing copy, 199
Servicing factor, 325
Servicing spirit, 357
Servicing-oriented factor, 367
Seven body system, 227, 298
Seven kins, 301
Seven members, 119, 338
Seven-fold consciousness, 159, 160
Seven-step freeriding process, 160
Seventeenth intellectual body, 331
Seventeenth member, 174
Seventeenth potential guider agent, 304
Seventeenth principal guider, 48
Seventh astral body, 298, 299, 317, 358
Seventh body, 298, 340, 341
Seventh brother, 286, 288
Seventh gamma photon, 307
Seventh infinity, 259
Seventh intellectual body, 327, 328, 359
Seventh member, 118, 119, 122, 173
Seventh principal guider, 48
Seventh sister, 288
Seventy bodies, 341
Sextuple bond, 165, 167

Sigma bond, 143, 144, 153, 154, 159, 165, 166
Sine, 156
Sister allele, 155, 197, 198
Sister copy, 313
Sister spirit, 313, 315, 326, 356, 357, 361, 365, 366
Six gamma photons, 307
Six triangles, 60, 61
Six-fold growth, 42, 56, 96, 99, 189, 191, 193–195, 285, 306, 310, 311
Sixteen brothers, 295
Sixteen effects, 300
Sixteen forces, 295
Sixteen four-body systems, 229
Sixteen light forces, 243, 295, 300
Sixteen sisters, 295, 296
Sixteenth astral body, 299
Sixteenth body, 341
Sixteenth brother, 297
Sixteenth intellectual body, 331
Sixteenth potential guider agent, 304
Sixteenth principal guider, 48
Sixteen-wheel system, 61, 62, 66, 67
Sixth astral body, 292
Sixth body, 293, 294, 340
Sixth brother, 286, 288, 289
Sixth copy, 194, 219, 225, 242–244, 292, 293
Sixth infinity, 253
Sixth intellectual body, 327
Sixth member, 122, 334
Sixth self-luminous entity, 193
Sixth sister, 287, 297
Skeletal muscle cells, 230
Skepticism, 82
Sleeping consciousness, 187
Sleeping entity, 19
Sleeping state, 176
Smooth muscle cells, 213
Social animal, 3, 9
Social benefit, 51, 112, 214, 262–264, 271, 272, 341
Social benefit-cost ratio, 214, 262–264, 341
Social body, 209
Social circle, 106, 109, 111
Social conscious well-being, 9
Social dispersion, 212

Social forces, 113
Social governance, 90, 92
Social network, 209, 211, 260, 279
Socialism, 73
Solar chakra, 184, 185
Solar cycle, 179, 183, 184
Solar energy, 227
Solar universe, 66, 72, 103, 241–243, 297
Solar wheel, 20, 179, 180, 183, 185–188, 190, 192, 201, 203, 204, 230, 231, 289
Solid earth, 69, 95, 215, 221, 227, 240, 241
Soliton, 147, 276, 345
Soliton crystal, 345
Somatic cell, 184
Somatic nervous system, 229
Somatosensory Cortex, 329, 330
Somatotrophs, 212
Somatotropin, 212, 213, 228
Someone known, 75
Someone unknown, 75
Son allele, 155, 198, 199
Son cell, 102, 181, 182, 222, 264, 327
Son consciousness, 91, 92, 105
Son soul, 218
Son spirit, 217
Sororal twin, 283
Soul family, 266, 267, 269, 275
Soul universe, 300, 302, 308
Sound echo, 363
Sound vibration, 201, 247, 248, 363
Sour acidic taste, 213
Southern light, 150
Space curvature, 44, 154, 171
Space cycle, 254
Space force, 98
Space realm, 131, 132
Space soul, 219
Space stock, 42, 43
Space vacuum, 305
Space-units rotation, 123
Span, 231
Spatial coordinate, 37, 38, 43–45, 269
Spatial development, 153
Spatial ellipse, 152
Spatial pie, 59, 126

Special seeing, 232
Specific living soul, 322
Speed of light, 5, 6, 35, 222, 304
Sphaleron, 301
Spherical pi, 56, 57, 128, 129
Spherical space, 125, 127, 129
Spider body, 93, 94
Spider breeds, 93
Spiderling, 93, 94
Spinnerets, 93
Spirit, 4, 9, 75, 76, 78, 79, 85, 97, 123, 135, 136, 151, 178, 202, 215–220, 238, 244, 254, 270–273, 276, 277, 279–281, 284, 285, 289–291, 295, 314, 315, 323, 346, 357, 366, 368
Spirit kingdom, 323
Spirit soul, 219
Spirit universe, 270
Spiritual body, 262–264, 278, 279
Spiritual code, 201
Spiritual consciousness, 196
Spiritual energy, 294
Spiritual experiences, 8
Spiritual flame, 352
Spiritual growth, 75, 302
Spiritual oneness, 162
Spiritual preceptor, 302
Spiritual realm, 98, 101
Spiritual sense, 320
Spiritual soul, 86
Spiritual union, 75, 76
Spiritual well-being, 326
Spiritualism, 111
Split photon, 262
Splitting force, 262
Splitting potential, 263
Splitting value, 263
Spontaneous fragmentation, 14
Spontaneous space, 338
Spontaneous time, 339
Spreading causation, 351, 354
Spreading path force, 363, 364
Square pi, 53, 60, 127, 144
Star of attraction, 96, 168
Starship, 354
Stem cell, 181–183

Index

Strange B-meson, 167
Strange D-meson, 167
Stress-energy tensor, 170
Strong extrinsic psychic force, 110
Strong intrinsic psychic force, 110, 363
Strong psychic force, 110, 146, 212
Strong psychic linkage, 145, 146, 220, 260, 275, 278, 282
Subconscious body, 351, 371
Subconscious impulse, 349, 350
Subconscious mind, 5, 57
Subjective consciousness, 368
Subliminal member, 121
Subordinate member, 119
Subtracted physical body, 307, 308
Sum of pie, 43
Super deity soul, 219, 347
Super friendly network, 319
Super ideal, 73
Super theory, 356
Super wisher, 366
Super-fertile, 99
Superimposed reality, 108
Superior consciousness, 364
Supernatural fragmentation, 15
Supernatural growth, 146, 148, 242, 251
Supernatural love, 256
Supernatural profiting, 312
Supernatural reality, 9, 278, 355
Supernormal energy, 225, 226
Supernormal mediation, 4
Superordinate member, 120
Superposition, 4, 46, 52, 166
Supplementary linkages, 328
Supra deity soul, 219
Supra deity universe, 346
Supreme Council, 243, 249–251, 253
Supreme deity soul, 219
Supreme deity universe, 345, 367
Supreme offering, 85
Supreme portal, 317, 318
Supreme sacrifice, 66, 85
Supreme self, 86, 332
Supreme Wisher, 265, 266
Sympathetic nervous system, 229

System-level consciousness, 351

T

Tachyon, 35
Tadpole, 91
Tandem copy, 311, 312
Tangent, 35, 76, 123, 129, 130, 136, 144, 164, 242
Tangent line, 242
Taurus, 63, 68, 70, 97, 98
Taurus house, 63, 68
Taurus zodiac, 97, 98
Taxonomical divergence, 14
T-cell receptors, 182, 183
Technological capability, 89, 210, 215, 231
Technological cost, 112, 204, 209, 246, 270, 273
Technological energy, 229
Technological exchange value, 253
Technological growth, 7, 116, 157, 227, 269–272, 321, 366
Technological input, 187
Technological oneness paradigm, 251
Technological servicing, 114, 116
Technological system, 229
Technological trading, 50, 114, 116
Temporal coordinate, 367
Temporal ellipse, 58, 152, 153
Temporal lobe, 327, 329
Temporary form, 63, 112
Temporary growth, 112, 197
Temporary joy, 85, 232, 369
Ten-body system, 66
Ten-face primordial illuminator, 275
Ten-fold growth, 35, 36, 105, 230, 237, 360
Tensor meson, 162
Tensor sequence, 163
Tenth body, 66, 67, 69, 243, 340
Tenth conscious light, 287
Tenth copy, 226, 245, 296
Tenth etheric body, 367
Tenth guider agent, 301
Tenth incarnation of the primeval perpetuator, 18
Tenth intellectual body, 328, 329
Tenth kin, 301
Tenth member, 119, 174, 335, 337

Tenth photon, 38, 139, 140, 274, 277
Tenth potential kin, 301
Tenth principal guider, 47
Tenth sister, 288
Tenth twin photon, 140
Tertiary causal body, 272, 274
Tertiary cause, 245, 246
Tertiary consciousness, 316
Tertiary growth, 156, 366
Tetra copy, 195
Tetraploid, 194, 196
Theoretical subject, 367
Theory force, 356
Theory-effect, 15, 22, 309, 324
Thermal force, 203, 230, 231
Thermal value, 204
Thermodynamic consciousness, 139
Thermodynamic energy, 342, 361
Thermodynamic entropy, 4, 66, 71, 202, 244
Thermodynamic equilibrium, 186
Thermodynamic force, 5, 166, 187, 204, 223, 239, 240, 317, 346, 362
Thermodynamic growth, 346
Thermodynamic limit, 137, 143, 145, 225, 336
Thermodynamic sensation, 363
Thermodynamics, 39, 168, 186, 187, 222–224, 231, 238
Third astral body, 290
Third brother, 286
Third copy, 237, 239, 272, 293
Third eye wheel, 179
Third infinity, 204, 207, 210
Third intellectual body, 192, 326, 346
Third member, 122, 333–335
Third sister, 285
Thirteen body system, 341
Thirteenth astral body, 299
Thirteenth body, 67, 69, 228, 229, 341
Thirteenth brother, 292
Thirteenth copy, 226, 296
Thirteenth entity group, 349
Thirteenth guider agent, 302
Thirteenth intellectual body, 330
Thirteenth kin, 302
Thirteenth member, 119

Thirteenth potential guider agent, 303
Thirteenth principal guider, 48
Thirteenth quarter, 58, 59
Thirteenth sister, 292, 296
Thirtieth astral body, 367
Thirtieth causal body, 366
Thirtieth etheric body, 367
Thirtieth intellectual body, 367
Thirtieth member, 334
Thirtieth mental body, 367
Thirtieth physical body, 367
Thirty bodies, 340, 341
Thirty entities, 349, 350
Thread of life, 323
Thread of sentient life, 201
Three body system, 282
Three energy system, 224, 328
Three-eyed child spirit, 216
Three-eyed holy spirit, 216
Three-faced system value, 163
Three-wheel system, 61, 65, 202
Throat wheel, 180, 183, 186
Thyroglobulin, 213, 228
Thyroid, 19, 179, 213, 227, 228
Time consciousness, 36, 96, 133, 135, 137
Time dimensions, 83
Time force, 60, 343
Time multiplier, 22, 24, 28, 30, 61, 66, 67, 74, 88, 91, 93, 97, 99, 100, 101, 103, 153, 165, 166, 192
Time realm, 134, 135
Time soul, 219
Time tensor, 41, 45–47, 49, 51, 53, 55, 57, 59, 61, 63, 65, 67, 69, 71, 73, 75, 77, 79, 81, 83, 85, 87, 89, 91, 93, 95, 97, 99, 101, 103, 105, 107, 109, 111, 113, 115, 117
Time value, 55, 57, 202, 354
Time wheel, 187
Time-dependent entropy, 214
Time-invariant, 15, 32, 319, 329, 345, 348, 359, 361
Time-units reflection, 123
Top Tau Meson, 167
Topological-effect, 147
Total absence, 321
Total presence, 321
Totipotency, 202

Index

Tradable physical body, 308
Traded allele, 199
Traded consciousness, 353
Traded physical body, 308
Trader class, 217
Trading factor, 7, 325
Trading photon, 156
Trading reality, 354
Trading spirit, 360, 361
Trading-effect, 68, 213, 238, 246
Trading-oriented factor, 366
Transcendental forces, 2
Transcendental form, 121, 221
Transcendental oneness, 286
Transcendental value, 93, 103, 138, 139, 174, 221, 222, 237, 238, 250, 255
Transcriptional machinery, 316
Transformation sequence, 330
Transformative development, 267, 328
Transformative exchange paradigm, 278
Transformative planning, 344
Transformative system, 229
Transgenic development, 48–50
Transient value, 56
Tree of creation, 357
Triangulate, 60–62, 137, 148, 230
Triangulating force, 61
Triangulation, 123, 124
Trifurcate, 154
Trillionaire, 7, 40, 369
Triple bond, 52, 53, 153
Triple chain, 161, 162
Triple conscious consciousness, 194
Triple copy culture, 141, 155, 156, 159, 163, 164, 272
Triple ellipse, 142, 151, 152, 164, 165
Triple family, 335
Triple octave, 91
Triple partial pi bond, 34–36, 38, 165
Triple population, 119
Triple radius, 337
Triple space, 176
Triple time, 46, 176
Triple twin quark, 138, 139
Triplets, 54, 302

Triploid, 193, 194, 199
Triploid-effect, 199
Trivial finding, 6
Tropical force, 83
True identity, 176
True nature, 174, 360, 362
True reality, 203
True trade, 271
True universe, 366
Trumpet, 257
Truth reality, 201
Tryptophan, 191, 207
Tryptophan hydroxylase, 191
Tun, 294
Twelfth body, 67, 69, 229, 340
Twelfth brother, 287–289, 292
Twelfth entity group, 349
Twelfth etheric body, 367
Twelfth guider agent, 302
Twelfth intellectual body, 329
Twelfth kin, 302
Twelfth member, 119
Twelfth potential guider agent, 302
Twelfth principal guider, 47, 48
Twelfth sister, 287, 288, 292
Twelve hours, 349, 351
Twelve octaves, 299
Twelve potential electrons, 323, 324
Twelve-fold growth, 285, 310
Twelve-wheel system, 202, 203, 230
Twentieth astral body, 302
Twentieth gamma photon, 306, 322
Twentieth guider agent, 303
Twentieth member, 174
Twentieth principal guider, 49
Twenty bodies, 340
Twenty quarters, 59, 60
Twenty-eighth astral body, 361
Twenty-eighth causal body, 360
Twenty-eighth etheric body, 361
Twenty-eighth intellectual body, 361
Twenty-eighth member, 334
Twenty-eighth mental body, 361
Twenty-eighth physical body, 362

Twenty-fifth astral body, 353
Twenty-fifth causal body, 354
Twenty-fifth etheric body, 353
Twenty-fifth intellectual body, 352
Twenty-fifth mental body, 352
Twenty-fifth physical body, 351
Twenty-first gamma photon, 306
Twenty-first guider agent, 303
Twenty-first member, 122, 333, 334
Twenty-four hours, 92, 105, 351
Twenty-fourth astral body, 349
Twenty-fourth etheric body, 349
Twenty-fourth intellectual body, 349
Twenty-fourth member, 333
Twenty-fourth mental body, 349
Twenty-fourth physical body, 349
Twenty-ninth astral body, 363
Twenty-ninth causal body, 363
Twenty-ninth etheric body, 363
Twenty-ninth intellectual body, 365
Twenty-ninth member, 334
Twenty-ninth mental body, 364
Twenty-ninth physical body, 365
Twenty-second astral body, 343
Twenty-second etheric body, 343, 344
Twenty-second member, 333
Twenty-second mental body, 342
Twenty-second physical body, 342
Twenty-seventh causal body, 357
Twenty-seventh etheric body, 358
Twenty-seventh member, 334
Twenty-seventh mental body, 359
Twenty-sixth astral body, 355
Twenty-sixth causal body, 355
Twenty-sixth etheric body, 355
Twenty-sixth mental body, 355
Twenty-third astral body, 345
Twenty-third causal body, 344
Twenty-third etheric body, 344
Twin mother cell, 267
Twin neutrino, 304, 305
Twin pair, 196
Twin person, 242, 348
Twin phi bond, 165, 166

Twin photon, 140
Twin pi bond, 159, 160, 166
Twin point fragment, 162
Twin point particle, 161, 162
Twin radius, 337
Twin sigma bond, 165, 166
Twin sister, 93, 99, 198, 285, 291, 292
Twin soul, 270
Twin space, 176
Twin time, 175
Twin triple copy, 120, 312
Twin zygote, 197
Two-wheel system, 61, 66

U

Ultraviolet, 35, 36, 42
Umbilical cord, 227
Umpire, 328
Uncertainty avoidance, 28
Unconditional confidence, 221
Unconditional radiant love, 221
Uncurved entity, 163, 164
Underflowing consciousness, 95
Undifferentiated four-body system, 295
Undifferentiated present, 331
Undivided astral body, 290
Undivided causal body, 267
Undivided etheric body, 279, 280
Undivided mental body, 309, 313
Undivided physical body, 301
Unfriendly network, 318
Unfulfilled subtle desire, 168, 215, 283
Unified consciousness, 162
Unified divinity, 368–370
Unique combination, 364
Unique deity, 178
Unique divine entity, 364
Unique entity, 296, 338
Uniqueness orientation, 28
Unison, 64
Universal conscious well-being, 189
Universal deity, 177
Universal sentient benefit, 65
Universal sentient wellness, 98, 112, 113
Universe dimension, 278

Index

Universe of alleles, 199, 314, 315
Universe of atoms, 86, 87
Universe of Blissful Self, 351
Universe of body organ systems, 188
Universe of bovines, 98
Universe of cells, 86, 87, 251
Universe of Child God, 345
Universe of child souls, 273, 319
Universe of child spirits, 280
Universe of children, 262, 278
Universe of copies, 90, 92, 93, 98, 100
Universe of departed entities, 344
Universe of departed souls, 277, 315
Universe of energy, 246
Universe of entities, 38, 123, 124, 257, 316
Universe of goals, 362, 363
Universe of granddaughter cells, 154, 269, 319, 327
Universe of ideal object, 314, 366
Universe of inanimate objects, 317, 318
Universe of light echoes, 154, 319
Universe of living child souls, 323
Universe of living entities, 344
Universe of living souls, 277, 315
Universe of luminous entity, 345
Universe of maternal souls, 273
Universe of maternal spirits, 279, 294
Universe of para deity, 280
Universe of para-conscious entities, 298, 299, 315
Universe of param children, 246
Universe of Paternal Godhead, 346
Universe of paternal spirit, 273, 280, 294
Universe of photons, 88, 101
Universe of potential electrons, 153, 154
Universe of primeval children, 246
Universe of principal guiders, 339
Universe of replication, 174, 339, 366
Universe of resident aliens, 324
Universe of self-luminous elements, 215
Universe of self-luminous entities, 270, 316
Universe of sentient entities, 216, 298, 315, 317
Universe of super wishers, 366
Universe of theoretical subject, 366
Universe of theory, 357

Universe of wishers, 217, 258, 318, 319
Unknown reality, 8, 82, 278, 314
Unsustainable value, 167
Up quark, 139, 141, 142, 263, 306, 322
Upper octave, 122, 164
Upsilon meson, 167
Uranus, 71, 72, 103, 104, 238, 241–244, 246, 247

V

Vacuum decay, 34
Vacuum point, 303
Vagaries, 232
Variable body, 310
Variable lifetime, 247, 248
Variable touch, 329
Vascular cells, 230
Vascular growth, 181
Vascular system, 227, 230
Vector D-meson, 167
Vector meson, 161, 162
Venus, 95, 238–241
Vertical chord, 335
Vibrational consciousness, 269, 291, 297
Vigesimal member, 122
Virgo house, 64, 68
Virgo zodiac, 94, 95
Virtual air, 304, 305
Virtual atom, 303
Virtual divine, 305
Virtual earth, 304
Virtual energy, 303
Virtual ether, 304
Virtual fire, 303
Virtual gluon, 303
Virtual guider, 304
Virtual light force, 304
Virtual particle, 302
Virtual photon, 304
Virtual quark, 303
Virtual water, 304
Visible horizon, 258, 259
Visual Cortex, 330
Visual information, 321, 322
Volatile, 209
Volunteer, 86

W

Waking consciousness, 187
Waking state, 176
Waning Crescent, 184, 186
Waste liquid air, 213
Waste solid earth, 213
Water force, 88, 240, 293
Water molecule, 291–294
Water wheel, 291
Water-effect, 208, 213, 221, 230, 241
Wavelet, 304
Wax body, 362
Weak extrinsic psychic force, 110
Weak intrinsic psychic force, 110
Weak point, 156
Weak psychic force, 110, 146
Weak psychic linkage, 3, 145, 146, 260, 275
Western culture-effect, 23
Western geography, 20, 23, 28
Western group, 23, 24
Wheel of causation, 248
Wheel of conception, 230
Wheel of creator, 230
Wheel of digestion, 297
Wheel of fire, 297
Wheel of innovation, 368
Wheel of joy, 232, 233
Wheel of life, 248
Wheel of psychic forces, 230, 361
Wheel of wheels, 182, 183, 201, 230, 237
Wheels of energy, 18
White blood cells, 182
White matter, 188, 189, 315
White star, 66, 67, 69, 71, 101, 148, 149, 239–243, 247, 305, 347
Whitened community, 272
Whole electron, 151
Whole ion, 160, 161
Whole photon, 159
Whole pie, 42–44
Whole spatial coordinate, 44
Whole sphere, 45, 52, 55–57
Whole value, 22, 23, 76
Wholeness Committee of Godheads, 327
Wholesome electron, 151
Wholesome ion, 160, 161
Wholesome knowing, 24
Wholesome pie, 42–44
Wholesomewhole ion, 160
Wholesomewhole pie, 42–44
Wisdom soul, 341
Wisher, 51–54, 74, 142, 152, 154–157, 176, 177, 180, 248, 260, 265, 275, 279, 319, 362
Wisher community, 362
Wisher entity, 279
Wisher paradigm, 156, 275
Wishing sequence, 214, 248, 252, 276, 277
Wishing tree, 365
Within affliction, 79
Within belief system, 80
Within force, 75, 76
Within God, 76
Without affliction, 80
Without belief system, 80
Without force, 75
Without oneness, 140, 142, 145, 146, 286, 313
Without religion, 83
Without system, 35, 237
Work energy, 231
Workculture, 23, 68–71, 115, 116, 203, 212, 215, 233, 298, 315, 367, 368
Workculture-effect, 68, 71, 212
Worker deity, 216, 217, 353
Worker social benefit, 64, 211
Worker social cost, 64
Worker-social-benefit-cost-ratio, 68
Workforce, 114, 183, 184, 188, 209, 214, 220, 228, 313, 314, 364
Workforce proficiency, 214
Workforce system, 188, 220, 228, 313, 314, 364
Workforce-effect, 184
Working agent, 220, 221
Working causation, 300, 308
Working consciousness, 215

X

X-chromosome, 343, 364
X-coordinate, 202
X-ray photon, 305

Y

Yang energy, 359

Y-chromosome, 343, 364
Y-coordinate, 202
Yin energy, 359
Y-linked, 154, 155, 198

Z

Zeal, 81
Zenith, 273
Zero electromagnetic mass, 37
Zero energy, 215, 231
Zero integrity, 270
Zero mass, 6
Zero morality, 113
Zeroth chromosome, 183
Zeroth finite points, 124
Zeroth gravitational constant, 55, 131
Zeroth member, 122
Zeroth metric tensor, 131, 163, 164
Zodiac element, 63
Zodiac entity, 63, 355
Zodiac house, 62, 70, 96
Zodiac system, 62, 256
Zodiacs, 70, 71, 97, 111
Zygote, 197
Zygote sequence, 197

Hindi Index

A

Aachar, 112
Aadesha, 121
Aap, 174, 261, 348
Aath, 358, 359
Aatm Sammaan, 215, 226
Aayojana, 345
Aayu, 214
Ab, 39, 51, 62, 154, 155, 158, 330, 331
Abadha, 64, 113, 178, 191, 195, 254, 289, 319, 326, 344, 347, 350
Abadi, 13, 14, 16, 46, 118, 338, 339
Abheda, 104, 321
Abhidheya, 191, 258, 333, 361, 368
Abhikarta, 221
Abhikha, 158
Abhilasha, 248, 252, 276, 277
Abhilasha karma, 252
Abhineta, 26, 265
Abhipraya, 64, 90, 232, 270
Abhyasa, 321
Achaarya, 108
Achaithanya, 317
Achetan, 365
Achitta shakti, 226, 325, 359
Adamiyah, 359, 360, 362
Adana vijnana, 250, 252, 339
Adarsha, 15, 22, 29, 63, 67, 73, 89, 102, 113, 253, 260, 320, 360
Adarshavadi, 22
Adarshikaran, 356
Adha, 262, 263
Adharma, 165, 308
Adhikriti, 84
Adhimoksha, 81
Adhirajya, 96
Adhmata, 221
Adhyapaka, 312
Adhyatamic sharira, 278
Adhyatma, 219, 256, 270, 274, 311, 313, 325, 334, 336

Adi, 36, 122, 164, 181, 204, 208, 214, 218, 224, 229, 252, 267, 274, 276, 283, 284, 325, 346, 349, 364
Adi Loka, 252, 283
Adi Marga, 364
Adi Para Atma, 36, 218, 224, 267, 274, 276, 349
Adi Para Shakti, 204, 208, 214
Adi Pitha, 122
Adi shakti, 181, 229
Adisthanam, 19
Aditi, 83, 161, 162, 164
Aditya, 153, 255
Adosha, 80
Adravya, 171, 208, 308
Adridalana, 263
Agaha, 46, 176
Agama, 26, 273
Agni, 19, 152, 154, 155, 204, 215, 221, 222, 259, 278, 297, 318, 326, 331, 333, 344, 361–363
Agni chakra, 297
Agni Loka, 363
Agnishvatta, 111
Agochar, 264
Agrata, 62, 239, 240
Agrima, 246
Aham, 24, 47, 73, 74, 190, 209, 220, 309, 362, 365, 366
Aham Brahma, 24
Aham shakti, 220
Ahamkara, 108
Ahkriya, 79
Aichchika, 307
Aikamatya, 24, 254, 279
AIM shakti, 186
Ajanman, 152, 153
Ajari, 54
Ajirna, 305
Ajitatma, 52, 88, 90, 91, 97, 156, 159, 160, 272
Ajiva, 249
Ajjhattikani, 64, 187, 321, 322
Ajna chakra, 19, 179, 231, 261, 290

Index

Ajnani, 201
Akala, 73, 105, 323, 360
Akalpa, 216, 298, 315, 317, 323
Akarta, 165, 167
Akasagarbha, 220
Akasha astikaya, 272
Akasha chakra, 183, 190, 231, 233
Akhanda, 34, 59, 153, 154, 228, 256, 293, 301, 303
Akhkhala, 189
Akrura, 267, 275
Aksha, 318
Akushala, 78
Alambusha Shakti, 273
Alaukika dharma, 361
Alaya vijnana, 190, 248
Alidha mandala, 151, 152, 302
Alobha, 80, 325
Amanaska, 368
Amara, 33, 63, 156, 159, 178, 231, 354, 356, 358
Amba, 77, 81, 87, 112
Ambalika, 329
Ambara, 38, 66, 83, 123, 321, 340, 341, 349, 350
Ambarant, 258
Ambhamsi, 362
Amnaya, 47, 55, 120, 174, 250
Amrina, 220
Amrit, 16, 18, 20, 22, 55, 57, 61, 91, 100, 104, 110, 149, 184, 189, 198, 201, 210, 226, 238, 263, 268, 273, 276, 293, 296, 316, 334–338, 352, 354, 356, 362, 364–366
Amsha, 64, 150, 165, 202, 213, 223, 231, 263, 294, 299, 312, 339, 343, 347, 361
Amukhikarana, 262, 278
Anagata, 15, 215, 289, 330
Anahata, 19, 179, 180, 226, 257, 291, 364
Anahata chakra, 19, 180, 226, 291
Anakala, 354
Anala, 359
Ananda, 83
Ananta, 14, 64, 156, 247
Anantariti, 94, 182, 337
Anapatrapya, 79
Anasuya, 285
Anataramsa, 16

Anda, 93, 225, 290, 297
Andesha, 291
Andha Kuan, 250, 253, 260, 325
Andhakara, 151, 195, 209, 309, 314
Aneka Rupa, 84, 87
Anga, 106, 190, 215, 221, 260, 287
Angana, 175, 176
Aniruddha, 352
Anishta, 224, 330
Aniyata, 20, 79
Annam, 21, 83
Antahkarana, 259, 269, 273, 274, 316
Antaka, 71, 95
Antakarana, 322
Antara, 64, 83, 85
Antaramsa, 49, 175, 258
Antaratma, 219, 293, 300, 347
Antarbhuta, 214
Antardasha, 23
Antardhyana, 82
Antarjata, 110, 114
Antarmukha, 20, 55, 100, 107, 109, 161, 162, 188, 224, 227, 256, 314, 316, 328, 366
Antavanta, 159
Antimansha, 342
Antra Tantra, 229
Anu, 76, 160, 225
Anubhav, 25, 75, 76, 121, 201, 269, 271, 272, 320, 353
Anubhuti, 71, 121, 352, 353
Anukampin Tantra, 229
Anuktasiddhi, 341
Anukulana, 14, 122, 335
Anumatra, 225
Anunaya, 211, 325, 330
Anuprastha Taranga, 60, 328
Anurenu, 89, 225
Anurupana, 14, 15
Anutrijya, 342
Anuvaka, 156
Anuvritta, 59, 111
Anuyoga, 75, 76
Anuyogakrit, 75, 302, 303
Anvayika, 60
Anyajatma, 194
Anyathasiddha, 259

Apacayana, 104, 190, 209

Apadharma, 244

Apahrtabhara, 213

Apamarjana, 355

Apana, 21, 23, 29, 193, 247

Apari, 150, 153, 154, 157–159

Apas, 72, 88, 221, 238, 239, 292, 293, 301, 327, 361

Apatrapya, 80

Appamanna, 80

Aprabuddha, 367

Aramadaya, 200

Aravu Maniyal, 163

Arcisa, 15, 25, 75, 78, 103, 104, 122, 220, 221, 223, 270, 271, 277, 283, 291, 307, 309, 350

Ardha tricone, 60

Ardhajya, 22, 35, 76, 164, 237, 242, 327

Ardhavakirna, 329

Arhat, 83

Arpita, 226, 357

Artharthi, 63, 224, 266

Arthatatma, 196

Arthatma, 35, 39, 196, 219

Artta, 55, 56, 60, 61, 160, 196, 305, 337

Arundhati, 87, 239

Aryaman, 220

Asahaya, 84

Asamakala, 198

Asamapat, 303

Asamjvala, 355

Asamprajanya, 79

Asamskrita, 100, 168

Asat, 187

Asathya, 79

Aseemita, 42

Asevana, 208, 217, 239

Asha, 254

Ashir, 361

Ashirwad, 18, 91, 260, 362

Ashobhana, 78

Ashrama, 81, 181, 191, 276, 286, 299

Ashtavakra, 15, 76, 121

Ashtottara-sata, 111, 152, 260

Ashtottara-sata lingam, 111, 260

Ashuddha, 203

Ashwini Kumaras, 182, 266, 300

Asmi siddhanta, 24

Asraddhya, 79, 106

Asthayi, 232

Asthayin, 63

Asthayitva, 63, 112

Asthiyantra, 111

Astikaya, 367

Asura, 34, 74, 76, 77, 107, 109, 111, 153, 163, 170, 176, 177, 183, 184, 189, 190, 192, 199, 200, 210, 217, 248, 276, 279, 288, 289, 318, 319, 339–341, 355, 365, 366

Asura shakti, 74, 109, 111, 153, 163, 170, 183, 184, 190, 192, 210, 276, 289, 319, 339–341

Asvaryogya, 151, 356

Atala Loka, 345, 346

Atisukshma, 64, 156

Atita, 100, 168

Atiyoga, 175, 286, 287

Atma, 65, 73, 77, 85, 86, 111, 151, 211, 216, 218, 246, 259, 260, 264, 277, 278, 280, 290, 306, 315, 317, 328, 329, 349

Atma bodha, 86, 328, 329

Atma Jagrook, 211

Atma sharira, 111, 277, 280, 315, 317

Atma vichara, 216

Atmata, 368

Atmatva, 23, 59, 222, 255, 275, 301

Atmavichara, 217, 276, 280, 290

Atmiya, 17

Atri, 355

Attan, 161, 165

Atthi, 264

Audavita, 56, 60, 362

Auddhyata, 79, 189, 190

Audumbra, 257

Augrya, 155, 197–199, 348, 350

AUM, 16, 25, 62, 72, 77, 86, 87, 100, 151, 152, 155, 156, 171, 181, 189–191, 208, 225, 227, 246, 259, 268, 274, 277, 292, 293, 296, 301, 318, 338–340, 352, 356, 370

AUM shakti, 155, 181, 189, 190, 208, 227, 268, 318

Aushadhi, 210

Auta-ghatakecem-rajya, 335, 336

Avachetan, 57, 349, 351

Avadata, 158

Avadhana, 17

Index

Avagraha, 342
Avahittha, 351
Avajati, 217, 291, 346
Avakirna, 240
Avanayaka, 188
Avarata, 187
Avartan, 191
Avasar, 111, 272, 282
Avashoshan, 307
Avasimiya, 121
Avatamsaka, 324
Avenika, 23
Avibhajya, 112, 347
Avibhakta, 263
Avidhi, 303
Avidya, 82
Avigata, 337
Avrit, 55, 57
Avritti, 213, 336, 337
Avyayatman, 64, 202, 217
Awaaz, 269
Ayam Atma Brahma siddhanta, 25
Ayanamsha, 240, 326
Ayatana, 213
Ayati, 64, 252
Ayati vela, 252
Ayoga, 286

B

Baap, 220, 250, 263, 264
Badalana, 250
Badan, 228–230, 293
Badha, 189
Badhabuddhi, 42, 59, 202, 209, 269, 301, 366
Badhabuddhi vadhartha, 209
Bagalamukhi, 102, 181, 182, 263, 267, 268, 317, 319, 322, 323, 326
Bahen, 94, 292, 296
Bahirani, 64, 322
Bahirbhuta, 213
Bahirini, 322
Bahirjata, 111, 114
Bahirmukha, 73, 265, 298, 325, 328
Bahutava, 196
Bahutva, 56
Baindava, 23
Balangi, 220
Balapradhamani, 362
Balukaprabha, 367
Balukrapha, 367
Bandhana shakti, 307, 315
Barah, 285, 310
Basera, 338
Beeja, 74, 84, 176, 214, 215, 361
Beeja dharma, 214, 361
Beeja-jagrat, 74, 176
Bekabu, 209
Beti, 220
Bha, 19, 21, 83, 101, 102, 269, 277
Bhaap, 271
Bhaavi, 287, 288, 290, 293, 307, 352, 353, 356, 364, 368
Bhagna, 262
Bhagnatma, 35, 36, 111, 179, 218, 279
Bhagwan, 159, 161, 176, 178, 253, 271, 289, 293, 345, 367
Bhagwat, 19, 101, 102, 269, 277
Bhagya, 250
Bhai, 292
Bhairava, 314, 343, 357
Bhairavi, 177
Bhakta, 24, 265, 273, 277
Bhakti Marga, 16, 33, 250
Bhakti Phal, 16
Bhakti shakti, 83
Bhakti Yogi, 85
Bhara, 159
Bharajaka pitta, 183
Bharana, 89, 101, 279
Bharata, 289
Bhashantara, 232
Bhaskara, 339, 340
Bhautika, 176, 214, 283
Bhautika Sharira, 214
Bhava, 65, 193, 209, 364
Bhavana, 327
Bhavanarama, 270, 271
Bhavanavasi, 339
Bhavartha, 37, 48, 71, 94–96, 100, 229, 245, 263, 272, 280, 283, 305, 316, 330, 347, 354, 356, 361
Bhavishyath Chaturananaya, 365

Bhavitatma, 219, 367
Bhaya, 83, 272
Bhayanvita, 276
Bheda, 104
Bhiksha, 211, 279
Bhima, 327, 328
Bhinbhinatma, 218
Bhoga sharira, 99, 212, 214, 216, 237, 245, 246, 261, 264, 279, 280, 282, 309, 325
Bhogabhumi, 261
Bhogi, 63, 255, 256
Bhojan, 230
Bhoyuojan, 88
Bhram, 260, 261
Bhram Sharira, 260, 261
Bhrigu, 179, 326
Bhringaraja, 19, 213, 221, 222
Bhu, 19, 92–94, 179, 221, 238, 245, 247, 261, 265, 297, 361
Bhu chakra, 179
Bhubhaga, 61
Bhujanka, 202
Bhulokasuranayaka, 277, 283, 284
Bhuta, 208, 215, 320, 325, 344
Bhuta Yajni, 320, 325, 344
Bhuva Loka, 367
Bhuvanesvari, 81, 220, 229, 288
Bhuvar Loka, 367
Bhuyobhava, 56, 157
Bhuyojan, 88, 95, 213, 215, 241
Bija, 23, 250
Bindu, 35, 59, 150, 224
Bodha, 276, 282
Bodhatma, 187
Bodhichitta, 162
Botala, 26
Brahli, 82, 181, 185, 354, 360, 361
Brahma, 13, 62, 87, 88, 177, 191, 250, 277, 291, 345, 367
Brahma Loka, 367
Brahmaastra, 316
Brahmani, 91, 181
Brahmin, 34, 164, 346
Brihadbala, 110, 150, 151, 276, 294, 335, 336
Brihaspati, 97, 240
Buddhi, 265

Budha, 105, 247
Bulava, 351
Buniyadi, 121

C

Cadhautari, 22
Caksur vijnana, 25, 171, 260, 276
Ceshta, 264
Cetana, 81
Chaahak, 74, 152, 154, 155, 176, 217, 248, 265, 275
Chaaripai, 56
Chabi, 244
Chah, 215, 248, 366
Chaithanya, 16, 17, 19, 25, 29, 30, 34–36, 42, 57, 59, 65, 77, 78, 88, 91, 92, 101, 104, 119, 123, 156, 169, 176, 177, 182, 189, 210, 212, 218, 240, 241, 243, 245, 248, 251, 255, 263, 265–268, 270, 284, 285, 291, 295–297, 299, 301, 309, 316, 324, 326, 335, 339, 342, 351, 355
Chakor, 61, 152, 268
Chakra, 55, 59, 100
Chakra Yukti, 55, 100
Chakraj, 59
Chakri, 113
Chakrika, 30, 90, 93, 98, 100, 166, 195, 226, 246, 342
Chalan, 119
Chamundeshwari, 257
Chand, 81
Chandra, 66, 92, 234, 239, 260, 272, 293, 296, 364
Chandra chakra, 272
Chandra Marga, 234
Chandramouli, 181, 182
Chapala, 209
Chara Paryaya dasha, 23, 203
Charanam, 193, 226, 298, 299, 340, 341
Charita, 255, 287
Charma, 18
Charya marga, 152, 304, 305, 326
Chathanya, 356
Chaturmukha lingam, 257
Chaudahbhuj, 354
Cheerna, 211, 247, 254, 357
Chela, 275, 358
Chelagiri, 346
Chetan, 215, 263, 297

Index 419

Chetan shakti, 215

Chetasika, 78

Chhaya, 81, 253

Chhayaputra, 340

Chini-Chini, 113

Chinta, 267

Chitra, 31, 47, 88, 119, 183, 189–194, 196, 197, 199, 200, 202, 207, 211, 213, 216, 218, 223, 225, 227, 230, 231, 237, 239, 243, 249, 263, 267, 268, 270, 284, 294, 299, 306, 310–313, 322, 348, 358

Chitralekha, 209

Chitrini, 53

Chitta, 29, 77, 98, 101, 175, 177, 195, 197, 209, 212, 231, 234, 239, 241, 247–249, 258, 261, 262, 266, 274, 276–278, 281, 282, 284, 285, 295, 299, 335–337, 339, 349, 355, 364, 365

Chitti, 187

Chulha, 254

Chunava, 75

Citraka, 83, 110, 220, 260, 275

D

Dada, 249, 281, 285, 350

Dadhikravan, 175, 271

Dadi, 103, 181, 182, 218, 220, 268, 320

Dahana, 158

Daihik tantrika, 229

Daivi, 155, 198, 208

Daksha, 263, 290

Dakshinashapati, 66, 67, 71, 87, 88, 95, 242, 245, 247, 291, 303

Dakshinya, 349

Dalana, 263

Dama, 290

Damodara, 184, 213, 238, 246, 253, 325, 366

Danda, 121, 173, 272

Dandanayaka, 85, 363, 364

Dandi, 176, 288

Dasha, 20, 23, 24, 71, 102, 171, 189, 217, 222, 223, 279, 280, 284, 290, 308, 317, 360

Dasharatha, 122

Dauhshilya, 113

Deha, 307

Deshanu, 357

Deva, 16, 64, 104, 176, 217, 249, 252, 253, 259, 260, 273, 275, 276, 279, 289, 321, 324, 344–346

Deva Loka, 275, 344, 345

Deva Rina, 252

Deva Yajni, 104

Devadata prana, 330

Devahuti, 211

Devata, 26, 34, 160, 161, 211

Devatamayi, 161

Devatatma, 304

Devatma, 219

Devendra, 156, 197, 199, 314

Deveshi, 197, 199, 200

Devi, 153, 176, 263, 318–320, 345, 346

Dhadakan, 213

Dhairya, 23

Dhamini, 93, 285, 291

Dhamma, 366

Dhanadhikara, 221

Dhanu Rashi, 103

Dharana, 64, 112, 201, 209, 319, 345

Dharati, 220, 251, 266

Dharitri, 330

Dharma, 82, 83, 85, 181, 217, 234, 367

Dharma Bhakshak, 83

Dharma Kaya, 367

Dharma puta, 181

Dharma Rakshak, 83

Dharma Sthapak, 82

Dharma yogi, 85, 217

Dharmasavarni, 212, 332

Dharmayuddha, 21

Dharshanatma, 219, 265, 321

Dhiratma, 219

Dhriti, 238, 273

Dhumavati, 102, 181, 182, 193, 261, 263, 265–267, 287, 316, 319

Dhumprabha, 367

Dhumra, 182

Dhumya, 210

Dhvani, 155, 201, 247, 363

Dhyamikrita, 210

Dhyana, 90, 169, 232, 240, 279, 326

Dhyana Marga, 279, 326

Digamabara, 323

Digant, 258

Dik, 56, 98, 244

Dikapala, 89

Dimbhak, 91
Dimbhakita, 91
Dina, 92, 351
Dipamala, 84
Dirghavritta, 56, 57, 149–152, 161, 164, 165
Dirghavrittaphala, 161, 162
Dishta, 14, 17, 18, 26, 30, 36, 42, 56, 67, 72, 82, 93, 95, 99, 100, 153, 156, 183, 189–191, 195, 215, 222, 267, 273, 284, 285, 288, 292, 295, 299, 306, 310, 311, 313, 317, 319, 321, 328, 338–341, 343, 365
Dishti, 366
Ditthigata, 357
Ditthujukamma, 256
Divangat, 122, 175, 176, 215, 281, 297, 321, 323, 324, 331, 344, 345, 349, 350, 355–357
Divya, 20, 65, 94, 208, 215, 216, 221, 222, 252, 254, 257, 316, 334, 361
Divya ratri, 20, 65
Divya shakti, 208, 215, 216, 221, 222, 254, 257, 316
Divyata, 15, 16, 47, 48, 64, 78, 219, 253, 258, 320
Doharana, 122, 360, 363, 365
Dokkalamma, 288
Doliltra, 320
Dono, 156
Dravya, 177, 217
Dridayudha, 36, 163, 318
Drishtanta, 238
Drishti, 82, 259
Drishti Mandala, 259
Drvaya, 325
Duguna, 60
Duniya, 17, 216, 217, 260, 318, 366
Durashravana, 342
Duratma, 203, 218, 290, 329
Durga, 21, 23, 66, 159, 162
Durgandha, 213
Dus, 360
Duschhaya ojas, 330
Dushkarakarana, 323
Dushtatma, 21, 219, 259, 280, 314, 364
Dvaiyogya, 151, 314, 333
Dvandva Brahma, 191
Dvara, 244
Dvidhruviya, 322

Dvijya, 156, 335
Dvilaya, 46, 175
Dvisaptati dasha, 269
Dvishakha, 154
Dvyartha, 278, 355
Dyumna, 151, 226, 240
Dyut, 65, 276

E

Eeshan, 297
Ek, 17, 352, 358, 359, 361
Ekadharma, 309
Ekagratha, 81, 224
Ekarajya, 64, 326
Ekarshi, 61, 71, 150
Ekartha, 17, 43, 57, 58, 92, 96, 101, 208, 215, 265
Ekarupa, 47, 182, 183, 267, 268
Ekashtaka, 219, 246
Ekatma, 162, 218, 338
Ekatva, 103
Ekavala, 273
Ekavali, 30, 114, 273, 354, 362
Eke, 17, 78, 228
Ekotrem, 307–309, 331, 340
Eroli, 113, 363
Evakara vadartha, 325, 355

F

Faraka, 17

G

Gabhira, 35, 60, 62, 63, 108, 228, 231, 237, 238, 248, 282, 284, 294, 295, 328, 340, 350, 351
Gabiratma, 351
Gaganaganja, 238
Gaja Mukha, 275
Gajakarani, 306, 322
Gana, 23, 47–50, 118, 216, 234, 255, 257, 260, 329, 331, 347, 350, 351
Gana Marga, 234, 255, 329
Ganapati, 18
Ganarajya, 20, 23, 48, 49, 163, 181, 182, 350, 351
Gandakanayaka, 100, 105, 112, 152, 156, 159, 224, 228, 305, 339
Gandha, 222
Ganesha, 244, 254, 259, 270, 271, 281, 315
Ganin, 119

Index   421

Garbha, 154, 226
Garbhadhana, 343
Garbhadharana, 61
Gardhaba, 230, 282, 295, 296, 302
Gardhabi mukha, 114, 328
Garjya, 65
Gat, 299
Gatr, 299
Gauranga, 285, 288, 290, 351, 364
Genda, 176, 212, 290, 293–296, 339, 340
Ghabira, 340
Ghadi, 47
Ghana, 166
Ghanabha, 154
Gharma, 240
Gharmatma, 353
Ghataana, 237, 325
Ghatanuka, 359
Ghataprakashaka, 166
Ghatika, 15, 31, 121, 155, 156, 176, 197, 199, 200, 226, 270, 279, 298, 313, 342, 347, 349, 352, 353, 357, 363, 364
Ghatikamandala, 59, 127, 199
Ghatna, 26, 82
Ghrinibhrit, 285
Gita, 101, 291
Glana, 354
Glani, 101, 338
Go Loka, 287, 339, 366
Gochar, 263
Goshthika, 33, 63
Gotra, 296
Griva, 162
Guna, 48, 61, 63, 67, 101, 208, 267, 272, 292, 318, 344, 355, 364
Guna chakra, 61, 67
Guna Sharira, 267, 272
Gunasa, 252, 346
Gunatipat, 354, 355
Gunja, 160
Gunya, 35
Guru, 17, 29, 57, 65, 75, 79, 80, 83, 96, 98, 177, 193, 209, 231, 252, 258, 259, 276, 290, 297, 303, 321, 323, 361
Guru dharma, 83
Guru Sthana, 17, 252, 321

H

Haisiyat, 179, 316
Ham, 92, 266, 273, 281
Hamsa chakra, 151
Hamsatma, 219
Hanuman, 81, 173, 222, 224, 263, 264, 276, 280, 281, 283, 285, 288, 300, 345, 346, 348, 358, 359
Haram, 104
Harbuddhi, 316
Hari, 22, 53, 82, 100, 156
Hariti, 25, 41, 78, 248, 350
Harkriya, 260, 317
Harsiddhi, 364
Haryasvas, 75
Haryyatma, 77, 78, 166, 219
Hasa, 178
Hasta, 187, 191, 195–197, 203, 204, 217, 281, 289, 323
Hasti, 16, 20, 23, 26, 29–31, 36, 39, 48, 67, 92, 100, 121, 155, 163, 164, 175, 176, 178, 193, 198, 200, 212, 219, 253, 259, 261, 279–281, 284, 313, 315–318, 321, 326, 338–340, 346–348, 350–352, 363, 364
HAUM shakti, 75, 77, 224, 225, 267, 269
HAUM shakti chakra, 267, 269
Hemarenu, 67, 161, 224
Heruka, 275
Hetavartha, 221
Hetu, 89, 93, 172, 175, 177, 178, 189, 190, 216, 248, 253, 267, 275, 276, 341, 356, 364
Himsa, 213
Hinayana, 25
Hiranya sharira, 309
Hiranyagarbha, 42, 57, 75, 76, 96, 182, 190, 216, 224, 227, 261, 263, 281, 338
Homa, 249, 274, 291, 295, 340
Hora, 182, 201, 237
Hora chakra, 182, 201, 237
Hradamana, 188, 313, 315
Hri, 80
Hridayesha, 119, 174, 192, 197, 308, 356, 358
Hridya Pranali, 122, 213, 230, 231, 300, 301, 314, 326
Hriti, 17
Hunduka, 71, 104, 108, 109, 189, 190, 362, 366
Hunduka shakti, 190

Hurupa, 253

I

Ibha, 93
Iccha, 215, 283
Ida, 159–161
Idam, 263, 264, 276, 289, 299
Ijara, 335, 337
Ijaradarita, 337
Indambara, 201
Indra, 74, 83, 85, 114, 249, 287, 367
Indra Loka, 287, 367
Indrani, 88, 98
Indriya, 367
Indu, 64
Insan, 18, 79, 87–89, 169, 171, 176, 216, 242, 259, 262, 264, 273, 276, 284, 315, 321, 324, 325, 327, 346, 348–350, 353, 356–360, 362–364, 368, 370
Insaniyat, 31, 275, 276, 341, 364
Irshya, 79
Isha, 47, 178, 291, 343, 364
Ishi shakti, 315
Ishta, 223, 367
Ishtartha, 201
Ishvara, 13, 28–30, 77, 78, 80, 83, 84, 99, 119, 177, 216, 249, 254, 257, 260, 265, 271, 277, 281–283, 285, 287, 289–291, 336, 345, 346
Ishvara Marga, 260
Ishvaratva, 85, 291

J

Jada, 211
Jadatva, 38, 62, 96
Jagadvinasa, 160, 161
Jagarya, 179
Jagath, 366
Jagatkritsna, 53
Jagrat, 176, 177, 257
Jagrit, 78, 329
Jagriti, 264
Jahar, 64
Jahatsvartha, 49, 50
Jaivik Anunaad, 320, 342
Jal, 19, 88, 221, 222, 291–293, 299, 304
Jala chakra, 291
Jalakrama, 209, 279
Jalandhara, 212
Jalaprana, 88, 208, 215, 240, 265, 293
Jaldi, 109, 220
Jalini Mukha, 191
Jamadagni, 303
Jambha, 93
Jambhaka, 222, 251, 293
Janabala, 114, 209
Janaka, 149, 255
Janana, 17, 22, 26, 30, 60, 71, 78, 83, 89, 119, 154, 169, 174, 186, 193, 196, 203, 268, 270, 285, 314, 315, 344
Janani, 246, 255
Jangamavisha, 182, 323
Jangha mukha, 20, 344
Janma yoga, 25, 159, 220
Jantu, 271, 284, 309, 315, 316, 318
Jantumati, 208, 239
Jathara, 27, 30, 272
Jimmedari, 21, 23, 203, 295
Jinpa, 65, 76, 302
Jism, 293
Jitatma, 218
Jiva, 26, 27, 150, 174, 176, 231, 286, 288, 289, 334, 346, 348
Jiva Astikya, 231
Jivagribh, 222
Jivan, 254, 277, 357, 358
Jivan mukta, 254
Jivan Rekha, 277
Jivatma, 273
Jivavijnana, 13
Jivita, 78, 129
Jnana, 16, 21, 24, 33, 85, 86, 90, 100, 102, 150, 154, 211, 212, 218, 249, 251, 268–271, 275, 282, 309, 338, 356, 365–367
Jnana Kaya, 366, 367
Jnana Marga, 16, 33
Jnana mimansiya, 365
Jnana Phal, 16
Jnana siddhi, 211, 212, 251, 270, 282
Jnana Yogi, 85, 86
Jnanadhikara, 234
Jnathru, 101
Jnyat, 278
Juta, 104
Juthalaya, 302
Jyotaranga, 121
Jyotir Loka, 367

Index

Jyotistava, 78, 91, 218, 264, 265, 306

K

Kaal, 17, 19, 20, 22, 37, 39, 46–48, 53, 54, 64, 77, 82, 86, 87, 94, 100, 113, 150, 154, 179, 187, 191, 193, 219, 252, 280, 340, 347, 349, 351, 354

Kaal chakra, 86, 187, 191

Kaal Sarp, 17, 19, 87, 113

Kaal Sarpa chakra, 17, 19, 87

Kacangala, 114

Kailasha, 252

Kaise, 218

Kaivalyashrama, 54, 55, 201

Kaki, 231

Kala astikaya, 273

Kala Devi, 288

Kalakaar, 187

Kalaptita, 354

Kalatma, 36, 96, 219

Kali, 16, 176, 178, 195, 197, 267, 272, 282–284, 331, 343, 345, 346, 355

Kali Shakti, 16

Kalika, 245, 271

Kalki, 18, 286, 323, 324

Kalpa, 17, 36, 181, 220, 301, 302, 313, 350

Kalpanik, 38, 75, 97, 150, 151, 167, 171, 175, 183, 223, 272, 274

Kalpavriksha, 365

Kama, 167

Kamagiri, 208

Kamartha, 110, 332

Kamarupitva, 351

Kamma, 83, 296

Kammannata, 80

Kana, 110, 228, 276, 277, 345, 363

Kanalakshamsha, 160, 197, 200, 342

Kantatma, 209

Kantida, 276

Kanya Rashi, 94, 95

Kapila Kumara, 18, 114, 273, 353, 362

Kapinjala, 87, 113, 216, 244, 246, 262, 264, 277, 296, 318

Kapisena Nayakaya, 221

Kapota, 309

Karak, 88, 335, 336

Karala, 101, 272

Karana, 17, 91, 101, 152, 153, 192, 211, 216, 238, 245, 248, 251, 252, 262, 264, 267, 270, 273, 274, 277, 282, 291, 299, 306, 310, 315, 320, 325, 334, 336, 340, 341, 357

Karana Sharira, 17, 274, 277, 299, 336

Karandi, 342

Kardama, 323

Karin, 17

Karka Rashi, 95

Karma, 16, 19, 26, 33, 85, 114, 178, 179, 188, 190, 217, 226, 233, 249, 250, 290, 292, 297, 298, 357

Karma chakra, 19, 179, 190, 290

Karma Marga, 16

Karma Phal, 16

Karma Yoga, 226, 298

Karma Yogi, 85

Karmana, 250

Karmi, 16, 192

Karna, 31

Kartavya, 277, 278

Karuna, 80

Karuyantra, 165

Kashi, 257, 316

Katai, 259

Katana, 113

Kathanayaka, 84

Kaumari, 256

Kaumaritantra, 367

Kaun, 287

Kaushalya, 209

Kausidya, 79

Kaya, 175, 181, 367

Kendrabhi mukha, 55, 71, 94, 354

Kendrabindu, 28, 169

Kendraja, 302

Kendrajita, 304

Kesari Nandan, 247, 277, 281, 297, 324

Kesarisutaya, 115

Keshava, 43, 56, 101, 222, 274

Ketu, 55, 71, 94, 95, 242, 243, 247

Kha, 152, 292

Khalnayaka, 31, 46, 223

Khanda, 350

Khandana, 14, 15

Khanijavarga, 324

Khara, 120, 310–313, 315, 316, 339–341, 348,

353
Kharcha, 278
Kharidi, 153, 250
Khatta, 213
Khayti, 343
Khojana, 218
Khola, 62
Khuda, 93, 103, 174, 220–222, 237, 238, 250, 255, 331
Khudrata, 156
Khyati, 83, 177, 343
Kincana, 208
Kincha, 75, 174, 302, 303
Kirt Karna, 271
Kismat, 25, 251
Klesha, 79, 80
KLIM, 17, 114, 155, 197, 275
KLIM shakti chakra, 17, 275
Kojyataranga, 158
Konasa, 75, 304
Konastha, 78, 109, 113, 159, 160, 196, 220, 244, 245, 256, 299, 308
Kramana, 36
Kramasuchaka, 121
Kranti, 57
Krantijya, 31
Kratu, 309
Krida, 231, 247, 279
KRIM shakti chakra, 230, 361
Kripita, 209
Krishna, 164, 177, 364
Krishnamurti, 47, 227, 248, 311
Krita, 161, 162
Kritajna, 78, 150, 151
Kriya shakti, 308
Kriyavasha, 306, 307
Krodha, 74, 79, 109
Kruralochana, 340
Kshatradharma, 102, 154, 181, 182, 268, 269, 271, 275, 319, 327
Kshatriya, 299, 341
Kshayaroga, 218
Kshema, 245, 308, 309, 326
Kshemya, 180, 191, 203, 328, 329
Kshemya shakti, 180, 191, 203
Kshepa, 368

Kshetra, 83, 152, 175, 177, 178, 251
Kshira sagar, 213
Kshiti, 93, 220, 221, 226
Kshitigarbha, 158, 174, 177, 287
Kshitij, 93, 259
Kud, 221
Kudrat, 15, 85, 101, 157, 219, 261, 267, 276, 288, 297, 334, 336, 361
Kuha, 62, 368
Kula, 16, 94, 98, 224, 276, 335, 337
Kumara, 18, 284, 285, 295, 301, 302, 305, 327, 350, 353
Kumari, 18, 285, 295, 296, 301, 327
Kumba Rashi, 100
Kumbhaka, 67, 75, 91, 122, 150, 151, 221, 274
Kundali, 355
Kundalini, 33, 46, 84, 93, 98, 99, 157, 175, 209, 219, 221, 259, 276, 279, 284, 360
Kundalini shakti, 221
Kunti, 84
Kuradana, 122, 209
Kuredana, 209
Kushala, 78
Kyun, 218, 287

L

Laal, 212, 213, 244–246, 281, 289
Labha, 63, 64, 112, 113, 234, 260
Ladhima, 354
Laganem, 331
Laghu, 57, 289
Lahari, 62, 118
Lahuta, 80
Lakshana, 25
Lakshmi, 84, 158, 218, 233, 261, 287, 288
Lakshya, 17, 76, 87–90, 93, 94, 232, 269, 274–276, 278, 288, 298, 300, 301, 305, 320, 342, 356, 359, 360, 362, 363
Lalita, 23, 55, 64, 67, 71, 97, 153, 189, 203, 213, 244–246, 276, 280, 287, 316
Lam, 42, 202
Lambila, 237, 238
Lasaka, 209, 221
Laseeka Tantra, 230
Laya, 219
Layanalika, 355
Lekhaka, 197
Lena, 17

Index

Liksha, 26, 27, 30, 90, 100, 105, 174

Lila, 218, 219

Linga, 47, 55, 100, 120, 189–191, 195, 207, 209, 210, 248, 274, 280, 281, 283–287, 289, 300, 309, 325, 340, 341, 347

Linga sharira, 100, 189, 190, 195, 207, 210, 248, 274, 280, 281, 283–287, 289, 300, 309, 325, 340, 341

Lingam, 331

Lingopahita-laingika, 73, 281

Lingopahita-laingika-vadartha, 73

Lobha, 167

Loka, 84, 174, 336, 344–347, 360, 363, 366

Lokanayaka, 262

Lokapurusha, 217

Lopamudra, 23, 326, 329

Lutashaav, 93

M

Mada, 79, 85, 109

Madahava, 269

Madan, 86, 278

Madhava, 269, 297

Madhusudan, 91, 111, 164, 177, 183, 241, 287, 308, 325, 333–335

Madhya, 108

Madhyasthahinta, 238, 260, 329

Madhyasthata, 329, 330

Maha, 15, 23, 55, 74, 81, 92, 96, 99, 112, 114, 158, 165, 176–178, 181, 187, 197, 198, 203, 222, 225, 228, 230, 232, 234, 244, 251, 256, 258, 259, 261, 263, 267–269, 272, 275, 278, 283, 284, 288–290, 292, 296, 297, 303, 305, 307, 309, 311–314, 316, 317, 319, 324, 325, 336–339, 341, 345, 346, 355–357, 361, 368

Maha Bhaavi, 23

Maha dasha, 15, 114, 222, 309, 324, 356, 234

Maha dharma, 23

Maha Durga, 288, 357, 361

Maha Gauri, 177, 288

Maha Gayatri, 244, 288

Maha Jagrat, 176, 251

Maha Kali, 112, 178, 197, 267, 268, 283, 288, 292, 296, 297, 317, 355

Maha Kalpa, 225, 228, 256, 313, 314

Maha Lakshmi, 178, 261, 288

Maha Nitya, 99, 181, 187, 251, 284, 311, 312, 313, 316, 345

Maha Pitha, 198

Maha Rasi, 263, 269

Maha Ratri, 81

Maha Saraswati, 177, 275, 288

Maha Shiva, 92, 165, 258, 275, 278, 290, 305, 307

Maha Siddha, 341

Maha Svapna, 74, 176, 203, 232, 272, 289, 325, 339

Maha Vibhu, 55, 96, 158, 178, 225, 230, 251, 259, 261, 319, 336–338, 368

Mahabala, 151, 300, 303

Mahabala Parakramaya, 151

Mahabhaumika, 78

Mahabhutta, 214

Mahachitti, 157

Mahakaal, 212, 214

Mahakriya, 163, 200, 201, 344, 345

Mahakriya marga, 163

Mahant, 109

Mahar Loka, 346, 347

Maharavana, 89, 187, 249, 292

Mahartha, 223

Mahasatta, 113, 347

Mahaspanda, 193

Mahat, 320

Mahatattva, 92, 212, 231, 245, 246, 253, 260, 283, 293

Mahatma, 73, 102, 189, 246, 262, 268, 280, 293, 309, 313, 328, 368

Mahavijya, 77

Mahayana, 67, 111, 275, 278, 318

Mahendrani, 15

Maheshwara, 177, 260, 261, 273, 289, 346

Mahika, 21, 260, 278, 318, 340, 366

Mahodbhava, 66, 67, 118, 169, 195, 200, 214, 222, 225, 239, 265, 271, 305–307, 318, 320

Mahurata, 307

Mahvijya, 223

Maitri Bhava, 328

Makadi, 93

Makar Rashi, 90, 91, 277, 315

Makuli, 327, 329

Malanga, 344

Malini, 196

Mallika Devi, 285

Malum, 278

Mamaka, 340

Mamaya, 90

Mamsaja, 308
Manas, 88, 219, 271, 295, 309, 313
Manasikara, 81
Manav, 150, 153–155, 198
Manava karaka, 209
Mandra, 36, 38, 163, 164, 177
Mandu, 288
Manduk, 92
Mandukak, 91
Mangala, 101, 103, 106, 240
Mangalnath, 86, 228
Manipura chakra, 19, 155, 180, 198, 226, 297
Manjovaya, 23
Manjushri, 264, 265
Manogata, 78
Manojava, 66, 235
Manomaya sharira, 188, 207, 209, 248, 309, 313, 325, 351
Manoratha, 357
Mansika, 355
Mantra, 331
Manushya, 239, 254, 262, 296, 297, 325, 364
Manushyagati, 301, 323
Manviya, 179, 212, 213, 324, 325, 334, 365
Manviya karak, 179, 212, 213, 325, 334, 365
Manyu, 102, 174, 181, 182, 222, 243, 247, 287, 293, 301, 327, 348
Mapanka, 155
Mara, 101
Mardangat, 155
Marga, 279, 364
Marichi, 57
Marji, 91
Markatesh, 35, 42, 222, 304
Martya, 113, 244, 279
Marudeva, 81
Maruti, 208, 209, 221, 239, 261, 266, 336
Masarya, 228
Mash, 160
Masika, 175
Masitasmrtita, 79
Mastishka, 330, 331
Mata, 220, 224, 250
Matali, 189
Mati, 353

Matri, 57, 362
Matsarya, 79, 160, 224
Matsya, 89
Maujud, 78, 150
Maujudgi, 30, 77, 81, 158, 268, 282, 343
Maya, 36, 38, 42, 43, 67, 79, 92, 97, 150, 156, 162, 169–191, 210, 215, 225, 249, 258, 262, 269, 270, 323
Maya shakti, 92, 97, 162, 190, 191, 210, 215
Medha, 23, 84, 215, 258
Meen Rashi, 87
Megha, 23
Mein, 35, 261, 262, 264, 294, 305, 306, 326, 330, 331, 348
Mena, 362
Mendaka, 92
Mera, 305, 325, 365
Mesha Rashi, 107
Midduta, 176
Milaana, 363, 364
Mimamsa, 29, 278, 332
Mithun Rashi, 106
Mitra, 228, 245, 256, 322, 323, 344, 351
Moha, 65, 203
Mohini, 60, 89, 267, 269, 275, 343
Moksha, 59, 85, 88–90, 232, 242, 253, 278, 298
Moksha Astikaya, 59
Mokshayitri, 192, 330
Mraksha, 79
Mridangapanava, 363
Mrig Trishna, 224, 357, 358
Mritya Ojas, 342, 343
Muddata, 176
Mudita, 80
Muduta, 80, 210
Mufta, 270
Muftakhori, 29, 101, 174, 297, 298, 344
Mukha, 122
Mukhya, 159, 292, 295, 300, 330, 331
Mukti, 85, 86, 91, 216, 233, 254, 281, 353
Muladhara Chakra, 18
Mulaguna, 323
Mulaklesa, 80
Mulaprakriti, 213
Mulya, 30, 56, 65, 149, 152, 202, 203, 212,

Index 427

251, 258, 329, 354
Muni, 275
Muthyalamma, 288, 294

N

Naad, 18, 155, 208, 214, 222, 231, 239, 247, 248, 258, 261, 266, 330, 363
Naadi, 177, 178
Nabhika, 122
Nagarajya, 98
Nagarika, 212
Nairatma, 269, 270, 291, 297
Nairitti, 297
Naitik, 113
Naitikta, 113
Naityaka, 259
Nakala, 15, 46, 48, 89, 119, 159, 174, 189–191, 198, 199, 202, 216, 222, 227, 241, 244, 249, 251, 252, 261, 284, 295, 310–312, 316, 322, 341, 346, 348, 358
Nakaratmak, 209, 325, 359
Nakaratmak shakti, 209, 359
Nakaratmak sharira, 325
Nakkhatta, 60, 222
Nanak, 281
Nanarupa, 63, 65, 78, 94, 100, 113, 153, 255, 320
Nandi, 221, 254
Nandini, 76
Napumsaka, 191
Nar, 250, 339
Narada, 55, 174, 223, 265, 286, 297, 348
Naraki, 270, 272, 312
Naran, 55, 74, 256, 334, 358, 359
Narayana, 159, 161, 257, 271, 308
Narayani, 174, 222, 227
Nari, 78, 156, 250
Naseeb, 251
Nashita artha, 175
Nastika, 368
Nataraja, 210, 261, 287
Natija, 297
Natini, 290, 295, 339
Nau, 359
Navgraha, 243, 250
Nayaka, 23, 31, 65, 223
Nayakatva shakti, 63, 64, 113, 159, 178, 357

Nayaki, 23, 47, 215, 233, 367
Nayana, 33, 121, 261, 282, 333
Nayanotsava, 289
Neta, 100, 108, 114, 191, 231, 357–359
Netritva, 85, 90, 104
Niayama, 20
Nidana, 17, 83, 86
Nidhi, 114
Nidra, 176
Nidramaya, 328
Nilalohita, 217
Nilanjana, 155
Nilirasa, 213
Nipaka, 194, 195
Nipparyaya, 147, 352
Nirbija, 24, 154, 299, 358
Nirbijayoga, 189
Nirdalana, 262
Nirdeshika, 152, 153
Nirdharma, 63, 83, 153
Nirguna, 63, 153, 359, 367
Nirharin, 257
Nirjara, 181, 182, 215, 259, 262, 263, 275, 296, 317, 321, 329
Nirman Puta, 181
Nirmana kaya, 325, 367
Nirmata, 251
Nirrti, 170, 171
Nirukta, 46
Nirvana, 85, 232
Nirvapana, 208
Nirvikalpana, 18, 26, 36, 57, 59, 67, 155, 196, 262, 273, 274, 322–324
Niryoga, 286
Nishchaya, 34, 357, 358
Nishedhavritti, 272
Nishkala, 279, 294
Nishkriyata, 200
Nishkriyatma, 357
Niskriya astikaya, 272
Nissara, 208, 218, 253, 292, 345, 347
Nitala Loka, 345
Nitya, 35, 108, 164, 166, 169, 171, 177, 227, 242, 258, 312, 313, 358, 359
Nitya ratri, 35, 108, 164, 166, 169, 171, 177, 227, 242, 258

Nityata, 362
Nivarana, 212
Nivas, 30, 78, 178, 216, 225, 273, 274, 315, 316
Nivasaniya, 167, 271, 274, 275, 296
Nivasi, 46, 118, 215, 263
Nivedana, 83
Nivritti, 252, 322
Nivritti marga, 252
Nivrtti dharma, 71, 212
Niyama, 15, 21, 23, 295, 302, 324, 356
Niyata, 22
Niyatamanasa, 252, 346
Niyatatma, 299, 341
Niyati, 64, 217, 250, 254
Niyogartha, 160, 199, 200, 208
Niyojana, 344, 349
Nyartha, 56, 66

O

Ojas, 98, 211, 351, 361
OM, 368, 370
Omkara, 208, 229, 299, 341, 355
Omkara vadartha, 208, 355
Ovada, 210

P

Padartha, 74, 86, 101, 170, 228, 240, 321, 323, 324, 336
Padavanita, 160
Padmanartteshvara, 292
Padoccaya, 365
Pagunnata, 80
Pahachaan, 26, 63, 107, 176, 251, 286
Pai, 42, 59, 61, 76
Pala, 210
Palala, 13, 177, 183, 243
Palanhaar, 78
Panava, 363
Panchajani, 48, 193, 263
Panchakone, 154, 158
Panchama, 297
Panchang Ganita, 297
Panchanguli, 209, 228, 276, 313, 314
Panchatattva, 214, 221, 226, 361
Panchtantra, 112
Pandrahbhuj, 154

Panva, 93
Papa astikaya, 272
Para, 54, 56, 72, 147, 152, 161, 175, 191, 207, 238, 243, 244, 255, 257, 259, 265, 266, 340, 342, 349
Para Ganesha, 56, 152, 161, 175, 244, 259
Para shakti, 191, 207
Parai, 304, 363
Parakarana, 64
Param, 42, 48–50, 52, 53, 86, 88, 98, 150, 157, 158, 161, 163, 165, 171, 193, 222, 224, 232, 233, 238, 242, 243, 252, 254–256, 259, 267, 269, 271, 273, 274, 278, 281, 282, 288, 289, 308, 327, 331, 338, 345, 354, 355
Param Ganesha, 254, 259, 267, 269, 271, 273, 274, 281, 288, 289, 327
Param Gauri, 289
Param Guru, 233, 282
Param Mukti, 98, 232, 256
Param Mukti Marga, 98
Param Parvati, 48–50, 157, 158, 242
Param Prapti, 224
Param rupa, 150
Param Shiva, 42, 161, 338
Param Shunya, 163, 165, 171
Param Vishnu, 252, 271, 273
Paramananda, 29
Paramanu, 56, 242
Paramartha, 240
Paramatma, 33, 81, 88, 102, 207, 208, 219, 234, 247, 253, 254, 278
Paramatma Marga, 33, 234, 254
Parameshthi, 158
Parameshwara, 328
Parampara, 234
Paramtattva, 345, 346
Paranukampi Tantra, 229
Pararabdh, 250
Parat, 26, 237
Paridhi, 340
Parijan, 63, 318, 356
Parijanana, 200
Pariksha, 63, 202
Parikshipta, 263
Parimandala, 42
Parimita, 90, 91
Parinama, 314

Index

Parinditartha, 204
Pariprashna, 75, 76
Pariprashnati, 76, 302
Parisar, 78
Pariseemita, 224
Parishkar, 252
Paritoshika, 200
Parivar, 14, 15, 46, 47, 118–121, 241, 262, 266, 305, 334–338, 356
Parjanya, 111
Parjanyatma, 347
Parshnisamasta, 81, 283, 324, 325, 358
Parvati, 18, 38, 49, 87, 160, 161, 178, 288, 293, 327, 334, 342, 343, 356, 360, 362, 364, 365
Paryaya, 23, 147, 352
Paryupti, 250
Pasaka, 64, 159, 209, 275, 358
Pasha, 16, 357
Pashu, 55, 256, 366
Pashyana, 232, 329
Passaddhi, 80
Patala Loka, 366
Pati, 250
Patni, 250
Pattanayaka, 223
Paturiya, 103
Pautra, 348, 349
Pautri, 348, 349
Pavamana, 102, 181, 182, 264, 321, 322, 326
Pavitratma, 29, 74, 171, 260, 273, 322
Payojya, 162
Peenugula Mallanna, 287
Pesha, 64
Phasava, 316
Pighal, 362
Pinakapani, 75, 301, 302
Pinda, 25
Pindajya, 84
Pipplayshraya, 93
Pitah, 218
Pitavasa, 86, 87, 280
Pitha, 122
Piti, 189, 268
Pitra, 24, 77, 87, 88, 91, 95, 101, 167, 190, 195, 218, 220, 249, 253, 264, 265, 280, 305–307, 321, 345–347

Pitri Loka, 120, 174, 367
Pivari, 286
Piyati, 55, 57, 60, 196
Pola, 121, 122
Porutham, 229, 294
Poshaka, 112, 348, 350
Poshan, 67
Poshita, 112
Poshitri, 338
Pota, 220
Poti, 220
Prabha, 38, 42, 47, 48, 65, 92, 96, 98, 101, 103, 158, 176, 177, 189, 239, 243, 244, 255, 260, 274, 310, 311, 339
Prabhaga, 316
Prabhasa, 17, 233, 358
Prabhasa Marga, 17, 233
Prabhasa Phal, 17
Prabhava, 91, 299
Prabhavi, 121
Prabhu, 34, 49, 113, 122, 253, 254, 281, 296, 358
Prachalita, 75, 78, 83
Prachar, 154
Pradarshan, 120, 166, 341
Pradasa, 79
Pradesha, 15, 42, 262, 300
Pradhi, 27
Pradyumna, 220, 247, 337, 340, 344
Pragami, 187
Prahar, 184, 350
Prajan, 47, 49, 89, 359
Prajana, 19, 88, 90, 323, 368
Prajanan, 30, 49, 63, 80, 88, 95, 156, 216, 221, 228, 250, 260, 271, 291, 335, 337, 351, 361
Prajanan Pranali, 30, 63, 156, 228, 250, 335
Prajaniyam, 86, 87, 112
Prajati, 14, 15, 31, 46–50, 110, 118, 120, 121, 174, 256, 257, 331, 360, 361
Prajatikaran, 359
Prajna, 20, 80
Prajnanam Brahma siddhanta, 24
Prajnatma, 23
Prakalpana, 77
Prakara, 14, 15, 46, 111, 119–121, 174, 221, 256, 331, 333
Prakarana, 19, 26, 107, 156, 224, 246, 354

Prakarana Vadartha, 107
Prakasha, 16, 84, 165, 196, 247, 260, 301
Prakashatma, 83, 102, 188, 190, 218, 281
Prakashisu, 61, 124
Prakatikarana, 110, 212, 260, 278
Prakramya, 90, 92
Prakrima, 312
Prakriti, 85, 166, 217, 250, 258, 268, 271, 293, 297, 298
Prakriti dharma, 268, 271
Prakriti-yogi, 166
Prakriya, 223
Prakshepanem, 221
Pramada, 79
Pramanita, 218
Pramanya, 342, 352, 353
Pramanya vadartha, 342
Pramathanaya, 303, 304
Pramatra, 30
Prameya, 23
Pramod, 15
Prana, 194, 263
Pranayama, 213, 328, 342
Prani, 161
Prapaka, 80
Prapta, 94
Prapti, 85
Prapya, 18, 224, 257, 272, 333, 356, 361, 363
Prarabdh, 249
Prasanskaran, 224
Prashana, 111
Prashant, 61
Prashantatma, 60, 77, 91, 108, 193, 219, 256, 333, 364
Prasthan, 62, 331
Prasushrut, 290
Prasuti, 275, 280
Pratana, 362, 363
Pratham Shakti, 173
Prati, 164, 177, 300
Pratigad, 111
Pratigha, 82
Pratikarak, 251, 270
Pratikriti, 122, 360
Pratilamb, 263

Pratinidhan, 278
Pratipad, 224
Pratirajya, 31
Pratirupa, 360
Pratisthapan, 278
Pratitya samutpada, 338
Pratiyaman, 18, 287, 293–295, 297, 303
Pratyabhijna, 80
Pratyach, 329
Pratyagatma, 82, 92, 95, 104, 194, 218, 249, 265, 299, 322, 356
Pratyalidha mandala, 112, 368
Pratyantara dasha, 222
Pratyayatma, 256
Pratyeka, 65, 101
Pravaha, 317
Pravritta, 62, 105, 202, 226, 228–230, 248, 279, 294, 295, 300, 314
Pravritta chakra, 202, 228–230
Pravritti, 97, 256, 257
Pravritti dharma, 256, 257
Pravrtti marga, 64
Prayasta, 26
Prayaya, 276, 363
Prayujya, 246, 361
Prema, 256, 317
Prerita mandala, 358, 359
Preshita, 255
Pretasharira, 350
Preya, 273, 332, 344, 345
Preyas, 220
Prithvi, 163, 245, 256, 318, 344
Priti, 77, 78, 87, 278, 307
Priyokti, 361
Pruthaktva, 112, 337
Pruthaktva Tejas, 112
Pruthvi, 364
Pudgala, 14, 272
Pudgala astikaya, 272
Pulastya, 55, 57, 163, 164, 202, 354
Puma, 341
Pumsavana, 182
Punargathan, 237
Punarjagran, 21
Punji, 251

Index 431

Punravritta, 265, 295, 336, 337
Punya astikaya, 272
Puraitri, 295, 300
Purani shakti, 214, 325, 359
Purejata, 202, 223, 354
Purna, 57, 256
Purna shakti, 256
Purnatavadi, 65
Purusha, 15, 25, 47, 62, 119, 192, 193, 250, 251, 284, 343, 355, 356, 367
Purushartha, 99, 174, 222, 251, 298, 318, 328
Purushayoni, 156
Purvabhasa, 237
Purvaja, 220, 339, 349
Puryashtaka, 86, 332
Pusalattu, 357
Pushtartha, 82, 108, 231, 319
Pushti, 73
Pustak, 248, 365, 366
Putana, 120
Putatma, 91, 92, 105, 218, 316
Puti mukha, 93
Putra, 212, 220
Putra dharma, 212, 220
Putri, 220, 250
Putri dharma, 220
Pyar, 257

R

Raasta, 42, 56, 57, 75, 176, 177, 282, 295, 311
Rab, 76, 77, 290
Rachana, 65, 112, 319
Rachanatmak, 337
Rachayita, 96, 255
Rachitartha, 34, 49, 82, 277, 280
Radha, 81, 183, 342, 352
Raga, 82, 110, 181, 257, 274–276, 362
Rahita, 228
Rahitatma, 219
Rahu, 71, 103, 104, 242, 243
Raivata, 83
Rajah, 33, 73, 321
Raja-rajesvari, 15, 324
Rajas, 22, 24, 83, 113, 161, 262, 312
Rajas marga, 113
Rajasva, 66, 67, 251
Rajayoga, 286
Rajju, 86
Rajni, 98
Rajo Guna, 15, 49, 318, 320, 342
Rajyaika-sheshena, 98, 101, 270, 271, 273, 280, 323
Rakshasa, 368
Rama, 194, 196, 260, 287, 288, 327
Ramakatha, 14, 23, 175, 339
Raman, 194
Ramarajya, 90, 92, 246
Rani, 103, 220, 221, 249, 347
Rasagni, 224
Rasayana, 112
Rashi, 63, 330, 331, 355
Rashtrikaran, 224, 309
Rasi, 17, 152, 280, 289
Rasika, 92
Rasiya, 25, 90
Rati, 64, 235, 278
Ratnaketu, 305
Ratnakosha, 179
Raub, 75, 270
Ravidina, 227
Ravinandana, 42, 74, 171, 339
Ribhu, 163, 164
Ridhyati, 21, 113
Riju, 202
Rikshanayaka, 251
Ripu, 33, 221
Rishta, 58, 60
Rochisha, 59, 63, 112, 113, 197, 202, 226, 227, 243, 263, 296, 298, 313, 317, 318, 320, 330, 334, 339, 345, 347, 355, 364
Rodayitri shakti, 103
Rodha, 18, 102, 166, 168, 169, 181, 182, 187, 193, 202–204, 208, 223, 239, 240, 247, 255, 280, 281, 291, 316, 327, 346, 354, 362
Rogapratirodhi Pranali, 230
Rohini, 20, 39, 42, 43, 62, 97, 100, 163, 211, 320, 325, 365
Rojamela, 188, 189
Roma, 20, 21
Ropan, 19
Ruah, 71, 75, 78, 85, 101, 151, 157, 178, 215, 216, 218, 238, 240, 244, 248, 272, 273, 277, 279–281, 284, 285, 295, 314, 315, 326, 356, 357, 361, 365, 366, 368

Rudhartha, 159, 160
Ruksha ojas, 322
Runasvara, 99
Rupa, 62, 78, 89, 90, 96, 120, 122, 163, 208, 221, 240, 266, 297, 322, 333–336, 338, 363
Rupa shakti, 335
Rupa siddhi, 96, 163
Rupatita, 351

S

Sab, 149, 303
Sabalasvas, 253, 354
Sabbacitta, 78, 232
Saboot, 80
Sacetana, 214
Sacheta, 78
Sachi, 99, 270
Sachivaya, 19, 279, 332
Sadachara Marga, 327
Sadagati, 186
Sadakhya, 16, 17, 23, 24, 26, 30, 36, 37, 47, 50, 62, 67, 73, 75, 95, 122, 152, 156, 161, 175, 243, 258, 259, 266, 268, 277, 334, 339, 349, 360
Sadakhya chakra, 36
Sadasat, 187
Sadashiva chakra, 368
Sadashiva nayaki, 23
Sadashya, 119
Sadasya, 47–49, 119–121, 174, 337, 356
Sadatma, 269
Sadgati, 107, 216
Sadhaka, 18, 20, 23, 295, 299, 300
Sadhana, 20, 193, 207, 249, 277, 353
Sadhaniya, 21, 65, 98, 113, 257, 270
Sadhanya, 257
Sadhibhuta, 17, 65, 98, 163, 275
Sadhya, 284, 307, 346
Sadhyata, 91, 238, 321, 325
Sadi, 365
Sagama, 269
Sagarottarakaya, 247
Saguna, 78, 95, 216, 221, 234, 269, 359
Sah, 59
Sahadeva, 245, 326, 331
Sahaja puta, 181
Sahajata, 222, 240
Sahajta, 286
Sahasra lingam, 42, 55, 257, 289, 350
Sahasrabda, 365
Sahasrara, 213, 244, 320
Sahasrara chakra, 213, 244
Sahastitva, 89
Sahatra, 365
Sahayaka, 156, 335
Sahokti, 247
Sahvaas, 55
Sajatya, 279, 283, 284
Sakala, 35, 323, 324
Sakaratmak shakti, 150, 209, 325, 359
Sakhya, 266
Sakriya, 15, 78, 105, 220, 272, 357
Sakriya astikaya, 272
Sakriya-Niskriya astikaya, 272
Sakriyata, 200
Sakriyatma, 357
Sakshatkar, 274
Sakshatkara, 113, 358
Salaam, 351
Salakegolake, 53
Sam, 228
Samaa, 94, 95, 157, 189, 192, 239, 271, 281, 289
Samabhasam, 260, 294
Samadhi, 114
Samagam, 103
Samajh, 320, 321
Samalamba, 316
Samalekha, 15, 30, 38, 119, 120, 122, 149, 158, 174, 177, 201, 238, 242, 244, 246, 247, 256, 264, 268, 279, 280, 283, 284, 287, 334, 360–363, 366, 368
Samana, 61, 105, 193, 365
Samanakala, 198
Samanantara, 65, 174, 201
Samanjasya, 64
Samanvaya, 120
Samanvayita, 120
Samanviya, 120
Samapurak, 287
Samarasya, 25, 320
Samashti, 323
Samasya, 211

Index

Samavaya, 59, 333
Samaya, 207, 233, 351
Samayam, 233
Sambahvi, 268
Sambandha, 62, 67
Sambhasam, 339
Sambhavi, 274
Sambhoga Kaya, 367
Samdesha, 151
Samghatartha, 50, 361
Samghatavigrhitartha, 18, 181, 201, 246
Samjna, 81
Samketana, 170
Samkshobha, 18
Samlagni, 35, 36, 38, 53, 160, 165
Samma Ajiva, 80
Samma Drishti, 80
Samma Jnana, 80
Samma Karma, 80
Samma Pasadana, 80
Samma Sankalpa, 80
Samma Sati, 80
Samma Vacha, 80
Samma Vayama, 80
Samma Vimukti, 81
Samposhaniyata, 111
Sampratapana, 304
Samrachana, 300
Samrajya, 81, 93
Samsarga, 96
Samskrita, 163, 164
Samskriti, 21
Samsrkita, 163, 164
Samtol, 189
Samudaya, 48, 352
Samunnati, 312
Samvarna, 358
Samvat shakti, 15, 89, 99, 203, 366
Samyama, 82, 90, 100, 254, 368
Samyoga, 286
Sanananda, 363
Sanatana, 233, 234, 269
Sanatana dharma, 234, 269
Sanatani, 256
Sandeha, 215

Sandesha, 174, 250, 260, 321–323
Sandhatre, 222
Sandra, 14, 47, 110, 222, 246, 340, 341, 350
Sangat, 290
Sangathan, 17, 18, 96, 200, 264, 269, 272, 307, 320, 326
Sangha, 358
Sangya, 81
Saniddhya shakti, 294
Sanjna, 269
Sankalpa, 224
Sankalpana, 367
Sankhara, 100, 213
Sankhya dharma, 273
Sanlagnata, 49, 119
Sanmati, 187
Sannati Chandralamba, 327, 328
Sanvaad, 209
Sanvardhan, 82
Saptarishi, 238
Sara, 17, 34, 35, 42, 59, 60, 62, 63, 67, 77, 78, 89–92, 95, 97, 101, 103, 104, 106, 111, 112, 122, 157–159, 178, 220, 228, 230, 247, 254, 255, 280, 289, 297, 300–302, 313, 324, 334, 359, 360, 362, 364, 368
Sara Kalpa, 17, 34, 35, 42, 59, 78, 89, 103, 104, 122, 220, 228, 230, 255, 301, 302, 313, 364, 368
Saranyu, 16–19, 22, 25, 28, 42, 55, 59, 84, 104, 119, 165, 174, 215, 216, 248, 249, 251, 254, 265, 281, 285, 286, 348, 356
Saraswata, 231, 237, 341
Saraswati, 275, 341
Sargam, 81, 91, 92, 118, 177, 223, 238, 243, 290, 294, 299, 311, 325, 338, 341, 348, 351, 363
Sarjana, 58
Sarpa, 86, 87
Sarthaka, 26, 81
Sarva shakti, 207
Sarva Vyapin, 304
Sarvabhadra Mahayoga, 264, 269
Sarvabhaumikaran, 94, 223
Sarvabhutatma, 219
Sarvadevatma, 188
Sarvalokacharine, 18, 181
Sarvam, 24, 100, 203, 209, 253, 277
Sarvanasha, 30, 97, 118, 162, 215, 243, 244, 254, 315, 355

Sarvashaktiman, 88, 90, 105, 241
Sarvatraga, 78, 112
Sarvavyapak, 88
Sarvodaya, 226, 334
Sasana, 42, 55, 57, 58, 129
Sashaktikaran, 156
Sasharira, 78, 81, 82
Sat, 187, 295
Sata, 114, 274, 353
Sataha, 30, 37, 158
Satan, 107, 277, 349
Satarupa, 83, 96, 99, 154, 166, 181, 196, 248, 284, 311–313, 316
Satha, 78, 150
Sathya, 26, 65, 187, 201, 259, 274
Sati, 80, 92, 228, 287, 325, 334, 335, 356, 358–360
Sati-Parvati, 92, 228, 287, 325, 334, 335, 356, 358–360
Satkayadrishti, 217, 221
Satma, 34, 35, 37, 41, 42, 58, 75, 178, 179, 226, 245, 356, 358, 364, 365
Sato Guna, 15, 48
Satrangi, 83
Satta, 30, 176, 282, 285, 291, 358
Sattva, 20, 83, 262
Sattvika, 201
Satvasva, 67, 303
Satya Loka, 270, 314, 316, 366
SAUM, 42, 51, 57, 88, 100, 215, 232, 241, 242, 253, 274, 277, 292, 303, 321, 332, 366, 368
Saundarya, 147, 352
Sauri, 105, 288, 293
Sautramani, 57–59, 202
Savahita, 65, 71
Savana, 110, 113, 332
Savanasha, 215
Seekh, 200
Seemandhara, 86
Sekha, 200
Seshi, 83, 255
Setu, 103, 113
Seva, 85
Sewadaar, 86
Shachi, 193, 208
Shadayatana, 213

Shaddhatma, 159, 160, 257
Shadyatana, 316
Shahidi, 272
Shakatasya mandala, 272
Shakrani, 213
Shakti, 17, 18, 29, 34, 38, 56, 67, 100, 102, 119, 122, 152, 161, 163, 169–171, 174, 188, 190, 191, 196, 198, 199, 201, 207, 209, 212, 216, 218, 222, 224, 225, 230, 231, 244, 245, 259, 265, 277, 293, 318, 358, 362
Shakti Bheda, 209, 212
Shakti chakra, 18
Shakti Pitha, 122
Shakti yogi, 218
Shakya, 274
Shamana, 245, 271, 296
Shamatha, 232
Shambhala, 367
Shaneeswar, 154, 155, 158, 198
Shani, 22, 24, 53, 85, 97, 100, 159–161, 171, 239, 259, 274, 276, 277, 286, 292, 293, 296, 339
Shani Bhagwan, 85, 159–161, 171, 239, 259, 276, 277, 286, 293
Shanideva, 194–196, 269, 287, 334–338, 340, 368
Shanka, 262
Shankha, 114, 332
Shanku, 60, 61
Sharira, 91, 231, 292, 306, 325
Shatakanta Mudapahartre, 115
Shayana, 202, 203
Sheetala, 283, 297
Shesha, 61, 62
Sheshanaga, 309
Shevalohita, 347
Shevetalohita, 350
Shevtah Mandala, 19
Shila, 113
Shilajit, 157, 271
Shinnan, 308
Shirati, 122
Shishya, 86
Shiva, 89, 90, 92, 93, 165, 177, 210, 253, 260, 261, 273, 275, 288, 290, 292, 305, 307, 338, 342, 356, 357
Shivadrishti, 42, 240, 241, 338
Shivagati, 94, 182, 263, 264, 266, 285, 292,

Index

300
Shivatma, 219, 279
Shlakshna, 24
Shobhana, 78
Shodashottari dasha, 114, 248
Shraddha, 80
Shradha charita, 228, 229
Shradhadeva, 213
Shrama shakti, 231
Shramika, 266
Shranta, 279
SHREEM, 187, 248, 264, 274, 279, 283, 284
Shreya, 63, 64, 113, 261, 276
Shri, 67, 82, 87, 177, 208, 214, 220, 244, 245, 257, 262–264, 272, 275, 287, 300, 327, 339, 342–344, 362–364
Shri Krishna, 82, 87, 177, 208, 257, 275, 287, 327, 343, 362–364
Shri Rama, 220, 300, 339
Shri Vishvakarma, 67
Shrimate, 23, 92, 320
Shrinkhala Bandhamochakaya, 269
Shubha, 211
Shuddhi, 48, 216, 239, 270, 273
Shudra, 16, 26, 34–36, 38, 42, 62, 121, 164, 165, 301, 328, 334, 336, 353
Shukra, 95, 96, 240
Shunya, 17, 34, 42, 63, 65, 76, 104, 171, 190, 222, 230, 255, 256, 293, 301, 303, 304, 313, 368
Shunya Kalpa, 17, 34, 104, 222, 230, 256, 293, 301, 304, 313
Shunya sthana, 303
Shunyata, 76, 263, 269, 273
Shurasena, 93
Shuru, 55
Shvakridin, 339
Shveta Varaha, 98
Shvetah, 19, 83
Shvetah Mandala, 19
Shymah, 83
Siddha, 17, 19, 20, 33, 65, 72, 93, 99, 100, 121, 173, 175, 176, 178, 189, 194, 200, 208, 212, 215, 216, 223, 233, 244, 245, 254, 259, 262, 267, 282, 284, 287, 298, 306, 307, 315, 317, 320, 326, 328, 336, 339, 352, 365
Siddha marga, 17, 33, 233, 254, 298
Siddha Phal, 17
Siddha ratri, 65
Siddhi, 15–17, 33, 113, 175, 184, 219, 249, 315, 341, 365
Siddhi chakra, 17
Siddhi Marga, 17, 33
Siddhi Phal, 17
Siddhi shakti, 184
Siddhi yogi, 219, 249
Sighraga, 314, 349
Sikha, 200
Sikhana, 200
Sikhya, 200
Simha Rashi, 105
Sirnapada, 243, 244, 246, 274, 281
Sitadevimudra Pradayakaya, 62
Sitanveshana Panditaya, 245
Skambha, 122
Smriti, 81, 200, 201, 327, 329
Soham Siddhanta, 281
Soma, 103, 239, 247, 250, 364
Somanasa Mahayuga, 176, 288
Somapa, 16, 175, 177, 178, 244, 343
Spanda shakti, 277
Sparsha, 81
Sparshi, 78
Sphul, 162
Sphurti, 19, 278, 279, 332, 342
Sprishti, 189, 297
Sraddhya, 106
Sri Loka, 362, 363
Sridhara, 191–193, 197, 329, 337
Srijak, 77, 82, 83, 175, 219, 256, 287, 323, 334, 335
Srijan, 23, 62, 65, 84, 233, 250, 278, 287, 367
SRIM shakti chakra, 248
Sthairya, 86
Sthanapatra, 278, 287
Sthanapatrata, 278
Sthanata, 14, 15
Sthavara, 78
Sthavaravisha, 196, 298, 299, 315, 317, 341, 368
Sthayitva, 112
Sthirikaran, 190
Sthiti, 85, 312

Sthula, 86, 87, 211, 220, 221, 300, 301, 308, 325, 331, 351
Sthula sharira, 87, 221, 300, 301, 308, 325, 331, 351
Sthuna, 335
Stree, 149, 260, 266
Stree Dirgha, 260
Stuti, 77
Styana, 79
Subheccha, 74
Suchaka, 121, 334
Suchan, 34, 47, 48, 120, 121, 164, 340
Suchanak, 340, 341
Suchi, 193, 194
Suddha, 87
Sudhanvan, 121, 316, 333, 345
Sudharatmak, 113
Sugriva, 16, 23, 27–30, 36, 41, 72, 119–122, 175–178, 322, 323, 340, 365
Sujata, 210
Sukara, 299
Sukha, 326
Sukshma, 188, 215, 246, 248, 310, 325, 331, 351
Sukshma sharira, 188, 215, 246, 248, 310, 325, 331, 351
Sunanayaka, 85, 255, 258
Sunayaka, 260
Sundari, 353
Supuma, 107, 299, 341
Sura, 25, 34, 42, 52, 58, 74, 76, 77, 85, 149, 150, 164, 168–170, 176, 177, 211, 215–217, 222, 225, 286, 312, 318, 336, 341, 349, 368
Sura shakti, 25, 170, 215, 225
Surari, 273
Surasa, 276, 326, 327, 330, 331
Surasa Chakra, 276
Surya, 17, 18, 20, 34, 66, 67, 72, 90–92, 95, 101, 103, 179, 187, 190, 227, 234, 239, 241, 243, 255, 289, 297
Surya chakra, 20, 179, 187, 190, 289
Surya mandala, 72, 103, 241, 243, 297
Surya Marga, 234
Surya shakti, 227
Suryaphani chakra, 61, 62, 66
Susana, 57, 58
Sushumna, 35, 75, 237, 347, 348, 365

Sushupti, 16, 33, 76, 89, 175, 176, 195, 197, 241, 282, 283, 285, 286, 331, 355
Sushuptivat, 123, 201, 211, 257
Suswani, 351
Sutala Loka, 360, 367
Suvarna, 347
Suvira, 275, 286, 356, 357, 359, 360
Sva, 15, 17, 20, 22, 23, 89, 91, 92, 94, 151, 175, 177, 213, 215, 252, 255, 273, 285, 288, 292, 294, 301, 340, 356
Svabhautik, 224, 315
Svabhav, 89, 90, 98, 99, 122, 157, 187, 284, 291, 334, 359–361, 366
Svadhina, 104, 201
Svadhinta, 222
Svadhisthan chakra, 19, 180, 210, 211
Svaha, 66, 85
Svahita, 191, 247, 351, 357
Svaiccchika, 307
Svajagrit, 268
Svaphalka, 21, 26, 222
Svapna, 176, 177, 203
Svapna-Jagrat, 176
Svapnil, 328
Svarajya, 39, 81, 87, 90
Svarga Loka, 354, 355
Svarochisha, 17, 33, 56, 63, 99, 121, 156, 167, 178, 200, 212, 248, 256, 262, 277, 290, 296–298, 303, 305, 306, 308, 310, 313, 314, 334, 346, 351, 363
Svartha, 201
Svarupanugata, 66
Svatantra, 113
Svayam, 33, 48, 49, 56, 168–170, 174, 190, 192, 212, 213, 215, 222, 226, 251, 254, 260, 262, 269, 271, 273, 279–284, 311, 313, 316, 329, 334, 343, 355, 363, 366, 368
Svayambhu, 267, 290, 291
Svayamseva, 226, 238, 251, 319, 339, 345
Svedana, 304
Swadha, 200, 278, 280
Swakshetra, 75, 162
Swamigata, 64
Swasthani, 326, 330, 331
Swati, 101

T

Taalmela, 202
Tadatma, 20, 23

Index

Taijaisa, 54, 160
Tajagi, 210
Tajaurba, 121, 353, 354, 358
Talaash, 272
Talatala Loka, 275, 280, 344, 345
Tamas, 21, 24, 48, 83, 113
Tamasika, 221
Tamo guna, 15, 16, 33, 93, 156, 212, 224, 238, 247, 280, 318
Tandarusti, 64, 210
Tantri, 194, 197
Tantu, 278
Tanunapat, 77
Tapo Loka, 275, 351
Tara, 17, 48, 49, 67, 95–97, 101, 120, 242, 243, 326, 331
Taraka, 62, 108, 292–295, 340
Taranga, 121, 232, 334
Tarashana, 17
Tarkshya, 78, 212, 219
Tat Tvam Asi siddhanta, 25
Tatkala, 339
Tatpraya, 81
Tatramajjhattata, 80
Teentarfa Vinamaya, 225
Tejas, 55, 57, 154, 215, 216, 222, 269, 278, 279, 323, 344, 362
Tejas yoga, 323
Tejomandala, 362
Thaili, 93
Thakaan, 210
Thikatala, 25
Tigma, 232
Tikari, 302, 303
Tikram, 153
Tikshana, 212
Tilaka, 59, 62, 112, 151, 199, 347
Tippa, 158, 354
Tirikarannasutti vrat, 216
Tiripurai, 155, 209
Tirthankara, 218
Tiryaggati, 323, 340
Titiksha, 94
Titimma, 356
Tittiri, 230
Trasa, 87, 122, 297

Trayastrimsha, 174, 223, 227
Trayi mukha, 25
Tribhajya, 42, 151, 179
Trichakra, 61, 65, 202
Trigunatma, 119, 161, 162, 211, 219, 255, 267
Trijya, 337
Trikonana, 61
Trinetra, 73, 216, 253, 270, 346
Tripta, 316
Tripura, 278
Tripurantaka, 120, 122, 303
Trishaka, 154
Trivikrama, 82, 92, 108, 155, 178, 179, 193, 198, 219, 253, 281
Triyancha, 93, 188, 249
Tula Rashi, 103, 104
Tulyartha, 104, 203, 204
Tulyatarka, 335
Tushti, 15, 55, 224
Tvam, 66, 72, 253

U

Ubhayato mukha, 253, 255
Ubhyato, 282
Uccitika, 103, 181, 182, 202, 316, 320
Udaharana, 238
Udana, 157, 159, 161, 294, 356
Udanavata, 343–345
Udbhava, 65
Uddalana, 263
Udita, 318, 324
Udu, 347
Udvaha, 57, 64, 102, 191, 263, 277, 299, 324, 341, 354
Ujukata, 80, 81
Ulat, 307
Ulatana, 307
Uljhana, 17, 98
Ulka, 212, 220
Uma, 17, 97, 153, 174, 178, 189, 222, 255, 278, 324
Umapati chakra, 229
Ummujjati, 15
Unha, 214, 337
Upa, 119, 122, 151, 164, 228
Upa Pitha, 119, 122
Upadhi, 95, 96

Upadishta, 246, 364
Upaja, 306
Upakarana artha, 55, 315
Upaklesa, 79
Upanaha, 79
Upanayana, 67, 270, 271, 284, 294
Upanishada, 232, 278, 287, 349, 356
Upanshanta, 328
Upantaran, 224
Upasampada, 273, 338
Upasarjana, 50, 119, 361
Upashanta, 60, 328
Upasravana, 93
Upaya, 24, 30, 35, 36, 42, 52, 60, 61, 66, 67, 82, 88, 91, 93, 97, 100, 101, 103, 153, 160, 165, 169, 192
Upjaana, 251
Urdhava, 273, 280
Urdhvagatma, 219
Urja, 214, 306, 307, 311–313, 347, 348
Usha, 83, 176, 251, 258, 296, 300, 301, 323
Uthapana, 269
Utsamskara, 30, 252, 329
Utsarajan, 19
Utsarjak, 19
Utsarji, 19
Uttama, 15, 55, 223, 224, 366
Uttanapada, 202, 224, 283, 346, 357
Uttarajya, 212
Uttaranga, 304
Uttardayitva, 211
Uttarmanasa, 23, 346, 347
Uttejana, 220

V

Vaas, 16, 24, 37, 39, 96, 98, 123, 151, 152, 168, 202, 211, 226, 268, 283, 292, 305, 321, 340, 357
Vafadaar, 79
Vahana mandapa, 52, 153, 160, 170
Vaicariki, 354
Vaidhatva, 94
Vaikari, 212, 224
Vaimitra, 76, 242, 283–285
Vaira Bhava, 328
Vaisha, 228, 309, 317
Vaishvanara, 249
Vaishvikrita, 42, 43, 55, 202
Vaishwadeva yajni, 153
Vaishya, 22, 24, 25, 27, 28, 47, 49, 74, 189–192, 196–200, 218, 226, 239, 243, 246, 306, 308, 311, 312, 318, 324, 335, 337, 343, 347, 348, 358, 359
Vajra, 218, 245, 315
Vajradhatu-mahamandala, 201
Vak, 320, 321
Vakra, 15
Vakradrishti, 63
Vakyartha, 212, 256
Valaya, 60, 62, 100, 106, 151, 154, 365
Vali, 274
Vam, 71, 72
Vamachara, 210
Vamaviddha mandala, 303
Vamsaa, 122
Vanara, 264, 288
Vanaspati, 322, 340
Vandana, 232
Vansha, 339
Vapu, 92, 293
Varakharcha, 83
Vardhimainaka Pujitaya, 248
Varga, 14, 15, 47, 49, 119, 174
Vargamula, 221
Varna, 14, 256, 257, 347, 351, 353
Varnatma, 353
Varshakritya Taranga, 37, 66, 121, 334
Vartaman, 169, 210, 211, 264
Vartula, 63, 78, 106, 156, 224, 304, 308, 309, 319, 365
Varuna, 273, 278, 317
Vasana, 74, 223, 224
Vasanatma, 275, 364
Vasiya, 222
Vastavika, 175, 223, 263
Vastavikta, 17, 24–26, 30, 43, 55, 71, 82, 99, 107, 129, 150, 154, 160, 164, 165, 169, 170, 174, 177, 178, 210, 219, 248, 249, 259, 261, 269, 270, 273, 278, 298, 314, 324
Vastu, 17, 24, 98, 168, 169, 238, 275, 282, 301, 322, 343–345, 347, 352
Vastunara, 277, 320
Vasudeva, 215, 285
Vataranga, 334

Index

Vatsanapat, 76
Vayu, 19, 208, 221, 297, 361
Vedana, 81, 83, 190, 297, 363
Vedatma, 188
Veerbhadra, 66, 67, 303
Veerya, 197
Vega, 100, 101, 270, 272
Vela, 257
Vevi, 224, 275
Veyyavacca, 204
Vibha, 254, 282
Vibhaga Tejas, 214, 276
Vibhajya, 112, 231
Vibhajyata, 220
Vibhasta, 23, 213
Vibhava Loka, 367
Vibhavana, 61
Vibodha, 223
Vichara, 79, 187
Vicikitsa, 82
Vicitra, 35, 36, 105, 230, 237, 267, 270, 360, 362
Vicitra mandala, 35, 36, 105, 230, 237, 267, 360, 362
Vidalana, 262
Vidhana, 99, 157, 194, 196, 214–216, 227, 231, 269–272, 321, 366
Vidhata, 255
Vidhi, 160, 294, 296
Vidhi Vadartha, 160
Vidhi Vidhartha, 296
Vidupatma, 194
Vidya, 20, 33, 54, 57, 101, 102, 152, 161, 269, 315, 344
Vidya ratri, 161, 315
Vidyadhara, 216, 217, 281
Vidyapati, 220
Viganana, 31, 75
Vignesh, 177
Vihimsa, 79
Vijatiya, 49
Vijnana, 21, 193, 272, 330, 353
Vijnanatma, 181, 219, 353, 354, 356
Vijnanik, 285
Vijujya, 105
Vikaranadharmitva, 86, 196

Vikatth, 321
Vikshepa, 79
Vilaga, 76, 85, 202, 343, 362
Vilatma, 335, 349
Vilayan, 23
Vimaleshvara, 22, 261, 365
Vimshottari dasha, 208
Vimukti, 81
Vinayaka, 179
Viniyojya, 112
Vipakya, 212
Viparyaya, 65, 82, 83, 112, 147, 267, 320, 329, 352
Vipashyana, 232
Vipluta, 349–351
Vira ratri, 66, 75, 93
Virasat, 34–36, 59, 105, 107, 112, 228
Virodhita, 163, 330
Virupaksha, 177, 280, 349
Visarajana, 18
Visaran, 256, 366
Visaranem, 57, 129
Visarjan, 343
Visarjana, 343
Visarjya, 105, 118, 340
Visata, 112, 347
Visayaniyata, 80
Vishesha, 302
Visheshana, 296
Vishnu nabhi, 253, 259, 262
Vishram, 191
Vishvabhava, 58, 59, 152, 153
Vishvagoptri, 88, 101
Vishvakarma, 248
Vishvamitra, 253, 261, 262
Vishvanatha, 252
Vishvasa, 221
Vishvavyapi, 224
Vishwikaran, 20, 222
Vismita, 280
Vistavikta, 268
Visuddha chakra, 19, 180, 291
Visukayita, 48, 277, 280
Vitala loka, 357
Vitasti, 231

Vitata, 20, 23, 54, 152, 209, 338, 344
Vithoba, 48, 264, 308
Vivasvan, 60, 61
Vividthta, 66
Viyojya, 99, 192, 211, 237, 238, 264, 282
Viyujya, 118
Vrajya, 94
Vrihat, 89
Vrishaba Rashi, 97, 98
Vruschik Rashi, 92, 93
Vutthana shakti, 224
Vyagrata, 62, 65, 154, 285, 299
Vyakti, 30, 98, 216, 242–244, 246, 276, 348, 353–356
Vyaktikrita, 113
Vyaktitva, 174, 360, 362
Vyakulatma, 219
Vyan Vata, 322
Vyanang, 93
Vyanavata, 184, 226, 244
Vyatikara, 87, 163
Vyavahar, 342
Vyavastha, 83
Vyayatman, 218

Y

Yaanshala, 175, 255, 278, 287, 354
Yahoodi, 15, 154, 171, 256
Yajna dharma, 77
Yajna Yoga, 286
Yajnopavita, 18, 201, 323
Yama, 66, 67, 71, 149, 186, 239, 241, 242, 245, 247, 254, 274, 288, 305, 347
Yama chakra, 254
Yamadeva, 149, 152, 304
Yamantaka, 265, 273
Yana, 354
Yanashakti, 358
Yantra, 277, 315
Yantratma, 187
Yatachittatma, 78
Yatamanasa, 160
Yathartha, 56, 257
Yathayogya, 111, 199, 314
Yaugika, 302
Yauvana dasha, 361
Yauvaniya, 367
Yava, 53, 54
Yavadartha, 14, 15
Yoga, 59, 61, 190, 286, 329
Yogataranga, 368
Yogatma, 162, 286
Yogihridaya, 13, 183, 228, 307, 317, 358
Yogini chakra, 62
Yojana, 65, 368
Yojya, 112, 209, 246, 270
Yoni, 225, 240, 331
Yuktartha, 241, 248, 320, 331, 332, 342
Yukti, 113, 212, 217, 246, 278, 319
Yukti yogi, 217
Yuvarajyarajya, 210

Z

Zindagi, 30, 75, 250, 334, 353, 357, 358

www.ingramcontent.com/pod-product-compliance
Lightning Source LLC
Chambersburg PA
CBHW071443220526
45472CB00003B/647